RELATIVITY:
MODERN LARGE-SCALE SPACETIME STRUCTURE OF THE COSMOS

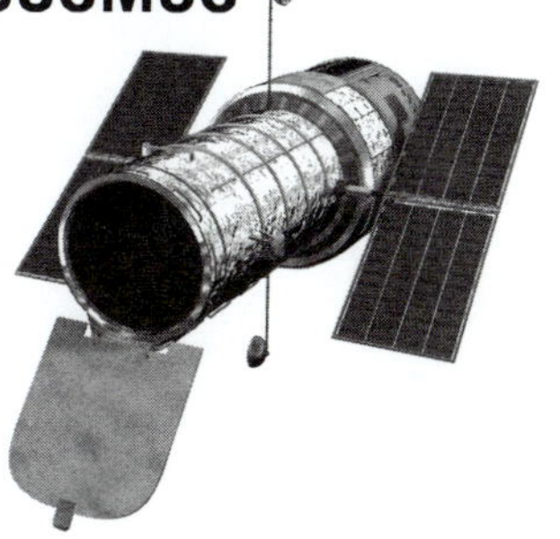

RELATIVITY:
MODERN LARGE-SCALE SPACETIME STRUCTURE OF THE COSMOS

Editor

Moshe Carmeli

Formerly of Ben Gurion University, Israel

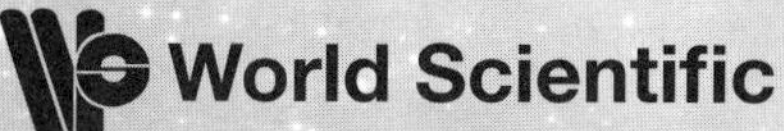

NEW JERSEY · LONDON · SINGAPORE · BEIJING · SHANGHAI · HONG KONG · TAIPEI · CHENNAI

Published by

World Scientific Publishing Co. Pte. Ltd.

5 Toh Tuck Link, Singapore 596224

USA office: 27 Warren Street, Suite 401-402, Hackensack, NJ 07601

UK office: 57 Shelton Street, Covent Garden, London WC2H 9HE

British Library Cataloguing-in-Publication Data
A catalogue record for this book is available from the British Library.

RELATIVITY
Modern Large-Scale Spacetime Structure of the Cosmos

ISBN-13 978-981-281-375-6
ISBN-10 981-281-375-6

Printed in Singapore by B & JO Enterprise

1 HAPPY IS the man that hath not walked in the counsel of the wicked, nor stood in the way of sinners, nor sat in the seat of the scornful.
2 But his delight is in the law of HaShem; and in His law doth he meditate day and night.
3 And he shall be like a tree planted by streams of water, that bringeth forth its fruit in its season, and whose leaf doth not wither; and in whatsoever he doeth he shall prosper...
Psalms, Chapter 1, lines 1 to 3, Ketuvim

Acknowledgements

We are hereby thankful for the permission given by the American Institute of Physics and Dr. Ray Sachs to use material from the published article in the *Journal of Mathematical Physics*: R. Kantowski and R.K. Sachs, *J. Math. Phys.* **7**, 443 (1966).

We are also thankful for the permission given by Cambridge University Press and to Prof. MacCallum to reproduce Tables 15.1 and 15.2 from the book by D. Kramer, M. MacCallum, H. Stephani and E. Herlt, entitled "Exact Solutions of Einstein's Field Equations", (1981).

Also we acknowledge the inclusion of material in the text on the biographies of Einstein, Lorentz, Minkowski, Riemann, Schwarzschild, Friedmann, Lie, Killing and Bianchi from articles by Profs. J.J. O'Connor and E.F. Robertson, to whom we highly express our gratitude.

We are particularly grateful to Dr. Julia Goldbaum for her constant assistance, in both technical and scientific aspects, and for her excellent job of typing, proof reading and stylistic corrections to this book.

Foreword

Late in 2002 a colleague of mine recommended I buy a book *Cosmological Special Relativity*. I must admit the combination of those words intrigued me. What could *Special Relativity* have to do with anything *cosmological* I thought? I just had to find out. So I bought the book and read it over one weekend – I just couldn't put it down. The idea was fascinating, that one could construct a theory for the large scale expansion of the cosmos by analogy to Einstein's *Special Relativity* theory.

I soon learned that the book was written by a leading theoretical physicist, Moshe Carmeli.[1] He had extended the concepts of *Special Relativity*

---

[1] Moshe (Ehezkel) Carmeli. Born: 15 June 1933 in Bagdad, Iraq. Died: 27 September 2007 in Beer Sheva, Israel.

Moshe Carmeli was born to Eliahu and Naima Carmeli, one of 7 children. He finished high school in 1948. Then for a year he remained at home and read most of Bagdad's library books.

In 1950 he immigrated to Israel through Iran. His parents came a year later. He enlisted in the Israeli Air Force. There he met people who told him a lot about physics. Moshe was fascinated by what he heard and quickly learned. Right away he knew that he wanted to study physics. In 1960 he received his Master Degree in Physics from the Hebrew University of Jerusalem. In 1961 he married Elisheva Cohen. At the same time he began to work on his PhD Thesis under supervision of Professor Nathan Rosen, who had been a student of Albert Einstein. Carmeli finished his D.Sc. in 1964.

In 1964 the Carmeli family moved to the USA, where Moshe carried out his research in general relativity and gravitation and lectured in Lehigh University, Temple University and then in the University of Maryland.

In 1972 the Carmeli family returned to Israel, and Moshe began to work as the Albert Einstein Professor at the Physics Department of Ben Gurion University, Beer Sheva. Between 1973–1977 he was the Head of the Department. He worked in the Physics Department as a researcher and lecturer until his retirement. After retiring he continued his scientific research. Carmeli's field included gravitation and gauge field theory, the theory of spinors as applied to physics, Einstein special and general relativity, and astrophysics. He developed his own cosmological relativity theory, in which the age of Universe is postulated as constant, just as the speed of light is in Einstein's theory, and

to the cosmos. Though he was in his early sixties, he developed a new description of the Universe around the kitchen table in his home in Beer Sheva, Israel. Being from the old school, he did most of his calculations on paper. That was in the mid-1990s. The book I found distilled the essence of his theory. And though it has yet to be fully recognized, I believe, the theory is a significant revolution in our description of the cosmos.

Carmeli spent most of his early years working on Einstein's Special and General theories and published a number of detailed books on relativity. In particular his *Classical Fields* is a standard text in many universities around the world. His new theory extends the Universe by one extra dimension — the velocity of the expansion of the Universe as measured by the redshifts of the receding galaxies in the Hubble expansion. In this theory spacetime is a brane in which particles interact and move but it is being "stretched" in relation to the velocity of the expansion. So spacevelocity is the relevant subspace to view the expansion in, in analogy to spacetime of relativity theory. And it follows that the Hubble law is fundamental.

The accelerating expansion, which he predicted in 1996, two years before the publication of high redshift supernova observations that indicated the acceleration, is due to a vacuum energy density that modifies the matter density of the Universe and is not due to dark energy. No cosmological constant appears in the field equations. Within his theory for bound systems around central potentials or for the cosmos as a whole there is no need to invoke dark matter either.

When I realized the potential of his theory I started to apply it to observations — spiral and elliptical galaxies and Hubble diagrams using type Ia supernova and even GRBs. It was able to fit the observational data without the need for dark matter or a cosmological constant. The new degree of freedom supplied what was needed. I have contributed a number of chapters to this book where my colleagues and I have applied his theory to these sorts of systems.

I am reminded of the era around the end of the 19th century when

---

the velocity of receding galaxies is considered as a new independent variable.

Carmeli was author and co-author to more than 120 research papers and 9 books. Professor Carmeli was the president of the Israel Physical Society as well as an elected Fellow of the American Physical Society, the American Association for the Advancement of Science and a Member of the New York Academy of Science. Professor Carmeli was a Visiting Professor and Member of the Institute for Theoretical Physics, SUNY, Stony Brook, a Visiting Professor and a Faculty Member at the University of Maryland, a Visiting Scientist at the Max-Planck Institute, Munich and the International Center for Theoretical Physics, Trieste.

Professor Carmeli left three children: Eli, Dorith and Yair, and six grandchildren.

the advance of the precession of the perihelion of the planet Mercury so troubled astronomers and physicists of the day they hypothesized "dark" matter in the form of an unseen planet Vulcan behind the Sun which always stayed out of sight of Earth observers. But by 1916 Einstein vanquished the illusionary stuff, by showing it wasn't unseen matter but new physics that was needed, with the Post-Newtonian development of his general relativity theory. General relativity is most significant where and when spacetime is warped by large masses and accelerations are large. The new theory, Cosmological General Relativity, applies on large scales where accelerations are very weak. This is where the gentle expansion of space is felt by particles inhabiting it.

Carmeli's work solves old problems with fresh ideas. You will find many new ideas in this book. Gravitational lensing has an extra term that is distance (hence epoch) dependent. Besides the Newtonian gravitation equation a new Post-Newtonian term arises in the dynamics within galaxies. The Tully–Fisher law for spiral galaxies results from the 5D theory. The new dimension with a timelike signature invokes particle production in the expanding Universe.

Though still much work needs to be done, this book is the last from Professor Carmeli due to his passing from this world on September 27th, 2007. He will be sadly missed. But I believe this book is only the beginning for his theory. I have had great pleasure knowing and working with Moshe and highly recommend this book to you.

Perth, W. Australia
May 2008

John Hartnett

Preface

This book is based on lectures given by the author throughout the years in Israel and outside Israel on relativity and cosmology. The book provides an extended text for researchers in the field and for graduate course on the subject.

Why does one need a new book on relativity almost one hundred years after it was first developed by Einstein? Is Einstein's theory no longer adequate? Or has it stopped providing the needed answers? In this book we show how a combination of cosmological relativity with Einstein's relativity provides new answers that have not been known before. We recall that the starting point of cosmological relativity is the Hubble expansion rather than the propagation of light as in Einstein's theory. In this sense cosmological relativity is a complimentary theory to Einstein's.

The biographies of Albert Einstein, Minkowski, Lorentz, Riemann, Friedmann, Schwarzschild, Sophus Lie, Killing and Bianchi are added to the book.

Beer Sheva, Israel
September 2007

Moshe Carmeli

Contents

Chapter 1

Special Relativity Theory

Moshe Carmeli

> *In September 1981 I asked* C.N. Yang *"What is more important, the special or the general theory of relativity." He answered: "The special theory is faaaaaaar more important than the general theory." I asked the same question of* Nathan Rosen, *and his answer was identical.*

In this chapter we present the theory of special relativity that fixes the structure of spacetime without gravitation. The subject has been discussed in many texts, but our approach here is less technical and goes along the original intuitive lines of Einstein. The chapter starts with the postulates of special relativity, namely the principles of relativity and the constancy of the speed of light. This is then followed by discussing the basic concept of coordinate systems, and particularly the inertial system. Simultaneity, an essential notion in special relativity theory, is subsequently analyzed. These basic concepts are then followed by the Galilean transformation and group, and the Lorentz transformation and group. Consequences of the Lorentz transformation are then drawn. A four-dimensional formulation of spacetime, following Minkowski, is subsequently given. The light cone structure, an important description of spacetime, is then given. The discussion on special relativity theory is concluded by giving the formula that relates the mass, energy and momentum, along with the introduction of the energy-momentum four-vector. Our presentation will be along the original lines of Einstein's theory rather than dealing with technicalities, in order to emphasize the deep analogy of special relativity to the cosmological special relativity given in Chapter 2.

1.1   Spacetime in Four Dimensions

In the following we give the basic principles of the special theory of relativity. These principles are needed to describe the electromagnetic field and other physical phenomena, and they constitute their spacetime symmetry background.

1.1.1   *Postulates of special relativity*

The special theory of relativity was developed by Einstein[1] in 1905 in order to overcome and correct certain basic concepts that were in use at that time, such as asymmetries in relative motion of bodies. Examples of relative motion in electrodynamics, and the unsuccessful attempt to detect the motion of the Earth by the experiment of Michelson and Morley, suggested that the phenomena of electrodynamics and mechanics do not depend on the Newtonian notion of absolute rest. Rather, the laws of electrodynamics should be valid in all frames of references in which the equations of mechanics are valid.

1.1.2   *The principle of relativity: Constancy of the speed of light*

Einstein raised the above observation to the status of a postulate and called it the *principle of relativity*. He also introduced another postulate (which is only *apparently* inconsistent with the former one) according to which light always propagates in empty space with a constant speed c independent of the motion of the emitting body and the measuring instrument.

---

[1] Albert Einstein (Born: 14 March 1879 in Ulm, Württemberg, Germany; Died: 18 April 1955 in Princeton, New Jersey, USA)

Around 1886 Albert Einstein began his school career in Munich. Two years later he entered the Luitpold Gymnasium. He studied mathematics, in particular calculus, beginning around 1891.

In 1894 Einstein's family moved to Milan but Einstein remained in Munich. In 1895 Einstein failed an examination that would have allowed him to study for a diploma as an electrical engineer at the Eidgenössische Technische Hochschule (ETH) in Zurich. Einstein then attended secondary school at Aarau planning to use this route to enter the ETH in Zurich. At Aarau he wrote an essay (for which he was only given a little more than half marks!) in which he wrote of his plans for the future:

"If I were to have the good fortune to pass my examinations, I would go to Zurich. I would stay there for four years in order to study mathematics and physics. I imagine myself becoming a teacher in those branches of the natural sciences, choosing the theoretical part of them. Here are the reasons which lead me to this plan. Above all, it is

The above two postulates were shown by Einstein to be enough for the development of a consistent theory of electrodynamics of moving charges, which is based on Maxwell's original theory that was assumed to be only

---

my disposition for abstract and mathematical thought, and my lack of imagination and practical ability."

Indeed Einstein succeeded with his plan graduating in 1900 as a teacher of mathematics and physics. One of his friends at ETH was Marcel Grossmann who was in the same class as Einstein. Einstein tried to obtain a post and wrote to universities in the hope of obtaining a job, but without success.

By mid 1901 he had a temporary job as a teacher of mathematics at the Technical High School in Winterthur. Around this time he wrote: "I have given up the ambition to get to a university ..."

Then Grossmann's father helped Einstein to get a job by recommending him to the director of the patent office in Bern; Einstein was appointed as a technical expert. Einstein worked in this patent office from 1902 to 1909, first holding a temporary post and then obtained a permanent position. While in the Bern patent office he completed an astonishing range of theoretical physics publications, written in his spare time without the benefit of close contact with scientific literature or colleagues.

Einstein earned a doctorate from the University of Zurich in 1905 for a thesis "On a new determination of molecular dimensions," which was dedicated to Grossmann.

In the first of three papers, all written in 1905, Einstein examined the phenomenon discovered by Max Planck, according to which electromagnetic energy seemed to be emitted from radiating objects in discrete quantities - quanta whose energy was proportional to the frequency of the radiation. Einstein used Planck's quantum hypothesis to describe the electromagnetic radiation of light.

Einstein's second 1905 paper proposed what is today called the special theory of relativity. He based his new theory on a reinterpretation of the classical principle of relativity, namely that the laws of physics should have the same form in any frame of reference. As a second fundamental hypothesis, Einstein assumed that the speed of light remained constant in all frames of reference, as required by Maxwell's theory. Later in 1905 Einstein showed how mass and energy were equivalent.

The third of Einstein's papers of 1905 concerned statistical mechanics, Brownian motion, a field of that had been studied by Ludwig Boltzmann and Josiah Gibbs.

After 1905 Einstein continued working in the areas described above. He made important contributions to quantum theory, but he sought to extend the special theory of relativity to phenomena involving acceleration. The key paper appeared in 1907 with the principle of equivalence, in which gravitational acceleration was held to be indistinguishable from acceleration caused by mechanical forces. Gravitational mass was therefore identical with inertial mass.

In 1908 Einstein became a lecturer at the University of Bern after submitting his Habilitation thesis "Consequences for the constitution of radiation following from the energy distribution law of black bodies." The following year he become professor of physics at the University of Zurich.

By 1909 Einstein was recognised as a leading scientific thinker and was appointed a full professor at the Karl-Ferdinand University in Prague in 1911. Also, in 1911 Einstein was able to make preliminary predictions about how a ray of light from a distant star, passing near the Sun, would appear to be bent slightly, in the direction of the Sun. This would be highly significant as it would lead to the first experimental evidence in favour of Einstein's theory.

valid in stationary systems. The theory did not require an "absolute stationary space."

About 1912, Einstein began a new phase of his gravitational research, with the help of his mathematician friend Marcel Grossmann, by expressing his work in terms of tensor calculus. Einstein called his new work the general theory of relativity.

Einstein returned to Germany in 1914. He accepted an impressive offer: a research position in the Prussian Academy of Sciences together with a chair at the University of Berlin. He was also offered the directorship of the Kaiser Wilhelm Institute of Physics in Berlin, which was about to be established.

Late in 1915 Einstein published the definitive version of general theory. Just before publishing this work he lectured on general relativity at Göttingen and he wrote: "To my great joy, I completely succeeded in convincing Hilbert and Klein." In fact Hilbert submitted for publication, a week before Einstein completed his work, a paper which contains the correct field equations of general relativity.

In 1920 Einstein's lectures in Berlin were disrupted by anti-Jewish demonstrations. During 1921 Einstein made his first visit to the United States. His main reason was to raise funds for the planned Hebrew University of Jerusalem. During his visit he received the Barnard Medal and lectured several times on relativity.

Einstein received the Nobel Prize in 1921, not for relativity but for his 1905 work on the photoelectric effect. Among further honors which Einstein received were the Copley Medal of the Royal Society in 1925 and the Gold Medal of the Royal Astronomical Society in 1926.

By 1930 he was making international visits again, back to the United States. A third visit to the United States in 1932 was followed by the offer of a post at Princeton. Einstein accepted it and left Germany in December 1932 for the United States. The following month the Nazis came to power in Germany and Einstein was never to return there.

During 1933 Einstein traveled in Europe visiting Oxford, Glasgow, Brussels and Zurich. Offers of academic posts, which he had found so hard to get in 1901, were plentiful. He received offers from Jerusalem, Leiden, Oxford, Madrid and Paris.

What was intended only as a visit became a permanent arrangement by 1935 when he applied and was granted permanent residency in the United States. In 1940 Einstein became a citizen of the United States, but chose to retain his Swiss citizenship. He made many contributions to peace during his life. In 1944 he made a contribution to the war effort by hand writing his 1905 paper on special relativity and putting it up for auction. It raised six million dollars, the manuscript today being in the Library of Congress.

By 1949 Einstein was unwell. He left his scientific papers to the Hebrew University in Jerusalem, the university which he had raised funds for on his first visit to the USA.

One more major event was to take place in his life. After the death of the first president of Israel in 1952, the Israeli government decided to offer the post of second president to Einstein. He refused but found the offer an embarrassment.

One week before his death Einstein signed his last letter. It was a letter to Bertrand Russell in which he agreed that his name should go on a manifesto urging all nations to give up nuclear weapons. It is fitting that one of his last acts was to argue, as he had done all his life, for international peace.

Einstein was cremated at Trenton, New Jersey at 4 pm on 18 April 1955 (the day of his death). His ashes were scattered at an undisclosed place.

The above report on Albert Einstein is based on the article by J J O'Connor and E F Robertson.

1.1.3 *Coordinates and the line element*

To describe the electromagnetic field, or any other classical field, one needs a system of coordinates in terms of which the fields are described. Such a coordinate system will include three *spatial* Cartesian coordinates x, y, z, to which we add the *time* coordinate t. In special relativity the united space and time are presented by the line element

$$ds^2 = c^2 dt^2 - \left(dx^2 + dy^2 + dz^2 \right), \qquad (1.1.1)$$

where c is the speed of light in vacuum. The signature is $(+ - - -)$. The four-dimensional spacetime is pseudo-Euclidean. It is assumed that the line element (1.1.1) is invariant under space and time inertial transformations, namely elements of the Lorentz transformation (see Section 1.2).

The three spatial coordinates will be denoted by x^k, where lower case Latin indices $k = 1, 2, 3$, and the time coordinate by $x^0 = ct$, where c is the speed of light in vacuum. The four coordinates will collectively be denoted by x^α, where Greek indices take the values $\alpha = 0, 1, 2, 3$.

1.1.4 *Inertial coordinate system*

A system of coordinates in which the law of inertia holds is called an *inertial coordinate system*. Hence Newton's laws of mechanics are valid only in inertial coordinate systems.

If K is an inertial coordinate system, then every other coordinate system K' is also an inertial system if it is in uniform motion with respect to K. Hence if, relative to K, K' is a uniformly moving coordinate system then the physical laws can be expressed with respect to K' exactly as with respect to K.

It will be seen in the sequel that one of the most important physical consequences of the special relativity theory is the existence of a maximum signal speed in nature, which coincides with the speed of light in empty space. It is therefore natural to define the same time at separate points by means of light signals. This then raises the problem of defining simultaneity.

1.1.5 *Simultaneity*

The definition of simultaneity is made as follows. If light requires the same time to pass across a path $A \to M$ as for a path $B \to M$, where M is in the middle of the distance AB, then we say that the light signals at A and

B started simultaneously if the observer at M detects the two light signals at the same time.

Will two events, which occur simultaneously in one system, also be simultaneous in another system moving with a velocity v with respect to the first one? The answer is *negative*; events which are simultaneous in one coordinate system are not necessarily simultaneous in others. This is so since every inertial system has its own particular time.

1.1.6 *The Galilean transformation*

We have seen that inertial coordinate systems are those which are in uniform, rectilinear, translational motion with respect to each other. Accordingly, inertial systems of coordinates differ from each other by orthogonal rotations, accompanied by translations of the origins of the systems, and by motion with uniform velocities. One can, furthermore, add the translation of the time coordinate thus enabling an arbitrary choice of the origin of time $t = 0$.

Counting the number of parameters which each system of coordinates has with respect to any other, we find that there are ten.

An example of such a transformation between two coordinate systems is given by (see Figure 1.1.1)

$$x' = x + Vt, \quad t' = t, \tag{1.1.2}$$

where the axes y and z were kept unchanged, and V is constant. If we assume that x and x' are the coordinates of a particle in the two systems, and taking now the time derivative of the above equation, we obtain

$$v' = v + V, \tag{1.1.3}$$

where v' and v are the velocities of the particle in the two systems of coordinates. As we see the velocity in the new coordinate system is equal to that in the old coordinate system with the addition of the constant velocity V. This is of course an expected result. There is no problem here. The velocities are added linearly.

1.1.7 *Difficulties with light*

But what happens if x and x' are assumed to be the coordinates of a light pulse? Instead of Eq. (1.1.3) we now obtain

$$c' = c + V, \tag{1.1.4}$$

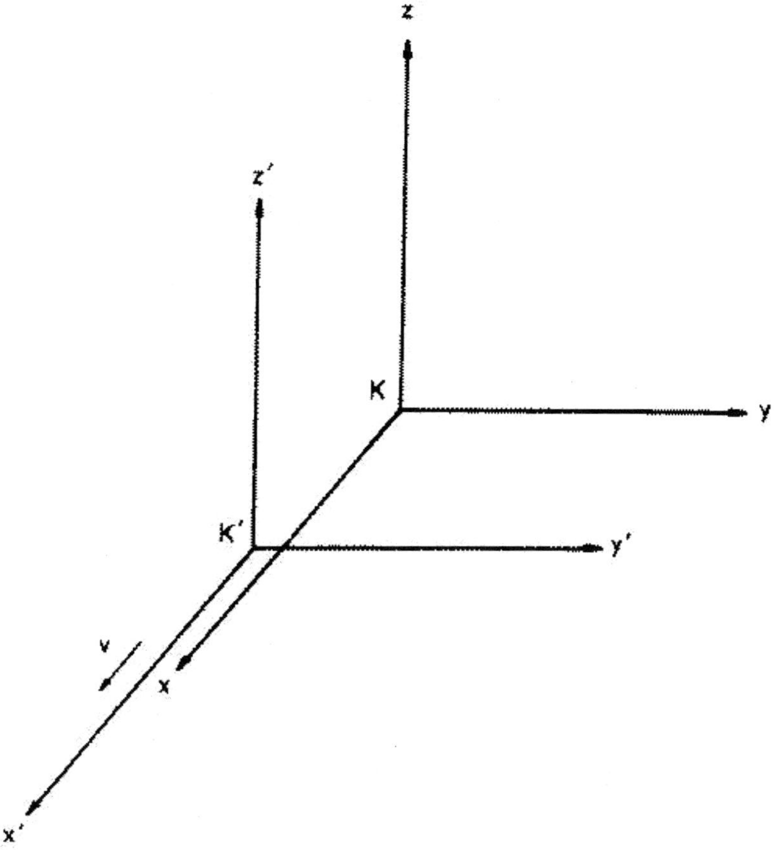

Fig. 1.1.1: Two coordinate systems K and K', one moving with respect to the other with a velocity v in the x-direction.

where c is the speed of light in the old system and c' is that in the new system. Thus we see that the speed of light c' is now larger than c. But this is impossible. It violates the assumption that the speed of light is constant with respect to all moving coordinate systems (one of the two postulates of Einstein's special relativity). Hence, the Galilean transformation is not adequate for light. It has to be generalized to accommodate the weird behavior of light. The generalization of the Galilean transformation then leads to the Lorentz transformation that connects space and time.

1.1.8 *Role of velocity in classical physics*

The role of the velocity in classical physics is not only important but also intrigues. Consider, for example, the Lagrangian of a free particle:

$$L = a\,\mathbf{R} \cdot \mathbf{R} + b\,\mathbf{R} \cdot \dot{\mathbf{R}} + c\,\dot{\mathbf{R}} \cdot \dot{\mathbf{R}},$$

where a, b and c are some constants, $\mathbf{R} = (x, y, z)$ are the Cartesian coordinates of the particle and the overdot denotes a time derivative. The first two terms depend on the location of the particle and hence they cannot represent a physical situation that gives a description of the particle. Only the third term can represent the motion of the particle. Using the Lagrange equation, then $(\partial/\partial\dot{x})L = 2c\dot{x}$. Consequently the Lagrange equation gives

$$2c\ddot{x} = 0,$$

and the same for the coordinates y and z. This is for a free particle. Accordingly, the velocity is constant. Thus when there are no external forces the particle will move with a constant velocity. This is the first of Newton's three laws of motion.

1.1.9 *The Galilean group*

A transformation between inertial coordinate systems which has ten parameters, as described above, is called a *Galilean transformation*. The aggregate of all Galilean transformations provides a group, called the *Galilean group*, which has ten parameters.

One can choose two inertial systems of coordinates so that their corresponding axes are parallel and coincide at $t = 0$. If v is the velocity of one inertial coordinate system with respect to the other, the Galilean transformation can then be reduced into a simple transformation as follows:

$$x' = x - v_x t, \qquad y' = y - v_y t, \qquad z' = z - v_z t, \qquad (1.1.5)$$

where v_x, v_y and v_z are the components of the velocity $\mathbf{v}$ along the x axis, y axis, and z axis, respectively. Of course the Newtonian laws of classical mechanics are invariant under the full ten-parameter Galilean group of transformations, and we have what can be called a *Galilean invariance*.

Finally we mention that if we have three systems of coordinates which move with constant velocities with respect to each other, then one can obtain the velocity between any two systems by adding or subtracting the appropriate relative velocities of the three systems. We shall see later on that this result, which expresses the law of the addition of velocities in classical mechanics, is valid only for velocities much smaller than that of light in the special theory of relativity.

In the next section the Lorentz transformation, a generalization of the Galilean transformation, will be derived.

1.2 The Lorentz Transformation

In the last section we presented the (nonrelativistic) Galilean transformation of spatial coordinates. We are now in a position to generalize this transformation to the relativistic case, where both the time and the spatial coordinates are involved.

1.2.1 *Measuring rods and clocks*

We notice that the assumptions of the existence of measuring rods and clocks are *not* independent of each other. This is so because a light signal, which is reflected back and forth, may provide an ideal clock, remembering that the speed of light in vacuum is constant. (For more details see Bondi.)

1.2.2 *Spatial coordinates and time*

We assume that there exists a rigid body of reference which is moving, and thus provides an inertial coordinate system. In this system, the spatial coordinates then denote the results of measurements that are made with stationary rods.

The time of an event can also be measured in an analogous way. One then needs a way to measure the time differences by a periodic process. A clock at rest in an inertial coordinate system records a local time. The local times of all points of an inertial coordinate system then give *the* time of that system, assuming that the clocks are at rest relative to each other. The times of different inertial coordinate systems are not necessarily identical if light is used to synchronize the clocks.

1.2.3 *Einstein's paradox*

The prerelativity difficulty with light was beautifully illustrated by Einstein ("a paradox upon which I had already hit at the age of sixteen") by the following *gedanken* experiment: If an observer follows a light beam with the velocity c, the beam would be observed as an electromagnetic field at rest which is spatially oscillating. However, this is impossible by Maxwell's equations. In fact, such an observer would see the same as another one who is not moving at all. The above difficulty exemplifies the essence of the special relativity theory. The difficulty is caused by the prerelativistic assumption of the absolute time.

1.2.4 *Apparent incompatibility of the special relativity postulates*

The above difficulty can be formulated in a different way as follows.

According to the Galilean transformation which relates the spatial coordinates and the time between inertial systems in prerelativity, the postulates of the constancy of the speed of light and of the principle of relativity are mutually incompatible, even though both are experimentally valid. The principle of relativity applies, in particular, to the propagation of light, constant speed of which is independent of the choice of inertial system.

The special theory of relativity resolves this impasse as follows.

The above two postulates will be compatible with each other if a new transformation relating the spatial coordinates and times of different inertial systems replaces the Galilean transformation. The new transformation, of course, turns out to be the Lorentz[2] transformation. This, subsequently,

---

[2]Hendrik Antoon Lorentz (Born: 18 July 1853 in Arnhem, Netherlands; Died: 4 Feb 1928 in Haarlem, Netherlands)

Hendrik Lorentz attended Mr Timmer's Primary School in Arnhem till the age of 13; then he entered the new High School there. He entered the University of Leiden in 1870 but, in 1872, he returned to Arnhem to take up teaching evening classes. He worked on his doctorate while holding the teaching post.

In his doctoral thesis, entitled "The theory of the reflection and refraction of light" (1875), Lorentz refined Maxwell's electromagnetic theory. He became professor of mathematical physics at Leiden University in 1878 and remained in this post until he retired in 1912, when Ehrenfest was appointed to his chair. After retiring he retained an honorary position at Leiden, where he continued to lecture.

Before the existence of electrons was proved, Lorentz proposed that light waves were due to oscillations of an electric charge in the atom. Lorentz developed his mathematical theory of the electron, and his student Peter Zeeman verified Lorentz's theoretical work experimentally, demonstrating the effect of a strong magnetic field on the oscillations by measuring the change in the wavelength of the light produced. For this work the Nobel Prize was awarded jointly to Lorentz and Zeeman in 1902.

Lorentz is famed for his work on the contraction in the length of an object at relativistic speeds, named also FitzGerald-Lorentz contraction. Lorentz transformations, introduced in 1904, describe the increase of mass, the shortening of length, and the time dilation of

requires certain behavior of the moving measuring rods and clocks.

The principle of relativity may, thus, alternatively be restated as follows: *The laws of physics should be covariant (or invariant) under the Lorentz transformations* relating different inertial coordinate systems. This Lorentz invariance is in accordance with the Michelson-Morley null experiment which showed that light on the moving Earth spreads with the same speed in all directions.

Consequently, the behavior of light is not incompatible with the principle of relativity. The incompatibility is only apparent.

1.2.5 *Derivation of the Lorentz transformation*

We now return to the problem of apparent disagreement between the law of propagation of light in vacuum and the principle of relativity, and how to resolve it. This then leads to the following question: Given the spatial coordinates and time of an event in an inertial coordinate system K, what are the corresponding quantities of the same event in another inertial system K', taking into account that light rays propagate with the same speed

---

a body moving at speeds close to the velocity of light. They form the basis of Einstein's special theory of relativity.

Lorentz was chairman of the first Solvay Conference in Brussels in 1911, which considered the problems of two approaches: that of classical physics and of quantum theory. However Lorentz never fully accepted quantum theory and always hoped that it would be possible to incorporate it back into the classical approach. He said in his presidential address at the opening ceremony of the conference:

"In this stage of affairs there appeared to us like a wonderful ray of light the beautiful hypothesis of energy elements which was first expounded by Planck and then extended by Einstein and Nernst, and others to many phenomena. It has opened for us unexpected vistas, even those, who consider it with a certain suspicion, must admit its importance and fruitfulness."

Some of Lorentz's numerous publications are: "Théorie Electromagnétique de Maxwell et son application auz Corps Mouvant" (1892) and "Versuch einer Theorie der Elektrischen und Optischen Erscheinungen in bewegten Körpern" (1895), the first systematic appearance of the electrodynamic principle of relativity; "Visible and Invisible Movements" (1901); "Theory of Electrons" (1909), based on a series of lectures at Columbia University; an account of statistical thermodynamic theories, published in French at Leipzig (1916), based on lectures delivered at the Collége de France in 1912; "Lessons on Theoretical Physics" (1919), an edition of his University lectures; "The Einstein Theory of Relativity: A Concise Statement" (1920); "Clerk Maxwell's Electromagnetic Theory" (1924). He was also the author of a textbook of the differential and integral calculus.

Lorentz received a great many honors for his outstanding work. He was elected a Fellow of the Royal Society in 1905. The Society awarded him their Rumford Medal in 1908 and their Copley Medal in 1918. He was given a great respect in The Netherlands.

The above report on Hendrik Antoon Lorentz is based on the article by J J O'Connor and E F Robertson.

c in both systems? The answer to this question leads to the Lorentz transformation determining the values t', x', y', z' of an event with respect to K' from the corresponding magnitudes t, x, y, z of the same event with respect to K, when the law of propagation of light is satisfied in both systems K and K'.

Consider two inertial coordinate systems K and K' whose origins coincide at time $t = 0$, and the events with respect to which are denoted by t, x, y, z and t', x', y', z', respectively. A light pulse emitted from the origin of K will be spread spherically with the speed c, according to the equation

$$x^2 + y^2 + z^2 = c^2 t^2. \tag{1.2.1}$$

Invariance of the speed of light tells us that an observer in K' will also see the light propagating from his origin spherically according to the equation

$$x'^2 + y'^2 + z'^2 = c^2 t'^2. \tag{1.2.2}$$

From Eqs. (1.2.1) and (1.2.2) one then obtains

$$c^2 t'^2 - \left(x'^2 + y'^2 + z'^2 \right) = c^2 t^2 - \left(x^2 + y^2 + z^2 \right), \tag{1.2.3}$$

or

$$\eta_{\mu\nu} x'^{\mu} x'^{\nu} = \eta_{\mu\nu} x^{\mu} x^{\nu}, \tag{1.2.4}$$

where x^{μ} and x'^{μ} are defined by

$$x^{\mu} = (ct,\ x,\ y,\ z),\ \ x'^{\mu} = (ct',\ x',\ y',\ z'), \tag{1.2.5}$$

and the symbol $\eta_{\mu\nu}$ (and later on $\eta^{\mu\nu}$) is the flat-space metric, given by the matrix

$$\eta = \begin{pmatrix} +1 & 0 & 0 & 0 \\ 0 & -1 & 0 & 0 \\ 0 & 0 & -1 & 0 \\ 0 & 0 & 0 & -1 \end{pmatrix}. \tag{1.2.6}$$

In the above equations, and throughout the following, repeated indices indicate the use of the summation convention.

We will seek a linear transformation of the form

$$x'^{\mu} = \Lambda^{\mu}{}_{\nu} x^{\nu} \tag{1.2.7}$$

between the times and spatial coordinates of the two inertial systems K and K'. Using matrix notation, Eqs. (1.2.4) and (1.2.7) can then be written in the form

$$x'^{t} \eta x' = x^{t} \eta x \tag{1.2.8}$$

and

$$x' = \Lambda x, \tag{1.2.9}$$

respectively. Here x and x' are the one-column matrices

$$x = \begin{pmatrix} x^0 \\ x^1 \\ x^2 \\ x^3 \end{pmatrix}, \ x' = \begin{pmatrix} x'^0 \\ x'^1 \\ x'^2 \\ x'^3 \end{pmatrix}, \tag{1.2.10}$$

x^t and x'^t are the transposed matrices to the matrices x and x', respectively, and Λ is the 4×4 matrix whose elements are $\Lambda^\mu{}_\nu$.

Using now Eq. (1.2.9) in Eq. (1.2.8) then gives

$$x^t \Lambda^t \eta \Lambda x = x^t \eta x, \tag{1.2.11}$$

from which we obtain the condition

$$\Lambda^t \eta \Lambda = \eta \tag{1.2.12}$$

that the 4×4 matrix Λ of the Lorentz transformation has to satisfy. Equation (1.2.12) is a generalization of the familiar relation

$$R^t I R = I, \tag{1.2.13}$$

which the 3×3 orthogonal matrix R, describing ordinary rotations of the spatial coordinates alone, satisfies. The essential difference between the two cases is in the replacement of the unit matrix I in the ordinary three-dimensional rotations by the matrix η in the four-dimensional Lorentz transformations.

Hence the transformation we are looking after is a "rotation" in a four-dimensional spacetime which consists of the time and the three dimensions of the ordinary space. Such a spacetime is usually called the *Minkowskian spacetime*.

The Lorentz transformation is thus the "orthogonal" transformation of the Minkowskian spacetime.

We now derive the Lorentz transformation connecting the two coordinate systems K and K' when they have the same orientations and their origins coincide at $t = 0$, but K' moves along the coordinate x with a velocity v.

The directions perpendicular to the motion are obviously left unaffected by the transformation. Hence

$$x'^2 = x^2, \ x'^3 = x^3, \tag{1.2.14}$$

and only the x^0 and x^1 coordinates require changes when transforming from one system to the other. One will therefore have the form

$$\Lambda = \begin{pmatrix} \Lambda^0{}_0 & \Lambda^0{}_1 & 0 & 0 \\ \Lambda^1{}_0 & \Lambda^1{}_1 & 0 & 0 \\ 0 & 0 & 1 & 0 \\ 0 & 0 & 0 & 1 \end{pmatrix} \tag{1.2.15}$$

for the matrix of the Lorentz transformation in our particular case.

The "orthogonality" condition (1.2.12) then yields the equation

$$\eta_{CD}\Lambda^C{}_A\Lambda^D{}_B = \eta_{AB}, \tag{1.2.16}$$

where the indices A, B, C, $D = 0$, 1, and $\eta_{00} = -\eta_{11} = 1$, $\eta_{01} = \eta_{10} = 0$. The above formula gives three relations connecting the four elements of the matrix (1.2.15):

$$\left(\Lambda^0{}_0\right)^2 - \left(\Lambda^1{}_0\right)^2 = 1,$$

$$\left(\Lambda^0{}_1\right)^2 - \left(\Lambda^1{}_1\right)^2 = -1, \tag{1.2.17}$$

$$\Lambda^0{}_0\Lambda^0{}_1 - \Lambda^1{}_0\Lambda^1{}_1 = 0.$$

The solution of these equations can therefore be determined up to an arbitrary parameter. One then finds that

$$\begin{aligned} \Lambda^0{}_0 &= \cosh\psi, \quad \Lambda^0{}_1 = \sinh\psi, \\ \Lambda^1{}_0 &= \sinh\psi, \quad \Lambda^1{}_1 = \cosh\psi. \end{aligned} \tag{1.2.18}$$

is such an appropriate solution. With these values for the four elements, we obtain

$$\Lambda = \begin{pmatrix} \cosh\psi & \sinh\psi & 0 & 0 \\ \sinh\psi & \cosh\psi & 0 & 0 \\ 0 & 0 & 1 & 0 \\ 0 & 0 & 0 & 1 \end{pmatrix} \tag{1.2.19}$$

for the matrix (1.2.15) of the Lorentz transformation.

The parameter ψ is related to the relative velocity v between the two inertial coordinate systems K and K'. The relationship between them is found by determining the motion of the origin of the coordinate system K as seen from K', for instance. This motion is determined by putting $x^1 = 0$

in the Lorentz transformation (1.2.9), using Eqs. (1.2.10) and (1.2.19). This gives

$$x'^0 = x^0 \cosh\psi,$$
$$x'^1 = x^0 \sinh\psi. \tag{1.2.20}$$

We therefore obtain

$$\frac{x'^1}{x'^0} = \frac{1}{c}\frac{x'}{t'} = -\beta = \tanh\psi, \tag{1.2.21}$$

where the parameter β is defined by

$$\beta = \frac{v}{c}. \tag{1.2.22}$$

Accordingly we obtain from Eq. (1.2.21)

$$\cosh\psi = \frac{1}{\sqrt{1-\beta^2}}, \tag{1.2.23a}$$

$$\sinh\psi = \frac{-\beta}{\sqrt{1-\beta^2}}. \tag{1.2.23b}$$

Using these results in Eq. (1.2.19) then yields

$$\Lambda = \begin{pmatrix} \dfrac{1}{\sqrt{1-\beta^2}} & \dfrac{-\beta}{\sqrt{1-\beta^2}} & 0 & 0 \\ \dfrac{-\beta}{\sqrt{1-\beta^2}} & \dfrac{1}{\sqrt{1-\beta^2}} & 0 & 0 \\ 0 & 0 & 1 & 0 \\ 0 & 0 & 0 & 1 \end{pmatrix} \tag{1.2.24}$$

for the matrix of the Lorentz transformation. We also obtain

$$\Lambda^{-1} = \begin{pmatrix} \dfrac{1}{\sqrt{1-\beta^2}} & \dfrac{\beta}{\sqrt{1-\beta^2}} & 0 & 0 \\ \dfrac{\beta}{\sqrt{1-\beta^2}} & \dfrac{1}{\sqrt{1-\beta^2}} & 0 & 0 \\ 0 & 0 & 1 & 0 \\ 0 & 0 & 0 & 1 \end{pmatrix} \tag{1.2.25}$$

for the inverse matrix describing the inverse Lorentz transformation.

The Lorentz transformation along the x axis is therefore given by

$$ct' = \frac{ct - \beta x}{\sqrt{1-\beta^2}}, \tag{1.2.26a}$$

$$x' = \frac{x - \beta ct}{\sqrt{1 - \beta^2}}, \qquad (1.2.26b)$$

$$y' = y, \; z' = z. \qquad (1.2.26c)$$

We also obtain

$$ct = \frac{ct' + \beta x'}{\sqrt{1 - \beta^2}}, \qquad (1.2.27a)$$

$$x = \frac{x' + \beta ct'}{\sqrt{1 - \beta^2}}, \qquad (1.2.27b)$$

$$y = y', \; z = z', \qquad (1.2.27c)$$

for the inverse transformation from the coordinates x'^{μ} back to x^{μ}. Different proofs to Eqs. (1.2.26) are given in Problem 1.4.1. They can also be proved by using an analogous method to that used in Section 2.11 for deriving the cosmological transformation.

Equations (1.2.27) show that the inverse transformation differs from Eqs. (1.2.26) only by a change in the sign of v. This result is obvious since the coordinate system K is moving relative to the system K' with the velocity $-v$.

A Lorentz transformation involving the time coordinate x^0 and one or more spatial coordinates x^k, such as that derived above, is often called a *boost*. A Lorentz transformation which keeps the time coordinate unchanged is, of course, just an ordinary three-dimensional rotation of the spatial coordinates.

In the same way one can find the other Lorentz transformations along the y axis and the z axis.

Finally we notice that if one neglects nonlinear terms in v/c in the Lorentz transformation (1.2.26), one obtains the *approximate Lorentz transformation* (see Problem 1.4.3):

$$\begin{aligned} x'^0 &= x^0 - \beta x^1, \\ x'^1 &= x^1 - \beta x^0, \\ x'^2 &= x^2, \\ x'^3 &= x^3, \end{aligned} \qquad (1.2.28)$$

in which the coordinates x^0 and x^1 appear on the same footing.

If one, in addition, neglects the term with β in the first of the above equations, one then obtains the nonrelativistic transformation

$$x' = x - vt, \qquad y' = y, \qquad z' = z, \qquad t' = t. \qquad (1.2.29)$$

Here the time and the coordinate x do not appear on the same footing.

Equations (1.2.29), of course, describe a Galilean transformation along the x axis. It is a particular case of the Galilean transformation, discussed in Section 1.1 and given by Eqs. (1.1.2), for which $v_x = v$ and $v_y = v_z = 0$.

In the next section the very important diagram, known as the light cone, that gives a geometrical presentation of the Universe in terms of past, present and future, is presented.

1.3   The Light Cone

In the four-dimensional spacetime, we choose an inertial coordinate system with coordinates $x^\alpha = (ct,\ x,\ y,\ z)$ and O describes the origin of the coordinate system. We now examine how other events are related to O. (See Figure 1.3.1.)

A finite-mass particle moving with a constant velocity and passing through O will be represented by a straight line. The inclination of this straight line with respect to the x axis is such that the cotangent of its angle is equal to v/c, where v is the ordinary velocity. But the maximum of such a velocity is c. Hence the minimum angle is $\pi/4$. Accordingly, we have a cone which is represented by the formula

$$c^2 t^2 - \left(x^2 + y^2 + z^2\right) = 0, \qquad (1.3.1)$$

called the *light cone*, whose symmetry axis coincides with the x^0 axis.

Figure 1.3.1 describes the light cone in two dimensions, one of which is the coordinate $x^0\,(= ct)$, whereas the other is the coordinate $x^1\,(= x)$. The propagation of two light signals in opposite directions passing through $x = 0$ at time $t = 0$, is represented by the two diagonal straight lines. The motion of finite-mass particles, on the other hand, is represented by straight lines in the interior of the light cone.

1.3.1   *Events and coordinate systems*

Let now two points in the four-dimensional Minkowskian spacetime be given by

$$\begin{aligned}
x_1^\alpha &= (ct_1,\ x_1,\ y_1,\ z_1), \\
x_2^\alpha &= (ct_2,\ x_2,\ y_2,\ z_2),
\end{aligned} \qquad (1.3.2)$$

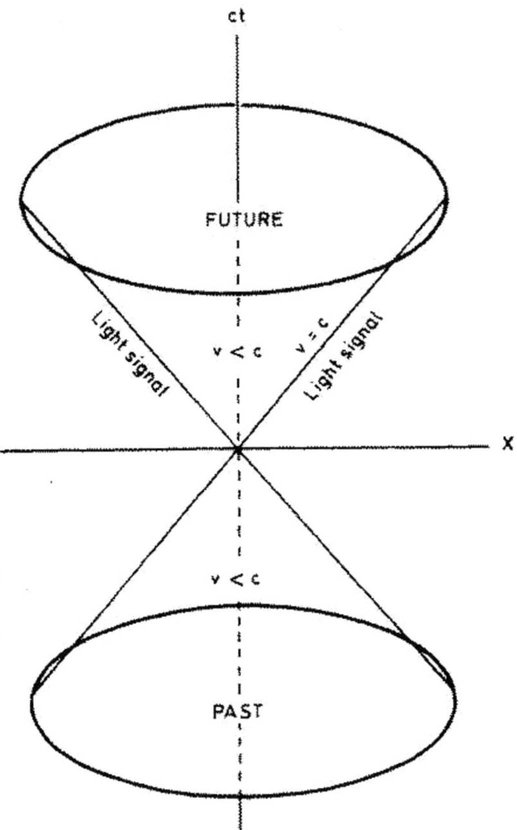

Fig. 1.3.1:   The light cone in two dimensions, $x^0 (= ct)$ and $x^1 (= x)$. The propagation of two light signals in opposite directions passing through $x = 0$ at time $t = 0$, is represented by the two diagonal straight lines. The motion of finite-mass particles, on the other hand, are represented by straight lines in the interior of the light cone. (Compare the galaxy cone given in Figure 2.3.1.)

in an inertial coordinate system K, and by

$$
\begin{aligned}
x_1'^{\alpha} &= (ct_1',\ x_1',\ y_1',\ z_1')\,, \\
x_2'^{\alpha} &= (ct_2',\ x_2',\ y_2',\ z_2')\,,
\end{aligned}
\qquad (1.3.3)
$$

in another system K'.

The *interval four-vectors* between the two points are defined by

$$X^\alpha = (cT,\ X,\ Y,\ Z)$$
$$= [c\,(t_2 - t_1),\ (x_2 - x_1),\ (y_2 - y_1),\ (z_2 - z_1)], \tag{1.3.4a}$$

and

$$X'^\alpha = (cT',\ X',\ Y',\ Z')$$
$$= [c\,(t'_2 - t'_1),\ (x'_2 - x'_1),\ (y'_2 - y'_1),\ (z'_2 - z'_1)], \tag{1.3.4b}$$

in the two coordinate systems K and K', respectively. The squares of the intervals between the two events in the systems K and K' are then given by

$$X_\alpha X^\alpha = c^2 T^2 - R^2,$$
$$X'_\alpha X'^\alpha = c^2 T'^2 - R'^2, \tag{1.3.5}$$

where

$$R^2 = X^2 + Y^2 + Z^2,$$
$$R'^2 = X'^2 + Y'^2 + Z'^2 \tag{1.3.6}$$

are the three-dimensional distances in K and K'. Lorentz invariance then requires that

$$X_\alpha X^\alpha = c^2 T^2 - R^2 = c^2 T'^2 - R'^2 = X'_\alpha X'^\alpha. \tag{1.3.7}$$

1.3.2 *Future and past*

Consider now the events within the light cone (see Figure 1.3.1) in which $c^2 t^2 - x^2 > 0$ and therefore the interval four-vector X^α is timelike. The upper part of the cone is called the *future* since all the events in it occur after O, and there is no coordinate system in which events can occur before O.

A similar analysis shows that all the events in the lower part of the light cone occur before O in all coordinate systems. This part is consequently called the *past*. For any event outside the light cone there exist coordinate systems in which the event occurs after and before O, and one system in which it occurs simultaneously with O.

In the next section the very important concept in theoretical physics, the Lorentz group, is presented in details.

1.3.3 *Problems*

P 1.3.1 Show that if the interval four-vector X^α between two events is timelike in an inertial coordinate system K, then there exists a coordinate system K' in which the two events occur at the same spatial point at different times. The time difference of the two events in K' is given by $T' = \sqrt{X_\alpha X^\alpha}/c$. Show also that if the interval four-vector X^α between two events in the system K is spacelike, then there exists a coordinate system K' in which the two events occur simultaneously at separate spatial points with the distance $R' = -\sqrt{X_\alpha X^\alpha}$ between them.

Solution: The solution is left for the reader.

1.4   The Lorentz Group

We now give a brief discussion on the groups which can be obtained from the Lorentz transformations. (For representations of the Lorentz group see M. Carmeli and S. Malin, *Representations of the Rotation and Lorentz Groups: An Introduction.*)

The Lorentz transformations form a group called the (homogeneous) *Lorentz group*. It is a subgroup of the *inhomogeneous Lorentz group*, also known as the *Poincaré group*. The latter group is formed from the *inhomogeneous* Lorentz transformations

$$x'^\mu = \Lambda^\mu{}_\nu x^\nu + x_0^\mu, \tag{1.4.1}$$

where x_0^μ describes *translations*.

The Lorentz group possesses four disconnected parts which arise as follows.

Equation (1.2.12) shows that $(\det\Lambda)^2 = 1$. Accordingly, the determinant of every Lorentz transformation is equal to either $+1$,

$$\det\Lambda = +1, \tag{1.4.2}$$

in which case the transformation is called *proper*, or to -1,

$$\det\Lambda = -1, \tag{1.4.3}$$

in which it is called *improper*.

From Eq. (1.2.12) when written with indices,

$$\eta_{\mu\nu}\Lambda^\mu{}_\alpha\Lambda^\nu{}_\beta = \eta_{\alpha\beta}, \tag{1.4.4}$$

and taking $\alpha = \beta = 0$, one obtains

$$\left(\Lambda^0{}_0\right)^2 - \left(\Lambda^1{}_0\right)^2 - \left(\Lambda^2{}_0\right)^2 - \left(\Lambda^3{}_0\right)^2 = 1. \tag{1.4.5}$$

Therefore $\left(\Lambda^0{}_0\right)^2 \geq 1$, and consequently we have either

$$\Lambda^0{}_0 \geq +1, \tag{1.4.6}$$

in which case the transformation is called *orthochronous*, or

$$\Lambda^0{}_0 \leq -1. \tag{1.4.7}$$

The aggregate of all orthochronous Lorentz transformations provides a subgroup of the Lorentz group.

The four parts of the Lorentz group are described as follows:

(1) $L_+^\uparrow$: $\det\Lambda = +1$, $\Lambda^0{}_0 \geq +1$. This part contains the identity element of the group. The aggregate of all proper, orthochronous, Lorentz transformations provides a group, which is a subgroup of the Lorentz group. It is called the *proper, orthochronous, Lorentz group.*

(2) $L_-^\uparrow$: $\det\Lambda = -1$, $\Lambda^0{}_0 \geq +1$. This part contains a *space inversion* element S which describes a reflection relative to the three spatial axes:

$$x'^0 = x^0, \quad x'^1 = -x^1, \quad x'^2 = -x^2, \quad x'^3 = -x^3. \tag{1.4.8}$$

(3) $L_-^\downarrow$: $\det\Lambda = -1$, $\Lambda^0{}_0 \leq -1$. This part contains a *time reversal* element T which describes a reflection relative to the time axis:

$$x'^0 = -x^0, \quad x'^1 = x^1, \quad x'^2 = x^2, \quad x'^3 = x^3. \tag{1.4.9}$$

(4) $L_+^\downarrow$: $\det\Lambda = +1$, $\Lambda^0{}_0 \leq -1$. This part contains the element ST.

As was mentioned before, from the above four parts of the Lorentz group one obtains the subgroup $L^\uparrow = L_+^\uparrow \bigcup L_-^\uparrow$ (the union of $L_+^\uparrow$ and $L_-^\uparrow$), called the *orthochronous Lorentz group.* Likewise, the subgroup $L_+ = L_+^\uparrow \bigcup L_+^\downarrow$, called the *proper Lorentz group*, is obtained.

Finally, we notice that every improper Lorentz transformation can be written in the form

$$\Lambda = S\Lambda_p, \tag{1.4.10}$$

where S is a space-inversion element and Λ_p is a proper Lorentz transformation.

In the next section some important consequences of the Lorentz group are drawn.

1.4.1 *Problems*

P 1.4.1 Let a light ray be emitted from the origin of a moving system K' at the time t_0' along the x axis to $\tilde{x} = x - vt$, where v is the velocity of K' relative to the "stationary" system K. Let the ray then be reflected back at the time t_1' to the origin of the coordinates, arriving there at time t_2'. We then have

$$\frac{1}{2}\left(t_0' + t_2'\right) = t_1'. \tag{1}$$

Derive the Lorentz transformation (1.2.26) by inserting in Eq. (1) the arguments of the function t', and applying the postulate of the constancy of the speed of light in the system K. (This is Einstein's historical derivation of the Lorentz transformation. The solution given below is a concise version of the original one.)

Solution: The function t' depends on the four coordinates of the system K, namely $t' = t'(t,\, x,\, y,\, z)$. Inserting these arguments in Eq. (1), and using the postulate of the constancy of the speed of light in the system K, one obtains after a lengthy but straightforward calculation,

$$\frac{v}{c^2 - v^2}\frac{\partial t'}{\partial t} + \frac{\partial t'}{\partial \tilde{x}} = 0. \tag{2}$$

Likewise, one obtains

$$\frac{\partial t'}{\partial y} = 0, \quad \frac{\partial t'}{\partial z} = 0, \tag{3}$$

since light always propagates along the y and z axes, when viewed from the stationary system K, with speed $\left(c^2 - v^2\right)^{1/2}$.

If one now assumes, furthermore, that t' is a *linear* function of its arguments, and using Eq. (2), then one finds

$$t' = a\left(t - \frac{v}{c^2 - v^2}\tilde{x}\right), \tag{4}$$

where a is a function of v (at present unknown), and it is assumed that at the origin of K', $t' = 0$ when $t = 0$.

Using Eq. (4) one can then determine the coordinates x', y', z', taking into account the fact that light propagates with the speed c when measured in the moving system K'. For a ray of light emitted at time $t' = 0$ along the x' axis, one accordingly has

$$x' = ct' = ca\left(t - \frac{v}{c^2 - v^2}\tilde{x}\right). \tag{5}$$

But the ray satisfies the propagation equation $x - ct$, and therefore

$$\tilde{x} = x - vt = (c - v)\,t, \tag{6}$$

or

$$t = \frac{\tilde{x}}{c - v}. \tag{7}$$

Using now this expression for t in Eq. (5) one obtains

$$x' = \frac{c^2 a\tilde{x}}{c^2 - v^2}. \tag{8}$$

Likewise, one can determine the coordinate y' and z' by considering rays moving along them. One obtains

$$y' = \frac{cay}{\sqrt{c^2 - v^2}}, \tag{9a}$$

$$z' = \frac{caz}{\sqrt{c^2 - v^2}}. \tag{9b}$$

Equations (4), (8) and (9) express the dependence of the coordinates t', x', y', z' on t, x, y, z, provided one substitutes for $\tilde{x}$ its value $x - vt$.

A straightforward calculation then gives:

$$\begin{aligned}
t' &= \phi(v)\,\gamma(v)\left(t - \frac{vx}{c^2}\right), \\
x' &= \phi(v)\,\gamma(v)\,(x - vt), \\
y' &= \phi(v)\,y, \\
z' &= \phi(v)\,z.
\end{aligned} \tag{10}$$

Here $\gamma(v)$ is defined by

$$\gamma(v) = \frac{1}{\sqrt{1 - \dfrac{v^2}{c^2}}}, \tag{11}$$

and $\phi(v)$ is a new unknown function of v, related to the function a (of v also) by

$$\phi(v) = \frac{a}{\sqrt{1 - \dfrac{v^2}{c^2}}}. \tag{12}$$

The function $\phi(v)$ is left, as yet, arbitrary.

In order to determine $\phi(v)$ one introduces a third coordinate system K'' which is moving with the velocity $-v$ relative to K', and writes the transformation law of its coordinates t'', x'', y'', z'' in terms of those of K and K'. One then obtains, after a lengthy but straightforward calculation,

$$\phi(v) = 1, \tag{13}$$

and

$$\begin{aligned}
t' &= \gamma\left(t - \frac{vx}{c^2}\right), \\
x' &= \gamma\,(x - vt), \\
y' &= y, \\
z' &= z,
\end{aligned} \tag{14}$$

for the transformation of the time and the spatial coordinates. In the above formulas γ is given by Eq. (11).

Equations (14) are, of course, those of the Lorentz transformation, Eqs. (1.2.26), given in the text.

P 1.4.2 Derive the Lorentz transformation by considering the transmission of light signals along the positive and negative parts of the x axis. (Also given by Einstein. The solution given below is a concise version of the original one.)

Solution: One expresses the coordinates x' and t' in terms of x and t, where unprimed and primed quantities refer to the systems K and K', respectively.

Let a light signal be transmitted along the positive x axis. Then its equation of propagation is given by

$$x = ct, \tag{1}$$

where c is the speed of light in vacuum. Relative to the system K', the light signal also propagates with the speed c. Accordingly,

$$x' - ct' = 0 \tag{2}$$

represents the propagation of light in the system K'. Spacetime events which satisfy Eq. (1) must also satisfy Eq. (2). This will, indeed, be the case if a relation of the form

$$(x' - ct') = \lambda (x - ct) \tag{3}$$

is fulfilled, where λ is a constant. Equation (3) shows that the vanishing of $(x - ct)$ yields the vanishing of $(x' - ct')$.

Light rays transmitted along the negative x axis, likewise, satisfy

$$(x' + ct') = \mu (x + ct), \tag{4}$$

where μ is a constant. From Eqs. (3) and (4) one then obtains a linear transformation between the variables ct and x,

$$ct' = act - bx, \tag{5a}$$

$$x' = ax - bct, \tag{5b}$$

where a and b are two new constants related to λ and μ by

$$a = \frac{1}{2} (\lambda + \mu), \tag{6a}$$

$$b = \frac{1}{2} (\lambda - \mu). \tag{6b}$$

It remains to determine the constants a and b in terms of the relative velocity v between the two systems K and K'.

A lengthy, but straightforward, calculation then yields

$$a = \frac{1}{\sqrt{1 - \dfrac{v^2}{c^2}}}, \quad b = \frac{v/c}{\sqrt{1 - \dfrac{v^2}{c^2}}}. \tag{7}$$

Using these expressions for the constants a and b in Eqs. (5) then gives

$$ct' = \frac{ct - \beta x}{\sqrt{1 - \beta^2}}, \quad x' = \frac{x - \beta ct}{\sqrt{1 - \beta^2}}, \tag{8a}$$

where $\beta = v/c$. When supplemented by the relations

$$y' = y, \; z' = z, \tag{8b}$$

one thus obtains the Lorentz transformation which was given in the text by Eqs. (1.2.26).

P 1.4.3 Find a nonrelativistic approximation to the Lorentz transformation which, unlike the Galilean one, contains at the same footing both the spatial coordinates and time.

Solution: Such an approximate transformation, when neglecting nonlinear terms in v/c, is given by

$$x'^0 = x^0 - \frac{v}{c}x^1,$$

$$x'^1 = x^1 - \frac{v}{c}x^0, \tag{1}$$

$$x'^2 = x^2, \; x'^3 = x^3,$$

if the coordinates x^2 and x^3 are kept untransformed. Here $x^0 = ct$, $x^1 = x$, $x^2 = y$, $x^3 = z$, and c is the speed of light in vacuum.

One then finds that (neglecting nonlinear terms in v/c)

$$c^2 t'^2 - \left(x'^2 + y'^2 + z'^2\right) = c^2 t^2 - \left(x^2 + y^2 + z^2\right). \tag{2}$$

Moreover, the aggregate of all the transformations (1) provides a group (keeping only linear terms in v/c).

1.5 Consequences of the Lorentz Transformation

We now draw some important consequences from the Lorentz transformation derived in the Section 1.2. These will include the Lorentz contraction of lengths, the dilation of time scales, and the law of addition of velocities. More implications of the Lorentz transformation will be discussed in the sequel.

1.5.1   *Nonrelativistic limit*

In the limit of small velocities relative to the speed of light c, namely $\beta \ll 1$, Eqs. (1.2.26) are easily seen to be reduced to the Galilean transformation.

1.5.2   *The Lorentz contraction of lengths*

Consider a rod whose length is one unit, placed along the x' axis of the system K' between the points $x' = 0$ and $x' = 1$. What is the length of this rod relative to the system K?

The above question is answered by determining the ends of the rod in K at a particular time t of K. Taking $t = 0$, for instance, and using the second formula of Eqs. (1.2.26), we obtain

$$x = x'\sqrt{1 - \beta^2}; \;\; \beta = v/c. \tag{1.5.1}$$

The values of the ends of the rod in K are then obtained from Eq. (1.5.1) by putting $x' = 0$ and $x' = 1$. This gives $x = 0$ and $x = \sqrt{1 - \beta^2}$, respectively. The length of the rod in K is thus $\sqrt{1 - \beta^2}$, rather than unit.

But the rod is moving relative to K with the velocity v. Hence the length of a rod moving with the velocity v is seen to be contracted to $\sqrt{1 - \beta^2}$ times its length; it is shorter when in motion than when at rest, a result known as the *Lorentz contraction of length*.

On the contrary, if the rod would have been placed at rest along the x axis in the coordinate system K, its length would also be contracted by the same factor $\sqrt{1 - \beta^2}$ as seen from K'. This result is, of course, in accordance with the principle of relativity.

Of course no contraction effect can be obtained using the Galilean transformation.

It will be seen that the phenomenon of the length contraction exists also in the cosmological special relativity as discussed in Subsection 2.4.2.

1.5.3   *The dilation of time*

The dilation of time scale can, similarly, be shown as follows.

Let a clock be placed at the origin ($x' = 0$) of the coordinate system K', and let $t' = 0$ and $t' = 1$ be two of its successive ticks. From the first two equations of the Lorentz transformation (1.2.26), when $x' = 0$, we obtain

$$t' = t\sqrt{1 - \beta^2}; \;\; \beta = \frac{v}{c}. \tag{1.5.2}$$

Hence for $t' = 0$ and $t' = 1$ we obtain

$$t = 0, \tag{1.5.3}$$

and

$$t = \frac{1}{\sqrt{1 - \beta^2}}, \tag{1.5.4}$$

respectively.

As observed from the coordinate system K, the clock is moving with the velocity v, and the time which elapses between two of its successive strokes is not one second but $1/\sqrt{1 - \beta^2}$ seconds, namely a longer time. Thus the clock runs *more slowly* when it is in motion than when it is at rest. Such a phenomenon is called the *dilation of time.*

The same conclusion would have been reached, of course, if the clock was placed in the system K and its time was judged from K'. Again the clock would be seen to run slower.

The analogous phenomenon to this in cosmological relativity is, of course, the velocity contraction given in Subsection 2.4.3.

1.5.4   *The addition of velocities law*

It might be thought possible to obtain a velocity greater than c by letting a particle move with a velocity w along the x' axis in K', and consider the motion from the system K with respect to which K' moves with the speed v along the x axis. Let the velocity of the particle be V with respect to K, and hence

$$x = Vt. \tag{1.5.5}$$

Classical mechanics, of course, gives $V = v + w$. However, this is not the case in special relativity. To see this we apply the Lorentz transformation, given by Eqs. (1.2.26), to the relation

$$x' = wt'. \tag{1.5.6}$$

A straightforward calculation then gives

$$x = \frac{(v + w)}{1 + \dfrac{vw}{c^2}} t = Vt, \tag{1.5.7}$$

and accordingly

$$V = \frac{v + w}{1 + \dfrac{vw}{c^2}}. \tag{1.5.8}$$

Equation (1.5.8) is called the *addition of velocities law.*

The same result is obtained by considering three coordinate systems, the second moves with the velocity v with respect to the first, and the third moves with the velocity w with respect to the second system. One may find the Lorentz transformation from the first to the third system directly by multiplying the matrices of the two separate transformations. It is then found that the total transformation corresponds to a velocity V given by Eq. (1.5.8), or equivalently by

$$\frac{V}{c} = \frac{\dfrac{v}{c} + \dfrac{w}{c}}{1 + \dfrac{vw}{c^2}}. \tag{1.5.9}$$

It is seen from Eq. (1.5.9) that V/c is always less than unity. The proof of Eq. (1.5.9) is left for the reader.

From the above discussion we also conclude that, in special relativity, the speed of light c is a limiting velocity. Namely, the velocity c can be neither reached nor exceeded by a finite-mass particle. The same conclusion clearly follows from the Lorentz transformation itself since it becomes meaningless for values of v larger than c.

The analogous law to this in cosmological special relativity is, of course, the law of addition of cosmic times given in Subsection 2.4.4.

In the next section the four-dimensional structure of spacetime is discussed.

1.5.5 *Problems*

P 1.5.1 Derive the formulas relating the velocity of a particle in one inertial coordinate system K to that in a second system K', where K' moves relative to K with the velocity V along the x axis. Use Eqs. (1.2.26) to show that (see Landau and Lifshitz, 1959)

$$v'_x = \frac{v_x - V}{1 - \dfrac{V v_x}{c^2}}, \tag{1a}$$

$$v'_y = \frac{v_y \sqrt{1 - \dfrac{V^2}{c^2}}}{1 - \dfrac{V v_x}{c^2}}, \tag{1b}$$

$$v_z' = \frac{v_z\sqrt{1 - \dfrac{V^2}{c^2}}}{1 - \dfrac{V v_x}{c^2}}, \tag{1c}$$

where $v_x = dx/dt$, etc. and $v_x' = dx'/dt'$, etc.

Use Eqs. (1.2.27), likewise, to show that the inverse transformation is given by

$$v_x = \frac{v_x' + V}{1 + \dfrac{V v_x'}{c^2}}, \tag{2a}$$

$$v_y = \frac{v_y'\sqrt{1 - \dfrac{V^2}{c^2}}}{1 + \dfrac{V v_x'}{c^2}}, \tag{2b}$$

$$v_z = \frac{v_z'\sqrt{1 - \dfrac{V^2}{c^2}}}{1 + \dfrac{V v_x'}{c^2}}. \tag{2c}$$

Solution: The solution is left for the reader.

P 1.5.2 Find the change in the direction of the velocity under the transition from one coordinate system K to another system K' moving with the velocity V with respect to K along the x axis. To this end, consider a particle moving in two dimensions. Then

$$v_x = v\cos\theta, \quad v_y = v\sin\theta, \tag{1}$$

$$v_x' = v'\cos\theta', \quad v_y' = v'\sin\theta', \tag{2}$$

for instance, in the two coordinate systems K and K'. Use Eqs. (1) and (2) in Eqs. (1) and (2) of the previous problem, and show that

$$\tan\theta' = \frac{v\sqrt{1 - \beta^2}\sin\theta}{v\cos\theta - V}, \tag{3a}$$

$$\tan\theta = \frac{v'\sqrt{1 - \beta^2}\sin\theta'}{v'\cos\theta' + V}, \tag{3b}$$

where $\beta = V/c$.

Solution: The solution is left for the reader.

P 1.5.3 Apply the results of Problems 1.5.1 and 1.5.2 to derive the aberration of light formula (see Landau and Lifshitz, 1959).

Solution: The apparent change in the direction of propagation of light under the transition from one inertial coordinate system to another can be obtained, using Eqs. (1) and (2) of Problem 1.5.1 and Eqs. (1) and (2) of Problem 1.5.2, by taking $v = v' = c$. One then gets

$$\cos \theta' = \frac{\cos \theta - \beta}{1 - \beta \cos \theta}, \tag{1}$$

$$\cos \theta = \frac{\cos \theta' + \beta}{1 + \beta \cos \theta'}. \tag{2}$$

Here the angles θ and θ' refer to the coordinate systems K and K', respectively, and $\beta = V/c$, where V is the velocity of K' with respect to K.

Assuming now that the relative velocity between the two coordinate systems is much smaller than the speed of light, $\beta \ll 1$, one easily finds, using Eq. (1),

$$\cos \theta' \approx \cos \theta - \beta \sin^2 \theta. \tag{3}$$

The *aberration angle* $\Delta\theta = \theta' - \theta$, using a straightforward calculation, is then given by

$$\Delta\theta \approx \cot \theta - \frac{\cos \theta'}{\sin \theta}. \tag{4}$$

Using Eq. (3) in Eq. (4) then gives

$$\Delta\theta \approx \beta \sin \theta = \frac{V}{c} \sin \theta, \tag{5}$$

which is the *aberration of light formula*.

1.6　The Structure of Spacetime

In the last section the foundations of special relativity were presented. These were based on the postulates of the principle of relativity and on the experimental fact that the speed of propagation of light in vacuum is constant. When put together, these two postulates then led us to the mixing of the time and the spatial coordinates. As a consequence, we obtained the Lorentz transformation as a generalization of the nonrelativistic Galilean transformation. Thus events are expressed in terms of the time t and the three spatial coordinates x, y, z. Consequently, we are naturally led to the notion of a four-dimensional spacetime. Such a spacetime was indeed introduced by Minkowski[3] in 1908. In this chapter the four-dimensional

---

[3]Hermann Minkowski (Born: 22 June 1864 in Alexotas, Russian Empire (now Kaunas, Lithuania); Died: 12 Jan 1909 in Göttingen, Germany)

Hermann Minkowski's parents were Lewin Minkowski, a businessman, and Rachel Taubmann, who were Germans. Hermann was his parents' second son, he was born while they were living in Russia. When Hermann was eight years old the family returned to Germany and settled in Königsberg where his father conducted his business.

Minkowski first showed his talent for mathematics while studying at the Gymnasium in Königsberg: he was already reading the work of Dedekind, Dirichlet and Gauss. He entered the University of Königsberg in April 1880. Hilbert was an undergraduate at the same time as Minkowski and they became close friends. The student Minkowski soon became close friends with academic Hurwitz, who was appointed to the staff in 1884.

From early university studies Minkowski became interested in quadratic forms. In 1883 the Academy of Sciences (Paris) granted the Grand Prix in Mathematics jointly to Minkowski and Smith for the solution to the problem of the number of representations of an integer as the sum of five squares. To produce this solution, Minkowski reconstructed Eisenstein's theory of quadratic forms.

Minkowski's doctoral thesis, entitled "Untersuchungen über quadratische Formen, Bestimmung der Anzahl verschiedener Formen, welche ein gegebenes Genus enthält" (1885), was a continuation of this prize winning work involving his natural definition of the genus of a form. After the award of his doctorate, he continued undertaking research at Königsberg.

In 1887 Minkowski began to teach at the University of Bonn. When he applied for the position, he submitted his original paper "Räumliche Anschauung und Minima positiv definiter quadratischer Formen" (Spatial visualization and minima of positive definite quadratic forms). It contained "the first example of the method which Minkowski would develop some years later in his famous 'geometry of numbers' ".

Minkowski was promoted to assistant professor in 1892. Two years later he moved back to Königsberg where he taught for two years before being appointed to the Eidgenössische Polytechnikum Zürich. There he became a colleague of his friend Hurwitz. Einstein was a student in several of the courses he gave, and later they became interested in similar problems in relativity theory.

Minkowski married Auguste Adler in Strasburg in 1897; they had two daughters, Lily born in 1898 and Ruth born in 1902. The family left Zürich in the year that their second daughter was born for Minkowski accepted a chair at the University of Göttingen in 1902. Hilbert arranged for the chair to be created specially for Minkowski, and he held it for the rest of his life.

At Göttingen he became interested in mathematical physics. He participated in a seminar on electron theory in 1905 and he learned the latest results and theories in electrodynamics.

Minkowski developed a new view of space and time and laid the mathematical foundation of the theory of relativity. By 1907 Minkowski realised that the work of Lorentz and Einstein could be best understood in a non-euclidean space. He considered space and time, which were formerly thought to be independent, to be coupled together in a four-dimensional 'space-time continuum'. Minkowski worked out a four-dimensional treatment of electrodynamics. His major works in this area are "Raum und Zeit" (1907) and "Zwei Abhand lungen über die Grundgleichungen der Elektrodynamik" (1909).

formulation of space and time is given.

1.6.1   *Special relativity as a valuable guide*

Before introducing the four-dimensional formulation of special relativity, it is worthwhile mentioning the following.

Special relativity is a kinematical rather than a dynamical theory. It is actually a skeleton theory and, as such, it provides a background and a guideline frame to the other dynamical theories of fields and matter. It imposes restrictions on the laws of physics which physical theories can have. It demands that every general law of physics, expressed in terms of the coordinates t, x, y, z of the system K, should be such that it can be transformed, by means of the Lorentz transformation, into a law of exactly the same form when expressed in terms of the new coordinates t', x', y', z' of the system K'. In other words, the laws of physics should be *invariant* under the Lorentz transformation. As a result, the theory is a valuable guide when looking for new laws of physics. A law of physics which cannot be written in a Lorentz-invariant form is simply not valid, or is only an approximate law.

This space-time continuum provided a framework for all later mathematical work in relativity. These ideas were used by Einstein in developing the general theory of relativity. In fact Minkowski had a major influence on Einstein as Corry points out in *Endeavor* **22** (1998), pp. 95-97:

"In the early years of his scientific career, Albert Einstein considered mathematics to be a mere tool in the service of physical intuition. In later years, he came to consider mathematics as the very source of scientific creativity. A main motive behind this change was the influence of two prominent German mathematicians: David Hilbert and Hermann Minkowski."

Minkowski's original mathematical interests were in pure mathematics and he spent much of his time investigating quadratic forms and continued fractions. His most original achievement, however, was his "geometry of numbers" which he initiated in 1890. "Geometrie der Zahlen" was first published in 1910 but the first 240 pages (of the 256) appeared as the first section in 1896. "Geometrie der Zahlen" was reprinted in 1953 by Chelsea, New York, and reprinted again in 1968. Minkowski published "Diophantische Approximationen: Eine Einführung in die Zahlentheorie" in 1907. It gave an elementary account of his work on the geometry of numbers and of its applications to the theories of Diophantine approximation and of algebraic numbers. Work on the geometry of numbers led on to work on convex bodies and to questions about packing problems, the ways in which figures of a given shape can be placed within another given figure.

At the young age of 44, Minkowski suddenly died from a ruptured appendix.

The above report on Hermann Minkowski is based on the article by J J O'Connor and E F Robertson.

1.6.2 *Four dimensions in classical mechanics*

It is well known that classical mechanics is based on a four-dimensional manifold of three-dimensional space and the time. However, there is an *essential* difference between the concepts of space and time in classical mechanics and the four-dimensional spacetime of special relativity.

In classical mechanics the three-dimensional subspace with constant t is absolute and is independent of the inertial coordinate system. This means one has a separate three-dimensional space, along with a one-dimensional time coordinate, i.e., $O(3) \times T(1)$. In special relativity, on the other hand, the spatial and time coordinates appear in the laws of physics at exactly the same footing, i.e., $O(1,3)$.

Without a four-dimensional formulation, one can also carry out a Lorentz transformation in order to check the invariance of a given law in special relativity. Indeed this was the case when Einstein first proved in his historical paper that Maxwell's equations are invariant under the Lorentz transformation. But such a procedure is quite lengthy, and it should be done for each and every field equation we have in physics. The four-dimensional formalism, on the other hand, provides a simple way to ensure Lorentz invariance by the form of the law itself.

1.6.3 *The Minkowskian spacetime*

If we use as coordinates of an event the quantities $x^0 = ct$, $x^1 = x$, $x^2 = y$, $x^3 = z$, then x^0, x^1, x^2, x^3 may be considered as the components of a vector in four dimensions. The four-dimensional space provided by these four-vectors is then called the *Minkowskian spacetime.* The square of the length of this four-vector,

$$\left(x^0\right)^2 - \left(x^1\right)^2 - \left(x^2\right)^2 - \left(x^3\right)^2 = c^2 t^2 - x^2 - y^2 - z^2, \qquad (1.6.1)$$

does not change under "rotations" of the four-dimensional coordinate system, that is under the Lorentz transformation. If Λ is the 4×4 matrix of the Lorentz transformation whose elements are $\Lambda^\alpha{}_\beta$, then the transformed coordinates are given by

$$x'^\alpha = \Lambda^\alpha{}_\beta x^\beta, \qquad (1.6.2)$$

where Greek indices take the values 0, 1, 2, 3. Invariance of the expression (1.6.1) then means that

$$\left(x'^0\right)^2 - \left(x'^1\right)^2 - \left(x'^2\right)^2 - \left(x'^3\right)^2 = \left(x^0\right)^2 - \left(x^1\right)^2 - \left(x^2\right)^2 - \left(x^3\right)^2, \qquad (1.6.3)$$

where x'^α are given by Eq. (1.6.2).

In general a set of four quantities V^α, which transform like the components of the coordinates x^α under the Lorentz transformation, is called a *four-vector*. One can extend the definition of vectors to *tensors* of any order in the Minkowskian spacetime. The event described by the *position four-vector* x^α is just an example of such quantities. A *scalar* is then a tensor of order 0, whereas a vector is a tensor of order 1. Under a Lorentz transformation a scalar is left *invariant*, a four-vector V^α transforms like the coordinates,

$$V'^\alpha = \Lambda^\alpha{}_\beta V^\beta, \tag{1.6.4}$$

whereas a tensor $T^{\alpha\beta}$ of order 2, for example, transforms like a product of coordinates,

$$T'^{\alpha\beta} = \Lambda^\alpha{}_\gamma \Lambda^\beta{}_\delta T^{\gamma\delta}, \tag{1.6.5}$$

and so on.

As was discussed in previous sections, the invariance of any physical law under the Lorentz transformation is evident once it is expressed in a *covariant* four-dimensional form. All terms of the law should then be four-tensors of the same order. There is also the possibility of spinor formulation, in addition to the four-tensor formalism. The physical law should again be formulated covariantly. (For the theory of spinors and their relation to the Lorentz group, see M. Carmeli, *Group Theory and General Relativity*; M. Carmeli, *Classical Fields: General Relativity and Gauge Theory*; E. Cartan, *The Theory of Spinors*.) A physical law which does not satisfy these requirements cannot be put in a covariant form. The four-dimensional transformation properties of the terms of a physical law, therefore, enable examining its relativistic validity. In the following some more details on four-vectors are given.

Four-vectors are natural generalization to the ordinary three-vectors of classical mechanics. However, use is made in the four-dimensional case of the flat-space metric

$$\eta_{\alpha\beta} = \begin{pmatrix} +1 & 0 & 0 & 0 \\ 0 & -1 & 0 & 0 \\ 0 & 0 & -1 & 0 \\ 0 & 0 & 0 & -1 \end{pmatrix} \tag{1.6.6}$$

and its inverse matrix $\eta^{\alpha\beta}$ (having the same expression as $\eta_{\alpha\beta}$), instead of the three-dimensional unit matrix. The metrics $\eta_{\alpha\beta}$ and $\eta^{\alpha\beta}$ can then be used to *lower* and *raise* the indices of four-quantities,

$$V_\alpha = \eta_{\alpha\beta} V^\beta, \quad V^\alpha = \eta^{\alpha\beta} V_\beta, \tag{1.6.7}$$

$$T_{\alpha\beta} = \eta_{\alpha\gamma}\eta_{\beta\delta}T^{\gamma\delta}, \quad T^{\alpha\beta} = \eta^{\alpha\gamma}\eta^{\beta\delta}T_{\gamma\delta}, \tag{1.6.8}$$

and so on for tensors of higher orders. Quantities with lower indices, like $T_{\alpha\beta}$, are called *covariant* whereas those with upper indices, such as $T^{\alpha\beta}$, are referred to as *contravariant*.

The *scalar product* of two four-vectors V^α and W^α is defined by

$$V_\alpha W^\alpha = \eta_{\alpha\beta}V^\beta W^\alpha = \eta^{\alpha\beta}V_\alpha W_\beta, \tag{1.6.9}$$

and it is a scalar (Lorentz invariant).

In analogy to the position four-vector x^α, the zeroth component of any four-vector is called *timelike* whereas the other three components are called *spacelike*. The square $V_\alpha V^\alpha$ of a four-vector V^α can be positive, zero, or negative. The four-vector is accordingly called *timelike*, *null*, or *spacelike*, respectively. A timelike vector is called *positive* or *negative* according to whether its timelike component is positive or negative, respectively. The manifold of all null vectors forms the *light cone* (a detailed discussion of which is given in Section 1.3).

The tensor δ_β^α is defined by

$$\delta_\beta^\alpha = \begin{cases} 1, \ \alpha = \beta \\ 0, \ \alpha \neq \beta \end{cases} \tag{1.6.10}$$

in all coordinate systems. It is called the *Kronecker delta*, and it satisfies

$$V_\alpha = \delta_\alpha^\beta V_\beta, \tag{1.6.11a}$$

$$V^\alpha = \delta_\beta^\alpha V^\beta \tag{1.6.11b}$$

for any vector V_α.

From any tensor $T_{\alpha\beta}$ of order 2 one can form the scalar

$$T_\alpha{}^\alpha = \delta_\beta^\alpha T_\alpha{}^\beta, \tag{1.6.12}$$

called the *trace* of the tensor.

A tensor is called *symmetric* with respect to two of its indices if their exchange does not affect the value of the tensor. Thus, for instance, the tensor $T_{\alpha\beta\gamma}$ of order 3 is symmetric with respect to the indices α and β if

$$T_{\beta\alpha\gamma} = T_{\alpha\beta\gamma}. \tag{1.6.13}$$

A tensor $A_{\alpha\beta\gamma}$ is called *antisymmetric* (or *skew-symmetric*) with respect to two of its indices α and β, for instance, if it satisfies

$$A_{\beta\alpha\gamma} = -A_{\alpha\beta\gamma}. \tag{1.6.14}$$

The diagonal components of an antisymmetric tensor $A_{\alpha\beta}$ of order 2, that is the components A_{00}, A_{11}, A_{22}, A_{33}, are all equal to zero since $A_{00} = -A_{00}$, and so on.

Properties of tensors are presented in more details in Section 3.1.

1.6.4   *The proper time*

When a particle moves in the ordinary three-dimensional space, it describes a path in the Minkowskian spacetime, called a *world line.* The four-vector dx^α represents the infinitesimal change in the position four-vector x^α, and it is a tangent vector to the world line.

The square of dx^α, namely $\eta_{\alpha\beta}dx^\alpha dx^\beta$, is a scalar, and it is denoted by ds^2. Accordingly we have

$$ds^2 = \eta_{\alpha\beta}dx^\alpha dx^\beta = c^2 dt^2 - dx^2 - dy^2 - dz^2. \tag{1.6.15}$$

The physical meaning of ds can best be understood if we evaluate it in an inertial coordinate system K' in which the particle is momentarily at rest. Denoting the coordinates in K' by x'^α, then in this system the timelike component of dx'^μ is $dx'^0 = cdt'$, whereas its spacelike components vanish, $dx'^k = 0$ ($k = 1,\ 2,\ 3$). Thus in the system K' one has

$$dx'^\alpha = (cdt',\ 0,\ 0,\ 0), \tag{1.6.16}$$

and consequently

$$ds^2 = c^2 dt^2 - dx^2 - dy^2 - dz^2 = c^2 dt'^2. \tag{1.6.17}$$

Accordingly, $d\tau = ds/c$ is the time interval as measured by a clock moving with the particle; it is called the *proper time.*

From Eq. (1.6.17) we obtain

$$d\tau = \frac{1}{c}\sqrt{c^2 dt^2 - dx^2 - dy^2 - dz^2}, \tag{1.6.18}$$

or

$$d\tau = dt\sqrt{1 - \frac{1}{c^2}\left[\left(\frac{dx}{dt}\right)^2 + \left(\frac{dy}{dt}\right)^2 + \left(\frac{dz}{dt}\right)^2\right]}. \tag{1.6.19}$$

But the expression in the square brackets on the right-hand side of Eq. (1.6.19) is simply v^2, where v is the velocity of the moving clock. Consequently we have

$$d\tau = dt\sqrt{1 - \frac{v^2}{c^2}}. \tag{1.6.20}$$

Equation (1.6.20) also follows from the time dilation formula given in Section 1.5.3, since $d\tau$ is the time interval of a clock moving with the particle and dt is the corresponding time interval as measured in a coordinate system from which the motion is observed.

The integration of Eq. (1.6.20) gives the time interval as measured by the moving clock,

$$\Delta\tau = \int_{t_1}^{t_2} \sqrt{1 - \frac{v^2}{c^2}}\, dt, \qquad (1.6.21)$$

whereas the corresponding time interval, as measured by a clock at rest, is given by $\Delta t = t_2 - t_1$.

Equation (1.6.21) shows that the proper time of a moving body is less than the corresponding time in the "rest" coordinate system; moving clocks go slower than those which are at rest.

Let us now have two inertial systems K and K', where K' is moving relative to K with the speed v. As viewed by an observer in the system K, the clocks in K' go slower than those in K. On the other hand, as judged by an observer in the coordinate system K', the clocks in K go slower than those in K'. It thus appears, at a first sight, that there is a contradiction. However, a careful analysis of the problem shows that there is no such a contradiction (see Problem 1.6.1).

A different problem, often referred to as the twins paradox, is that in which we have two clocks one of which goes along a *closed* path, returning to the starting point where the other clock is left behind at rest. Then clearly the moving clock must go slower than the one at rest. A converse reasoning, according to which the moving clock should be considered to be at rest and the other one as moving, is not possible; the clock going along the closed trajectory does not move with a constant velocity, and therefore a coordinate system that is attached to it cannot be inertial along the entire path. Indeed an experiment has been carried out with particles that decay, and the result predicted by special relativity was confirmed.

One can therefore conclude that the time interval shown by a clock is given by the integral

$$\int_{a}^{b} d\tau, \qquad (1.6.22)$$

where $d\tau = ds/c$ and the integration is carried out along the world line of the clock. For a clock at rest, the world line is a straight line parallel to the x^0 axis. If the clock goes along a closed path in the ordinary three-dimensional space, on the other hand, then its world line is a curve passing through the initial and the final points of the motion.

Figure 1.6.1 gives the Minkowskian spacetime diagram describing the world lines of two clocks one at rest, while the other moves along a closed curve, in the three-dimensional space. The world lines are described by the

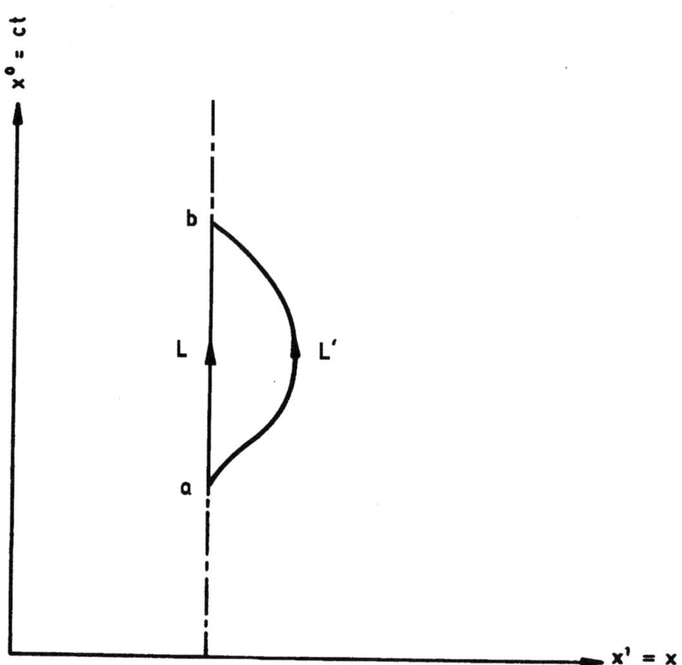

Fig. 1.6.1:   The Minkowskian spacetime diagram describing the world lines of two clocks one at rest, while the other moves along a closed curve, in the three-dimensional space.

straight line L and the curve L' between the initial point a and the final point b of the motion. (The x^2 and x^3 axes are omitted for brevity.)

Finally, it will also be noted that the clock at rest always shows a longer time interval than that of the moving clock. In other words, the integral (1.6.22) becomes maximum if the integration is carried out along the straight world line connecting the points a and b.

1.6.5   *Velocity and acceleration four-vectors*

We continue our four-dimensional analysis by forming the velocity and acceleration four-vectors from their corresponding ordinary three-dimensional vectors.

The velocity four-vector of a particle is defined by

$$u^\alpha = \frac{dx^\alpha}{ds},$$
(1.6.23)

and its relation to the ordinary three-dimensional velocity $\mathbf{v}$ can be obtained by expressing ds in terms of dt, using Eq. (1.6.20). We then have

$$u^\alpha = \frac{1}{c\sqrt{1 - \dfrac{v^2}{c^2}}} \frac{dx^\alpha}{dt}. \tag{1.6.24}$$

The timelike component of u^α is consequently given by

$$u^0 = \frac{1}{\sqrt{1 - \dfrac{v^2}{c^2}}}, \tag{1.6.25}$$

whereas its spacelike components

$$\mathbf{u} = \left(u^1,\ u^2,\ u^3\right) \tag{1.6.26}$$

are given by

$$\mathbf{u} = \frac{1}{\sqrt{1 - \dfrac{v^2}{c^2}}} \frac{\mathbf{v}}{c}. \tag{1.6.27}$$

Here

$$\mathbf{v} = (v_x,\ v_y,\ v_z) = \left(\frac{dx}{dt},\ \frac{dy}{dt},\ \frac{dz}{dt}\right) \tag{1.6.28}$$

is the ordinary three-dimensional velocity vector.

It will be noted that the velocity four-vector u^α is dimensionless. Moreover, by Eq. (1.6.15),

$$u_\alpha u^\alpha = 1, \tag{1.6.29}$$

i.e., its length is unity.

The acceleration four-vector of a particle is subsequently defined by

$$\frac{du^\alpha}{ds} = \frac{d^2 x^\alpha}{ds^2}, \tag{1.6.30}$$

which, by Eq. (1.6.29), satisfies the orthogonality condition

$$u_\alpha \frac{du^\alpha}{ds} = 0. \tag{1.6.31}$$

Using Eqs. (1.6.26), (1.6.27) and (1.6.29) we then find for the components of the acceleration four-vector the following:

$$\frac{du^0}{ds} = \frac{\gamma}{c} \frac{d\gamma}{dt}, \tag{1.6.32a}$$

$$\frac{d\mathbf{u}}{ds} = \frac{\gamma}{c^2} \frac{d\left(\gamma \mathbf{v}\right)}{dt}, \tag{1.6.32b}$$

where

$$\gamma = \frac{1}{\sqrt{1 - \dfrac{v^2}{c^2}}}. \tag{1.6.33}$$

In the next section the kinematics of mass, energy, etc. is given in almost the original way that was given by Einstein.

1.6.6 *Problems*

P 1.6.1 Given two inertial systems K and K', where the latter is moving relative to K with the speed v. As judged by an observer in K, the clocks in the system K' go slower than those in K. On the other hand, as viewed by an observer in K', the clocks in the system K go more slowly than those in K'. Show that there is really no contradiction between the above two observations.

Solution: The solution is left for the reader.

P 1.6.2 Find the timelike component of the acceleration four-vector of a particle in a coordinate system in which the particle is momentarily at rest. Express the result in terms of the ordinary velocity $\mathbf{v}$ and acceleration $\mathbf{a} = d\mathbf{v}/dt$ for the two limiting cases where in one v changes only in direction and in the other v changes only in magnitude. Use Eqs. (1.6.32) to show that one obtains $\gamma^2|\mathbf{a}|$ and $\gamma^3|\mathbf{a}|$ for these limiting cases, where $\gamma = 1/\sqrt{1-\beta^2}$ and $\beta = v/c$.

Solution: The solution is left for the reader.

1.7 Mass, Energy and Momentum

We conclude the fundamentals of the special relativity theory by discussing the *dynamical* concepts of mass, energy and momentum. It follows that one of the most important consequences of the special theory of relativity is the relationship between the mass and energy, stated in the form of a simple and universal law, $E = mc^2$. Before the advent of special relativity, there were two separate conservation laws of great importance. These were the conservation laws of mass and of energy, and they were independent of each other. The special theory of relativity united them into one law. In this chapter the relationship between the mass and energy is presented. This is done by introducing the momentum four-vector whose timelike component is the energy whereas its spatial components are those of the ordinary momentum.

1.7.1 *Preliminaries*

We start our discussion by presenting some physical comments about the relationship between the mass and energy.

According to the theory of special relativity, the total energy of a particle is not given by the familiar Newtonian kinetic energy $mv^2/2$, where m is the mass and v is the velocity of the particle. Rather, it is given by

$$\frac{mc^2}{\sqrt{1-\dfrac{v^2}{c^2}}},\qquad(1.7.1)$$

where c is the speed of light.

Expanding (1.7.1) into a power series in v/c, we obtain

$$\frac{mc^2}{\sqrt{1-\dfrac{v^2}{c^2}}}=mc^2+\frac{1}{2}mv^2+\frac{3}{8}m\frac{v^4}{c^2}+\cdots.\qquad(1.7.2)$$

The second term on the right-hand side of Eq. (1.7.2) is, of course, the Newtonian kinetic energy. The first term, mc^2, is called the *rest energy* of the particle.

1.7.2   *Relationship between mass, energy and momentum*

We now derive the relationship between the mass, energy and momentum.

Starting with the relation between the energy and momentum of the photon,

$$E=cp,\qquad(1.7.3)$$

one can associate the inertial mass

$$m=\frac{E}{c^2}\qquad(1.7.4)$$

to the photon. Hence, the inertial mass of a photon with the energy E is given by

$$m=\frac{p}{c}.\qquad(1.7.5)$$

Equation (1.7.5) is a particular case of the formula

$$m=\frac{p}{v}\qquad(1.7.6)$$

for a particle, familiar from the Newtonian mechanics, for $v=c$.

From Eqs. (1.7.4) and (1.7.6) we obtain

$$\frac{v}{c}=\frac{cp}{E},\qquad(1.7.7)$$

expressing the relationship between the energy, momentum and velocity.

From classical mechanics we know that the increase of kinetic energy due to work done by an external force f is given by

$$dE = f dx = \frac{dp}{dt} dx = v dp. \tag{1.7.8}$$

Equations (1.7.7) and (1.7.8) then yield

$$E dE = c^2 p dp, \tag{1.7.9}$$

the integration of which gives

$$E^2 - c^2 p^2 = E_0^2, \tag{1.7.10}$$

where E_0^2 is a constant of integration. Equation (1.7.10) is invariant under the Lorentz transformation as will be shown in the sequel.

To determine the constant E_0 we use Eqs. (1.7.7) and (1.7.10), getting

$$E = \frac{E_0}{\sqrt{1 - \dfrac{v^2}{c^2}}}. \tag{1.7.11}$$

Expanding now the right-hand side of this equation into a power series in v/c then gives

$$E = E_0 + \frac{1}{2}\left(\frac{E_0}{c^2}\right) v^2 + \frac{3}{8}\left(\frac{E_0}{c^2}\right)\frac{v^4}{c^2} + \cdots. \tag{1.7.12}$$

This should now be compared to Eq. (1.7.2), giving

$$\frac{E_0}{c^2} = m_0. \tag{1.7.13}$$

Here m_0 is a constant, called the *rest mass* of the particle (previously denoted by m).

Using Eq. (1.7.13) in Eqs. (1.7.10) and (1.7.11) then gives

$$E^2 - c^2 p^2 = m_0^2 c^4 \tag{1.7.14}$$

and

$$E = \frac{m_0 c^2}{\sqrt{1 - \dfrac{v^2}{c^2}}}, \tag{1.7.15}$$

respectively. Equation (1.7.4) can now be written in the form

$$m = \frac{m_0}{\sqrt{1 - \dfrac{v^2}{c^2}}}, \tag{1.7.16}$$

where m is the inertial mass of the particle. The rest mass m_0 is, consequently, equal to the inertial mass for $v = 0$.

From Eqs. (1.7.15) and (1.7.16) one obtains

$$E = mc^2. \tag{1.7.17}$$

This is Einstein's famous formula expressing the total energy of the particle in terms of its inertial mass. Moreover, the three-dimensional momentum $\mathbf{p}$ is now given by

$$\mathbf{p} = m\mathbf{v}, \tag{1.7.18}$$

where $\mathbf{v} = d\mathbf{x}/dt$ is the three-dimensional velocity and m is the inertial mass rather than the rest mass m_0.

It should be emphasized that none of the equations (1.7.11) and (1.7.14)$-$(1.7.18) demands that the velocity v of the particle should be constant even though they were all derived from special relativity, a theory based on transformations between inertial systems moving relative to each other with constant velocities. In fact, these formulas are valid for nonconstant velocities v as well.

Physical phenomena described within the framework of special relativity are by no means restricted to processes with constant velocities. The velocity v can be replaced by any other variable related to it, such as the momentum p or the angular momentum J, for instance. All one needs is expressing the ratio $\beta = v/c$ in the Lorentz contraction factor $\sqrt{1 - \beta^2}$, appearing in the above equations of the particle, by its value in terms of the desired new variable.

Let us, for instance, express the Lorentz contraction factor as a function of the momentum p. Using Eq. (1.7.18), we then obtain

$$\frac{1}{1 - \dfrac{v^2}{c^2}} = 1 + \frac{p^2}{p_0^2}, \tag{1.7.19}$$

where

$$p_0 = m_0 c = \frac{E_0}{c} \tag{1.7.20}$$

is a characteristic constant of the momentum of the particle. As v takes the values $0 \le v < c$, the momentum goes from zero to infinity, $0 \le p < \infty$, and consequently the expressions on both sides of Eq. (1.7.19) take the values 1 to ∞, as expected. Notice that for a photon $p_0 = 0$ since its rest mass m_0 equals to zero.

Equations (1.7.15) and (1.7.16) will consequently assume the forms

$$E = E_0 \sqrt{1 + \frac{p^2}{p_0^2}} = \frac{p_0^2}{m_0} \sqrt{1 + \frac{p^2}{p_0^2}} \qquad (1.7.21)$$

and

$$m = m_0 \sqrt{1 + \frac{p^2}{p_0^2}}, \qquad (1.7.22)$$

respectively, when the new expression of the Lorentz contraction factor is used. Notice that Eq. (1.7.21) is completely equivalent to the energy-momentum formula (1.7.14).

Expanding now Eq. (1.7.21) in powers of p/p_0, we then obtain for the energy of the particle as a function of its momentum,

$$E\,(p) = E_0 \left(1 + \frac{1}{2} \frac{p^2}{p_0^2} - \frac{1}{8} \frac{p^4}{p_0^4} + \frac{1}{16} \frac{p^6}{p_0^6} + \cdots \right). \qquad (1.7.23)$$

This should then be compared with its equivalent expression in terms of the velocity,

$$E\,(v) = E_0 \left(1 + \frac{1}{2} \frac{v^2}{c^2} + \frac{3}{8} \frac{v^4}{c^4} + \frac{5}{16} \frac{v^6}{c^6} + \cdots \right), \qquad (1.7.24)$$

obtained by expanding Eq. (1.7.15) in powers of v/c.

The first terms on the right-hand sides of Eqs. (1.7.23) and (1.7.24) are equal to the rest energy of the particle. However, the second term on the right-hand side of Eq. (1.7.23) is not equal to the second term of Eq. (1.7.24) since, by Eq. (1.7.18), p is proportional to m rather than to m_0. The third terms on the right-hand sides of Eqs. (1.7.23) and (1.7.24) appear with opposite signs.

1.7.3 *Angular-momentum representation*

Likewise, use can be made of the angular momentum J of the particle by expressing the ratio p^2/p_0^2 in terms of J (Carmeli 1983). Using Eq. (1.7.20) one then finds

$$\frac{p^2}{p_0^2} = \frac{p^2}{m_0^2 c^2} = \frac{1}{m_0 c^2} \frac{p^2}{m_0} = \frac{1}{m_0 c^2} \frac{J^2}{I_0}, \qquad (1.7.25)$$

where I_0 is the *rest moment of inertia* of the particle (calculated, using the rest mass). The last equality in the above equation is a result of expressing the kinetic energy in terms of p and J. Accordingly one obtains

$$\frac{p^2}{p_0^2} = \frac{J^2}{J_0^2}, \qquad (1.7.26)$$

where

$$J_0 = \sqrt{m_0 c^2 I_0} \tag{1.7.27}$$

is a characteristic constant of the angular momentum of the particle. Notice that for a photon $J_0 = 0$, since both m_0 and I_0 are equal to zero in this case.

Equation (1.7.19) then gives

$$\frac{1}{1 - \dfrac{v^2}{c^2}} = 1 + \frac{J^2}{J_0^2}, \tag{1.7.28}$$

and, as a consequence, Eqs. (1.7.21)–(1.7.23) give

$$E = E_0 \sqrt{1 + \frac{J^2}{J_0^2}} = \frac{J_0^2}{I_0} \sqrt{1 + \frac{J^2}{J_0^2}}, \tag{1.7.29}$$

$$m = m_0 \sqrt{1 + \frac{J^2}{J_0^2}}, \tag{1.7.30}$$

$$E(J) = E_0 \left(1 + \frac{1}{2}\frac{J^2}{J_0^2} - \frac{1}{8}\frac{J^4}{J_0^4} + \frac{1}{16}\frac{J^6}{J_0^6} + \cdots \right). \tag{1.7.31}$$

Finally, just as Eq. (1.7.31) is completely equivalent to the energy-momentum formula (1.7.14), we can rewrite Eq. (1.7.29) as an energy-angular-momentum formula. We then have

$$E^2 - \gamma^2 J^2 = I_0^2 \gamma^4, \tag{1.7.32}$$

where

$$\gamma = c \sqrt{\frac{m_0}{I_0}} \tag{1.7.33}$$

is a natural angular velocity of the particle, and the expression on the right-hand side of Eq. (1.7.32) is equal to $m_0^2 c^4$. The equivalent formulas to Eqs. (1.7.15)–(1.7.18) and (1.7.7) will then have the forms:

$$E = \frac{I_0 \gamma^2}{\sqrt{1 - \dfrac{\Omega^2}{\gamma^2}}}, \tag{1.7.34}$$

$$I = \frac{I_0}{\sqrt{1 - \dfrac{\Omega^2}{\gamma^2}}}, \tag{1.7.35}$$

$$E = I\gamma^2, \tag{1.7.36}$$

$$\mathbf{J} = I\mathbf{\Omega}, \tag{1.7.37}$$

$$\frac{\Omega}{\gamma} = \frac{\gamma J}{E}, \tag{1.7.38}$$

where Ω is the angular velocity of the particle satisfying the condition $0 \leq \Omega < \gamma$. We may also add the relation

$$\frac{1}{1 - \dfrac{\Omega^2}{\gamma^2}} = 1 + \frac{J^2}{J_0^2}, \tag{1.7.39}$$

which is equivalent to Eq. (1.7.19).

1.7.4 *Energy-momentum four-vector*

To conclude this chapter we introduce the *energy-momentum four-vector* defined by

$$p^\alpha = m_0 c^2 u^\alpha, \tag{1.7.40}$$

where m_0 is the rest mass of the particle, and u^α is the velocity four-vector defined by Eq. (1.6.23),

$$u^\alpha = \frac{dx^\alpha}{ds}. \tag{1.7.41}$$

Using Eqs. (1.6.24)–(1.6.28) one then finds that the components of p^α are given by

$$p^0 = m_0 c^2 u^0 = \frac{m_0 c^2}{\sqrt{1 - \beta^2}} = mc^2 = E, \tag{1.7.42a}$$

$$p^k = m_0 c^2 u^k = \frac{m_0 c}{\sqrt{1 - \beta^2}} \frac{dx^k}{dt} = mc \frac{dx^k}{dt}, \tag{1.7.42b}$$

for $k = 1,\ 2,\ 3$. Thus the energy-momentum four-vector is given by

$$p^\alpha = \left(p^0,\ p^k\right) = \left(E,\ c\mathbf{p}\right), \tag{1.7.43a}$$

$$p_\alpha = \left(p_0,\ p_k\right) = \left(E,\ -c\mathbf{p}\right), \tag{1.7.43b}$$

where $\mathbf{p} = m\, d\mathbf{x}/dt$ is the ordinary three-dimensional momentum given by Eq. (1.7.18).

The square of p^α is consequently given by

$$p_\alpha p^\alpha = m_0^2 c^4 u_\alpha u^\alpha. \tag{1.7.44}$$

Thus using Eqs. (1.6.29) and (1.7.43), we obtain

$$E^2 - c^2 \mathbf{p}^2 = m_0^2 c^4, \tag{1.7.45}$$

that is, the relationship (1.7.14) between mass, energy and momentum.

With the above discussion on mass, energy and momenta we end this part on the fundamentals of the special theory of relativity.

In the next chapter cosmological relativity theory is presented and its important consequences are described.

1.7.5 *Problems*

P 1.7.1 Assume that $J_0 = \hbar$ for particles with *intrinsic* angular momentum (spin), such as electrons, where $\hbar = h/2\pi$ and h is Planck's constant. Use Eq. (1.7.27) to calculate the tangential velocity of a material point in such bodies due to their internal rotation.

Solution: The tangential velocity of a material point in such a body is defined by

$$v_g = \omega_0 r_g, \tag{1}$$

where ω_0 is the angular velocity of the internal rotation of the body, and r_g, defined by

$$I_0 = m_0 r_g^2, \tag{2}$$

is its *radius of gyration*.

Equation (1.7.27) with $J_0 = \hbar$, along with the ordinary relationship among angular momentum, moment of inertia and angular velocity, then give

$$J_0 = \sqrt{m_0 c^2 I_0} = I_0 \omega_0 = \hbar. \tag{3}$$

Using now Eqs. (1)–(3) one then obtains

$$v_g = \omega_0 r_g = \frac{\hbar}{I_0}\sqrt{\frac{I_0}{m_0}} = \frac{\hbar}{\sqrt{m_0 I_0}} = \frac{\hbar c}{J_0} = c, \tag{4}$$

that is, the speed of light.

P 1.7.2 Use the special relativistic line element

$$c^2 dt^2 - \left(dx^2 + dy^2 + dz^2\right) = ds^2 \tag{1}$$

to derive the relationship between the energy and the momentum.

Solution: A straightforward calculation, using Eq. (1), gives

$$c^2 dt^2 \left(1 - \frac{v^2}{c^2}\right) = ds^2, \tag{2}$$

thus

$$\frac{dt}{d\tau} = \frac{1}{\sqrt{1 - \frac{v^2}{c^2}}}, \tag{3}$$

where $d\tau$ is related to ds by $ds = c\,d\tau$.

Multiplying now Eq. (2) by $m_0^2 c^4/ds^2$, and using Eq. (3), one then obtains

$$\left(1 - \frac{v^2}{c^2}\right)^{-1} \left(m_0^2 c^4 - m_0^2 c^2 v^2\right) = m_0^2 c^4. \tag{4}$$

Using now the relationship between the inertial mass and the rest mass,

$$m = \frac{m_0}{\sqrt{1 - \frac{v^2}{c^2}}}, \tag{5}$$

in Eq. (4), we obtain

$$m^2 c^4 - c^2 \mathbf{p}^2 = m_0^2 c^4, \tag{6}$$

where $\mathbf{p} = m\mathbf{v}$.

1.8 Suggested References

D. Bohm, *The Special Theory of Relativity* (Benjamin, New York, 1965).

H. Bondi, Some special solutions of the Einstein equations, in: *1964 Brandeis Summer Institute in Theoretical Physics*, Vol. 1 (Prentice-Hall, Englewood Cliffs, New Jersey, 1965), pp. 379-406.

M. Born, *Einstein's Theory of Relativity* (Dover, New York, 1962).

M. Carmeli, *Group Theory and General Relativity* (McGraw-Hill, New York, 1977; reprinted by Imperial College Press, London, 2000).

M. Carmeli, *Classical Fields: General Relativity and Gauge Theory* (John Wiley, New York, 1982; reprinted by World Scientific, Singapore, 2001).

M. Carmeli, Extension of the principle of minimal coupling to particles with magnetic moments, *Nuovo Cimento Lett.* **37**, 205 (1983).

M. Carmeli, *Cosmological Special Relativity: The Large-Scale Structure of Space, Time and Velocity*, Second Edition (World Scientific, Singapore, 2002).

M. Carmeli and S. Malin, *Representations of the Rotation and Lorentz Groups: An Introduction* (Marcel Dekker, New York and Basel, 1976).

E. Cartan, *The Theory of Spinors* (The M.I.T. Press, Cambridge, Massachusetts, 1966).

A. Einstein, *Ann. Phys.* **17**, 891 (1905); English translation in: *The Principle of Relativity* (Dover, New York, 1923), p. 35.

A. Einstein, *Relativity: The Special and General Theory* (Crown Publishers, New York, 1931).

A. Einstein, *The Meaning of Relativity* (Princeton University Press, Princeton, N.J., 1955).

A. Einstein, *Autobiographical Notes*, P.A. Schilpp, Editor (Open Court Publishing Company, La Salle and Chicago, Illinois, 1979).

A.P. French, *Special Relativity* (W.W. Norton, New York and London, 1968).

L.D. Landau and E.M. Lifshitz, *The Classical Theory of Fields* (Addison-Wesley, Reading, Massachusetts, 1959).

A.I. Miller, *Albert Einstein's Special Theory of Relativity* (Addison-Wesley, Reading, Massachusetts, 1981).

H. Minkowski, Space and time (an address delivered at the 80th Assembly of German Natural Scientists and Physicians, at Cologne, 21 September, 1908); English translation in: *The Principle of Relativity* (Dover, New York, 1923), p.73.

Chapter 2

Cosmological Special Relativity

Moshe Carmeli

In this chapter we present the cosmological special relativity theory along the lines of Einstein's special relativity. While Einstein's special relativity deals with the continuum of space and time, cosmological special relativity deals with distances and velocities of celestial bodies. The role of the propagation of light will consequently be replaced by Hubble's law, with negligible gravitation, and in the next chapters with gravitation. The relative quantity here is the cosmic time as compared to the velocity in Einstein's theory. In other words, cosmic times are assumed to be relative in this theory. The chapter starts by reviewing the present day status of cosmology. The postulates of the theory are given and the notion of cosmic frame is introduced. Spacevelocity and relative cosmic time are subsequently discussed. The inadequacy of the classical transformation in physics is discussed and a comparison of the Universe expansion to the light propagation is given. The transformation between spacevelocity coordinates at different cosmic times, called the cosmological transformation, is derived and its physical interpretation is given. Consequences of the cosmological transformation are then drawn. The inflation of the Universe at the very early stage is discussed in details and is shown to be of the order of magnitude similar to that first predicted by Guth. Also discussed is the temperature of the Universe. It is shown that in the Universe there is a minimal acceleration, always in the direction of the motion of particles. An interesting formula that relates the cosmological redshift to the cosmic time without any extra parameters is given. This is like the Doppler shift that is well known in physics. The theory is then extended into five dimensions by adding the time coordinate, so as to make it spacetime and velocity theory but still without gravitation.

This fact demands a special treatment for Maxwell's equations which are now extended into five dimensions by adding to them a new interaction, the Higgs interaction. The generalized Maxwell's equations in five dimensions are written down explicitly in the standard three-dimensional notation, and it is shown how the Higgs interaction is unified with the electromagnetic theory. Kinematical topics in the Universe, such as velocity, acceleration and cosmic distances, are then discussed in details and some formulas are obtained within this theory and compared with those in special relativity theory.

2.1 Spacevelocity in Four Dimensions

In cosmology one also needs spatial coordinates x, y, z and velocity v to describe the location and the velocity of an object in an expanding Universe. The velocity v is considered to be independent of the coordinates x, y, z, as the theory of Hamilton in mechanics has shown. The four coordinates x, y, z and v also provide a pseudo-Euclidean manifold just as the coordinates x, y, z and t in ordinary special relativity (see Section 1.1). The line element is now given by

$$ds^2 = \tau^2 dv^2 - \left(dx^2 + dy^2 + dz^2\right). \qquad (2.1.1)$$

Remarks on the constant τ:

The constant τ is a universal constant (called the Hubble-Carmeli constant). It is actually the inverse of the Hubble constant h in empty space (in vacuum). It is also the Big Bang time and hence it is the age of the Universe. Its numerical value will be shown in Chapter 4 to be $\tau = 13.56 \pm 0.48 \mathrm{Gyr}$. The constant τ is just as the constant c is in special relativity, even though it is well known that both the speed of light and the rate of expansion of the Universe change their values due to gravity. This is possible since local measurements of both the speed of light and the rate of expansion of the Universe (at the present epoch of time) always yield constants c and τ, respectively. Except for Newton's gravitational constant G (or Einstein's gravitational constant κ) the constant τ is the only constant that appears in cosmological special relativity and cosmological general relativity.

2.1.1 *Present-day cosmology*

We wish to point out that at present we have a similar situation in cosmology to that existed in prerelativistic times with respect to space and (not velocity but) cosmic time, in conjunction with the constancy of expansion of the Universe (and not propagation of light). If we make the convention according to which cosmic time, denoted by t, is measured *backward*, then our present time ($t = 0$) is a preferred time with respect to which all cosmological physical phenomena are referred. This is exactly analogous to the prerelativity assumption that physical phenomena are referred to only one "stationary" ($v = 0$) system (see M. Carmeli, *Found. Phys.* **25**, 1029 (1995); **26**, 413 (1996); M. Carmeli, *Intern. J. Theor. Phys.* **36** (1997)).

Actually space has no such a preference: When we consider an astronomical object and say that it is, let us say, at $t = \tau/2$, that faraway object has the same right to say that it is at cosmic time zero ($t = 0$) and we are at $t = \tau/2$ with respect to him, exactly as in relativistic physics but with the roles of cosmic time and velocity exchanged. We will assume that such a reciprocity relationship between cosmological objects is a universal property of space and cosmic time just as Einstein did with respect to space and velocity in special relativity.

2.1.2 *Postulates*

In addition, we will make two assumptions which will be elevated to postulates. These are:
(1) The *principle of the constancy of the expansion of the Universe* at all cosmic times (analogous to the principle of the constancy of propagation of light in all moving frames); and
(2) The *principle of cosmological relativity* (analogous to the principle of special relativity) according to which the laws of physics are the same at all cosmic times (as moving frames in special relativity).

2.1.3 *The cosmic frames*

In this way the Universe has *cosmic frames of reference* located at fixed cosmic times and differ from each other by relative constant cosmic times, similar to the situation in special relativity but now cosmic times replace velocities (see Figure 2.1.1). Observers in each cosmic frame are equipped with rulers to measure distances (like in special relativity) and with small radar devices (similar to those used by highway patrol) for velocity measure-

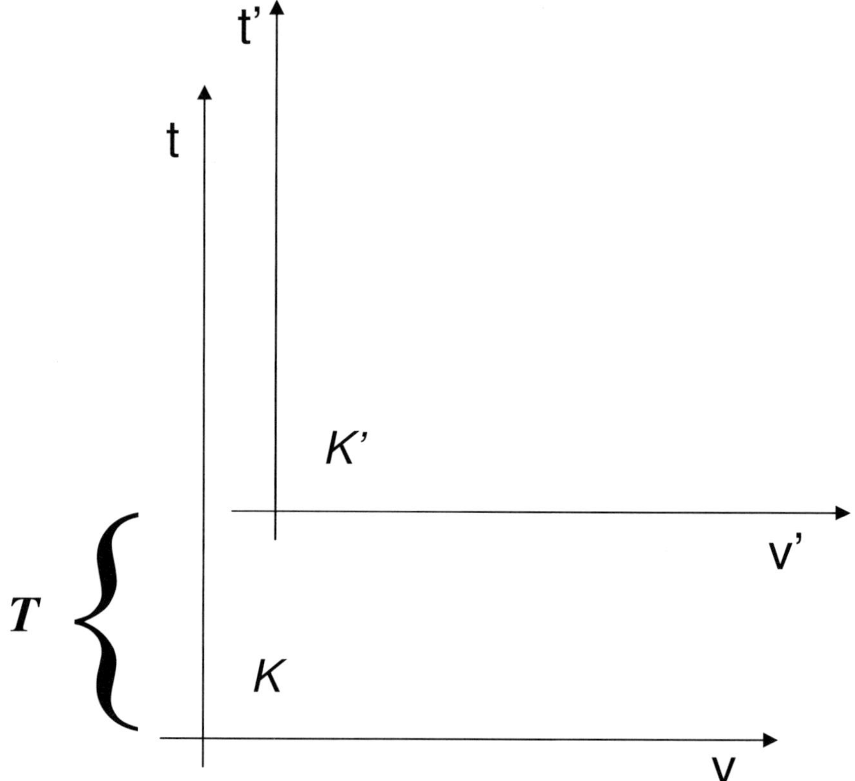

Fig. 2.1.1:   Two cosmic frames K and K' with relative cosmic time T.

ments (instead of clocks in special relativity). Notice the analogy between the relation $[\tau] = $ distance/velocity in the present theory and $[c] = $ distance/time in special relativity, which suggests the choice of distance and velocity as our fundamental variables as compared to distance and time in special relativity.

2.1.4 *Spacevelocity in cosmology*

With the above postulates, and by comparison with special relativity, it is obvious that space and velocity cannot be independent if Hubble's law is to be preserved at all cosmic times. In fact this will enable us to derive a transformation that relates space points and velocities (and other quantities) measured in different cosmic frames of reference that differ in relative cosmic times just like the Lorentz transformation which relates space points and times (and other quantities) measured in different inertial frames that differ in relative velocities. Space coordinates and velocities become unified in cosmology just as space and time are unified in local (noncosmological) physics.

2.1.5 *Pre-special relativity*

With the above preliminaries we are now in a position to develop the theory. To begin with we repeat very briefly what preceded to special relativity. The Galilean transformation between two inertial systems K and K', where K' moves relative to K with a constant velocity v along the x axis, is given by

$$x' = x - vt, \quad t' = t, \quad y' = y, \quad z' = z.$$

Here x and x' represent the coordinates of a particle in the systems K and K', respectively.

The trouble with the Galilean transformation is its incompatibility with the equation of propagation of light which satisfies

$$c^2 t'^2 - x'^2 = c^2 t^2 - x^2, \quad y' = y, \quad z' = z.$$

Hence the Galilean transformation should be abolished in favor of a new one that relates not only x' to x leaving t unchanged but relates x' *and* t' to x *and* t. And this immediately leads to the familiar Lorentz transformation.

2.1.6 *The relative cosmic time*

In cosmology one is not interested in comparing quantities at two reference frames moving with a constant velocity with respect to each other. Rather, one is interested in comparing quantities at two different cosmic times. For example, one often asks what was the density of matter or the temperature of the Universe at an earlier cosmic time t as compared to the values of these quantities at our present time now ($t = 0$). The backward time t is the *relative* cosmic time with respect to our present time.

The concept of the relative cosmic time is not restricted only to the backward cosmic time t with respect to the present time ($t = 0$). Every two observers with cosmic times t_1 and t_2 with respect to us are related to each other by a relative cosmic time t. Thus t plays the role of the velocity

v in special relativity and we will see in the sequel that t has an upper limit which is the Hubble-Carmeli time τ just as the maximum velocity permitted in special relativity is c.

The variables (coordinates) in this theory are naturally the Hubble variables, i.e. the velocity v and the distance x. To derive the transformation between these variables in the systems K and K', where K' has a relative cosmic time t with respect to K, we proceed as follows.

2.1.7   *Inadequacy of the classical transformation*

We first do it classically, and for simplicity it is assumed that the motion is one-dimensional. Denoting the coordinates and velocities in the systems K and K' by x, v and x', v', respectively, then

$$x' = x - tv, \qquad v' = v, \qquad y' = y, \qquad z' = z,$$

where v was assumed to be constant. The x's and v's in these equations represent the coordinates and velocities not for just one particle but for as many as one wishes, with t the same for all of them.

The above transformation does not satisfy the equation of expansion of the Universe which, according to the principle of the constancy of expansion of the Universe and the principle of cosmological relativity demands the laws of physics (and in particular Hubble's law) to be valid at all cosmic times, satisfies

$$\tau^2 v'^2 - x'^2 = \tau^2 v^2 - x^2, \qquad y' = y, \qquad z' = z.$$

The situation here is similar to what we had at the beginning of the century where the Galilean transformation could not accommodate both of the principle of special relativity and the principle of the constancy of the speed of light, whence leading to the Lorentz transformation. A new transformation here also has to be found, which relates not only x' to x leaving v unchanged but relates x' *and* v' to x *and* v.

2.1.8   *Nonrelativistic cosmological transformation*

The nonrelativistic transformation between two cosmic systems of coordinates is given by

$$x' = x + Tv, \qquad v' = v, \tag{2.1.2}$$

where v is a parameter which has the dimension of velocity and T is the relative cosmic time between the two coordinate systems. Taking now the derivative of the above equation with respect to v we obtain

$$\frac{dx'}{dv} = \frac{dx}{dv} + T. \qquad (2.1.3)$$

But $dx'/dv = t'$ and $dx/dv = t$, where t' and t are the cosmic times at the two coordinate systems. Thus

$$t' = t + T. \qquad (2.1.4)$$

2.1.9 *Difficulties at the Big Bang*

In cosmology, observers are supposed to be in different cosmic times rather than moving with respect to each other at constant velocities. Equation (2.1.4) is the analogue of Eq. (1.1.3) in Einstein's special relativity. There is no problem here also for finite times t and t'. But what happens if we take t to be equal to the Big Bang time (which is now assumed, in analogy to Einstein's special relativity, to be constant at all cosmic times). We obtain

$$\tau' = \tau + T, \qquad (2.1.5)$$

where τ denotes the Big Bang time (τ is equal to $13.56 \pm 0.48\text{Gyr}$) and τ' is the new Big Bang time. Thus the cosmic time now is larger than the Big Bang time and that is impossible by our assumption. Therefore the above nonrelativistic transformation should be replaced by a larger transformation that involves both space and velocity. This is similar to the situation in Einstein's special relativity. In the previous case that leads to the Lorentz transformation and in cosmology it leads to the cosmological transformation.

In cosmology as we have seen, we have to deal with quantities that depend on the velocity rather than on time. If R represents the location of a galaxy with respect to an observer and v is the receding velocity of the galaxy then one can, in analogous to what was done in the preceding section write the Lagrangian: $L = (dR/dv)^2$, neglecting terms that depend on the location. From the Lagrange equation one then has $(\partial/\partial(dR/dv))L = 2(dR/dv)$. And the Lagrange equation gives $d^2R/dv^2 = 0$. Thus $dR/dv = \text{const}$, and $R = \text{const} \times v$. One then easily identify the constant with the reciprocal of the Hubble parameter H_0 for the case of negligible gravity (empty space).

In the next two sections the Lorentz transformation and its extension to cosmology are presented.

2.1.10 *Universe expansion versus light propagation*

Under the assumption that Hubble constant is constant in cosmic time, there is an analogy between the propagation of light, $x = ct$, and the expansion of the Universe, $x = \tau v$, where τ is the Hubble-Carmeli time, a constant which is also the age of the Universe under the above assumption, and c is the speed of light in vacuum. Thus one can express the expansion of the Universe, assuming that it is homogeneous and isotropic, in terms of the null vector $(v,\ x,\ y,\ z)$ satisfying

$$\tau^2 v^2 - \left(x^2 + y^2 + z^2\right) = 0, \tag{2.1.6}$$

where v is the receding velocity of the galaxies. Equation (2.1.6), in the 4-dimensional flat space of the Cartesian 3-space and the velocity, is similar to

$$c^2 t^2 - \left(x^2 + y^2 + z^2\right) = 0 \tag{2.1.7}$$

for the null propagation of light in Minkowskian spacetime. We assume, furthermore, that a relationship of the form (2.1.6) is valid at all cosmic times. Thus, at a cosmic time t' at which the coordinates and velocity are labeled with primes, we have

$$\tau^2 v'^2 - \left(x'^2 + y'^2 + z'^2\right) = 0 \tag{2.1.8}$$

with the same τ, just as for light emitted from a source with velocity v with respect to the first one,

$$c^2 t'^2 - \left(x'^2 + y'^2 + z'^2\right) = 0. \tag{2.1.9}$$

Accordingly, we have a four-dimensional space with zero curvature of x, y, z, v just as the Minkowskian spacetime of x, y, z, t.

We now assume that at two cosmic times t and t' we have

$$\tau^2 v'^2 - \left(x'^2 + y'^2 + z'^2\right) = \tau^2 v^2 - \left(x^2 + y^2 + z^2\right), \tag{2.1.10}$$

in analogy to the special relativistic formula

$$c^2 t'^2 - \left(x'^2 + y'^2 + z'^2\right) = c^2 t^2 - \left(x^2 + y^2 + z^2\right). \tag{2.1.11}$$

The question is then what is the transformation between x', y', z', v' and x, y, z, v that satisfies the invariance formula (2.1.10).

In the next section the cosmological transformation of space and velocity, the analogue of the Lorentz transformation of space and time, is presented.

2.2 The Cosmological Transformation

The transformation analogous to the Lorentz transformation is the *cosmological transformation* that gives the relationship between physical quantities at different cosmic times. The derivation of the cosmological transformation is just like that of the derivation of the Lorentz transformation.

We start with

$$x'^2 - \tau^2 v'^2 = x^2 - \tau^2 v^2, \tag{2.2.1}$$

assuming $y' = y$ and $z' = z$, the solution gives us

$$x' = x \cosh \psi - \tau v \sinh \psi,$$
$$\tau v' = \tau v \cosh \psi - x \sinh \psi. \tag{2.2.2}$$

At $x' = 0$ we obtain

$$\tanh \psi = \frac{x}{\tau v} = \frac{t}{\tau}, \tag{2.2.3}$$

and therefore

$$\sinh \psi = \frac{\dfrac{t}{\tau}}{\sqrt{1 - \dfrac{t^2}{\tau^2}}}, \tag{2.2.4a}$$

$$\cosh \psi = \frac{1}{\sqrt{1 - \dfrac{t^2}{\tau^2}}}, \tag{2.2.4b}$$

which lead to the transformations

$$x' = \frac{x - tv}{\sqrt{1 - \dfrac{t^2}{\tau^2}}}, \tag{2.2.5a}$$

$$v' = \frac{v - \dfrac{xt}{\tau^2}}{\sqrt{1 - \dfrac{t^2}{\tau^2}}}, \tag{2.2.5b}$$

$$y' = y, \ z' = z, \tag{2.2.5c}$$

and

$$x = \frac{x' + tv'}{\sqrt{1 - \dfrac{t^2}{\tau^2}}}, \tag{2.2.6a}$$

$$v = \frac{v' + \dfrac{x't}{\tau^2}}{\sqrt{1 - \dfrac{t^2}{\tau^2}}}, \qquad\qquad (2.2.6b)$$

$$y = y', \; z = z', \qquad\qquad (2.2.6c)$$

for the inverse transformation. Equations (2.2.5) and (2.2.6) will be referred to as the *cosmological transformation.*

Equations (2.2.5) give the transformed values of x and v as measured in the system K' with a relative cosmic time t with respect to K. The roles of the time and the velocity are *exchanged* as compared to special relativity. This fits our needs in cosmology where one measures distances and velocities at different cosmic times in the past. The parameter t/τ replaces v/c of special relativity.

Remark:

It should be emphasized that the transformation (2.2.5) is not a trivial exchange of v/c, appearing in the Lorentz transformation, and t/τ here. For example, the redshift $z = v/c$ at low velocities, but is certainly not equal to t/τ for small t/τ.

In the next section the galaxy cone, a presentation of galaxies in empty spacetime, and which is completely analogous to the light cone in ordinary special relativity is presented.

2.2.1 *Problems*

P 2.2.1. Derive the cosmological transformation assuming that the motion is along the x axis.

Solution: Hubble's law in the systems K and K' is given by

$$x = \tau v, \; x' = \tau v', \qquad\qquad (1)$$

where x, v and x', v' are measured in K and K'. Assuming now that x, v and x', v' transform linearly, then

$$x' = ax - bv, \qquad\qquad (2)$$

$$x = ax' + bv', \qquad\qquad (3)$$

where a and b are some variables which are independent of the coordinates.

At $x' = 0$ and $x = 0$, Eqs. (2) and (3) yield, respectively,

$$\frac{b}{a} = \frac{x}{v} = t, \tag{4}$$

and

$$\frac{b}{a} = -\frac{x'}{v'} = t. \tag{5}$$

Using now Eqs. (1), (2) and (3) we obtain

$$\tau v = x = ax' + bv' = a\tau v' + bv' = (a\tau + b)\, v', \tag{6a}$$

and similarly

$$\tau v' = (a\tau - b)\, v. \tag{6b}$$

Eliminating v and v' from Eqs. (6), and using $b = at$ from Eq. (4), we get

$$\tau^2 = a^2 \left(\tau^2 - t^2\right), \tag{7}$$

or

$$a = \frac{1}{\sqrt{1 - \dfrac{t^2}{\tau^2}}}, \tag{8}$$

and therefore

$$b = \frac{t}{\sqrt{1 - \dfrac{t^2}{\tau^2}}}. \tag{9}$$

Inserting these results in Eqs. (2) and (3) we obtain Eqs. (2.2.5) and (2.2.6).

2.3   The Galaxy Cone

The invariant equation (2.2.1), describing the distribution of galaxies in the Universe at any cosmic time, has a very simple geometrical interpretation. It enables one to present the locations of galaxies as a cone in the space of distance and velocity. One then has a *galaxy cone*, similar to the familiar light cone in special relativity (see Figure 1.3.1). The symmetry axis of the cone coincides with the x^0 axis which extends from $-\tau c$ to $+\tau c$.

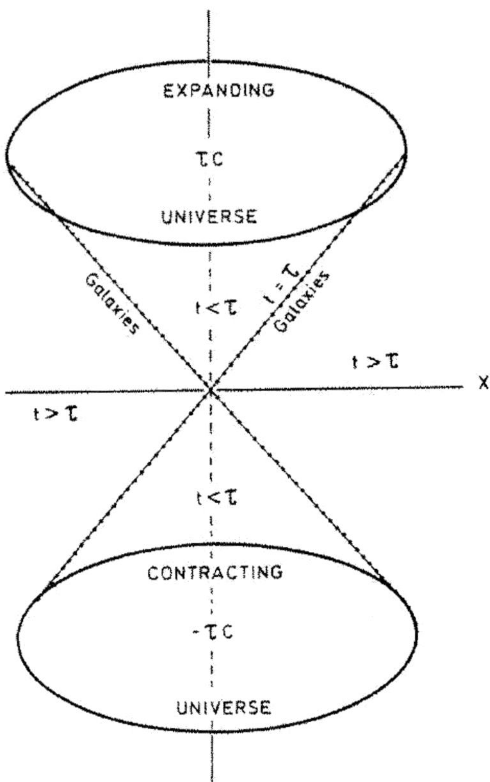

Fig. 2.3.1:   The galaxy cone in the $x - v$ space satisfying $x^2 - \tau^2 v^2 = 0$, where x represents the three-dimensional space. The heavy dots describe galaxies. The galaxy cone represents the locations of the galaxies at a given time rather than their path of motion in the real space. (Compare the light cone given in Figure 1.3.1.)

Figure 2.3.1 describes the galaxy cone in cosmological relativity. It gives the description of the cone in the $x - v$ space satisfying $x^2 - \tau^2 v^2 = 0$, where x stands for the three-dimensional space. The heavy dots describe galaxies. The galaxy cone represents the locations of the galaxies at a certain cosmic time rather than their path of motion in the real space (as is the case for light in the light cone). While points at the surface of the cone represent bodies which follow the Hubble expansion, those in its interior represent all other bodies.

In the next section some important consequences that follow from the cosmological transformation are presented.

2.4 Consequences of the Cosmological Transformation

In the following we draw some consequences of the cosmological transformations (2.2.5) and (2.2.6).

2.4.1 *The classical limit*

Assuming that t is much smaller than τ, one can neglect t^2 with respect to τ^2, and the transformation (2.2.5) gives

$$x' = x - tv, \quad v' = v, \quad y' = y, \quad z' = z, \qquad (2.4.1)$$

which is exactly the transformation obtained from classical mechanics.

2.4.2 *The length contraction*

Suppose there is a rod located in the K system parallel to the x axis. Let its length, measured in this system, be $\Delta x = x_2 - x_1$, where x_1 and x_2 are the coordinates of the two ends of the rod. To determine the length of this rod as measured in the K' system we must find the coordinates of the two ends of the rod x_1' and x_2' in this system at the same velocity v'. From Eqs. (2.2.6) we have

$$x_1 = \frac{x_1' + tv'}{\sqrt{1 - \dfrac{t^2}{\tau^2}}},$$

$$x_2 = \frac{x_2' + tv'}{\sqrt{1 - \dfrac{t^2}{\tau^2}}}.$$

The length of the rod in the K' system is $\Delta x' = x_2' - x_1'$, thus

$$\Delta x = \frac{\Delta x'}{\sqrt{1 - \dfrac{t^2}{\tau^2}}}.$$

The *proper* length of a rod is its length in a system in which it is located. Let us denote it by $L_0 = \Delta x$ and the length of the rod in any other system K' by L. Then

$$L = L_0 \sqrt{1 - \frac{t^2}{\tau^2}}. \qquad (2.4.2)$$

Thus a rod has its greatest length in the system in which its relative cosmic time with respect to the system is zero; its length in a system in which it is located at a relative cosmic time t with respect to that system is decreased by the factor $\left(1 - t^2/\tau^2\right)^{1/2}$. This result of the present theory is exactly similar to the familiar Lorentz contraction with the factor $\left(1 - v^2/c^2\right)^{1/2}$ in special relativity given in Subsection 1.5.2.

2.4.3   *The velocity contraction*

Suppose a velocity measuring instrument is located at $x' = 0$ in the K' system. Then from Eqs. (2.2.6) we have

$$v = \frac{v'}{\sqrt{1 - \dfrac{t^2}{\tau^2}}}. \qquad (2.4.3)$$

Denoting now v by v_0 and v' by v we obtain

$$v = v_0 \sqrt{1 - \frac{t^2}{\tau^2}}. \qquad (2.4.4)$$

The above result is like the time dilation in special relativity (see Subsection 1.5.3) and was expected since time in special relativity goes over to the velocity in the present theory. The velocity measured by an observer with a relative cosmic time t with respect to us is smaller by the factor $\left(1 - t^2/\tau^2\right)^{1/2}$ than what is observed by us at $t = 0$.

2.4.3.1   *Remark on dark matter*

As is well known much of the support for the existence of the dark matter is due to higher than expected observed velocities of galaxies measured from type Ia supernovae. Equation (2.4.4) clearly shows that the velocity observed by us is not the velocity measured by a local observer at a relative time t with respect to us. He would measure a smaller velocity, and the more back in time the more that velocity decreases. Does this mean that the hypothetical dark matter can be abolished just as the "luminiferous ether" was proved to be superfluous by special relativity?

2.4.4 *The law of addition of cosmic times*

Dividing the first of Eqs. (2.2.6) by the second we find, choosing $t = t_1$,

$$\frac{x}{v} = \frac{x' + t_1 v'}{v' + \dfrac{t_1}{\tau^2} x'}, \tag{2.4.5}$$

or, dividing the numerator and the denominator of the right-hand side of this equation by v', we obtain

$$t = \frac{t_1 + t_2}{1 + \dfrac{t_1 t_2}{\tau^2}}, \tag{2.4.6}$$

where $t_2 = x'/v'$ and $t = x/v$.

Equation (2.4.6) determines the transformation of cosmic time and describes the law of composition of cosmic times in the Universe. In the limiting case of t much smaller than the Hubble-Carmeli time τ, Eq. (2.4.6) goes over to the formula $t = t_1 + t_2$ of classical physics.

We see that the simple law of adding and subtracting cosmic times is no longer valid or, more precisely, is only approximately valid for recent times with respect to us, but not for those near the Hubble-Carmeli time, which is also the age of Universe in this case. If two consecutive events occur at $t_1 = (9/10)\tau$ and $t = (180/181)\tau$ both with respect to us (at $t = 0$), for example, then with respect to the first event the second one does not occur at $t - t_1 \approx \tau/10$ but rather at

$$t_2 = \frac{t - t_1}{1 - \dfrac{t t_1}{\tau^2}} = \frac{9}{10}\tau, \tag{2.4.7}$$

which is much longer than $t - t_1$ and happens to be exactly equal to t_1. We also notice that the past cosmic time cannot be greater than τ, the age of Universe. This is similar to what we have in special relativity where the velocity cannot exceed c (see Subsection 1.5.4). It will be noted that one may add as many successive time intervals as one wishes without ever reaching the age of the Universe τ.

2.4.5 *The inflation of the Universe*

The line element is given by

$$\tau^2 dv^2 - \left(dx^2 + dy^2 + dz^2\right) = ds^2. \tag{2.4.8}$$

Hence

$$\tau^2 \left(\frac{dv}{ds}\right)^2 - \left[\left(\frac{dx}{dv}\right)^2 + \left(\frac{dy}{dv}\right)^2 + \left(\frac{dz}{dv}\right)^2\right]\left(\frac{dv}{ds}\right)^2$$

$$= \left(\tau^2 - t^2\right)\left(\frac{dv}{ds}\right)^2 = 1. \tag{2.4.9}$$

Multiplying now this equation by ρ_0^2, square of the matter density of the Universe at the present time, we obtain for the matter density at a past time t

$$\rho = \tau\rho_0\frac{dv}{ds} = \frac{\rho_0}{\sqrt{1 - \dfrac{t^2}{\tau^2}}}. \tag{2.4.10}$$

Since the volume of the Universe is inversely proportional to its density, it follows that the ratio of the volumes at two backward cosmic times t_1 and t_2 with respect to us is given by $(t_2 < t_1)$.

$$\frac{V_2}{V_1} = \sqrt{\frac{1 - \dfrac{t_2^2}{\tau^2}}{1 - \dfrac{t_1^2}{\tau^2}}} = \sqrt{\frac{(\tau - t_2)(\tau + t_2)}{(\tau - t_1)(\tau + t_1)}}. \tag{2.4.11}$$

For times t_1 and t_2 very close to τ we can assume that $\tau + t_2 \approx \tau + t_1 \approx 2\tau$. Hence

$$\frac{V_2}{V_1} = \sqrt{\frac{T_2}{T_1}}, \tag{2.4.12}$$

where $T_1 = \tau - t_1$ and $T_2 = \tau - t_2$. For $T_2 - T_1 \approx 10^{-32}$ sec and $T_2 \ll 1$ sec, we then have

$$\frac{V_2}{V_1} \approx \sqrt{1 + \frac{10^{-32}}{T_1}} \approx \sqrt{\frac{10^{-32}}{T_1}} = \frac{10^{-16}}{\sqrt{T_1}}. \tag{2.4.13}$$

For $T_1 \approx 10^{-132}$ sec we obtain $V_2 \approx 10^{50} V_1$.

The above result conforms with inflationary Universe theory without assuming any model, such as the Universe is propelled by a sort of antigravity (A.H. Guth, and A. D. Linde).

2.4.6 *Minimal acceleration in the expansion of the Universe*

From Eq. (4) of Problem 2.2.1 we have

$$t = \frac{x}{v} = \frac{dx}{dv} = \frac{v}{a},$$
(2.4.14)

where a is the acceleration. Hence

$$t_{max} = \tau = \left(\frac{v}{a}\right)_{max} = \frac{c}{a_{min}}.$$
(2.4.15)

It thus appears that in the expansion of the cosmos there is a minimal acceleration

$$a_{min} = \frac{c}{\tau} = \frac{3 \times 10^{10} cm/s}{4.28 \times 10^{17} s} = 0.7 \times 10^{-7} \frac{cm}{s^2}.$$
(2.4.16)

2.4.7 *The cosmological redshift*

By analogy with Doppler theory in special relativity the wavelength of light emitted by a source at cosmic time t received by the observer at cosmic time $t = 0$, can be written as

$$\frac{\lambda}{\lambda_0} = \sqrt{\frac{1+\beta}{1-\beta}},$$
(2.4.17)

where $\beta = t/\tau$ instead of v/c. And for $t/\tau \ll 1$ we have

$$z = \frac{\lambda}{\lambda_0} - 1 \approx \frac{t}{\tau} + \frac{1}{2}\frac{t^2}{\tau^2}.$$
(2.4.18)

Sources further back in time are more redshifted. See section 2.4.9 for more details.

2.4.8 *The temperature of the Universe*

Denote the temperature of the Universe at a cosmic time t by T and that at present by T_0 (=2.73K), we then have

$$T = \frac{T_0}{\sqrt{1 - \frac{t^2}{\tau^2}}},$$
(2.4.19)

where t is the cosmic time measured with respect to us now and $\tau = 13.56 \pm 0.48$Gyr. For temperatures T at very early times, we can use the approximation $t \approx \tau$, thus

$$1 - \frac{t^2}{\tau^2} = \left(1 + \frac{t}{\tau}\right)\left(1 - \frac{t}{\tau}\right) \approx 2\left(1 - \frac{t}{\tau}\right)$$

$$= \frac{2}{\tau}\left(\tau - t\right) = 2\frac{\tilde{t}}{\tau}, \tag{2.4.20}$$

where $\tilde{t}$ is the cosmic time with respect to the Big Bang. Using this result we obtain

$$T = T_0 \sqrt{\frac{\tau}{2\tilde{t}}}. \tag{2.4.21}$$

The thermodynamical formula that relates the temperature to the cosmic time with respect to the Big Bang is well known and given by

$$T = \left(\frac{45\hbar^3}{32\pi^3 k^4 G}\right)^{1/4} \tilde{t}^{-1/2}, \tag{2.4.22}$$

where k is Boltzmann's constant, G is Newton's gravitational constant, $\hbar = h/2\pi$ and h is Planck's constant. As is seen, both equations show that the temperature T depends on $\tilde{t}^{-1/2}$. The coefficients appearing before the $\tilde{t}^{-1/2}$, however, are not identical. A simple calculation shows

$$\left(\frac{45\hbar^3}{32\pi^3 k^4 G}\right)^{1/4} = 1.52 \times 10^{10} \mathrm{Ks}^{1/2}, \tag{2.4.23}$$

and

$$T_0\sqrt{\frac{\tau}{2}} = 1.21 \times 10^9 \mathrm{Ks}^{1/2}. \tag{2.4.24}$$

In the above we have used $\hbar = 1.05 \times 10^{-34}\,\mathrm{Js}$, $k = 1.38 \times 10^{-23}\,\mathrm{J/K}$, $G = 6.67 \times 10^{-11}\,\mathrm{m}^3/\mathrm{s}^2\mathrm{Kg}$, $T_0 = 2.73\mathrm{K}$, $\tau = 13.56\mathrm{Gyr}$. Accordingly we can write for the temperatures in both cases

$$T \approx 1.5 \times 10^{10}\mathrm{Ks}^{1/2}\tilde{t}^{-1/2}, \tag{2.4.25}$$

and

$$T \approx 1.2 \times 10^9\mathrm{Ks}^{1/2}\tilde{t}^{-1/2}. \tag{2.4.26}$$

The ratio between them is approximately 13.

It thus appears that the dominant part of the plasma energy of the early Universe has gone to the creation of matter appearing now in the Universe, and only a small fraction of it was left for the background cosmic radiation.

2.4.9 *The relationship between redshift and cosmic time*

Problems with redshift and cosmic time are of considerable importance in cosmology. In particular one would like to know if there is any direct relationship between these two physical quantities. We know that in electrodynamics when a charged particle moves with acceleration and radiates electromagnetic waves there is a redshift when the particle is moving away from the observer. This is known as the Doppler shift. Is there any similar relationship in cosmology?

In this subsection we derive the formula, first given by Carmeli, Hartnett and Oliveira (see reference),

$$T = \frac{2h^{-1}}{1 + (1 + z)^2} \approx \frac{28}{1 + (1 + z)^2} \, \text{Gyr}, \qquad (2.4.27)$$

where h is the Hubble constant for the Universe with negligible gravity, and might be taken as $h = 72.17 \pm 0.84 \text{km/s-Mpc}$, and z is the redshift. The time T is now measured from the Big Bang onward, thus $T = \tau - t$, where t is measured backward. The formula is valid for all z.

The Universe expands, of course, by the Hubble law $x = H_0^{-1} v$, where H_0 is the Hubble constant at the present time. But one cannot use this law directly to obtain a relation between z and t. So we start by assuming that the Universe is empty of gravity. As we have seen, one can describe the property of expansion as a null-vector in the flat four dimensions of space and the expanding velocity v.

As was shown in Section 2.2, the cosmological line element was given by Eq. (2.2.1), where τ is the Big Bang time, the reciprocal of the Hubble constant H_0 in the limit of zero gravity, and it is a constant in this epoch of time. When $ds = 0$ one gets the Hubble expansion with no gravity.

Space and time coordinates transform according to the Lorentz transformation given by Eqs. (1.2.26) in ordinary physics. In cosmology the coordinates transform by the cosmological transformation given by Eqs. (2.2.5), where t is the cosmic time with respect to us now.

Comparing the Lorentz transformation and the cosmological transformation shows that the cosmological one can formally be obtained from the Lorentz transformation by changing v to t and c to τ $(v/c \rightarrow t/\tau)$. Thus the transfer from ordinary physics to the expanding Universe, under the above assumption of empty space, for null four-vectors is simply achieved by replacing v/c with t/τ, where t is the cosmic time measured with respect to us now.

We now use the above description as follows. In electrodynamics the electromagnetic radiation is described by its frequency ω and the wave vector $\mathbf{k}$. Using the wave four-vector $(\omega, \mathbf{k})$ one can easily derive the transformation of ω and $\mathbf{k}$ from one coordinate system to another. This then gives the Doppler effect.

A charged particle receding from the observer with a velocity v and emitting electromagnetic waves will experience a frequency shift given by

$$\omega = \omega' \sqrt{\frac{1 + \dfrac{v}{c}}{1 - \dfrac{v}{c}}}, \tag{2.4.28}$$

where ω' and ω are the frequencies of the emitted radiation recieved from the particle at velocity v and at rest, respectively. And thus a redshift is obtained from

$$1 + z = \sqrt{\frac{1 + \dfrac{v}{c}}{1 - \dfrac{v}{c}}}. \tag{2.4.29}$$

In our case τ replaces c and t replaces v (v/c goes over to t/τ), thus getting

$$1 + z = \sqrt{\frac{1 + \dfrac{t}{\tau}}{1 - \dfrac{t}{\tau}}}. \tag{2.4.30}$$

Rearranging, we get

$$\frac{t}{\tau} = \frac{(1 + z)^2 - 1}{(1 + z)^2 + 1}. \tag{2.4.31}$$

By using $\tau = 1/h$, and $T = \tau - t$, where now T is measured with respect to the Big Bang time, in the above formula, one easily obtains Eq. (2.4.27) (see Figure 2.4.1 for the case of backward time and Figure 2.4.2 for forward time). The two figures give the dependence of the redshift z on the cosmic time, in the two options of forward time T beginning with the Big Bang and the backward cosmic time t starting now.

In this section we have derived a simple formula valid for the case of negligible gravity and all redshift values in the Universe. The formula relates the cosmic time t since the Big Bang, for an Earth observer at present epoch, to the measured redshift z of light emitted at time t. The formula could be useful for identifying objects at the early Universe since we can go back in time as far as we desire but not to the Big Bang event itself at which the redshift becomes infinity.

In the next section the important kinematical concepts of velocity, acceleration and cosmic distances are brought in details.

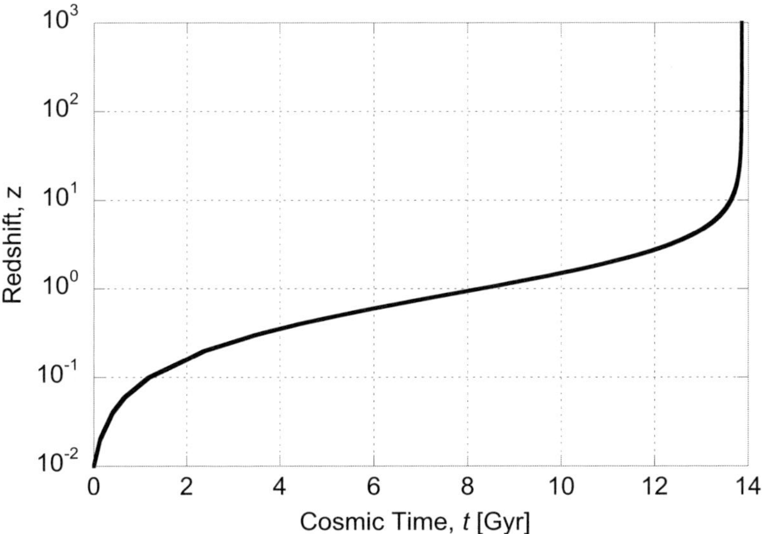

Fig. 2.4.1: Redshift z vs. backward cosmic time t where the time now is zero and is approximately 13.6 Gyr at the Big Bang. (Source: Carmeli, Hartnett & Oliveira, 2006)

2.4.10 *Problems*

P 2.4.1. Find the lengths of the 24 hour days since the early Universe, day by day, from the first day after the Big Bang up to our present time. Show that the first day actually lasted the Big Bang time τ. If T_n denotes the length of the n-th day in time units of the early Universe, then we have a very simple relation

$$T_n = \frac{\tau}{2n - 1}. \tag{1}$$

Hence we obtain for the first few days the following lengths of time:

$$T_1 = \tau,\ T_2 = \frac{\tau}{3},\ T_3 = \frac{\tau}{5},\ T_4 = \frac{\tau}{7},\ T_5 = \frac{\tau}{9},\ T_6 = \frac{\tau}{11}. \tag{2}$$

It also follows that the accumulation of time from the first day to the second, third, fourth, ..., up to now is just exactly the Big Bang time τ. The Big Bang time in the limit of zero gravity is the maximum time allowed in nature ($\tau = 13.56\text{Gyr}$).

Solution: We assume that the Big Bang time with respect to us now was $t_0 = \tau$, the time of the first day after that was t_1, the time of the second day

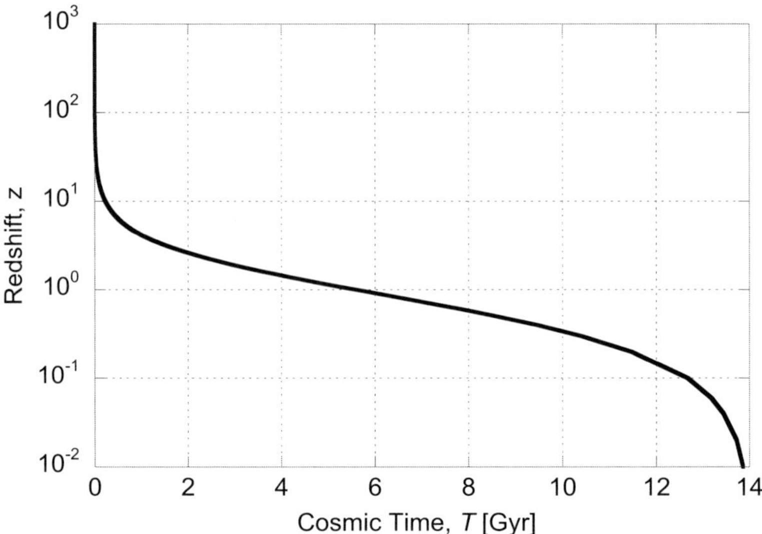

Fig. 2.4.2:   Redshift z vs. cosmic time T forward from the Big Bang. (Source: Carmeli, Hartnett & Oliveira, 2006)

was t_2, and so on. In this way the time scale is progressing in units of one day (24 hours) in our units of present time. The time difference between t_0 and t_1, denoted by T_1, is the time as measured at the early Universe and is by no means equal to one day of our time. In this way we denote the times elapsed from the Big Bang to the end of the first day t_1 by T_1, between the first day t_1 and the second day t_2 by T_2 and so on. According to the rule of the addition of cosmic times one has, for example,

$$t_6 + 1(\text{day}) = \frac{t_6 + T_6}{1 + \dfrac{t_6 T_6}{\tau^2}}. \tag{3}$$

A straightforward calculation then shows that

$$T_6 = \frac{\tau^2}{\tau^2 - (\tau - 6)(\tau - 5)} = \frac{\tau^2}{11\tau - 30}. \tag{4}$$

In general one finds that

$$T_n = \frac{\tau^2}{\tau^2 - (\tau - n)(\tau - n + 1)}, \tag{5}$$

or

$$T_n = \frac{\tau}{n + (n - 1) - n(n - 1)/\tau}. \tag{6}$$

As is seen from the last formula one can neglect the last term in the denominator in the first approximation and we get the simple Eq. (1).

From the above one reaches the conclusion that the age of Universe exactly equals the Big Bang time τ, and it is a universal constant regardless of epoch. This means that the age of Universe tomorrow will be the same as it was yesterday or today.

2.5 Velocity, Acceleration and Cosmic Distances

In this section we conclude the fundamentals of the cosmological special relativity (CSR) by discussing the *dynamical* concepts of *velocity, acceleration* and *cosmic distances* in spacevelocity. These concepts occur in CSR just as those of mass, linear momentum and energy appear in Einstein's special relativity (ESR) of spacetime (see Chapter 1).

2.5.1 *Preliminaries*

The most important result of Einstein's special relativity is probably the relationship between mass and energy (see Section 1.6). How it happened that the mass became so critical in this theory? The answer is very simple. The theory involves the square of the speed of light, c^2. What physical quantity incorporates the square of velocity? It is the energy, $mv^2/2$. Hence it is the mass that goes with v^2 and c^2. Thus m is taken as an invariant under the Lorentz transformation. This becomes the rest mass m_0. The inertial mass m follows to depend on the velocity.

What is the comparable physical quantity in cosmological special relativity? Certainly not the mass. Here we have τ^2, the square of the Big Bang time. What physical quantity goes with the square of time? It is the acceleration $\mathbf{a}$ because $\mathbf{a}t^2/2$ describes the distance a particle makes at time t when it is subject to acceleration $\mathbf{a}$. Hence $\mathbf{a}$ should be taken as the invariant quantity under the cosmological transformation. But then the acceleration depends on the cosmic time just as the mass depends on the velocity.

In this section these concepts are explored.

2.5.2 *Velocity and acceleration four-vectors*

We start our four-dimensional spacevelocity analysis by defining the velocity and acceleration (see Subsection 1.6.5 for the special relativistic treatment in spacetime).

The *velocity* four-vector of a particle in spacevelocity is defined as the dimensionless quantity

$$u^\mu = \frac{dx^\mu}{ds}, \tag{2.5.1}$$

where $\mu = 0, 1, 2, 3$, $x^\mu = (x^0, x^1, x^2, x^3) = (\tau v, x, y, z)$, and τ is the Big Bang time, the inverse of the Hubble parameter (constant) in the limit of zero gravity, a universal constant whose value is 13.56 Gyr (see M. Carmeli, 2006).

In flat spacevelocity one has for the line element,

$$\tau^2 dv^2 - \left(dx^2 + dy^2 + dz^2\right) = ds^2, \tag{2.5.2}$$

thus

$$\tau^2 \left(\frac{dv}{ds}\right)^2 \left(1 - \frac{dx^2 + dy^2 + dz^2}{\tau^2 dv^2}\right) = 1. \tag{2.5.3}$$

This gives

$$\tau^2 \left(\frac{dv}{ds}\right)^2 \left(1 - \frac{t^2}{\tau^2}\right) = 1, \tag{2.5.4}$$

and therefore

$$\frac{dv}{ds} = \frac{1}{\tau\sqrt{1 - \dfrac{t^2}{\tau^2}}}. \tag{2.5.5}$$

The *velocity* four-vector in spacevelocity can thus be expressed as

$$u^\mu = \frac{dx^\mu}{ds} = \frac{dx^\mu}{dv}\frac{dv}{ds} = \frac{1}{\tau\sqrt{1 - \dfrac{t^2}{\tau^2}}}\frac{dx^\mu}{dv}. \tag{2.5.6}$$

The velocitylike component of u^μ is therefore given by

$$u^0 = \gamma, \tag{2.5.7}$$

whereas its spatial components

$$\mathbf{u} = u^k = \left(u^1, u^2, u^3\right) \tag{2.5.8}$$

are given by

$$u^k = \frac{\gamma}{\tau}\frac{dx^k}{dv}, \tag{2.5.9}$$

where

$$\gamma = \frac{1}{\sqrt{1 - \dfrac{t^2}{\tau^2}}}. \tag{2.5.10}$$

It will be noted that, by Eq. (2.5.1),

$$u_\alpha u^\alpha = 1, \tag{2.5.11}$$

namely, the length of u^μ is unity.

The *acceleration* four-vector of a particle in spacevelocity is defined by

$$\frac{du^\mu}{ds} = \frac{d^2 x^\mu}{ds^2}. \tag{2.5.12}$$

By differentiating Eq. (2.5.11) we find that the acceleration four-vector satisfies the orthogonality condition

$$u_\alpha \frac{du^\alpha}{ds} = 0. \tag{2.5.13}$$

The components of the acceleration four-vector of a particle in spacevelocity are, by Eqs. (2.5.5) and (2.5.7)–(2.5.9), then given by

$$\frac{du^0}{ds} = \frac{du^0}{dv}\frac{dv}{ds} = \frac{\gamma}{\tau}\frac{d\gamma}{dv}, \tag{2.5.14}$$

$$\frac{du^k}{ds} = \frac{du^k}{dv}\frac{dv}{ds} = \frac{\gamma}{\tau^2}\frac{d}{dv}\left(\gamma\frac{dx^k}{dv}\right), \tag{2.5.15}$$

where γ is given by Eq. (2.5.10).

2.5.3   *Acceleration and distances*

Multiplying Eq. (2.5.4) by $\mathbf{a}_0^2 \tau^4$, where $\mathbf{a}_0$ is the ordinary three-vector acceleration as measured in the cosmic frame of reference (see Subsection 2.1.3) at cosmic time $t = 0$ (i.e. now), we obtain

$$\mathbf{a}_0^2 \tau^6 \left(\frac{dv}{ds}\right)^2 \left(1 - \frac{t^2}{\tau^2}\right) = \mathbf{a}_0^2 \tau^4. \tag{2.5.16}$$

Using now Eq. (2.5.5) in Eq. (2.5.16) the latter then yields

$$\left(1 - \frac{t^2}{\tau^2}\right)^{-1} \left(\mathbf{a}_0^2 \tau^4 - \mathbf{a}_0^2 t^2 \tau^2\right) = \mathbf{a}_0^2 \tau^4. \tag{2.5.17}$$

We now define the acceleration $\mathbf{a}$ at an arbitrary cosmic time t by

$$\mathbf{a} = \frac{\mathbf{a}_0}{\sqrt{1 - \dfrac{t^2}{\tau^2}}}, \tag{2.5.18}$$

then Eq. (2.5.17) will have the form

$$\mathbf{a}^2\tau^4 - \mathbf{a}^2 t^2\tau^2 = \mathbf{a}_0^2\tau^4, \tag{2.5.19}$$

or

$$\mathbf{a}^2\tau^4 - \tau^2\mathbf{v}^2 = \mathbf{a}_0^2\tau^4, \tag{2.5.20}$$

where $\mathbf{v} = \mathbf{a}t$ is the ordinary three-dimensional velocity. Equation (2.5.20) is the analog to

$$m^2 c^4 - c^2\mathbf{p}^2 = m_0^2 c^4, \tag{2.5.21}$$

in ESR, where $\mathbf{p} = m\mathbf{v}$ is the linear momentum (see Subsection 1.7.2). Thus $\mathbf{a}$ reduces to $\mathbf{a}_0$ when the cosmic time $t = 0$ (i.e. at present).

The comparable to Eq. (2.5.18) in ESR is, of course,

$$m = \frac{m_0}{\sqrt{1 - \dfrac{v^2}{c^2}}}, \tag{2.5.22}$$

where m and m_0 are the inertial mass and the rest mass of the particle, with m reduces to m_0 at $v = 0$. If we multiply Eq. (2.5.22) by c^2 and expand both sides in v/c, we obtain

$$mc^2 = m_0 c^2 + \frac{m_0}{2}v^2 + \cdots. \tag{2.5.23}$$

Doing the same with Eq. (2.5.18) but multiplication by τ^2, and expanding in t/τ, we obtain

$$\mathbf{a}\tau^2 = \mathbf{a}_0\tau^2 + \frac{\mathbf{a}_0}{2}t^2 + \cdots. \tag{2.5.24}$$

Equation (2.5.24) in CSR is of course the analog to Eq. (2.5.23) in ESR.

2.5.4   *Energy in ESR versus cosmic distance in CSR*

While Eq. (2.5.23) yields

$$E = E_0 + \frac{m_0}{2}v^2 + \cdots, \tag{2.5.25}$$

Eq. (2.5.24) gives

$$\mathbf{S} = \mathbf{S}_0 + \frac{\mathbf{a}_0}{2}t^2 + \cdots. \tag{2.5.26}$$

In the above equations E is, of course, the energy of the particle, whereas $\mathbf{S}$ is the cosmic distance. $E_0 = m_0 c^2$ is the rest energy of the particle whereas $m_0 v^2/2$ is the Newtonian kinetic energy. What about the terms in Eq. (2.5.26)? The term $\mathbf{a}_0 t^2/2$ is, of course, the Newtonian distance the particle makes due to the acceleration $\mathbf{a}_0$, the term $\mathbf{S}_0 = \mathbf{a}_0 \tau^2$ is unique to CSR, and it might be called the *intrinsic* cosmic distance of the particle.

Equation (2.5.20) can now be written as

$$\mathbf{S}^2 - \tau^2 \mathbf{v}^2 = \mathbf{S}_0^2, \tag{2.5.27}$$

in complete analogy to

$$E^2 - c^2 \mathbf{p}^2 = E_0^2 \tag{2.5.28}$$

in ESR.

2.5.5 *Distance-velocity four-vector*

We now define the cosmic distance-velocity four-vector. It is the analogous to the energy-momentum four-vector in ESR (see Subsection 1.7.4). It is defined by

$$\mathbf{v}^\mu = \mathbf{a}_0 \tau^2 u^\mu, \tag{2.5.29}$$

where u^μ has been defined in Subsection (2.5.2) We have

$$\mathbf{v}^0 = \mathbf{a}_0 \tau^2 u^0, \tag{2.5.30a}$$

$$\mathbf{v}^k = \mathbf{a}_0 \tau^2 u^k, \tag{2.5.30b}$$

where u^0 and u^k are given by Eqs. (2.5.7)–(2.5.10), with

$$\mathbf{v}_0 = \mathbf{v}^0, \qquad \mathbf{v}_k = -\mathbf{v}^k. \tag{2.5.31}$$

Accordingly we have

$$\mathbf{v}^2 = \mathbf{v}_0 \cdot \mathbf{v}^0 + \mathbf{v}_k \cdot \mathbf{v}^k = \mathbf{a}_0^2 \tau^4 \left(u_0 u^0 + u_k u^k \right)$$

$$= \mathbf{a}_0^2 \tau^4 u_\alpha u^\alpha = \mathbf{a}_0^2 \tau^4 = \mathbf{S}_0^2. \tag{2.5.32}$$

But, using Eqs. (2.5.7), (2.5.9) and (2.5.18),

$$\mathbf{v}^0 = \mathbf{a}_0 \tau^2 \gamma = \mathbf{a} \tau^2 = \mathbf{S}, \tag{2.5.33a}$$

$$\mathbf{v}^k = \mathbf{a}_0 \tau \gamma \frac{dx^k}{dv} = \mathbf{a} \tau \frac{dx^k}{dv} = \mathbf{a} \tau \frac{dx^k}{dt} \frac{dt}{dv} = \tau \mathbf{v}, \tag{2.5.33b}$$

where $\mathbf{v}$ is the three-dimensional velocity. Hence

$$\mathbf{v}_\alpha \cdot \mathbf{v}^\alpha = \mathbf{v}_0 \cdot \mathbf{v}^0 + \mathbf{v}_k \cdot \mathbf{v}^k = \mathbf{a}_0^2 \tau^4 \gamma^2 - \mathbf{a}_0^2 \tau^2 \gamma^2 \left(\frac{dx^k}{dv} \right)^2$$

$$= \mathbf{a}^2\tau^4 - \tau^2\mathbf{v}^2 = \mathbf{S}^2 - \tau^2\mathbf{v}^2, \tag{2.5.34}$$

and accordingly, using (2.5.32),

$$\mathbf{S}^2 - \tau^2\mathbf{v}^2 = \mathbf{S}_0^2, \tag{2.5.35}$$

which is exactly Eq. (2.5.20) with $\mathbf{S} = \mathbf{a}\tau^2$ and $\mathbf{S}_0 = \mathbf{a}_0\tau^2$. The above analysis also shows that Eq. (2.5.35) is covariant under spacevelocity transformation.

Equation (2.5.35) is the analog to

$$E^2 - c^2\mathbf{p}^2 = E_0^2 \tag{2.5.36}$$

in ESR, where E and E_0 are the energy and rest energy, respectively, $\mathbf{S}$ and $\mathbf{S}_0$ are the cosmic distances at cosmic time t and present time $(t = 0)$, respectively.

Finally, in ESR when the rest mass is zero (like the photon), one then has

$$E = cp, \tag{2.5.37}$$

which is valid at the light cone (see Section 1.3). In the case of CSR one also has, when $\mathbf{S}_0 = 0$,

$$\mathbf{S} = \tau\mathbf{v}, \tag{2.5.38}$$

now valid at the galaxy cone (see Section 2.3).

2.5.6 *Conclusions*

In this section it has been shown that the comparable quantity to the mass in Einstein's special relativity is the ordinary acceleration three-vector in cosmological special relativity. They both have similar behavior, one with respect to v/c and the other with t/τ:

$$m = \frac{m_0}{\sqrt{1 - \dfrac{v^2}{c^2}}},$$

and

$$\mathbf{a} = \frac{\mathbf{a}_0}{\sqrt{1 - \dfrac{t^2}{\tau^2}}}.$$

Furthermore, the role of the energy in Einstein's theory is being taken by the cosmic distance,

$$E = mc^2,$$

and

$$\mathbf{S} = \mathbf{a}\tau^2.$$

Finally, the analog of the energy formula

$$E^2 - c^2\mathbf{p}^2 = E_0^2$$

in ordinary special relativity is

$$\mathbf{S}^2 - \tau^2\mathbf{v}^2 = \mathbf{S}_0^2$$

in cosmological special relativity.

With this we finished our flat spacetime discussion of cosmological special relativity. In the next chapter, gravitation is invoked and the full theory of general relativity is presented and applied to standard topics in that theory.

2.6 Suggested References

R.R. Caldwell and P.J. Steinhardt, *Phys. Rev. D* **57**, 6057 (1998).

M. Carmeli, *Cosmological Relativity: The Special and General Theories for the Structure of the Universe* (World Scientific, Singapore, 2006).

M. Carmeli, J.G. Hartnett and F. Oliveira, The cosmic time in terms of the redshift, *Found. Phys. Lett.* **19**, 276-283 (2006); gr-qc/0506079

N. Dauphas, The U/Th production ratio and the age of the Milky Way from meteorites and galactic halo stars, *Nature* **435**, 1203-1205 (2005).

E. Egami , J.-P. Kneib , G. H. Rieke , R. S. Ellis , J. Richard , J. Rigby , C. Papovich, D. Stark , M. R. Santos , J.-S. Huang , H. Dole , E. Le Floch and P. G. Pérez-González , Spitzer and Hubble Space Telescope constraints on the physical properties of the $z{\approx}7$ galaxy strongly lensed by A2218, *Astrophys. J.* **618**, L5-L8 (2005).

A. Einstein, *Ann. Physik* (Germany) **17**, 891 (1905); English translation in: A. Einstein *et al.: The Principle of Relativity* (Dover, New York, 1923).

A. Einstein, *Autobiographical Notes*, translated and edited by P.A. Schilpp (Open Court Publishing Co., La Salle and Chicago, 1979).

C.H. Gibson, Turbulent mixing, viscosity, diffusion and gravity in the formation of cosmological structures: The fluid mechanics of dark matter, *J. Fluid Eng.* **122**, 830-835 (2000).

A.H. Guth, Inflationary Universe: A possible solution to the horizon and flatness problems, *Phys. Rev. D* **23**, 347 (1982).

L. Hernquist and V. Springel, An analytical model for the history of cosmic star formation, *MNRAS* **341**, 1253-1267 (2003).

A.D. Linde, Scalar field fluctuations in the expanding Universe and the new inflationary Universe scenario, *Phys. Lett. B* **116**, 335 (1982).

A.M. Öztaş and M.L. Smith, Elliptical solutions to the standard cosmology model with realistic values of matter density, *Inter. J. Theor. Phys.* **45**, 925-936 (2006).

P.J.E. Peebles, *Principles of Physical Cosmology,* Princeton Series in Physics (Princeton University Press, Princeton, NJ, 1993).

S. Perlmutter *et al.*, *Astrophys. J.* **517**, 565 (1999).

A.G. Riess *et al.*, *Astron. J.* **116**, 1009 (1998).

A.G. Riess, L.-G. Strolger, J. Tonry, S. Casertano, H.C. Ferguson, B. Mobasher, P. Challis, A.V. Filippenko, S. Jha, W. Li, R. Chornock, R.P. Kirchner, B. Leibundgut, M. Dickinson, M. Livio, M. Giavalisco, C.C. Steidel, N. Benitez and Z. Tsvetanov, Type Ia Supernova discoveries at $z > 1$ from the Hubble Space Telescope: Evidence for past deceleration and constraints on dark energy evolution, *Astrophys. J.* **607**, 665-687 (2004).

J.L. Sievers, J.K. Cartwright, C.D. Contaldi, B.S. Mason, S.T. Myers, S. Padin, T.J. Pearson, U.-L. Pen, D. Pgosyan, S. Prunet, A.C.S. Readhead, M.C. Sheppard, P.S. Udonpraesert, L. Bronfman, W.L. Holzapfel and J. May, Cosmological parameters from cosmic background imager observations and comparisons with BOOMERANG, DASI, and MAXIMA, *Astrophys. J.* **591**, 599-622 (2003).

M.L. Smith, A.M. Öztaş and J. Paul, A model of light from ancient blue emissions, *Int. J. Theor. Phys.* **45**, 937-952 (2006).

M.L. Smith, A.M. Öztaş and J. Paul, Estimation of redshifts from early galaxies, *Annales de la Foundation Louis de Broiglie* **32**(1) (2007).

V. Springel, S.D. White, A. Jenkins, C.S. Frenk, N. Yosida, L. Gao, J. Navarro, R. Thacker, D. Croton, J. Helly, J.A. Peacock, S. Cole,P. Thomas, H. Couchman, A. Evrard, J. Colberg and F. Pearce, Simulations of the formation, evolution and clustering of galaxies and quasars, *Nature* **435**, 629-636 (2005).

D. Watson, J. Hjorth, P. Jakobsson, D. Xu, J.P.U. Fynbo, J. Sollerman, C.C. Thöne and K. Pedersen, Are short gamma-ray bursts collimated? GRB 050709, a flare but no break, *Astron. Astrophys.* **454**, L123-L126 (2006).

Chapter 3

General Relativity Theory

Moshe Carmeli

In this chapter a short review of general relativity theory is given. We begin the discussion with a brief review of Riemannian geometry, including transformation of coordinates, covariant and contravariant vectors, tensors, the metric tensor, Christoffel symbols, covariant differentiation, the Riemann, Einstein and Ricci tensors, geodesics, and the Bianchi identities. This followed by a description of the physical foundations of general relativity. These are the principle of equivalence and the principle of general covariance. A detailed description is given of the Eötvös experiment and its new version, the Dicke experiment that improves its accuracy. The gravitational field equations of Einstein are then derived in a tensorial form. The basis of these equations is the Newtonian field equations and this fact is shown in details. The Schwarzschild solution of Einstein's field equations is derived. Although this is the simplest solution of the Einstein's field equations, its astrophysical importance is very high since it leads to the concept of black hole. Experimental verification of general relativity is subsequently given. The chapter is then concluded with a review of the problem of motion in general relativity theory. This problem is unique in general relativity theory since the equations of motion are consequence of the field equations. The Bianchi identities play an important role in the singularities at the locations of particles. The Einstein-Infeld-Hoffmann equations of motion, which are post-Newtonian, are derived. These equations include nonlinear terms that lead to the description of the perihelion advance of planets.

3.1 Riemannian Geometry

3.1.1 *Transformation of coordinates*

Any four independent variables x^μ, where Greek letters take the values 0, 1, 2, 3, may be considered as the coordinates of a four-dimensional space V_4 (Figure 3.1.1). Each set of values of x^μ defines a point of V_4. Let there be another set of coordinates x'^μ related to the first set x^ν by

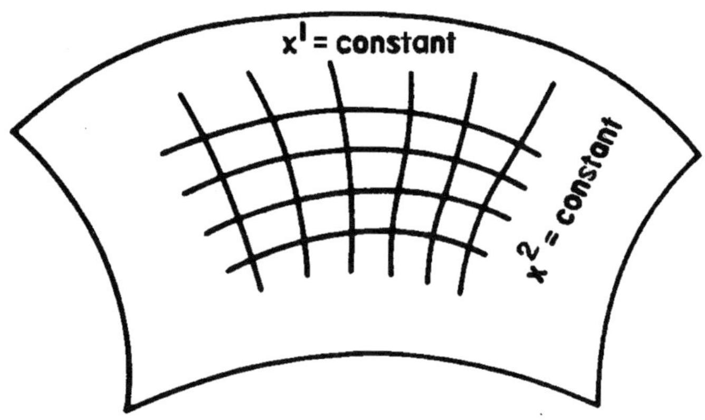

Fig. 3.1.1: Coordinate system in curved Riemannian space of dimension two. The coordinates are denoted by x^1 and x^2.

$$x'^\mu = f^\mu\left(x^\nu\right),\qquad(3.1.1)$$

where f^μ are four independent real functions of x^ν. A necessary and sufficient condition that f^μ be independent is that their Jacobian,

$$\left|\frac{\partial f^\mu}{\partial x^\nu}\right| = \begin{vmatrix} \dfrac{\partial f^0}{\partial x^0} & \cdots & \dfrac{\partial f^3}{\partial x^0} \\ \vdots & & \\ \dfrac{\partial f^0}{\partial x^3} & \cdots & \dfrac{\partial f^3}{\partial x^3} \end{vmatrix},\qquad(3.1.2)$$

does not vanish identically. Equation (3.1.1) defines a *transformation of coordinates* in the space V_4. When the Jacobian is different from zero, one can also write x^μ in terms of x'^ν as

$$x^\mu = g^\mu\left(x'^\nu\right).\qquad(3.1.3)$$

A *direction* at a point P in the space V_4 is determined by the differential dx^μ. The same direction is determined in another set of coordinates x'^μ by the differential dx'^μ. The two differentials are related, using Eq. (3.1.1), by

$$dx'^\mu = \frac{\partial x'^\mu}{\partial x^\nu} dx^\nu = \frac{\partial f^\mu}{\partial x^\nu} dx^\nu. \qquad (3.1.4)$$

Here the Einstein summation convention is used, according to which repeated Greek indices are summed over the values 0, 1, 2, 3.

3.1.2 Contravariant vectors

Let two sets of functions V^μ and V'^μ be related by

$$V'^\mu = \frac{\partial x'^\mu}{\partial x^\nu} V^\nu, \qquad (3.1.5)$$

similar to the way the differentials dx'^μ and dx^μ are related. V^μ and V'^μ are then called the *components* of a *contravariant vector* in the coordinate systems x^μ and x'^μ, respectively. Hence any four functions of the x's in one coordinate system can be taken as the components of a contravariant vector whose components in any other coordinate system are given by Eq. (3.1.5).

A contravariant vector determines a direction at each point of the space V_4. Let V^μ be the components of a contravariant vector and let dx^μ be a displacement in the direction of V^μ. Then $dx^0/V^0 = \cdots = dx^3/V^3$. This set of equations admits three independent functions $f^k(x^\mu) = c^k$, where $k = 0, 1, 2$, and the c's are arbitrary constants and the matrix $\partial f^k/\partial c^\mu$ is of rank three. The functions f^k are solutions of the partial differential equation $V^\nu \partial f^k/\partial x^\nu = 0$. Hence using the transformation laws (3.1.1) and (3.1.3) one obtains $V'^k = 0$ for $k = 0, 1, 2$, and $V'^3 \neq 0$. Hence a system of coordinates can be chosen in terms of which all components but one of a given contravariant vector are equal to zero.

3.1.3 Invariants. Covariant vectors

Two functions $f(x)$ and $f'(x')$ define an *invariant* if they are reducible to each other by a coordinate transformation.

Let f be a function of the coordinates. Then

$$\frac{\partial f}{\partial x'^\mu} = \frac{\partial f}{\partial x^\nu} \frac{\partial x^\nu}{\partial x'^\mu}. \qquad (3.1.6)$$

Two sets of functions V_μ and V'_μ are called the *components* of a *covariant vector* in the systems x and x', respectively, if they are related by the transformation law of the form (3.1.6),

$$V'_\mu = \frac{\partial x^\nu}{\partial x'^\mu} V_\nu. \tag{3.1.7}$$

For example, if f is a scalar function, then $\partial f/\partial x^\mu$ is a covariant vector. It is called the *gradient* of f. The product $V^\mu W_\mu$ is an invariant if V is a contravariant vector and W is a covariant vector. Conversely, if the quantity $V^\mu W_\mu$ is an invariant and either V^μ or W_μ are arbitrary vectors, then the other set is a vector.

3.1.4 *Tensors*

Tensors of any order are defined by generalizing Eqs. (3.1.5) and (3.1.7). Thus the equation

$$T'^{\mu_1\cdots\mu_m}_{\nu_1\cdots\nu_n} = \frac{\partial x'^{\mu_1}}{\partial x^{\rho_1}} \cdots \frac{\partial x'^{\mu_m}}{\partial x^{\rho_m}} \frac{\partial x^{\sigma_1}}{\partial x'^{\nu_1}} \cdots \frac{\partial x^{\sigma_n}}{\partial x'^{\nu_n}} T^{\rho_1\cdots\rho_m}_{\sigma_1\cdots\sigma_n} \tag{3.1.8}$$

defines a mixed tensor of order $m + n$, contravariant of the mth order and covariant of the nth order. If the Kronecker delta function is taken as the components of a mixed tensor of the second order in one set of coordinates, for example, then it defines the components of a tensor in any set of coordinates. An invariant is a tensor of zero order and a vector is a tensor of order one.

When the relative position of two indices, either contravariant or covariant, is immaterial, the tensor is called *symmetric* with respect to these indices. When the relative position of two indices of a tensor is interchanged and the tensor obtained differs only in sign from the original one, the tensor is called *skew-symmetric* with respect to these indices. The process by means of which one obtains a tensor of order $r - 2$ from a mixed tensor of order r is called *contraction*.

3.1.5 *The metric tensor*

Let $g_{\mu\nu}$ be the components of the *metric tensor*, i.e., a symmetric covariant tensor, which is a function of coordinates, and let $g = \det g_{\mu\nu}$. The quantity $g^{\mu\nu}$, denoting the *cofactor* of $g_{\mu\nu}$ divided by g, is a symmetric contravariant tensor and satisfies

$$g^{\mu\rho} g_{\nu\rho} = \delta^\mu_\nu. \tag{3.1.9}$$

The element of length is defined by means of a quadratic differential form $ds^2 = g_{\mu\nu}dx^\mu dx^\nu$. By means of the tensors $g_{\mu\nu}$ and $g^{\mu\nu}$ one can lower or raise tensor indices:

$$T^\mu{}_{\nu\rho} = g^{\mu\sigma}T_{\sigma\nu\rho}, \qquad (3.1.10a)$$

$$T_\alpha{}^{\beta\gamma} = g_{\alpha\rho}T^{\rho\beta\gamma}. \qquad (3.1.10b)$$

Other certain quantities transform according to the law

$$T'^{\mu\cdots}_{\alpha\cdots} = J^N \frac{\partial x'^\mu}{\partial x^\rho}\frac{\partial x^\beta}{\partial x'^\alpha}\cdots T^{\rho\cdots}_{\beta\cdots}. \qquad (3.1.11)$$

Here J is the Jacobian determinant $|\,\partial x^\alpha/\partial x'^\beta\,|$. The superscript N is the power to which J is raised. $T^{\mu\cdots}_{\nu\cdots}$ is called a *tensor density* of weight N. For example, if g' denotes det $g'_{\mu\nu}$ then $g' = J^2 g$, where $g = $ det $g_{\mu\nu}$. Hence for the four-dimensional elements in two coordinate systems one has the equality:

$$\sqrt{-g}d^4x = \sqrt{-g'}d^4x'. \qquad (3.1.12)$$

Tensor densities and their properties are considered in details in Subsection 3.1.11.

3.1.6 *The Christoffel symbols*

From the two tensors $g_{\mu\nu}$ and $g^{\mu\nu}$ one can define the two functions

$$\Gamma_{\alpha\rho\sigma} = \frac{1}{2}\left(\frac{\partial g_{\rho\alpha}}{\partial x^\sigma} + \frac{\partial g_{\sigma\alpha}}{\partial x^\rho} - \frac{\partial g_{\rho\sigma}}{\partial x^\alpha}\right), \qquad (3.1.13)$$

$$\Gamma^\mu_{\rho\sigma} = g^{\mu\alpha}\Gamma_{\alpha\rho\sigma}. \qquad (3.1.14)$$

They are symmetric in ρ and σ, and are called the *Christoffel symbols* of the *first* and *second* kind, respectively.

Both kinds of Christoffel symbols are not components of tensors. By starting with the differential transformation law for $g_{\mu\nu}$ it is not too difficult to show that $\Gamma_{\alpha\rho\sigma}$ transforms according to the following relation:

$$\Gamma'_{\nu\mu\alpha} = \frac{\partial x^\beta}{\partial x'^\mu}\frac{\partial x^\gamma}{\partial x'^\alpha}\frac{\partial x^\delta}{\partial x'^\nu}\Gamma_{\delta\beta\gamma} + g_{\beta\gamma}\frac{\partial x^\beta}{\partial x'^\nu}\frac{\partial^2 x^\gamma}{\partial x'^\mu\partial x'^\alpha}. \qquad (3.1.15)$$

Using the transformation law for $g^{\alpha\beta}$ then leads to the transformation law of $\Gamma^\delta_{\beta\nu}$ as

$$\Gamma'^\delta_{\beta\nu} = \frac{\partial x'^\delta}{\partial x^\alpha}\frac{\partial x^\mu}{\partial x'^\beta}\frac{\partial x^\sigma}{\partial x'^\nu}\Gamma^\alpha_{\mu\sigma} + \frac{\partial x'^\delta}{\partial x^\sigma}\frac{\partial^2 x^\sigma}{\partial x'^\beta x'^\nu}. \qquad (3.1.16)$$

From Eq. (3.1.13) we obtain

$$\Gamma^{\mu}_{\alpha\mu} = \frac{1}{2} g^{\mu\nu} \frac{\partial g_{\mu\nu}}{\partial x^{\alpha}}. \tag{3.1.17}$$

This equation can be rewritten in terms of the determinant g of $g_{\mu\nu}$. The rule for expansion of a determinant leads to the formula

$$\frac{\partial g}{\partial g_{\mu\nu}} = \Delta^{\mu\nu}, \tag{3.1.18}$$

where $\Delta^{\mu\nu}$ is the cofactor of the element $g_{\mu\nu}$. From the law for obtaining the inverse of a determinant, and from the definition of $g^{\mu\nu}$, Eq. (3.1.18) may be written as

$$\frac{\partial g}{\partial g_{\mu\nu}} = g g^{\mu\nu}, \tag{3.1.19}$$

and consequently

$$dg = g g^{\mu\nu} dg_{\mu\nu} = -g g_{\mu\nu} dg^{\mu\nu}. \tag{3.1.20}$$

Hence we have

$$\partial_{\alpha} g = g g^{\mu\nu} \partial_{\alpha} g_{\mu\nu} = -g g_{\mu\nu} \partial_{\alpha} g^{\mu\nu}. \tag{3.1.21}$$

The use of Eq. (3.1.21) enables us to write Eq. (3.1.17) in the form

$$\Gamma^{\mu}_{\alpha\mu} = \frac{1}{2g} \frac{\partial g}{\partial x^{\alpha}} = \frac{1}{\sqrt{-g}} \frac{\partial \sqrt{-g}}{\partial x^{\alpha}}. \tag{3.1.22}$$

3.1.6.1 *Geodesic coordinates*

We prove below the following very useful lemma concerning the vanishing of the Christoffel symbols under certain conditions.

Lemma *It is always possible to choose a coordinate system in which all the components of the Christoffel symbols vanish at a given point. Such a coordinate system is called a geodesic system.*

The proof of this lemma is quite simple. Suppose that the Christoffel symbols do not vanish at the given point A in a coordinate system x^{α}. We then introduce a new coordinate system x'^{α} by carrying out the coordinate transformation

$$x'^{\alpha} = x^{\alpha} - x^{\alpha}_{A} + \frac{1}{2} \Gamma^{\alpha}_{\beta\gamma}(A) \left(x^{\beta} - x^{\beta}_{A}\right) \left(x^{\gamma} - x^{\gamma}_{A}\right). \tag{3.1.23}$$

Here the subscript A indicates the value at a given point A. The new coordinate system and the coordinate transformation (3.1.23) certainly have a meaning in a sufficiently small region around the point A. Of course we have $x'^\alpha_A = 0$.

We now calculate the transformed Christoffel symbols in the new coordinate system x'^α, using the law of transformation (3.1.16). The coordinate transformation (3.1.23) then yields

$$\frac{\partial x'^\alpha}{\partial x^\mu}\Big|_A = \delta^\alpha_\mu,$$

$$\frac{\partial x'^\alpha}{\partial x'^\mu} = \delta^\alpha_\mu = \frac{\partial x^\alpha}{\partial x'^\mu} + \Gamma^\alpha_{\beta\gamma}(A)\frac{\partial x^\beta}{\partial x'^\mu}\left(x^\gamma - x^\gamma_A\right).$$

Therefore

$$\frac{\partial x^\alpha}{\partial x'^\mu}\Big|_A = \delta^\alpha_\mu.$$

Furthermore,

$$\frac{\partial^2 x'^\alpha}{\partial x'^\nu \partial x'^\mu} = 0$$

$$= \frac{\partial^2 x^\alpha}{\partial x'^\nu \partial x'^\mu} + \Gamma^\alpha_{\beta\gamma}(A)\frac{\partial^2 x^\beta}{\partial x'^\nu \partial x'^\mu}\left(x^\gamma - x^\gamma_A\right) + \Gamma^\alpha_{\beta\gamma}(A)\frac{\partial x^\beta}{\partial x'^\mu}\frac{\partial x^\gamma}{\partial x'^\nu}.$$

At A we then have

$$0 = \frac{\partial^2 x^\alpha}{\partial x'^\mu \partial x'^\nu} + \Gamma^\alpha_{\mu\nu}(A),$$

thus

$$\frac{\partial^2 x^\alpha}{\partial x'^\mu \partial x'^\nu}\Big|_A = -\Gamma^\alpha_{\mu\nu}(A).$$

Using these results in Eq. (3.1.16), we find the transformed Christoffel symbols at a point A in the new coordinate system x'^α:

$$\Gamma'^\lambda_{\rho\sigma}(A) = \delta^\lambda_\beta \delta^\mu_\rho \delta^\nu_\sigma \Gamma^\beta_{\mu\nu}(A) - \delta^\lambda_\beta \Gamma^\beta_{\sigma\rho}(A)$$

$$= \Gamma^\lambda_{\rho\sigma}(A) - \Gamma^\lambda_{\sigma\rho}(A)$$

$$= 0.$$

The above lemma can be generalized to the case for which one can always find a coordinate system in which the Christoffel symbols vanish at all points of a given curve in the spacetime and not just at one point alone.

The proof of this important extension of the above lemma is due to Fermi (Fermi 1922; Levi-Civita 1927) and will not be given in this book.

The possibility of choosing the Christoffel symbols as zeros at a certain point in spacetime has a very interesting and deep physical meaning which is worth mentioning here. In the next chapter we see that the acceleration of a particle, moving in the gravitational field, is proportional to the Christoffel symbols. The possibility of choosing the Christoffel symbols to be zeros at a preferred point, therefore, means that we can always choose a coordinate system in which the acceleration of a given particle at a given spacetime point vanishes. This is one way of saying that the gravitational force acting on the particle vanishes there. In other words, the gravitational field acts as if it were eliminated at the arbitrary point by means of a coordinate transformation. We have to remember, however, that this is possible only at a point and not in a finite region of spacetime. The above considerations are intimately related to the principle of equivalence, which is discussed in the next section, and confirm its mathematical foundations beyond the Newtonian limit.

3.1.7 *Covariant differentiation*

We have seen that the derivatives of an invariant are the components of a covariant vector. This is the only case for a general system of coordinates in which the derivative of a tensor is a tensor. However, there are expressions involving first derivatives which are components of a tensor. To see this we proceed as follows.

Let V^μ and V'^ν be a contravariant vector in two coordinate systems x and x'. Then

$$V^\mu = V'^\nu \frac{\partial x^\mu}{\partial x'^\nu}.$$

Differentiating this equation with respect to x^α and using Eq. (3.1.16) gives (see Problem 3.9.2):

$$\frac{\partial V^\mu}{\partial x^\alpha} = \left(\frac{\partial V'^\rho}{\partial x'^\nu} + V'^\sigma \Gamma'^\rho_{\sigma\nu} \right) \frac{\partial x'^\nu}{\partial x^\alpha} \frac{\partial x^\mu}{\partial x'^\rho} - V^\rho \Gamma^\mu_{\rho\alpha}. \tag{3.1.24}$$

Hence if we define a *covariant derivative* of V^μ by

$$\nabla_\alpha V^\mu = \partial_\alpha V^\mu + \Gamma^\mu_{\rho\alpha} V^\rho, \tag{3.1.25}$$

then Eq. (3.1.24) can be written as

$$\nabla_\alpha V^\mu = \nabla_\nu V'^\rho \frac{\partial x'^\nu}{\partial x^\alpha} \frac{\partial x^\mu}{\partial x'^\rho}. \tag{3.1.26}$$

Therefore $\nabla_\alpha V^\mu$ is a mixed tensor of second order.

In the same way one shows that the covariant derivative of a covariant vector V_μ is given by:

$$\nabla_\alpha V_\mu = \partial_\alpha V_\mu - \Gamma^\rho_{\mu\alpha} V_\rho, \tag{3.1.27}$$

From the above equation one has for the *curl* of a vector V_μ:

$$\nabla_\beta V_\alpha - \nabla_\alpha V_\beta = \partial_\beta V_\alpha - \partial_\alpha V_\beta. \tag{3.1.28}$$

Hence the necessary and sufficient condition that the first covariant derivative of a covariant vector be symmetric is that the vector be a gradient.

It is easily seen, using the law of covariant differentiation of tensors (see Problem 3.9.3), that

$$\nabla_\rho g^{\mu\nu} = 0, \tag{3.1.29a}$$

$$\nabla_\rho g_{\mu\nu} = 0, \tag{3.1.29b}$$

$$\nabla_\rho \, \delta^\mu_\nu = 0. \tag{3.1.29c}$$

Other properties of covariant differentiation can be established (see Problem 3.9.4).

3.1.8 *The Riemann, Ricci and Einstein tensors*

If we differentiate covariantly the tensor $\nabla_\alpha V_\mu$, given by Eq. (3.1.27), we obtain

$$\left(\nabla_\gamma \nabla_\beta - \nabla_\beta \nabla_\gamma\right) V_\alpha = R^\delta{}_{\alpha\beta\gamma} V_\delta, \tag{3.1.30}$$

where $R^\delta{}_{\alpha\beta\gamma}$ is called the *Riemann tensor* [1] and is given by

$$R^\delta{}_{\alpha\beta\gamma} = \partial_\beta \Gamma^\delta_{\alpha\gamma} - \partial_\gamma \Gamma^\delta_{\alpha\beta} + \Gamma^\mu_{\alpha\gamma} \Gamma^\delta_{\mu\beta} - \Gamma^\mu_{\alpha\beta} \Gamma^\delta_{\mu\gamma}. \tag{3.1.31}$$

---

[1] Georg Friedrich Bernhard Riemann (Born: 17 Sept 1826 in Breselenz, Hanover (now Germany); Died: 20 July 1866 in Selasca, Italy)

Bernhard Riemann's father, Friedrich Bernhard Riemann, was a Lutheran minister. He married Charlotte Ebell, and Bernhard was the second of their six children: two boys and four girls. Friedrich Riemann acted as teacher to his children, and he taught Bernhard until he was ten. A teacher from a local school assisted in Bernhard's education.

In 1840 Bernhard entered the third class at the Lyceum in Hannover. While at the Lyceum he lived with his grandmother. When she died in 1842, Bernhard moved to the Johanneum Gymnasium in Lüneburg. Bernhard showed a particular interest in mathematics, and the director of the Gymnasium allowed him to study mathematics texts from his own library. On one occasion he lent Bernhard Legendre's book on the theory of numbers and Bernhard read the 900 page book in six days.

In the spring of 1846 Riemann enrolled at the University of Göttingen. Encouraged by his father, he entered the theology faculty, but attended some mathematics lectures.

After getting permission from his father, he transferred to the faculty of philosophy and took courses in mathematics from Moritz Stern and Gauss.

At this time the University of Göttingen was a rather poor place for mathematics. So Riemann moved to Berlin University in 1847 to study under Steiner, Jacobi, Dirichlet and Eisenstein.

This was an important time for Riemann. He learnt much from Eisenstein and discussed using complex variables in elliptic function theory. The main person to influence Riemann at this time, however, was Dirichlet. During this time at the University of Berlin Riemann worked out his general theory of complex variables that formed the basis of some of his most important work.

In 1849 he returned to Göttingen, and submitted his Ph.D. thesis, supervised by Gauss, in 1851. By this time Weber had returned to a chair of physics at Göttingen, and Riemann was his assistant for 18 months. Also Listing had been appointed as a professor of physics in Göttingen in 1849. Through Weber and Listing, Riemann gained a strong background in theoretical physics and important ideas in topology which were to influence his ground breaking research.

Riemann's thesis studied the theory of complex variables and, in particular, what we now call Riemann surfaces. It introduced topological methods into complex function theory. The work builds on Cauchy's foundations of the theory of complex variables and on Puiseux's ideas of branch points. However, Riemann's thesis is a strikingly original piece of work which examined geometric properties of analytic functions, conformal mappings and the connectivity of surfaces. In proving some of the results in his thesis Riemann used a variational principle. Riemann's thesis was examined on 16 December 1851. In his report on the thesis Gauss described Riemann as having "a gloriously fertile originality."

On Gauss's recommendation Riemann was appointed to a post in Göttingen. To become a lecturer, he worked for thirty months on his Habilitation dissertation, which was on the representability of functions by trigonometric series. He gave the conditions of a function to have an integral, what we now call the condition of Riemann integrability. In the second part of the dissertation he examined the problem which he described as following:

"While preceding papers have shown that if a function possesses such and such a property, then it can be represented by a Fourier series, we pose the reverse question: if a function can be represented by a trigonometric series, what can one say about its behaviour."

To complete his Habilitation Riemann had to give a lecture. He prepared three lectures, two on electricity and one on geometry. Gauss had to choose one of the three for Riemann to deliver and chose the lecture on geometry. Riemann's lecture "Über die Hypothesen welche der Geometrie zu Grunde liegen" (On the hypotheses that lie at the foundations of geometry), delivered on 10 June 1854, became a classic of mathematics.

There were two parts to Riemann's lecture. In the first part he posed the problem of how to define an n-dimensional space and ended up giving a definition of what today we call a Riemannian space. In fact the main point of this part was the definition of the curvature tensor. The second part of Riemann's lecture posed deep questions about the relationship of geometry to the world we live in. He asked what the dimension of real space was and what geometry described real space. The lecture was not appreciated by most scientists of that time. It was not fully understood until sixty years later, when:

"The general theory of relativity splendidly justified his work. In the mathematical apparatus developed from Riemann's address, Einstein found the frame to fit his physical ideas, his cosmology, and cosmogony: and the spirit of Riemann's address was just what physics needed: the metric structure determined by data."

This brilliant work entitled Riemann to begin to lecture. However,

"Not long before, in September, he read a report "On the Laws of the Distribution of Static Electricity" at a session of the Göttingen Society of Scientific researchers and Physicians. In a letter to his father, Riemann recalled, among other things, "the fact that I spoke at a scientific meeting was useful for my lectures." In October he set to work on his lectures on partial differential equations. Riemann's letters to his dearly-loved father were full of recollections about the difficulties he encountered. Although only eight students attended the lectures, Riemann was completely happy. Gradually he overcame his natural shyness and established a rapport with his audience." (M Monastyrsky, *Rieman, Topology and Physics* (Boston-Basel, 1987)).

In 1857 he was appointed as professor and in the same year another of his masterpieces was published. The paper "Theory of abelian functions" was the result of work carried out over several years and contained in a lecture course he gave to three people, one of who was Dedekind, in 1855-56.

The abelian functions paper continued where his doctoral dissertation had left off and developed further the idea of Riemann surfaces and their topological properties. He examined multi-valued functions as single valued over a special Riemann surface and solved general inversion problems which had been solved for elliptic integrals by Abel and Jacobi. The Dirichlet Principle used by Riemann in his doctoral thesis was used by him again for the results of this paper. Weierstrass, however, showed that there was a problem with the Dirichlet Principle. The problem of the use of Dirichlet's Principle in Riemann's work was, however, sorted out.

In 1859 Dirichlet died and Riemann was appointed to the chair of mathematics at Göttingen on 30 July. A few days later he was elected to the Berlin Academy of Sciences. He had been proposed by Kummer, Borchardt and Weierstrass, whose proposal read:

"Prior to the appearance of his most recent work [Theory of abelian functions], Riemann was almost unknown to mathematicians. This circumstance excuses somewhat the necessity of a more detailed examination of his works as a basis of our presentation. We considered it our duty to turn the attention of the Academy to our colleague whom we recommend not as a young talent which gives great hope, but rather as a fully mature and independent investigator in our area of science, whose progress he in significant measure has promoted."

As a newly elected member of the Berlin Academy of Sciences, Riemann sent a report on "On the number of primes less than a given magnitude" where he examined the zeta function (earlier considered by Euler)

$$\zeta(s) = \sum (1/n^s) = \prod (1 - p^{-s})^{-1}$$

(the sum is over all natural numbers n while the product is over all prime numbers). Riemann looked at the zeta function as a complex function rather than a real one. Except for a few trivial exceptions, the roots of (s) all lie between 0 and 1. In the paper he stated that the zeta function had infinitely many nontrivial roots and that it seemed probable that they all have real part 1/2. This famous Riemann hypothesis remains today one of the most important of the unsolved problems of mathematics.

Riemann studied the convergence of the series representation of the zeta function and found a functional equation for the zeta function. The main purpose of the paper was to give estimates for the number of primes less than a given number. Many of the results which Riemann obtained were later proved by Hadamard and de la Vallée Poussin.

In June 1862 Riemann married Elise Koch. They had one daughter. In the autumn of the year of his marriage Riemann caught a heavy cold which turned to tuberculosis.

A generalization of Eq. (3.1.30) to an arbitrary tensor can be made (see Problem 3.9.5).

One can show that in order that there can exist a coordinate system in which the first covariant derivatives reduce to ordinary ones at every point in space, it is necessary and sufficient that the Riemann tensor be zero and that the coordinates be those in which the metric is constant.

One notices that the Riemann tensor satisfies

$$R_{\alpha\beta\gamma\delta} = -R_{\beta\alpha\gamma\delta} = -R_{\alpha\beta\delta\gamma}, \tag{3.1.32a}$$

$$R_{\alpha\beta\gamma\delta} = R_{\gamma\delta\alpha\beta}, \tag{3.1.32b}$$

$$R_{\alpha\beta\gamma\delta} + R_{\alpha\gamma\delta\beta} + R_{\alpha\delta\beta\gamma} = 0. \tag{3.1.32c}$$

Moreover, counting the number of components, one finds that in a four-dimensional space the Riemann tensor has 20 components.

From the Riemann tensor one can define the *Ricci tensor* and the *Ricci scalar* by

$$R_{\mu\nu} = R^{\alpha}{}_{\mu\alpha\nu} = \frac{1}{\sqrt{-g}}\left(\sqrt{-g}\,\Gamma^{\alpha}_{\mu\nu}\right)_{,\alpha} - \left(\ln\sqrt{-g}\right)_{,\mu\nu} - \Gamma^{\alpha}_{\mu\beta}\Gamma^{\beta}_{\nu\alpha}, \tag{3.1.33}$$

$$R = R^{\mu}{}_{\mu}, \tag{3.1.34}$$

respectively. Here a comma denotes partial differentiation, $f_{,\alpha} = \partial_{\alpha} f$. The *Einstein tensor* is then defined by

$$G_{\mu\nu} = R_{\mu\nu} - \frac{1}{2}g_{\mu\nu}R. \tag{3.1.35}$$

The last important tensor constructed from the Riemann tensor is the *Weyl conformal tensor* (see Section 3.8):

$$C_{\rho\sigma\mu\nu} = R_{\rho\sigma\mu\nu} - \frac{1}{2}\left(g_{\rho\mu}R_{\nu\sigma} - g_{\rho\nu}R_{\mu\sigma} - g_{\sigma\mu}R_{\nu\rho} + g_{\sigma\nu}R_{\mu\rho}\right)$$

He had never had good health, and in fact his mother had died when Riemann was 20 while his brother and three sisters all died young. Riemann tried to fight the illness by going to the warmer climate of Italy.

The winter of 1862-63 was spent in Sicily and he then traveled through Italy, spending time with Italian mathematicians who had earlier visited him in Göttingen. He returned to Göttingen in June 1863 but had to return to Italy because of health. Having spent from August 1864 to October 1865 in northern Italy, Riemann once again returned to Göttingen for the winter of 1865-66, but then returned to Selasca on the shores of Lake Maggiore on 16 June 1866. Dedekind writes that "...the day before his death, resting under a fig tree, his soul filled with joy at the glorious landscape, he worked on his final work which unfortunately, was left unfinished."

The above report on Bernhard Riemann is based on the article by J J O'Connor and E F Robertson.

$$-\frac{1}{6}\left(g_{\rho\nu}g_{\mu\sigma} - g_{\rho\mu}g_{\nu\sigma}\right)R. \tag{3.1.36}$$

It has the special property that

$$C^{\rho}{}_{\mu\rho\nu} = 0. \tag{3.1.37}$$

Furthermore, if the Weyl tensor vanishes everywhere, then the metric is said to be *conformally flat*. (Two spaces V and $\tilde{V}$ are called conformal spaces if their metric tensors $g_{\mu\nu}$ and $\tilde{g}_{\mu\nu}$ are related by $\tilde{g}_{\mu\nu} = e^{\beta}g_{\mu\nu}$, where β is a function of the coordinates.) That is, there exists a mapping such that $g_{\mu\nu}$ can be diagonalized, with $\pm\beta(x)$ appearing in the diagonal positions, and where $\beta(x)$ is some function. This follows from the fact that the Weyl tensor can be expressed entirely in terms of the density $\tilde{g}_{\mu\nu} = (-g)^{-1/4}g_{\mu\nu}$ and its inverse, and is equal to the Riemann tensor formed by replacing $g_{\mu\nu}$ by $\tilde{g}_{\mu\nu}$, $R_{\alpha\beta\gamma\delta}(\tilde{g}_{\mu\nu}) = C_{\alpha\beta\gamma\delta}(g_{\mu\nu})$.

Consequently, the vanishing of the Weyl tensor implies the vanishing of $R_{\alpha\beta\gamma\delta}(\tilde{g}_{\mu\nu})$. This implies in turn that there exists a mapping such that $\tilde{g}_{\mu\nu}$ is diagonal everywhere, with ± 1 appearing along the diagonal; only g is arbitrary and $\pm(-g)^{1/4}$ appears along the diagonal of $g_{\mu\nu}$.

3.1.9 *Geodesics*

The differential equations of the curves of extremal length are called *geodesic equations*. To find their equations we seek the relations which must be satisfied to give a stationary value to the integral $\int ds$. Hence we have to find the solution of the variational problem (see Figure 3.1.2)

$$\delta \int L\, ds = 0, \tag{3.1.38}$$

where the Lagrangian L is given by

$$L = \sqrt{g_{\mu\nu}\frac{dx^{\mu}}{ds}\frac{dx^{\nu}}{ds}}. \tag{3.1.39}$$

Accordingly we have

$$\delta \int L\, ds = \int \left[\frac{\partial L}{\partial x^{\mu}}\delta x^{\mu} + \frac{\partial L}{\partial\left(\dfrac{dx^{\mu}}{ds}\right)}\delta\left(\frac{dx^{\mu}}{ds}\right)\right]ds. \tag{3.1.40}$$

The second term of the integrand may be written in the form of the following difference

$$\frac{d}{ds}\left[\frac{\partial L}{\partial\left(\dfrac{dx^{\mu}}{ds}\right)}\delta x^{\mu}\right] - \frac{d}{ds}\left[\frac{\partial L}{\partial\left(\dfrac{dx^{\mu}}{ds}\right)}\right]\delta x^{\mu}. \tag{3.1.41}$$

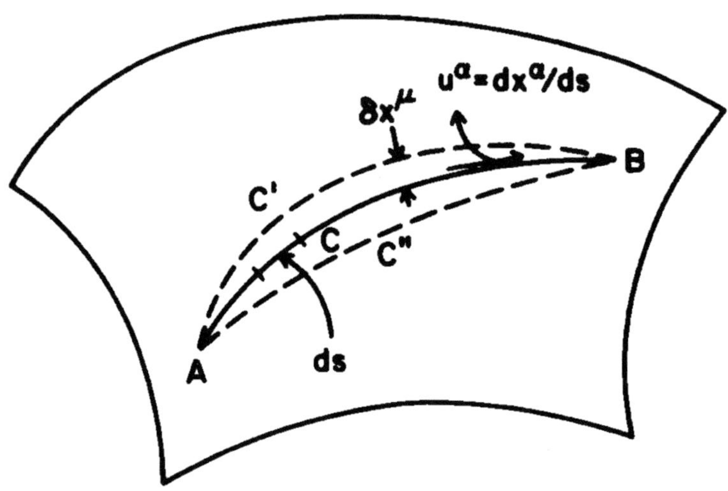

Fig. 3.1.2:   Description of geodesic line connecting points A and B. Curve C is the geodesic line. Curves C' and C" are nongeodesic lines which deviate from the geodesic line C by the variation δx^μ. The geodesic line can be described in a parametric way by $x^\mu = x^\mu(s)$ or $x^\mu = x^\mu(\sigma)$, where s and σ are parameters along the curve, and $ds^2 = g_{\mu\nu}dx^\mu dx^\nu$. The unit vector $u^\alpha = dx^\alpha/ds$ is the tangent vector to the curve $x^\alpha = x^\alpha(s)$. The two points A and B are assumed not to be varied.

Integrating, one obtains that the first of these expressions contributes nothing, since the variations are assumed to vanish at the end points of the curve.

As expected, the equation obtained is the usual Lagrange equation:

$$\frac{d}{ds}\frac{\partial L}{\partial\left(\dfrac{dx^\mu}{ds}\right)} - \frac{\partial L}{\partial x^\mu} = 0. \qquad (3.1.42)$$

A simple calculation then gives, using the Lagrangian given by Eq. (3.1.39),

$$\frac{d^2 x^\mu}{ds^2} + \Gamma^\mu_{\alpha\beta}\frac{dx^\alpha}{ds}\frac{dx^\beta}{ds} = 0. \qquad (3.1.43)$$

3.1.10 *The Bianchi identities*

In Subsection 3.1.6 it was shown that it is always possible to choose a geodesic coordinate system in which all the Christoffel symbols vanish at a point A, by means of the coordinate transformation

$$x'^{\alpha} = x^{\alpha} - x^{\alpha}_{A} + \frac{1}{2}\Gamma^{\alpha}_{\beta\gamma}\left(A\right)\left(x^{\beta} - x^{\beta}_{A}\right)\left(x^{\gamma} - x^{\gamma}_{A}\right), \qquad (3.1.44)$$

where the subscript A indicates to the value at the point A. By Eq. (3.1.16) one finds that the Christoffel symbols in the new coordinate system vanish at the point A.

If we choose a geodesic coordinate system at a point A, then at A one has

$$\nabla_{\nu}R^{\mu}{}_{\delta\beta\gamma} = \partial_{\beta}\partial_{\nu}\Gamma^{\mu}_{\delta\gamma} - \partial_{\gamma}\partial_{\nu}\Gamma^{\mu}_{\delta\beta}. \qquad (3.1.45)$$

Consequently, at the point A one has:

$$\nabla_{\nu}R^{\mu}{}_{\delta\beta\gamma} + \nabla_{\gamma}R^{\mu}{}_{\delta\nu\beta} + \nabla_{\beta}R^{\mu}{}_{\delta\gamma\nu} = 0. \qquad (3.1.46)$$

Since the terms of this equation are components of a tensor, this equation holds for any coordinate system and at each point. Hence Eq. (3.1.46) is an identity throughout the space. It is known as the *Bianchi identities*.

Multiplication of Eq. (3.1.46) by $g^{\delta\beta}\delta^{\gamma}_{\mu}$ gives

$$g^{\delta\beta}\left(\nabla_{\nu}R^{\gamma}{}_{\delta\beta\gamma} + \nabla_{\gamma}R^{\gamma}{}_{\delta\nu\beta} + \nabla_{\beta}R^{\gamma}{}_{\delta\gamma\nu}\right) = 0. \qquad (3.1.47)$$

Using the symmetry properties of the Riemann tensor, one has for the last equation:

$$\nabla_{\nu}\left(R_{\gamma}{}^{\nu} - \frac{1}{2}\delta^{\nu}_{\gamma}R\right) = \nabla_{\nu}G_{\gamma}{}^{\nu} = 0. \qquad (3.1.48)$$

Equation (3.1.48) is called the *contracted Bianchi identity*.

3.1.11 *Tensor densities*

3.1.11.1 *Definition of a tensor density*

The concept of a tensor, introduced and discussed in the last few sections, is not the only mathematical quantity that transforms under a coordinate transformation with a linear, homogeneous law of transformation. Tensors provide, in fact, a subclass of a more general class of quantities called *tensor densities*. A tensor density transforms like a tensor, except for the appearance of an extra factor, which is the Jacobian of the transformation of the coordinates, raised to some power, in transformation law.

Thus a tensor density $\mathfrak{S}^{\mu\cdots}_{\nu\cdots}$ transforms according to the following rule of transformation:

$$\mathfrak{S}'^{\mu\cdots}_{\nu\cdots} = \left|\frac{\partial x}{\partial x'}\right|^{W} \frac{\partial x'^{\mu}}{\partial x^{\beta}} \cdots \frac{\partial x^{\alpha}}{\partial x'^{\nu}} \cdots \mathfrak{S}^{\beta\cdots}_{\alpha\cdots}. \tag{3.1.49}$$

In the above equation $|\partial x/\partial x'|$ denotes the Jacobian of the transformation from the x^{α} to the x'^{β} coordinate system, and W is a constant, a positive or a negative integer, called the *weight* of the tensor density.

A tensor density of order 1 is called a *vector density*, and a tensor density of order 0 is called a *scalar density*. Of course, ordinary tensors are tensor densities of weight 0.

An example of a scalar density is provided by the determinant of an ordinary covariant tensor $T_{\alpha\beta}$ of order 2. The transformation rule for such a tensor $T_{\alpha\beta}$ can be regarded as a matrix equation:

$$T'_{\alpha\beta} = \frac{\partial x^{\mu}}{\partial x'^{\alpha}} T_{\mu\nu} \frac{\partial x^{\nu}}{\partial x'^{\beta}}. \tag{3.1.50}$$

Using the rule for the determinants of a product of matrices, we find that

$$\det T'_{\alpha\beta} = \left|\frac{\partial x}{\partial x'}\right|^{2} \det T_{\alpha\beta}. \tag{3.1.51}$$

Hence the determinant of a covariant tensor of order 2 is a scalar density of weight 2. In particular, Eq. (3.1.51) can be applied to the metric tensor $g_{\mu\nu}$. One then finds that, under a coordinate transformation,

$$g' = \left|\frac{\partial x}{\partial x'}\right|^{2} g. \tag{3.1.52}$$

From the theory of differential and integral calculus one knows that under a general coordinate transformation of the form given by Eq. (3.1.1), the volume element d^4x transforms into

$$d^4x' = \left|\frac{\partial x'}{\partial x}\right| d^4x. \tag{3.1.53}$$

Using Eq. (3.1.52) in the latter equation, we obtain the following relation for the four-dimensional volume element:

$$\sqrt{-g'}\,d^4x' = \sqrt{-g}\,d^4x. \tag{3.1.54}$$

Hence the expression $\sqrt{-g}\,d^4x$ is an invariant volume element.

Equation (3.1.52) can be used in the law of transformation of tensor densities (3.1.49). The latter equation can then be written in the form

$$(-g')^{-W/2}\,\mathfrak{S}'^{\mu\cdots}_{\nu\cdots} = \frac{\partial x'^{\mu}}{\partial x^{\lambda}} \cdots \frac{\partial x^{\kappa}}{\partial x'^{\nu}} \cdots (-g)^{-W/2}\,\mathfrak{S}^{\lambda\cdots}_{\kappa\cdots}. \tag{3.1.55}$$

We thus see, from the above equation, that the quantity

$$(-g)^{-W/2}\,\mathfrak{S}^{\lambda\cdots}_{\kappa\cdots} \tag{3.1.56}$$

has the same rule of transformation as an ordinary tensor $T^{\lambda\cdots}_{\kappa\cdots}$. Consequently, a tensor density of weight W, multiplied by the factor $(-g)^{-W/2}$, is an ordinary tensor, namely, a tensor density of weight 0. By the same token, an ordinary tensor, when multiplied by $(-g)^{W/2}$, becomes a tensor density of weight W. In particular, the quantity $\sqrt{-g}\,T^{\alpha\cdots}_{\beta\cdots}$ is a tensor density of weight $W = 1$ if $T^{\alpha\cdots}_{\beta\cdots}$ is an ordinary tensor.

It thus follows that raising and lowering indices, by means of the metric tensor, of tensor densities do not change the weight of the tensor density. Moreover, the symmetry properties of tensor densities are defined in exactly the same way as those of ordinary tensors.

3.1.11.2 *Levi-Civita tensor densities*

Symmetry properties of tensors tell us that the only nonvanishing components of a totally skew-symmetric tensor of order 4, $A_{\alpha\beta\gamma\delta}$, are those for which its four indices are permutations of 0,1,2,3. Hence all of the components of $A_{\alpha\beta\gamma\delta}$ are equal to either $+A_{0123}$ or $-A_{0123}$, depending upon whether α, β, γ, δ is an even or an odd permutation of 0,1,2,3, and zero otherwise. As a result we found that a completely skew-symmetric tensor of order 4 has actually only one independent component. Of course, the same reasoning is valid for contravariant tensors of order 4.

Let us now define the quantity $\varepsilon^{\alpha\beta\gamma\delta}$ by the following:

$$\varepsilon^{\alpha\beta\gamma\delta} = \begin{cases} +1, & \alpha\beta\gamma\delta \text{ is an even permutation of } 0123 \\ -1, & \alpha\beta\gamma\delta \text{ is an odd permutation of } 0123 \\ \;\;0, & \text{otherwise.} \end{cases} \tag{3.1.57}$$

The discussion presented above shows that any totally contravariant skew-symmetric tensor of order 4 should be proportional to $\varepsilon^{\alpha\beta\gamma\delta}$. In particular, the quantity $\varepsilon^{\alpha\beta\gamma\delta}$ itself, defined by Eq. (3.1.57), can be taken as a contravariant tensor density of weight $W = +1$.

The tensor density $\varepsilon^{\alpha\beta\gamma\delta}$ has a useful property, namely, if its components are given by Eq. (3.1.57) in one coordinate system, then the values of these components are unchanged in any other coordinate system (see Problem 3.1.8). The tensor density $\varepsilon^{\alpha\beta\gamma\delta}$ is called the *Levi-Civita contravariant tensor density.*

In the same way we can also define the covariant tensor density $\varepsilon_{\alpha\beta\gamma\delta}$, whose weight is $W = -1$. The components of $\varepsilon_{\alpha\beta\gamma\delta}$ are obtained from $\varepsilon^{\mu\nu\rho\sigma}$

by lowering the indices in the usual way, and multiplying it by $(-g)^{-1}$:

$$\varepsilon_{\alpha\beta\gamma\delta} = g_{\alpha\mu}g_{\beta\nu}g_{\gamma\rho}g_{\delta\sigma}\,(-g)^{-1}\,\varepsilon^{\mu\nu\rho\sigma}. \tag{3.1.58}$$

The components of $\varepsilon_{\alpha\beta\gamma\delta}$ can easily be found. For instance, we have

$$\varepsilon_{0123} = g_{0\mu}g_{1\nu}g_{2\rho}g_{3\sigma}\,(-g)^{-1}\,\varepsilon^{\mu\nu\rho\sigma}$$
$$= (-g)^{-1}\det g_{\mu\nu}$$
$$= -1. \tag{3.1.59}$$

Accordingly, one has for the *covariant Levi-Civita tensor density* of weight $W = -1$ the following:

$$\varepsilon_{\alpha\beta\gamma\delta} = \begin{cases} -1, & \alpha\beta\gamma\delta \text{ is an even permutation of } 0123 \\ +1, & \alpha\beta\gamma\delta \text{ is an odd permutation of } 0123 \\ 0, & \text{otherwise.} \end{cases} \tag{3.1.60}$$

Again, the components of this tensor density do not change under a coordinate transformation (see Problem 3.1.9).

In addition to the tensor densities $\varepsilon^{\alpha\beta\gamma\delta}$ of weight $W = +1$ and $\varepsilon_{\alpha\beta\gamma\delta}$ of weight $W = -1$ described above, we can also define ordinary tensors. The contravariant tensor is defined by

$$\epsilon^{\alpha\beta\gamma\delta} = \frac{1}{\sqrt{-g}}\varepsilon^{\alpha\beta\gamma\delta}, \tag{3.1.61}$$

whereas the covariant tensor is given by

$$\epsilon_{\alpha\beta\gamma\delta} = \sqrt{-g}\,\varepsilon_{\alpha\beta\gamma\delta}. \tag{3.1.62}$$

It is left for the reader to show that $\epsilon^{\alpha\beta\gamma\delta}$ and $\epsilon_{\alpha\beta\gamma\delta}$ are indeed ordinary contravariant and covariant tensors (see Problem 3.1.11). Of course we now have the following relations between them:

$$\epsilon^{\alpha\beta\gamma\delta} = g^{\alpha\mu}g^{\beta\nu}g^{\gamma\rho}g^{\delta\sigma}\epsilon_{\mu\nu\rho\sigma}. \tag{3.1.63}$$

and

$$\epsilon_{\alpha\beta\gamma\delta} = g_{\alpha\mu}g_{\beta\nu}g_{\gamma\rho}g_{\delta\sigma}\epsilon^{\mu\nu\rho\sigma}. \tag{3.1.64}$$

Finally, if $F_{\mu\nu}$ is a skew-symmetric tensor, then the *pseudotensor*

$$\star\Im^{\alpha\beta} = \frac{1}{2}\varepsilon^{\alpha\beta\mu\nu}F_{\mu\nu} \tag{3.1.65}$$

is said to be the *dual* to $F_{\mu\nu}$. The product of $\star\Im^{\alpha\beta}F_{\alpha\beta}$ is obviously a *pseudoscalar*, and we have

$$\star\Im_{\alpha\beta} = \frac{1}{2}\varepsilon_{\alpha\beta\mu\nu}F^{\mu\nu}. \tag{3.1.66}$$

The *dual ordinary tensors* are then given by

$$\star F^{\alpha\beta} = \frac{1}{2}\epsilon^{\alpha\beta\mu\nu}F_{\mu\nu} \tag{3.1.67}$$

$$\star F_{\alpha\beta} = \frac{1}{2}\epsilon_{\alpha\beta\mu\nu}F^{\mu\nu}. \tag{3.1.68}$$

In the next sections the physical foundation of general relativity, the principle of equivalence and the principle of general covariance, are given.

3.1.12 *Problems*

P 3.1.1. The symmetric and antisymmetric parts of the tensor $T_{\alpha\beta\gamma}$ of order 3 are given by

$$T_{(\alpha\beta\gamma)} = \frac{1}{3!}\left(T_{\alpha\beta\gamma} + T_{\beta\gamma\alpha} + T_{\gamma\alpha\beta} + T_{\beta\alpha\gamma} + T_{\alpha\gamma\beta} + T_{\gamma\beta\alpha}\right), \qquad (1)$$

$$T_{[\alpha\beta\gamma]} = \frac{1}{3!}\left(T_{\alpha\beta\gamma} + T_{\beta\gamma\alpha} + T_{\gamma\alpha\beta} - T_{\beta\alpha\gamma} - T_{\alpha\gamma\beta} - T_{\gamma\beta\alpha}\right), \qquad (2)$$

respectively. Show that the tensor $T_{\alpha\beta\gamma}$ satisfies the following conditions:
(i) If $T_{[\alpha\beta]\gamma} = 0$ and $T_{\alpha(\beta\gamma)} = 0$, then

$$T_{\alpha\beta\gamma} = 0. \qquad (3)$$

(ii) If $T_{[\alpha\beta]\gamma} = 0$, then

$$T_{(\alpha\beta\gamma)} = \frac{1}{3}\left(T_{\alpha\beta\gamma} + T_{\beta\gamma\alpha} + T_{\gamma\alpha\beta}\right). \qquad (4)$$

(iii) If $T_{(\alpha\beta)\gamma} = 0$, then

$$T_{[\alpha\beta\gamma]} = \frac{1}{3}\left(T_{\alpha\beta\gamma} + T_{\beta\gamma\alpha} + T_{\gamma\alpha\beta}\right). \qquad (5)$$

Solution: To prove Eq. (3) we use the properties of $T_{\alpha\beta\gamma}$ outlined above. Hence we have

$$T_{\alpha\beta\gamma} = -T_{\alpha\gamma\beta} = -T_{\gamma\alpha\beta} = +T_{\gamma\beta\alpha} = +T_{\beta\gamma\alpha} = -T_{\beta\alpha\gamma} = -T_{\alpha\beta\gamma},$$

which proves Eq. (3).

Likewise, the proofs of Eqs. (4) and (5) are straightforward if we use the definitions (1) and (2) for the symmetrization and antisymmetrization of tensors of order 3.

P 3.1.2. Show that if $T_{\alpha\beta\gamma\delta}$ is a tensor of order 4, then

$$T_{\alpha[[\beta\gamma]\delta]} = T_{\alpha[\beta\gamma\delta]}. \qquad (1)$$

Solution: The proof of Eq. (1) is straightforward:

$$T_{\alpha[[\beta\gamma]\delta]} = \frac{1}{6}\left(T_{\alpha[\beta\gamma]\delta} + T_{\alpha[\gamma\delta]\beta} + T_{\alpha[\delta\beta]\gamma} - T_{\alpha[\gamma\beta]\delta} - T_{\alpha[\beta\delta]\gamma} - T_{\alpha[\delta\gamma]\beta}\right)$$

$$= \frac{1}{12}\Big(T_{\alpha\beta\gamma\delta} - T_{\alpha\gamma\beta\delta} + T_{\alpha\gamma\delta\beta} - T_{\alpha\delta\gamma\beta} + T_{\alpha\delta\beta\gamma} - T_{\alpha\beta\delta\gamma}$$

$$-T_{\alpha\gamma\beta\delta} + T_{\alpha\beta\gamma\delta} - T_{\alpha\beta\delta\gamma} + T_{\alpha\delta\beta\gamma} - T_{\alpha\delta\gamma\beta} + T_{\alpha\gamma\delta\beta}\Big)$$

$$= T_{\alpha[\beta\gamma\delta]}.$$

P 3.1.3. The symmetric and antisymmetric parts of a tensor of order 2 are given by

$$T_{(\alpha\beta)} = \frac{1}{2}\left(T_{\alpha\beta} + T_{\beta\alpha}\right), \tag{1}$$

$$T_{[\alpha\beta]} = \frac{1}{2}\left(T_{\alpha\beta} - T_{\beta\alpha}\right), \tag{2}$$

respectively. Find the transformed symmetric part $T'_{(\alpha\beta)}$ and the transformed antisymmetric part $T'_{[\alpha\beta]}$ of a tensor $T_{\alpha\beta}$ from its transformed components $T'_{\alpha\beta}$. Show that $T'_{(\alpha\beta)}$ is a function of $T_{(\alpha\beta)}$ alone and $T'_{[\alpha\beta]}$ is a function of $T_{[\alpha\beta]}$ alone, also.

Solution: Using the definition (1) for $T_{(\alpha\beta)}$ and the transformation rule (3.1.8) for a tensor, we find

$$T'_{(\alpha\beta)} = \frac{1}{2}\left(T'_{\alpha\beta} + T'_{\beta\alpha}\right)$$

$$= \frac{1}{2}\left(\frac{\partial x^\mu}{\partial x'^\alpha}\frac{\partial x^\nu}{\partial x'^\beta}T_{\mu\nu} + \frac{\partial x^\nu}{\partial x'^\beta}\frac{\partial x^\mu}{\partial x'^\alpha}T_{\nu\mu}\right)$$

$$= \frac{\partial x^\mu}{\partial x'^\alpha}\frac{\partial x^\nu}{\partial x'^\beta}T_{(\mu\nu)}. \tag{3}$$

In the same way we find for the skew-symmetric tensor the following:

$$T'_{[\alpha\beta]} = \frac{\partial x^\mu}{\partial x'^\alpha}\frac{\partial x^\nu}{\partial x'^\beta}T_{[\mu\nu]} \tag{4}$$

P 3.1.4. A quantity that can be decomposed into parts, which transform among themselves, is called *reducible*. If such a decomposition is impossible, the quantity is called *irreducible*.

Show that a mixed tensor of order 2, $T_\alpha{}^\beta$, with a nonvanishing trace, is reducible. Show that the irreducible parts of $T_\alpha{}^\beta$ are its trace $T = T_\rho{}^\rho$ and the tracefree tensor $S_\alpha{}^\beta$ obtained from $T_\alpha{}^\beta$ by

$$S_\alpha{}^\beta = T_\alpha{}^\beta - \frac{1}{4}\delta_\alpha^\beta T. \tag{1}$$

Solution: From the rule of transformation of tensors (3.1.8) one obtains

$$T'_\alpha{}^\beta = \frac{\partial x^\mu}{\partial x'^\alpha}\frac{\partial x'^\beta}{\partial x^\nu}T_\mu{}^\nu.$$

Hence the trace of $T_\alpha{}^\beta$ transforms as follows:

$$T' = T'_\alpha{}^\alpha = \frac{\partial x^\mu}{\partial x'^\alpha}\frac{\partial x'^\alpha}{\partial x^\nu}T_\mu{}^\nu = T_\mu{}^\mu = T. \tag{2}$$

Consequently the trace of $T_\alpha{}^\beta$ transforms into itself, and therefore the tensor $T_\alpha{}^\beta$ is reducible. In the same way one finds

$$S'_\mu{}^\nu = T'_\mu{}^\nu - \frac{1}{4}\delta'^\nu_\mu T$$

$$= \frac{\partial x^\alpha}{\partial x'^\mu}\frac{\partial x'^\nu}{\partial x^\beta}\left(T_\alpha{}^\beta - \frac{1}{4}\delta^\beta_\alpha T\right)$$

$$= \frac{\partial x^\alpha}{\partial x'^\mu}\frac{\partial x'^\nu}{\partial x^\beta}S_\alpha{}^\beta. \tag{3}$$

The tensor $S_\alpha{}^\beta$ is irreducible, since it cannot be decomposed any more as it has no trace. Obviously the trace $T = T_\rho{}^\rho$ is also irreducible.

P 3.1.5. Let A_α, B_α and C_α be three linearly independent vectors, and define the tensor

$$A_{\alpha\beta\gamma} = \begin{vmatrix} A_\alpha & B_\alpha & C_\alpha \\ A_\beta & B_\beta & C_\beta \\ A_\gamma & B_\gamma & C_\gamma \end{vmatrix}. \tag{1}$$

Show that $A_{\alpha\beta\gamma}$ is completely skew-symmetric.

Solution: The completely antisymmetric property of the tensor $A_{\alpha\beta\gamma}$ is an immediate consequence of the definition of determinants.

P 3.1.6. Generalizations of the Kronecker delta δ^α_β can be constructed as follows:

$$\delta^{\alpha\beta}_{\mu\nu} = \begin{vmatrix} \delta^\alpha_\mu & \delta^\beta_\mu \\ \delta^\alpha_\nu & \delta^\beta_\nu \end{vmatrix}, \tag{1}$$

$$\delta^{\alpha\beta\gamma}_{\mu\nu\rho} = \begin{vmatrix} \delta^\alpha_\mu & \delta^\beta_\mu & \delta^\gamma_\mu \\ \delta^\alpha_\nu & \delta^\beta_\nu & \delta^\gamma_\nu \\ \delta^\alpha_\rho & \delta^\beta_\rho & \delta^\gamma_\rho \end{vmatrix}, \tag{2}$$

$$\delta^{\alpha\beta\gamma\delta}_{\mu\nu\rho\sigma} = \begin{vmatrix} \delta^\alpha_\mu & \delta^\beta_\mu & \delta^\gamma_\mu & \delta^\delta_\mu \\ \delta^\alpha_\nu & \delta^\beta_\nu & \delta^\gamma_\nu & \delta^\delta_\nu \\ \delta^\alpha_\rho & \delta^\beta_\rho & \delta^\gamma_\rho & \delta^\delta_\rho \\ \delta^\alpha_\sigma & \delta^\beta_\sigma & \delta^\gamma_\sigma & \delta^\delta_\sigma \end{vmatrix}. \tag{3}$$

Show that these quantities are tensors, and that they cannot be extended beyond the tensor of order eight $\delta^{\alpha\beta\gamma\delta}_{\mu\nu\rho\sigma}$. Show that these tensors have the property that for arbitrary tensors $T_{\mu\nu}$, $T_{\mu\nu\rho}$, and $T_{\mu\nu\rho\sigma}$ one has

$$T_{[\alpha\beta]} = \frac{1}{2!} T_{\mu\nu} \delta^{\mu\nu}_{\alpha\beta} \tag{4a}$$

$$T_{[\alpha\beta\gamma]} = \frac{1}{3!} T_{\mu\nu\rho} \delta^{\mu\nu\rho}_{\alpha\beta\gamma} \tag{4b}$$

$$T_{[\alpha\beta\gamma\delta]} = \frac{1}{4!} T_{\mu\nu\rho\sigma} \delta^{\mu\nu\rho\sigma}_{\alpha\beta\gamma\delta}. \tag{4c}$$

Finally show that the following relationships hold:

$$\delta^{\alpha\beta\gamma\tau}_{\mu\nu\rho\tau} = \delta^{\alpha\beta\gamma}_{\mu\nu\rho} \tag{5a}$$

$$\delta^{\alpha\beta\tau}_{\mu\nu\tau} = 2\delta^{\alpha\beta}_{\mu\nu} \tag{5b}$$

$$\delta^{\alpha\tau}_{\mu\tau} = 3\delta^{\alpha}_{\mu} \tag{5c}$$

$$\delta^{\tau}_{\tau} = 4. \tag{5d}$$

Solution: The tensor character of $\delta^{\alpha\beta}_{\mu\nu}$ follows from its definition, since

$$\delta^{\alpha\beta}_{\mu\nu} = \delta^{\alpha}_{\mu}\delta^{\beta}_{\nu} - \delta^{\beta}_{\mu}\delta^{\alpha}_{\nu}.$$

The same holds for $\delta^{\alpha\beta\gamma}_{\mu\nu\rho}$ and $\delta^{\alpha\beta\gamma\delta}_{\mu\nu\rho\sigma}$. From their definitions one also finds that $\delta^{\alpha\beta}_{\mu\nu}$ can be expressed as follows:

$$\delta^{\alpha\beta}_{\mu\nu} = \begin{cases} +1, & \alpha \neq \beta, \quad \alpha\beta = \mu\nu \\ -1, & \alpha \neq \beta, \quad \alpha\beta = \nu\mu \\ 0, & \text{otherwise,} \end{cases} \tag{6}$$

whereas $\delta^{\alpha\beta\gamma}_{\mu\nu\rho}$ can be expressed in the form

$$\delta^{\alpha\beta\gamma}_{\mu\nu\rho} = \begin{cases} +1, \alpha \neq \beta \neq \gamma, \; \alpha\beta\gamma \text{ is an even permutation of } \mu\nu\rho \\ -1, \alpha \neq \beta \neq \gamma, \; \alpha\beta\gamma \text{ is an odd permutation of } \mu\nu\rho \\ 0, \quad \text{otherwise,} \end{cases} \tag{7}$$

Likewise, the tensor $\delta^{\alpha\beta\gamma\delta}_{\mu\nu\rho\sigma}$ cam be expressed in a similar way.

We thus see why such tensors do not exist in orders higher than eight. Likewise, Eqs. (4) are easily seen to be consequences of Eqs. (6) and (7).

To prove the relations (5) we use the definitions given by Eqs. (1) and (3). For instance, we have

$$\delta^{\alpha\beta\tau}_{\mu\nu\tau} = \begin{vmatrix} \delta^{\alpha}_{\mu} & \delta^{\beta}_{\mu} & \delta^{\tau}_{\mu} \\ \delta^{\alpha}_{\nu} & \delta^{\beta}_{\nu} & \delta^{\tau}_{\nu} \\ \delta^{\alpha}_{\tau} & \delta^{\beta}_{\tau} & \delta^{\tau}_{\tau} \end{vmatrix}$$

$$= \delta^\alpha_\tau \begin{vmatrix} \delta^\beta_\mu & \delta^\tau_\mu \\ \delta^\beta_\nu & \delta^\tau_\nu \end{vmatrix} - \delta^\beta_\tau \begin{vmatrix} \delta^\alpha_\mu & \delta^\tau_\mu \\ \delta^\alpha_\nu & \delta^\tau_\nu \end{vmatrix} + \delta^\tau_\tau \begin{vmatrix} \delta^\alpha_\mu & \delta^\beta_\mu \\ \delta^\alpha_\nu & \delta^\beta_\nu \end{vmatrix}$$

$$= \begin{vmatrix} \delta^\beta_\mu & \delta^\alpha_\mu \\ \delta^\beta_\nu & \delta^\alpha_\nu \end{vmatrix} - \begin{vmatrix} \delta^\alpha_\mu & \delta^\beta_\mu \\ \delta^\alpha_\nu & \delta^\beta_\nu \end{vmatrix} + 4 \begin{vmatrix} \delta^\alpha_\mu & \delta^\beta_\mu \\ \delta^\alpha_\nu & \delta^\beta_\nu \end{vmatrix}$$

$$= 2 \begin{vmatrix} \delta^\alpha_\mu & \delta^\beta_\mu \\ \delta^\alpha_\nu & \delta^\beta_\nu \end{vmatrix} = 2\delta^{\alpha\beta}_{\mu\nu}. \tag{8}$$

P 3.1.7. Use the results of Problem 3.1.6 in order to express the values of the determinants of the matrices A, B, and C of orders 2×2, 3×3 and 4×4, respectively, in terms of their traces.

Solution: We first notice that the determinants can be written in terms of generalized Kronecker deltas as follows:

$$\det A = \frac{1}{2!} A_\alpha{}^\mu A_\beta{}^\nu \delta^{\alpha\beta}_{\mu\nu} \tag{1}$$

$$\det B = \frac{1}{3!} B_\alpha{}^\mu B_\beta{}^\nu B_\gamma{}^\rho \delta^{\alpha\beta\gamma}_{\mu\nu\rho} \tag{2}$$

$$\det C = \frac{1}{4!} C_\alpha{}^\mu C_\beta{}^\nu C_\gamma{}^\rho C_\delta{}^\sigma \delta^{\alpha\beta\gamma\delta}_{\mu\nu\rho\sigma}. \tag{3}$$

Using now the definitions of $\delta^{\alpha\beta}_{\mu\nu}$, $\delta^{\alpha\beta\gamma}_{\mu\nu\rho}$, and $\delta^{\alpha\beta\gamma\delta}_{\mu\nu\rho\sigma}$, given by Eqs. (1) - (3) of Problem 3.1.6, in the above equations we then obtain the desired results:

$$\det A = \frac{1}{2!} \left[(\mathrm{Tr}A)^2 - \mathrm{Tr}A^2 \right] \tag{4}$$

$$\det B = \frac{1}{3!} \left[(\mathrm{Tr}B)^3 - 3\mathrm{Tr}B\mathrm{Tr}B^2 + 2\mathrm{Tr}B^3 \right] \tag{5}$$

$$\det C = \frac{1}{4!} \left[(\mathrm{Tr}C)^4 - 6 (\mathrm{Tr}C)^2 \mathrm{Tr}C^2 + 8\mathrm{Tr}c\mathrm{Tr}C^3 - 6\mathrm{Tr}C^4 + 3 (\mathrm{Tr}C^2)^2 \right]. \tag{6}$$

P 3.1.8. Prove Eqs. (3.1.15) and (3.1.16) for the transformation laws of the Christoffel symbols.

Solution: The solution is left for the reader.

P 3.1.9. Show that the components of the Levi-Civita contravariant and covariant tensor densities $\varepsilon^{\alpha\beta\gamma\delta}$ and $\varepsilon_{\alpha\beta\gamma\delta}$ of weights $W = +1$ and $W = -1$, respectively, are unchanged under a coordinate transformation.

Solution: In the transformed coordinate system we have, for instance, for the transformed component ε'^{0123},

$$\varepsilon'^{0123} = \left|\frac{\partial x}{\partial x'}\right| \frac{\partial x'^0}{\partial x^\alpha} \frac{\partial x'^1}{\partial x^\beta} \frac{\partial x'^2}{\partial x^\gamma} \frac{\partial x'^3}{\partial x^\delta} \varepsilon^{\alpha\beta\gamma\delta}. \tag{1}$$

Accordingly we obtain

$$\varepsilon'^{0123} = \left|\frac{\partial x}{\partial x'}\right| \left|\frac{\partial x'}{\partial x}\right| \varepsilon^{0123}. \tag{2}$$

But

$$\left|\frac{\partial x}{\partial x'}\right| \left|\frac{\partial x'}{\partial x}\right| = 1,$$

thus we obtain the desired result:

$$\varepsilon'^{0123} = \varepsilon^{0123}. \tag{3}$$

In the same way we find that all the other components of $\varepsilon^{\alpha\beta\gamma\delta}$ do not depend on the particular coordinate system in which it has been defined.

Likewise, for the covariant tensor density $\varepsilon_{\alpha\beta\gamma\delta}$ of weight $W = -1$ we find

$$\varepsilon'_{0123} = \left|\frac{\partial x}{\partial x'}\right|^{-1} \frac{\partial x^\alpha}{\partial x'^0} \frac{\partial x^\beta}{\partial x'^1} \frac{\partial x^\gamma}{\partial x'^2} \frac{\partial x^\delta}{\partial x'^3} \varepsilon_{\alpha\beta\gamma\delta}$$

$$= \left|\frac{\partial x}{\partial x'}\right|^{-1} \left|\frac{\partial x}{\partial x'}\right|$$

$$= \varepsilon_{0123}. \tag{4}$$

P 3.1.10. Let $A_{\alpha\beta\gamma\delta}$ and $A_{\alpha\beta\gamma}$ be two completely skew-symmetric tensors of orders 4 and 3, respectively. Find the quantities obtained by contracting these tensors with the Levi-Civita tensor density $\varepsilon^{\alpha\beta\gamma\delta}$ of weight $+1$.

Solution: For $A_{\alpha\beta\gamma\delta}$ we obtain

$$\varepsilon^{\alpha\beta\gamma\delta} A_{\alpha\beta\gamma\delta} = 4! A_{0123}, \tag{1}$$

which is a scalar density of weight $+1$. Likewise, we denote the contraction of $\varepsilon^{\alpha\beta\gamma\delta}$ with $A_{\alpha\beta\gamma}$ by $3! V^\alpha$,

$$\varepsilon^{\alpha\beta\gamma\delta} A_{\alpha\beta\gamma} = 3! V^\alpha. \tag{2}$$

Then V^α is a contravariant vector density of weight $+1$. The components of V^α can be taken, for instance, as

$$V^\alpha = (A_{023}, A_{031}, A_{012}, A_{123}). \tag{3}$$

P 3.1.11. Show that the quantities defined by Eqs. (3.1.61) and (3.1.62),

$$\epsilon^{\alpha\beta\gamma\delta} = \frac{1}{\sqrt{-g}} \varepsilon^{\alpha\beta\gamma\delta} \tag{1}$$

and

$$\epsilon_{\alpha\beta\gamma\delta} = \sqrt{-g}\,\varepsilon_{\alpha\beta\gamma\delta}, \tag{2}$$

where $\varepsilon^{\alpha\beta\gamma\delta}$ and $\varepsilon_{\alpha\beta\gamma\delta}$ are respectively the Levi-Civita contravariant and covariant tensor densities of weights $+1$ and -1, are ordinary contravariant and covariant tensors.

Solution: The solution is left for the reader.

P 3.1.12. Let $\mathfrak{I}^{\alpha\beta}$ be a tensor density of weight W. Show that

$$\det \mathfrak{I}^{\alpha\beta} = \frac{1}{4!} \varepsilon_{\alpha\beta\gamma\delta} \varepsilon_{\mu\nu\rho\sigma} \mathfrak{I}^{\alpha\mu} \mathfrak{I}^{\beta\nu} \mathfrak{I}^{\gamma\rho} \mathfrak{I}^{\delta\sigma} \tag{1}$$

$$\det \mathfrak{I}^{\beta}_{\alpha} = -\frac{1}{4!} \varepsilon_{\alpha\beta\gamma\delta} \varepsilon^{\mu\nu\rho\sigma} \mathfrak{I}_{\mu}^{\ \alpha} \mathfrak{I}_{\nu}^{\ \beta} \mathfrak{I}_{\rho}^{\ \gamma} \mathfrak{I}_{\sigma}^{\ \delta} \tag{2}$$

$$\det \mathfrak{I}_{\alpha\beta} = \frac{1}{4!} \varepsilon^{\alpha\beta\gamma\delta} \varepsilon^{\mu\nu\rho\sigma} \mathfrak{I}_{\alpha\mu} \mathfrak{I}_{\beta\nu} \mathfrak{I}_{\gamma\rho} \mathfrak{I}_{\delta\sigma}. \tag{3}$$

Show that these determinants are scalar densities of weight $4W - 2$, $4W$, and $4W + 2$, respectively. Show also that if $\wp_{\mu\nu}$ is a skew-symmetric tensor density, then

$$\det \wp^{\mu\nu} = \left\{ \frac{1}{8} \varepsilon_{\alpha\beta\gamma\delta} \wp^{\alpha\beta} \wp^{\gamma\delta} \right\}^2 \tag{4}$$

$$\det \wp_{\mu\nu} = \left\{ \frac{1}{8} \varepsilon^{\alpha\beta\gamma\delta} \wp_{\alpha\beta} \wp_{\gamma\delta} \right\}^2 \tag{5}$$

Solution: The solution is left for the reader.

P 3.1.13. Show that the product of the Levi-Civita tensor densities of weights $+1$ and -1 satisfies the following equation:

$$\varepsilon^{\alpha\beta\gamma\delta} \varepsilon_{\mu\nu\rho\sigma} = -\delta^{\alpha\beta\gamma\delta}_{\mu\nu\rho\sigma}, \tag{1}$$

where $\delta^{\alpha\beta\gamma\delta}_{\mu\nu\rho\sigma}$ is the ordinary tensor of order 8 whose definition is given in Problem 3.1.6.

Solution: The solution is left for the reader.

P 3.1.14. Show that the cofactor $\Delta^{\mu\nu}$ of the element $g_{\mu\nu}$ of the determinant g can be written in the form

$$\Delta^{\mu\nu} = \frac{\partial g}{\partial g_{\mu\nu}} = \frac{1}{3!}\varepsilon^{\mu\alpha\beta\gamma}\varepsilon^{\nu\rho\sigma\tau}g_{\alpha\rho}g_{\beta\sigma}g_{\gamma\tau},\tag{1}$$

where $\varepsilon^{\alpha\beta\gamma\delta}$ is the Levi-Civita contravariant tensor density of weight $W = +1$, defined by Eq. (3.1.57).

Solution: Equation (1) is a consequence of the definition of a cofactor and Eq. (3) of Problem 3.1.12.

P 3.1.15. Define a two-dimensional infinitesimal surface element in a four-dimensional curved spacetime.

Solution: An infinitesimal element of a two-dimensional surface, spanned by the two infinitesimal displacements d_1x^μ and d_2x^ν, is given by the following skew-symmetric tensor of order 2:

$$d\tau^{\alpha\beta} = \delta^{\alpha\beta}_{\mu\nu}d_1x^\mu d_2x^\nu = \begin{vmatrix} d_1x^\alpha & d_2x^\alpha \\ d_1x^\beta & d_2x^\beta \end{vmatrix}.\tag{1}$$

We can also define the dual $^\star d\tau_{\alpha\beta}$ to the tensor $d\tau^{\alpha\beta}$ by

$$^\star d\tau_{\alpha\beta} = \frac{1}{2}\varepsilon_{\alpha\beta\gamma\delta}d\tau^{\gamma\delta},\tag{2}$$

using Eq. (3.1.66), which is a tensor density of weight $W = -1$ and satisfies $d\tau^{\alpha\beta}\,^\star d\tau_{\alpha\beta} = 0$.

If we choose the two-dimensional surface to be given by $x^0 = x^3 = 0$, for instance, and the two vectors d_1x^α and d_2x^α are taken along the coordinates x^1 and x^2 at the chosen point, respectively, then the only nonvanishing components of the tensor $d\tau^{\alpha\beta}$ are $d\tau^{12} = -d\tau^{21} = dx^1 dx^2$.

P 3.1.16. Define a three-dimensional infinitesimal "area" of a hypersurface in a four-dimensional curved spacetime.

Solution: The element of "area" in a curved spacetime of a hypersurface, which is spanned by the three infinitesimal vectors d_1x^α, d_2x^α, and d_3x^α, is defined as the completely skew-symmetric contravariant tensor of order 3 given by

$$d\tau^{\alpha\beta\gamma} = \delta^{\alpha\beta\gamma}_{\mu\nu\rho}d_1x^\mu d_2x^\nu d_3x^\rho = \begin{vmatrix} d_1x^\alpha & d_2x^\alpha & d_3x^\alpha \\ d_1x^\beta & d_2x^\beta & d_3x^\beta \\ d_1x^\gamma & d_2x^\gamma & d_3x^\gamma \end{vmatrix}.\tag{1}$$

As an element of integration over the hypersurface it is more convenient sometimes to use the vector dual to $d\tau^{\alpha\beta\gamma}$. This vector is obtained from $d\tau^{\alpha\beta\gamma}$ by

$$dS_\alpha = -\frac{1}{3!}\varepsilon_{\alpha\beta\gamma\delta}d\tau^{\beta\gamma\delta}, \tag{2}$$

and its components are explicitly given by

$$dS_0 = d\tau^{123}, \quad dS_1 = d\tau^{023}, \quad dS_2 = d\tau^{031}, \quad dS_3 = d\tau^{012}. \tag{3}$$

Geometrically, dS_α is a vector density of weight $W = -1$. It is equal in magnitude to the element of "area" of the hypersurface, and is perpendicular to it. In particular, the vector dS_α can be taken to have the following components:

$$dS_\alpha = (dx^1 dx^2 dx^3, dx^0 dx^2 dx^3, dx^0 dx^1 dx^3, dx^0 dx^1 dx^2). \tag{4}$$

Thus, for example, the component $dS_0 = dx^1 dx^2 dx^3$ is the element of the three-dimensional spatial infinitesimal volume element on the hypersurface $x^0 = $ constant.

P 3.1.17. Define a four-dimensional volume element, and then generalize the Gauss and Stokes theorems.

Solution: In four dimensions, the infinitesimal volume element is given by

$$d\tau^{\alpha\beta\gamma\delta} = \delta^{\alpha\beta\gamma\delta}_{\mu\nu\rho\sigma}d_0 x^\mu d_1 x^\nu d_2 x^\rho d_3 x^\sigma. \tag{1}$$

The dual to $d\tau^{\alpha\beta\gamma\delta}$ is then defined by

$$dS = -\frac{1}{4!}\varepsilon_{\alpha\beta\gamma\delta}d\tau^{\alpha\beta\gamma\delta}, \tag{2}$$

In particular, the volume element dS will be given by the simple expression

$$dS = dx^0 dx^1 dx^2 dx^3 \tag{3}$$

if the four infinitesimal vectors $d_0 x^\mu$, $d_1 x^\mu$, $d_2 x^\mu$, and $d_3 x^\mu$ are chosen to be directed along the coordinates x^0, x^1, x^2, and x^3, respectively.

The integral over a closed hypersurface can be transformed into an integral over the four-volume contained in it. This can be done by the substitution $dS_\alpha \to dS\partial/\partial x^\alpha$, where dS_α is defined in Problem 3.1.16. Thus, for example, the integration over a vector V^α can be written in the form

$$\oint V^\alpha dS_\alpha = \int \frac{\partial V^\alpha}{\partial x^\alpha}dS. \tag{4}$$

The above equation can be considered as a generalization of the *Gauss theorem*.

The integral over a two-dimensional surface can also be transformed into an integral over a hypersurface by the following substitution:

$$^{\star}d\tau_{\alpha\beta} \rightarrow dS_{\alpha}\frac{\partial}{\partial x^{\beta}} - dS_{\beta}\frac{\partial}{\partial x^{\alpha}}, \tag{5}$$

where $^{\star}d\tau_{\alpha\beta}$ is defined by Eq. (2) of Problem 3.1.15. Thus, for instance, we can write for the integral over a skew-symmetric tensor of order $A^{\alpha\beta}$ the following:

$$\frac{1}{2}\int A^{\alpha\beta}\,^{\star}d\tau_{\alpha\beta} = \frac{1}{2}\int\left(dS_{\alpha}\frac{\partial A^{\alpha\beta}}{\partial x^{\beta}} - dS_{\beta}\frac{\partial A^{\alpha\beta}}{\partial x^{\alpha}}\right)$$

$$= \int dS_{\alpha}\frac{\partial A^{\alpha\beta}}{\partial x^{\beta}}. \tag{6}$$

Finally, the integral over a four-dimensional closed curve can be transformed into an integral over the surface spanned by it. This can be done by the substitution $dx^{\alpha} \rightarrow d\tau^{\beta\alpha}\partial/\partial x^{\beta}$. For example, we have

$$\oint A_{\alpha}dx^{\alpha} = \int d\tau^{\beta\alpha}\frac{\partial A_{\alpha}}{\partial x^{\beta}} = \frac{1}{2}\int d\tau^{\alpha\beta}\left(\frac{\partial A_{\beta}}{\partial x^{\alpha}} - \frac{\partial A_{\alpha}}{\partial x^{\beta}}\right). \tag{7}$$

Equation (7) is a generalization of the *Stokes theorem*.

P 3.1.18. Show that the difference between two affine connections is a tensor.

Solution: If in a given spacetime two connections are defined, let us say $_{1}\Gamma^{\lambda}_{\rho\sigma}$ and $_{2}\Gamma^{\lambda}_{\rho\sigma}$, then the transformed components are given by Eq. (3.1.16) for both of them. If we denote their difference by $T^{\lambda}_{\rho\sigma}$,

$$T^{\lambda}_{\rho\sigma} = {}_{1}\Gamma^{\lambda}_{\rho\sigma} - {}_{2}\Gamma^{\lambda}_{\rho\sigma}, \tag{1}$$

then, by Eq. (3.1.16), $T^{\lambda}_{\rho\sigma}$ is transformed into

$$T'^{\lambda}_{\rho\sigma} = \frac{\partial x'^{\lambda}}{\partial x^{\beta}}\frac{\partial x^{\mu}}{\partial x'^{\rho}}\frac{\partial x^{\nu}}{\partial x'^{\sigma}}T^{\beta}_{\mu\nu}. \tag{2}$$

Hence $T^{\lambda}_{\rho\sigma}$ is a tensor of order 3, contravariant in its index λ and covariant in the indices ρ and σ.

P 3.1.19. Prove the formulas for the covariant derivative of a covariant vector V_{μ}.

Solution: The solution is left for the reader.

P 3.1.20. Prove the formulas for the laws of covariant differentiation of tensors of order 2.

Solution: The solution is left for the reader.

P 3.1.21. Calculate the covariant derivative of the Levi-Civita contravariant tensor density of weight $W = +1$, and show that it is equal to zero,

$$\nabla_\mu \varepsilon^{\alpha\beta\gamma\delta} = 0. \tag{1}$$

Solution: The solution is left for the reader.

P 3.1.22. Show that the covariant derivative of the Levi-Civita covariant tensor density of weight $W = -1$ is equal to zero,

$$\nabla_\mu \varepsilon_{\alpha\beta\gamma\delta} = 0. \tag{1}$$

Solution: Equation (1) follows from Eq. (3.1.58) and the fact that the covariant derivatives of both the metric tensor and the Levi-Civita contravariant tensor density of weight $W = +1$ vanish.

P 3.1.23. Calculate the expression $\nabla_\beta T_\alpha{}^\beta$ for a symmetric tensor $T_{\alpha\beta}$.

Solution: We have

$$\nabla_\beta T_\alpha{}^\beta = \frac{\partial T_\alpha{}^\beta}{\partial x^\beta} + \Gamma^\beta_{\beta\rho} T_\alpha{}^\rho - \Gamma^\gamma_{\alpha\beta} T_\gamma{}^\beta.$$

Using Eq. (3.1.22), and combining the first two terms on the right-hand side of the above equation, we obtain

$$\nabla_\beta T_\alpha{}^\beta = \frac{1}{\sqrt{-g}} \frac{\partial \left(\sqrt{-g}\, T_\alpha{}^\beta\right)}{\partial x^\beta} - \Gamma^\gamma_{\alpha\beta} T_\gamma{}^\beta.$$

The second term on the right-hand side of this equation can be written in the form

$$-\Gamma^\gamma_{\alpha\beta} T_\gamma{}^\beta = -\frac{1}{2} \left(\frac{\partial g_{\lambda\alpha}}{\partial x^\beta} + \frac{\partial g_{\lambda\beta}}{\partial x^\alpha} - \frac{\partial g_{\alpha\beta}}{\partial x^\lambda} \right) T^{\lambda\beta}.$$

Since the tensor $T^{\lambda\beta}$ is symmetric, the first and third terms on the right-hand side of the above equation cancel out and, as a result, we obtain

$$\nabla_\beta T_\alpha{}^\beta = \frac{1}{\sqrt{-g}} \frac{\partial \left(\sqrt{-g}\, T_\alpha{}^\beta\right)}{\partial x^\beta} - \frac{1}{2} T^{\lambda\beta} \frac{\partial g_{\lambda\beta}}{\partial x^\alpha}. \tag{1}$$

Equation (1) can also be written in a somewhat different form. By differentiating the relation

$$g_{\lambda\beta} g^{\lambda\mu} = \delta^\mu_\beta,$$

we obtain

$$g^{\lambda\mu} \frac{\partial g_{\lambda\beta}}{\partial x^\alpha} = -g_{\lambda\beta} \frac{\partial g^{\lambda\mu}}{\partial x^\alpha}.$$

Hence we obtain for the last term on the right-hand side of Eq. (1) the following:

$$-\frac{1}{2}T^{\lambda\beta}\frac{\partial g_{\lambda\beta}}{\partial x^\alpha} = -\frac{1}{2}g^{\lambda\mu}g^{\beta\nu}T_{\mu\nu}\frac{\partial g_{\lambda\beta}}{\partial x^\alpha}$$

$$= \frac{1}{2}T_{\mu\nu}g^{\beta\nu}g_{\lambda\beta}\frac{\partial g^{\lambda\mu}}{\partial x^\alpha}$$

$$= \frac{1}{2}T_{\mu\nu}\frac{\partial g^{\nu\mu}}{\partial x^\alpha}.$$

Accordingly one obtains for the covariant divergence of the tensor $T_\alpha^{\ \beta}$ the following second form:

$$\nabla_\beta T_\alpha^{\ \beta} = \frac{1}{\sqrt{-g}}\frac{\partial\left(\sqrt{-g}T_\alpha^{\ \beta}\right)}{\partial x^\beta} + \frac{1}{2}T_{\mu\nu}\frac{\partial g^{\mu\nu}}{\partial x^\alpha}. \tag{2}$$

P 3.1.24. Show that in the Minkowskian spacetime with $ds^2 = \eta_{\mu\nu}dx^\mu dx^\nu$, where $\eta_{\mu\nu}$ is the flat-space metric given by $\eta_{00} = -\eta_{11} = -\eta_{22} = -\eta_{33} = +1$, with $\eta_{\mu\nu} = 0$ for $\mu \neq \nu$, every *null curve* can be presented in the parametric form

$$x^0 = \int \sin\psi\, ds$$

$$x^1 = \int \sin\psi\sin\theta\cos\phi\, ds$$

$$x^2 = \int \sin\psi\sin\theta\sin\phi\, ds$$

$$x^3 = \int \sin\psi\cos\theta\, ds. \tag{1}$$

Here ψ, θ, and ϕ are functions of the proper time parameter s.

Solution: The solution is left for the reader.

P 3.1.25. Find the Hamilton-Jacobi equation for a particle moving in the gravitational field.

Solution: The motion of a test particle in a gravitational field is determined by the principle of the least action, where the action is given by

$$I = mc\int L\, ds. \tag{1}$$

The four-momentum of a test particle moving in a gravitational field may be defined by

$$p^\alpha = mcu^\alpha = mc\frac{dx^\alpha}{ds} \tag{2}$$

whose square is given by

$$p_\alpha p^\alpha = g^{\alpha\beta} p_\alpha p_\beta = m^2 c^2. \tag{3}$$

Substituting now $-\partial I/\partial x^\alpha$ for p_α we obtain

$$g^{\alpha\beta}\frac{\partial I}{\partial x^\alpha}\frac{\partial I}{\partial x^\beta} = m^2 c^2. \tag{4}$$

Equation (4) is the Hamilton-Jacobi equation for a test particle moving in a gravitational field.

P 3.1.26. Show that the Ricci tensor can be written in the form

$$R_{\mu\nu} = \frac{1}{\sqrt{-g}}\frac{\partial\left(\sqrt{-g}\Gamma^\alpha_{\mu\nu}\right)}{\partial x^\alpha} - \frac{\partial^2 \ln\sqrt{-g}}{\partial x^\mu \partial x^\nu} - \Gamma^\alpha_{\mu\beta}\Gamma^\beta_{\nu\alpha}. \tag{1}$$

Solution: By a direct calculation, using Eq. (3.1.22), we show that the above expression (1) can be reduced to the standard form given by Eqs. (3.1.31) and (3.1.33):

$$R_{\mu\nu} = \frac{\partial\Gamma^\alpha_{\mu\nu}}{\partial x^\alpha} + \Gamma^\alpha_{\mu\nu}\frac{1}{\sqrt{-g}}\frac{\partial\sqrt{-g}}{\partial x^\alpha} - \frac{\partial}{\partial x^\nu}\left(\frac{1}{\sqrt{-g}}\frac{\partial\sqrt{-g}}{\partial x^\mu}\right) - \Gamma^\alpha_{\mu\beta}\Gamma^\beta_{\nu\alpha}$$

$$= \frac{\partial\Gamma^\alpha_{\mu\nu}}{\partial x^\alpha} - \frac{\partial\Gamma^\alpha_{\mu\alpha}}{\partial x^\nu} + \Gamma^\alpha_{\mu\nu}\Gamma^\beta_{\alpha\beta} - \Gamma^\alpha_{\mu\beta}\Gamma^\beta_{\nu\alpha}. \tag{2}$$

P 3.1.27. The *Gaussian curvature K* (also known as the *mean curvature*) of a two-dimensional space (a surface) is defined by

$$R_{ABCD} = K\left(g_{AC}g_{BD} - g_{AD}g_{BC}\right) \tag{1}$$

where the indices A, B, ... take the values 1, 2, and

$$K = \frac{1}{R_1 R_2}. \tag{2}$$

Here R_1 and R_2 are the principal radii of curvature of the surface at the point. The signs of R_1 and R_2 are assumed to be the same if the corresponding centers of curvature are on the same side of the surface, thus $K > 0$ in this case. The signs are opposite if the centers of curvature are on the opposite sides of the surface, thus $K < 0$.

Find the expressions for the Riemann curvature tensor and the Gaussian curvature for a surface. [See D.J. Struik, *Lectures on Classical Differential Geometry*, Addison-Wesley, Reading, MA, 1961.]

Solution: Let us denote the Riemann tensor by R_{ABCD} and the metric tensor by g_{AB}. Since R_{ABCD} is antisymmetric in the indices A, B and C, D, and is symmetric under the exchange of AB with CD, we see that all the nonvanishing components of the Riemann tensor are equal in magnitude and either coincide or differ in sign:

$$R_{1212} = -R_{2112} = -R_{1221} = R_{2121}$$

$$R_{1111} = R_{1122} = R_{2211} = R_{2222} = 0. \tag{3}$$

If we denote by g the determinant of the metric tensor,

$$g = g_{11}g_{22} - g_{12}g_{21}, \tag{4}$$

with $g_{12} = g_{21}$, we may then write Eq. (1) in the form

$$R_{ABCD} = (g_{AC}g_{BD} - g_{AD}g_{BC})\frac{R_{1212}}{g}. \tag{5}$$

Using Eq. (5), we obtain for the Ricci tensor and the Ricci scalar curvature the following expressions:

$$R_{BD} = g^{AC}R_{ABCD} = g_{BD}\frac{R_{1212}}{g}, \tag{6}$$

$$R = g^{BD}R_{BD} = 2\frac{R_{1212}}{g}. \tag{7}$$

Substituting the above expression for the Ricci scalar curvature in Eq. (5), we obtain the following expression for the Riemann tensor in two dimensions:

$$R_{ABCD} = \frac{1}{2}(g_{AC}g_{BD} - g_{AD}g_{BC})R. \tag{8}$$

Comparing the latter expression for the Riemann tensor with that given by Eqs. (1) and (2), we obtain the following for the Gaussian curvature:

$$K = \frac{1}{R_1 R_2} = \frac{R}{2}. \tag{9}$$

P 3.1.28. Decompose the Riemann tensor into its irreducible components for a general n-dimensional space.

Solution: For such a decomposition let us assume the following formula:

$$R_{\mu\nu\rho\sigma} = C_{\mu\nu\rho\sigma} + A\left(g_{\mu\rho}R_{\nu\sigma} - g_{\mu\sigma}R_{\nu\rho} - g_{\nu\rho}R_{\mu\sigma} + g_{\nu\sigma}R_{\mu\rho}\right)$$

$$+B\left(g_{\mu\sigma}g_{\nu\rho} - g_{\mu\rho}g_{\nu\sigma}\right)R. \tag{1}$$

Here A and B are some numerical constants to be determined, and $R_{\mu\nu\rho\sigma}$ is the Riemann tensor, $C_{\mu\nu\rho\sigma}$ is the Weyl tensor, $R_{\alpha\beta}$ is the Ricci tensor, $g_{\alpha\beta}$ is the metric tensor, and R is the Ricci scalar.

Contracting Eq. (1) with respect to the two indices μ and ρ, and taking into account that the Weyl tensor is traceless, namely, $C^\rho{}_{\nu\rho\sigma} = 0$, we obtain

$$R^\rho{}_{\nu\rho\sigma} = R_{\nu\sigma} = A\left[(n-2)R_{\nu\sigma} + g_{\nu\sigma}R\right] + B\left(1-n\right)g_{\nu\sigma}R. \tag{2}$$

Here n is the number of dimensions of the space. By equating the coefficients of $R_{\nu\sigma}$ and $g_{\nu\sigma}R$ in Eq. (2), we can now determine the values of the constants A and B. We obtain

$$A = \frac{1}{n-2}, \qquad B = \frac{1}{(n-1)(n-2)}. \tag{3}$$

We therefore obtain for the decomposition of the Riemann tensor, in an n-dimensional space, the following:

$$R_{\mu\nu\rho\sigma} = C_{\mu\nu\rho\sigma} + \frac{1}{n-2}\left(g_{\mu\rho}R_{\nu\sigma} - g_{\mu\sigma}R_{\nu\rho} - g_{\nu\rho}R_{\mu\sigma} + g_{\nu\sigma}R_{\mu\rho}\right)$$

$$+\frac{1}{(n-1)(n-2)}\left(g_{\mu\sigma}g_{\nu\rho} - g_{\mu\rho}g_{\nu\sigma}\right)R. \tag{4}$$

From its structure we see that the above equation is only valid for spaces of dimensions n higher than 2, $n > 2$.

P 3.1.29. Find the number of independent components of both the Riemann and the Ricci tensors in three dimensions.

Solution: Let us denote the metric tensor, the Riemann tensor, and the Ricci tensor by g_{ab}, R_{abcd}, and R_{ab}, respectively, where the indices $a,b, \ldots$ take the values 1, 2, 3. The pairs of indices ab and cd of the Riemann tensor can then take on the values 23, 31, and 12. Hence R_{abcd} behaves like a symmetric matrix in three dimensions. The Riemann tensor, therefore, has only six independent components. The Ricci tensor $R_{ab} = g^{cd}R_{cadb}$ is symmetric. It therefore has also only six independent components. Since the decomposition of the Riemann tensor into the Weyl tensor and the Ricci tensor is valid for any dimension higher that 2 (see previous problem), it follows that the Weyl tensor vanishes in the three-dimensional spaces.

Since both the Riemann tensor and the Ricci tensor have the same number of independent components, we expect that these two tensors are related to each other. Let us, therefore, assume that we have the following relationship between them:

$$R_{abcd} = g_{ac}S_{bd} - g_{ad}S_{bc} - g_{bc}S_{ad} + g_{bd}S_{ac}. \tag{1}$$

The right-hand side of Eq. (1) satisfies the symmetry properties of the Riemann tensor. Here S_{ab} is some symmetric tensor whose explicit form has to be determined.

Contracting the indices a and c in Eq. (1), gives

$$R_{bd} = S_{bd} + g_{bd}S, \tag{2}$$

where $S = g^{ad}S_{ad}$ is the trace of the tensor S_{ab}. Contracting now the indices bd in Eq. (2) gives $R = 4S$. Hence we obtain for the tensor S_{ab}, when expressed in terms of the Ricci tensor R_{ab}, the following:

$$S_{ab} = R_{ab} - \frac{1}{4}g_{ab}R. \tag{3}$$

Substituting the above expression for S_{ab} in Eq. (1), we finally obtain

$$R_{abcd} = (g_{ac}R_{bd} - g_{ad}R_{bc} - g_{bc}R_{ad} + g_{bd}R_{ac}) + \frac{1}{2}(g_{ad}g_{bc} - g_{ac}g_{bd})R. \tag{4}$$

It will be noted that Eq. (4) is a particular case of the general decomposition formula, obtained in the previous problem for the Riemann tensor, for a three-dimensional space (in which case the Weyl tensor vanishes, as has been shown above.)

P 3.1.30. Show that a necessary and sufficient condition for the Weyl conformal tensor to vanish everywhere is that the spacetime should be conformally flat, for a spacetime of dimension $n > 3$. [See L.P. Eisenhart, *Riemannian Geometry*, Princeton University Press, Princeton, NJ, 1949.]

Solution: The proof is left for the reader.

P 3.1.31. Show that the symmetry relation for the Riemann tensor, Eq. (3.1.32b), can be obtained as a consequence of Eqs. (3.1.32a) and (3.1.32c).

Solution: The solution is left for the reader.

P 3.1.32. Derive the Bianchi identities without using a geodesic coordinate system.

Solution: The solution is left for the reader.

3.2   The Principle of Equivalence

3.2.1   *Null experiments: Eötvös experiment*

One of the most interesting *null experiments* in physics is due to Eötvös, first performed in 1890 and recently repeated by Dicke. The experiment showed, in great precision, that all bodies fall with the same acceleration. The roots of the experiment go back to Newton and Galileo, who demonstrated experimentally that the gravitational acceleration of a body is independent of its composition.

The importance of the Eötvös experiment is in the fact that the null result of the experiment is a *necessary* condition for the theory of general relativity to be valid.

Eötvös employed a static torsion balance, balancing a component of the Earth's gravitational pull on the weight against the centrifugal force field of the Earth acting on the weight. He employed a horizontal torsion beam, 40 cm long, suspended by a fine wire. From the ends of the torsion beam two masses of different compositions were suspended, one lower than the other. A lack of exact proportionality between the inertial and gravitational masses of the two bodies would then lead to a torque tending to rotate the balance. There appears to be no need for one mass to be suspended lower than the other.

The experiment of Eötvös showed, with an accuracy of a few parts in 10^9, that inertial and gravitational masses are equal.

In the experiment performed by Dicke, the gravitational acceleration toward the Sun of small gold and aluminum weights were compared and found to be equal with an accuracy of about one part in 10^{11}. Hence the necessary condition to be satisfied for the validity of general relativity theory seems to be rather satisfactorily met.

The question therefore arises as to what extent is this experiment also a sufficient condition to be satisfied in order that general relativity theory be valid.

It has been emphasized by Dicke that gold and aluminum differ from each other rather greatly in several important aspects. First, the neutron to proton ratio is quite different in the two elements, varying from 1.08 in aluminum to 1.50 in gold. Second, the electrons in aluminum move with nonrelativistic velocities, but in gold the k-shell electrons have a 15 per cent increase in their masses as a result of their relativistic velocities. Third, the electromagnetic negative contribution to the binding energy of the nucleus varies as Z^2 (Z is the proton number) and represents $\frac{1}{2}$ per cent of the total mass of a gold atom, whereas it is negligible in aluminum. Fourth, the virtual pair field and other fields would be expected to be different in the two atoms. We thus conclude that the physical aspects of gold and aluminum differ substantially, and consequently the equality of their accelerations represents an important condition to be satisfied by any theory of gravitation.

Since the accuracy of the Eötvös experiment is very high, the question arises as to whether it implies that the equivalence principle is very nearly valid. This is true in a limited sense; certain aspects of the equivalence principle are not supported in the slightest by the Eötvös experiment.

3.3 The Principle of General Covariance

We have seen in the preceding section that a gravitational field can be considered locally equivalent to an accelerated frame. This implies that the special theory of relativity (see Section 1.3) cannot be valid in an extended region where gravitational fields are present. A curved spacetime is needed and all laws of nature should be covariant under the most general coordinate transformations.

The original formulation of general relativity by Einstein was based on two principles: (1) the principle of equivalence (discussed in detail in the last section); and (2) the principle of general covariance.

The principle of general covariance is often stated in one of the following forms, which are not exactly equivalent:

(1) All coordinate systems are equally good for stating the laws of physics, and they should be treated on the same footing.

(2) The equations of physics should have tensorial forms.

(3) The equations of physics should have the same form in all coordinate systems.

According to the principle of general covariance, the coordinates become nothing more than a bookkeeping system to label the events. The principle is a valuable guide to deducing correct equations.

It has been pointed out that any spacetime physical law can be written in a covariant form and hence the principle of general covariance has no necessary physical consequences, and Einstein concurred with this view.

In spite of Einstein's acceptance of this objection, it appears that the principle of general covariance was introduced by Einstein as a generalization of the principle of special relativity and he often referred to it as the principle of general relativity. In fact the principle of equivalence (which necessarily leads to the introduction of a curved spacetime), plus the assumption of general covariance, is most of what is needed to generate Einstein's theory of general relativity. They lead directly to the idea that gravitation can be explained by means of Riemannian geometry. This is done in the next section.

3.4 Gravitational Field Equations

We have seen in Section 3.1 that the Riemannian geometry is characterized by a geometrical metric, i.e., a symmetric tensor $g_{\mu\nu}$ from which one can construct other quantities. Classical general relativity theory identifies this tensor as the gravitational potential. Hence in general relativity there are ten components to the gravitational potential, as compared with the single potential function in the Newton theory of gravitation.

3.4.1 *The Einstein field equations*

In trying to arrive at the desired gravitational field equations that the metric tensor has to satisfy, we are guided by the requirement that, in an appropriate limit, the theory should reduce to the Newton gravitational theory. In the latter theory, the gravitational potential ϕ is determined by the Poisson equation:

$$\nabla^2\phi = 4\pi G\rho, \tag{3.4.1}$$

where G $(= 6.67\times10^{-8}$ cm^3 g^{-1} sec$^{-2})$ is the Newton gravitational constant and ρ is the mass density of matter. Hence $g_{\mu\nu}$ should satisfy second order partial differential equations. The equations should then be linearly related to the energy-momentum tensor $T_{\mu\nu}$. Such equations are

$$R_{\mu\nu} - \frac{1}{2}g_{\mu\nu}R = \kappa T_{\mu\nu}, \tag{3.4.2a}$$

or their equivalent

$$R_{\mu\nu} = \kappa\left(T_{\mu\nu} - \frac{1}{2}g_{\mu\nu}T\right), \tag{3.4.2b}$$

where T is the trace of $T_{\mu\nu}$, $T = T_{\mu\nu}g^{\mu\nu}$, and κ is some constant, called the Einstein gravitational constant, to be determined. In cosmology theory, one sometimes adds an additional term, $\lambda g_{\mu\nu}$, to the left-hand side of Eq. (3.4.2a). The constant λ is known as the cosmological constant.

But the contracted Bianchi identities, Eq. (3.1.48), show that the covariant divergence of the left-hand side of Eq. (3.4.2a) vanishes. Hence

$$\nabla_\nu T_\mu{}^\nu = 0, \tag{3.4.3}$$

which expresses the covariant conservation of energy and momentum. The constant κ can be determined by going to the limit of weak gravitational field (see Problem 3.9.11). Its value is $\kappa = 8\pi G/c^4$.

3.4.2 *Problems*

P 3.4.1. Find the Newtonian potential produced by a system of masses at distances that are large compared to the dimensions of the system.

Solution: The Newtonian theory of gravitation is still a useful theory in spite of all the advances made by Einstein.

The Newtonian potential is the solution of the Poisson equation

$$\nabla^2 \phi(x) = 4\pi G \rho(x), \tag{1}$$

where $\rho(x)$ is the mass density of the system, and G is Newton's gravitational constant. In Eq. (1) the variable x denotes the three spatial coordinates x, y, z.

The solution of Eq. (1) is given by

$$\phi(x) = -G \int \frac{\rho(x')}{|\mathbf{r} - \mathbf{r}'|} d^3 x', \tag{2}$$

where $\mathbf{r}=(x^1, x^2, x^3)$ is the radius vector of the point where the potential is being calculated, and $\mathbf{r}'=(x'^1, x'^2, x'^3)$ is the radius vector of an arbitrary point at the mass distribution of the matter.

The potential $\phi(x)$ can be expanded in powers of $1/r$, thus getting

$$\phi = -G \left[\frac{m}{r} + \frac{1}{6} D_{ij} \frac{\partial^2}{\partial x^i \partial x^j} \left(\frac{1}{r} \right) + \cdots \right], \tag{3}$$

where

$$m = \int \rho d^3 x \tag{4}$$

is the total mass of the system. The missing $1/r^2$ term, corresponding to the dipole moment of the system of masses, vanishes identically by virtue of the choice of the origin of the coordinates at the center of mass. The quantity

$$D_{ij} = \int \rho \left(3 x^i x^j - r^2 \delta^{ij} \right) d^3 x \tag{5}$$

is called the *mass quadrupole moment tensor*, and is related to the *moment of inertia tensor*

$$J_{ij} = \int \rho \left(r^2 \delta^{ij} - x^i x^j \right) d^3 x \tag{6}$$

by

$$D_{ij} = J_{kk} \delta_{ij} - 3 J_{ij}, \tag{7}$$

where $J_{kk} = J_{11} + J_{22} + J_{33}$. Notice that, by definition, the mass quadrupole moment tensor is traceless, $D_{kk} = D_{11} + D_{22} + D_{33} = 0$.

P 3.4.2. Calculate the mass quadrupole moment tensor of a homogeneous body having the shape of an ellipsoid.

Solution: Let the surface of the ellipsoid be given by the equation

$$\frac{x^2}{a^2} + \frac{y^2}{b^2} + \frac{z^2}{c^2} = 1. \tag{1}$$

By introducing the new coordinates $x' = x/a$, $y' = y/b$, and $z' = z/c$, the volume integration over the ellipsoid reduces to that over the unit sphere. Hence we have, for example,

$$D_{11} = \int \int \int \rho \left(3x^2 - r^2\right) dx\,dy\,dz$$

$$= \int \int \int \rho \left(2x^2 - y^2 - z^2\right) dx\,dy\,dz$$

$$= \int \int \int \rho abc \left(2a^2 x'^2 - b^2 y'^2 - c^2 z'^2\right) dx'\,dy'\,dz'$$

$$= \rho abc \left(2a^2 - b^2 - c^2\right) \int \int \int z'^2 dx'\,dy'\,dz'$$

$$= \rho abc \left(2a^2 - b^2 - c^2\right) \int_0^{2\pi} \int_0^{\pi} \int_0^1 r^4 dr \cos^2 \theta \sin \theta d\theta d\phi$$

$$= \frac{m}{5} \left(2a^2 - b^2 - c^2\right), \tag{2}$$

where $m = 4\pi abc\rho/3$ is the mass of the ellipsoid. Likewise, the other nonvanishing components of the mass quadrupole moment tensor are given by

$$D_{22} = \frac{m}{5} \left(-a^2 + 2b^2 - c^2\right) \tag{3}$$

$$D_{33} = \frac{m}{5} \left(-a^2 - b^2 + 2c^2\right). \tag{4}$$

P 3.4.3. Write the general term in the expansion of the Newtonian potential using spherical harmonics.

Solution: We expand the expression $1/\left|\mathbf{r} - \mathbf{r}'\right|$ into *spherical harmonics*:

$$\frac{1}{\left|\mathbf{r} - \mathbf{r}'\right|} = \frac{1}{\left(r^2 + r'^2 - 2rr' \cos \beta\right)^{1/2}} = \sum_{l=0}^{\infty} \frac{r'^l}{r^{l+1}} P_l\left(\cos \beta\right). \tag{1}$$

Here β is the angle between vectors $\mathbf{r}$ and $\mathbf{r}'$ (for notation see Problem 3.4.1). Using now the addition theorem for the spherical harmonics, we obtain

$$P_l\left(\cos\beta\right) = \sum_{m=-l}^{l} \frac{\left(l-|m|\right)!}{\left(l+|m|\right)!} P_l^{|m|}\left(\cos\theta\right) P_l^{|m|}\left(\cos\theta'\right) e^{-im(\phi-\phi')}, \qquad (2)$$

where the spherical angles θ, ϕ and θ', ϕ' denote the directions of the vectors $\mathbf{r}$ and $\mathbf{r}'$, respectively, with respect to the fixed coordinate system. The functions $P_l^m\left(\cos\theta\right)$ are the associated Legendre polynomials.

Introducing now the *spherical functions* defined by

$$Y_{lm}(\theta,\phi) = (-1)^{m_i l}\left[\frac{\left(2l+1\right)\left(l-m\right)!}{2\left(l+m\right)!}\right]^{1/2} P_l\left(\cos\theta\right) e^{im\phi}, \qquad (3)$$

for $m \geq 0$, and

$$Y_{l,-|m|}(\theta,\phi) = (-1)^{l-m}\bar{Y}_{l,|m|}, \qquad (4)$$

the expansion given above can then be written as

$$\frac{1}{|\mathbf{r}-\mathbf{r}'|} = \sum_{l=0}^{\infty}\sum_{m=-l}^{l} \frac{4\pi}{2l+1}\frac{r'^l}{r^{l+1}} Y_{lm}\left(\theta',\phi'\right) \bar{Y}_{lm}\left(\theta,\phi\right). \qquad (5)$$

If we now write the Newtonian potential in the form

$$\phi\left(x\right) = -G\int \frac{\rho\left(x'\right) d^3 x'}{|\mathbf{r}-\mathbf{r}'|} = \sum_{l=0}^{\infty}\phi_l\left(x\right), \qquad (6)$$

then the lth term will have the form

$$\phi_l = \frac{-G}{r^{l+1}}\sum_{m=-l}^{l}\left(\frac{4\pi}{2l+1}\right)^{1/2} Q_l^m\bar{Y}_{lm}\left(\theta,\phi\right), \qquad (7)$$

where use has been made of the notation

$$Q_l^m = \left(\frac{4\pi}{2l+1}\right)^{1/2}\int \rho\left(x'\right) r'^l Y_{lm}\left(\theta',\phi'\right) d^3 x'. \qquad (8)$$

The $2l+1$ quantities Q_l^m, with $m = -l, -l+1, \ldots, l$, describe the 2^l-pole moment of the mass system. The quantity $Q_0^0 = 2\pi^{1/2}m$, where m is the total mass of the system. The quantities Q_2^m, with $m = -2, -1, 0, 1, 2$, are related to the components of the mass quadrupole moment tensor D_{ij} by

$$Q_2^0 = -\frac{1}{2}D_{33} \qquad (9a)$$

$$Q_2^{\pm 1} = \pm\frac{1}{\sqrt{6}}\left(D_{13}\pm iD_{23}\right) \qquad (9b)$$

$$Q_2^{\pm 2} = -\frac{1}{2\sqrt{6}} \left(D_{11} - D_{22} \pm 2i D_{12} \right). \tag{9c}$$

P 3.4.4. The Newtonian field equation is given by the Poisson equation. At points where there is no matter, that is, at points of space where $\rho(x, y, z) = 0$ the Poisson equation can be replaced by the Laplace equation,

$$\nabla^2 \phi(x, y, z) = 0. \tag{1a}$$

Solve the Laplace equation in cylindrical coordinates, using the method of the separation of variables.

Solution: In cylindrical coordinates ρ, z and ϕ, the Laplace equation takes the form

$$\frac{\partial^2 f}{\partial \rho^2} + \frac{1}{\rho} \frac{\partial f}{\partial \rho} + \frac{1}{\rho^2} \frac{\partial^2 f}{\partial \phi^2} + \frac{\partial^2 f}{\partial z^2} = 0. \tag{1b}$$

The separation of variables can then be achieved by the following substitution:

$$f(\rho, z, \phi) = R(\rho) Z(z) \Phi(\phi). \tag{2}$$

Using the solution (2) in the Laplace equation (1) then yields the following three differential equations:

$$\frac{d^2 R}{d\rho^2} + \frac{1}{\rho} \frac{dR}{d\rho} + \left(k^2 - \frac{\nu^2}{\rho^2} \right) R = 0 \tag{3a}$$

$$\frac{d^2 Z}{dz^2} - k^2 Z = 0 \tag{3b}$$

$$\frac{d^2 \Phi}{d\phi^2} + \nu^2 \Phi = 0. \tag{3c}$$

Here k^2 and ν^2 are separation constants.

The solutions of the last two equations are elementary and are given by

$$Z(z) = e^{\pm kz} \tag{4a}$$

$$\Phi(\phi) = e^{\pm i\nu\phi}. \tag{4b}$$

In order that the potential f be single valued, ν must be an integer. The parameter k, on the other hand, is arbitrary and may be assumed to be real. By changing variables from ρ into $x = k\rho$, the radial equation (3a) becomes

$$\frac{d^2 R}{dx^2} + \frac{1}{x} \frac{dR}{dx} + \left(1 - \frac{\nu^2}{x^2} \right) R = 0. \tag{5}$$

Equation (5) is the familiar *Bessel equation* whose solutions are *Bessel functions* of order ν.

We assume that the solution of the Bessel equation can be written in the form of a power series as

$$R(x) = x^{\alpha} \sum_{k=0}^{\infty} a_k x^k. \tag{6}$$

Then we find that $\alpha = \pm\nu$, and the coefficients a_k are given by

$$a_{2j-1} = 0 \tag{7a}$$

$$a_{2j} = -\frac{1}{4j(j+\alpha)} a_{2j-2}, \tag{7b}$$

for $j = 1, 2, 3, \ldots$. Hence the coefficients of the odd powers of x vanish. Iterating the recursion formula then yields

$$a_{2j} = \frac{(-1)^j \, \Gamma(\alpha+1)}{2^{2j} j! \Gamma(j+\alpha+1)} a_0. \tag{8}$$

If we choose the coefficient $a_0 = 1/2^{\alpha}\Gamma(\alpha+1)$, then the two solutions, corresponding to $\alpha = \pm\nu$, are given by

$$J_{\nu}(x) = \left(\frac{x}{2}\right)^{\nu} \sum_{j=0}^{\infty} \frac{(-1)^j}{j! \Gamma(j+\nu+1)} \left(\frac{x}{2}\right)^{2j} \tag{9a}$$

$$J_{-\nu}(x) = \left(\frac{x}{2}\right)^{-\nu} \sum_{j=0}^{\infty} \frac{(-1)^j}{j! \Gamma(j-\nu+1)} \left(\frac{x}{2}\right)^{2j} \tag{9b}$$

These are *Bessel functions of the first kind* of order $\pm\nu$. The series converge for all finite values of x. If we assume that $\nu = m$ is an *integer*, the above two solutions are then nearly *dependent*, and we have

$$J_{-m}(x) = (-1)^m J_m(x). \tag{10}$$

If ν is taken to be *not* an integer, however, the two solutions $J_{\pm\nu}(\alpha)$ are then linearly *independent*.

We may replace the two solutions (9) by $J_{\nu}(x)$ and $N_{\nu}(x)$, where

$$N_{\nu}(x) = \frac{J_{\nu}(x)\cos\nu\pi - J_{-\nu}(x)}{\sin\nu\pi} \tag{11}$$

is a *Newmann function*, or a *Bessel function of the second kind*. The function $N_{\nu}(x)$ is linearly *independent* of $J_{\nu}(x)$, both when ν is not an integer and in the limit $\nu \to$ integer.

Finally, *Bessel functions of the third kind*, called *Hankel functions*, are defined as linear combinations of $J_\nu(x)$ and $N_\nu(x)$ by

$$H_\nu^{(1)}(x) = J_\nu(x) + iN_\nu(x) \tag{12a}$$

$$H_\nu^{(2)}(x) = J_\nu(x) - iN_\nu(x). \tag{12b}$$

Hankel functions also provide independent solutions to the Bessel equation, just as $J_\nu(x)$ and $N_\nu(x)$ do.

P 3.4.5. The Newtonian potential energy produced by a distribution of masses with mass density ρ is given by

$$U = \frac{1}{2} \int \rho\phi d^3x, \tag{1}$$

where ϕ is the Newtonian potential.

Find the Newtonian potential energy produced by a homogeneous ellipsoidal body. (See Landau and Lifshitz 1975.)

Solution: The solution is left for the reader.

3.4.3 *The Newtonian limit in general relativity*

The Einstein gravitational field equations were represented in Subsection 3.4.1. We now apply them to the case of weak gravitational field. This is important in two aspects, the first is to find the link to the Newtonian gravitational theory along with its equation of motion, and secondly it will give us the possibility of fixing the value of Einstein's gravitational constant κ. Obviously the Newtonian theory should be obtained as a limiting case of a weak gravitational field in Einstein's theory. In the sequel we will use approximation methods that will give us post-Newtonian equations of motion that are very important in the experimental verification of general relativity theory. A similar procedure of obtaining the Newtonian limit in cosmological general relativity, that deals with the expansion of the Universe, will be given in Subsection 4.1.5. To obtain the Newtonian limit we proceed as follows.

In this regard we employ the geodesic equation to give us a description of the motion of particles. We also have to find out how the ten components of the metric tensor are related to the single function that appears in the Newtonian theory as a potential function. And in this regard the ten components of the metric tensor are also considered the potential functions of gravitation in Einstein's general relativity theory. We will have to use an approximation method that fits the theory of general relativity in considering the velocities of particles to be much smaller than the speed of light in vacuum c, $v \ll c$. Hence we proceed as follows.

We first examine the line element

$$ds^2 = g_{\mu\nu}dx^\mu dx^\nu. \tag{3.4.4}$$

We notice that the different elements in the sum on the right-hand side of the above equation depend on the ratio of the velocity to the speed of light in vacuum c quite differently. For example the term $g_{00}dx^0 dx^0 = g_{00}c^2 dt^2$ (since $dx^0 = cdt$) is one order of magnitude larger than the term $2g_{0k}dx^0 dx^k = 2g_{0k}cdtdx^k$, where $k = 1, 2, 3$. The latter term, in turn, is one order of magnitude larger than $g_{kl}dx^k dx^l$.

It thus appears that the term $g_{00}dx^0 dx^0$ is the dominant term in the line element. Again we have to emphasize that this approximation is valid only for the case where the velocities of the masses producing the gravitational field are much smaller than that of the speed of light in vacuum. This approximation fits, for example, for describing the motion of the planets around the Sun. In order to obtain the appropriate Einstein corrections to the Newtonian theory one has to go to the second, third, forth orders and higher ones. It is the combination of the metric tensor approximation with the geodesic equation that will enable us to obtain Newton's second law of motion.

The geodesic equation is usually written in terms of the independent parameter of length ds,

$$\frac{d^2 x^\mu}{ds^2} + \Gamma^\mu_{\alpha\beta}\frac{dx^\alpha}{ds}\frac{dx^\beta}{ds} = 0. \tag{3.4.5}$$

Since Newton's law of motion is written with the time independent parameter, we will have to rewrite the above equation in terms of the time coordinate. Changing the variable s to a new variable σ, we obtain the alternative form

$$\frac{d^2 x^\mu}{d\sigma^2} + \Gamma^\mu_{\alpha\beta}\frac{dx^\alpha}{d\sigma}\frac{dx^\beta}{d\sigma} = -\frac{d^2\sigma/ds^2}{(d\sigma/ds)^2}\frac{dx^\mu}{d\sigma} \tag{3.4.6}$$

for the geodesic equation. We can now choose the parameter σ to be equal x^0, where x^0 is the time coordinate. The latter equation can therefore be written as

$$\ddot{x}^\mu + \Gamma^\mu_{\alpha\beta}\dot{x}^\alpha \dot{x}^\beta = -\frac{d^2 x^0/ds^2}{(dx^0/ds)^2}\dot{x}^\mu. \tag{3.4.7}$$

In the above equation an overdot denotes differentiation with respect to the coordinate x^0.

The zero component of Eq. (3.4.7) is

$$\ddot{x}^0 + \Gamma^0_{\alpha\beta}\dot{x}^\alpha\dot{x}^\beta = -\frac{d^2x^0/ds^2}{(dx^0/ds)^2}\dot{x}^0. \tag{3.4.8}$$

We notice that $\dot{x}^0 = dx^0/dx^0 = 1$, and therefore $\ddot{x}^0 = 0$. Accordingly we obtain

$$\frac{d^2x^0/ds^2}{(dx^0/ds)^2} = -\Gamma^0_{\alpha\beta}\dot{x}^\alpha\dot{x}^\beta. \tag{3.4.9}$$

Inserting Eq. (3.4.9) in Eq. (3.4.7) we obtain for the geodesic equation

$$\ddot{x}^\mu + \left(\Gamma^\mu_{\alpha\beta} - \dot{x}^\mu\Gamma^0_{\alpha\beta}\right)\dot{x}^\alpha\dot{x}^\beta = 0, \tag{3.4.10}$$

now written in terms of the coordinate x^0. The zero component of Eq. (3.4.10) is now an identity since $\ddot{x}^0 = 0$ and $\dot{x}^0 = 1$. Thus the formula has actually only three independent components, namely

$$\ddot{x}^k + \left(\Gamma^k_{\alpha\beta} - \dot{x}^k\Gamma^0_{\alpha\beta}\right)\dot{x}^\alpha\dot{x}^\beta = 0, \tag{3.4.11}$$

where $k = 1, 2, 3$. This is the equation of motion we now have to work with. We will have to find the lowest approximation of this equation and also of the Einstein field equations.

To find the lowest approximation of Eq. (3.4.11), we assume the particles move with velocities much smaller than the speed of light in vacuum c, $v \ll c$. Thus for example

$$\Gamma^k_{\alpha\beta} \gg \Gamma^0_{\alpha\beta}\dot{x}^k,$$

and as a result the term $\Gamma^0_{\alpha\beta}\dot{x}^k$ can be neglected in Eq. (3.4.11). Proceeding in this way all terms with velocities can be neglected and the geodesic equation yields

$$\ddot{x}^k \approx -\Gamma^k_{00}, \tag{3.4.12}$$

in the lowest approximation. Thus Γ^k_{00} acts like the Newtonian force per mass unit.

And in terms of the metric tensor we have

$$\Gamma^k_{00} = \frac{1}{2}g^{k\lambda}\left(2\frac{\partial g_{\lambda 0}}{\partial x^0} - \frac{\partial g_{00}}{\partial x^\lambda}\right)$$

$$\approx -\frac{1}{2}\eta^{k\lambda}\frac{\partial g_{00}}{\partial x^\lambda}$$

$$\approx -\frac{1}{2}\eta^{kl}\frac{\partial g_{00}}{\partial x^l}$$

$$= \frac{1}{2}\delta^{kl}\frac{\partial g_{00}}{\partial x^l}$$

$$= \frac{1}{2}\frac{\partial g_{00}}{\partial x^k}. \tag{3.4.13}$$

In the above equations $\eta^{\mu\nu}$ is the Minkowskian metric

$$\eta^{\mu\nu} = \begin{pmatrix} +1 & 0 & 0 & 0 \\ 0 & -1 & 0 & 0 \\ 0 & 0 & -1 & 0 \\ 0 & 0 & 0 & -1 \end{pmatrix}.$$

In order to show the similarity of the equation of motion, obtained from the geodesic equation in the slow-motion approximation, to the Newtonian limit, we write

$$\ddot{x}^k = -\frac{1}{2}\frac{\partial g_{00}}{\partial x^k}, \tag{3.4.14}$$

and introduce a new function $\phi(x)$ by means of

$$g_{00}(x) = 1 + \frac{2}{c^2}\phi(x). \tag{3.4.15}$$

In terms of the new function $\phi(x)$ we now have

$$\ddot{x}^k = -\frac{1}{c^2}\frac{\partial \phi(x)}{\partial x^k}. \tag{3.4.16}$$

This equation still does not resemble the Newton law of motion since a dot means differentiation with $x^0 = ct$, and in terms of a time coordinate we now have

$$\frac{d^2 x^k}{dt^2} = -\frac{\partial \phi(x)}{\partial x^k}. \tag{3.4.17}$$

We now have to relate the function $\phi(x)$ to the Einstein field equations. It still has to be shown that the function $\phi(x)$ satisfies the Poisson equation as the Newtonian theory requires.

We will use the Einstein field equations

$$R_{\mu\nu} = \kappa\left(T_{\mu\nu} - \frac{1}{2}g_{\mu\nu}T\right). \tag{3.4.18}$$

In the above equation T is the trace of $T_{\mu\nu}$, and in the lowest approximation we have

$$T = T_{\mu\nu}g^{\mu\nu} \approx T_{\mu\nu}\eta^{\mu\nu} \approx T_{00}\eta^{00} = T_{00}. \tag{3.4.19}$$

We do not need all the ten equations of Einstein. It is sufficient to use the 00 component. This gives

$$R_{00} = \kappa \left(T_{00} - \frac{1}{2} g_{00} T \right)$$

$$\approx \kappa \left(T_{00} - \frac{1}{2} \eta_{00} T \right)$$

$$= \frac{1}{2} \kappa T_{00}$$

$$= \frac{1}{2} \kappa c^2 \rho\left(x\right). \tag{3.4.20}$$

In the last equation use has been made of

$$T_{\mu\nu} = \rho_0 u_\mu u_\nu, \tag{3.4.21}$$

where $u^\mu = dx^\mu/ds$, and

$$\rho\left(x\right) = \rho_0 \left(\frac{dt}{ds} \right)^2. \tag{3.4.22}$$

In the above equations $\rho\left(x\right)$ is the mass density of the matter distribution that produces the gravitational field.

Using the slow-motion approximation to the Ricci tensor R_{00} gives

$$R_{00} \approx \frac{\partial \Gamma^s_{00}}{\partial x^s}. \tag{3.4.23}$$

And to the lowest approximation

$$R_{00} \approx \frac{1}{c^2} \nabla^2 \phi\left(x\right), \tag{3.4.24}$$

where ∇^2 is the ordinary three-dimensional Laplace operator

$$\nabla^2 = \frac{\partial^2}{\partial x^1 \partial x^1} + \frac{\partial^2}{\partial x^2 \partial x^2} + \frac{\partial^2}{\partial x^3 \partial x^3}. \tag{3.4.25}$$

Equating now the two expressions given by Eqs. (3.4.20) and (3.4.24) for R_{00} then gives the differential equation that the function $\phi\left(x\right)$ has to satisfy

$$\nabla^2 \phi\left(x\right) = \frac{1}{2} \kappa c^4 \rho\left(x\right). \tag{3.4.26}$$

Equation (3.4.26) can thus be identified with the Newtonian Poisson equation for the gravitational potential if we identify the term $\frac{1}{2}\kappa c^4$ with the Newtonian expression $4\pi G$, where G is the Newton gravitational constant. Such an identification gives the equation

$$\kappa = \frac{8\pi G}{c^4}, \tag{3.4.27}$$

for the Einstein gravitational constant. Equation (3.4.26) then becomes

$$\nabla^2 \phi\left(x\right) = 4\pi G \rho\left(x\right). \tag{3.4.28}$$

In Subsection 4.1.5 we will find the Newtonian limit in cosmological general relativity.

3.4.4   *Derivation of the Einstein equations from variational principle*

We start with the action integral

$$I = \int \sqrt{-g}\,(L_G - 2\kappa L_F)\,d^4x, \tag{3.4.29}$$

and demand its variation to be zero. Here L_G and L_F are the Lagrangians for the gravitational and other fields, respectively. We take $L_G = R$, where R is the Ricci scalar, $R = R_{\mu\nu}g^{\mu\nu}$.

The first part of the integral (3.4.29) gives

$$\delta \int \sqrt{-g}R\,d^4x = \int \sqrt{-g}g^{\mu\nu}\delta R_{\mu\nu}d^4x + \int R_{\mu\nu}\delta\left(\sqrt{-g}g^{\mu\nu}\right)d^4x. \tag{3.4.30}$$

To find $\delta R_{\mu\nu}$ we note that in a geodesic coordinate system one has

$$\delta R_{\mu\nu} = \nabla_\alpha\left(\delta\Gamma^\alpha_{\mu\nu}\right) - \nabla_\nu\left(\delta\Gamma^\alpha_{\mu\alpha}\right). \tag{3.4.31}$$

But the latter is a tensorial equation. Hence it is valid in all coordinate systems. Consequently, the first integral on the right-hand side of Eq. (3.4.30) can be written as

$$\int \sqrt{-g}g^{\mu\nu}\delta R_{\mu\nu}d^4x = \int \sqrt{-g}\nabla_\alpha\left(g^{\mu\nu}\delta\Gamma^\alpha_{\mu\nu} - g^{\mu\alpha}\delta\Gamma^\beta_{\mu\beta}\right)d^4x, \tag{3.4.32}$$

and hence (by Problem 3.9.7) is equal to

$$\int \partial_\alpha\left[\sqrt{-g}\left(g^{\mu\nu}\delta\Gamma^\alpha_{\mu\nu} - g^{\mu\alpha}\delta\Gamma^\beta_{\mu\beta}\right)\right]d^4x. \tag{3.4.33}$$

This integral, however, vanishes, since by Gauss' theorem it is equal to a surface integral, which is equal to zero in consequence of the vanishing of the variations at the boundary.

The second integral on the right-hand side of Eq. (3.4.30) gives, by Eq. (3.1.21),

$$\int R_{\mu\nu}\delta\left(\sqrt{-g}g^{\mu\nu}\right)d^4x = \int \sqrt{-g}\left(R_{\mu\nu} - \frac{1}{2}g_{\mu\nu}R\right)\delta g^{\mu\nu}d^4x. \tag{3.4.34}$$

The second part of the integral (3.4.29) leads to (see Problem 3.9.12)

$$\delta \int \sqrt{-g}L_F d^4x = -\frac{1}{2}\int \sqrt{-g}T_{\mu\nu}\delta g^{\mu\nu}d^4x, \tag{3.4.35}$$

Here $T_{\mu\nu}$ is the energy-momentum tensor and is given by

$$T_{\mu\nu} = \frac{-2}{\sqrt{-g}}\left[\left(\frac{\partial\left(\sqrt{-g}L_F\right)}{\partial g^{\mu\nu}_{,\alpha}}\right)_{,\alpha} - \frac{\partial\left(\sqrt{-g}L_F\right)}{\partial g^{\mu\nu}}\right], \tag{3.4.36}$$

and a comma denotes partial differentiation, $f_{,\alpha} = \partial_\alpha f$. Combining Eqs. (3.4.29), (3.4.34) and (3.4.35) then leads to the field equations (3.4.2):

$$R_{\mu\nu} - \frac{1}{2}g_{\mu\nu}R = \kappa T_{\mu\nu}. \tag{3.4.37}$$

3.4.5 *The electromagnetic energy-momentum tensor*

The energy-momentum tensor $T_{\mu\nu}$ for the electromagnetic field is obtained from the general expression (3.4.36) with the field Lagrangian L_F. The latter is given by the first part of the Lagrangian density (5.22) of Chapter 5 of Carmeli's and Malin's *Theory of Spinors* (see Suggested References), namely,

$$L_F = -\frac{1}{16\pi} g^{\alpha\mu} g^{\beta\nu} f_{\alpha\beta} f_{\mu\nu}. \tag{3.4.38}$$

It can easily be shown to be given by

$$T_{\rho\sigma} = \frac{1}{4\pi} \left(\frac{1}{4} g_{\rho\sigma} f_{\alpha\beta} f^{\alpha\beta} - f_{\rho\alpha} f_{\sigma}{}^{\alpha} \right). \tag{3.4.39}$$

If we calculate the trace of the energy-momentum tensor (3.4.39), we find that it vanishes,

$$T = T_{\rho}{}^{\rho} = g^{\rho\sigma} T_{\rho\sigma} = 0. \tag{3.4.40}$$

Using now $R = -\kappa T$ then leads to the vanishing of the Ricci scalar curvature, $R = 0$. We therefore obtain

$$R_{\mu\nu} = \kappa T_{\mu\nu} \tag{3.4.41}$$

for the Einstein field equations in the presence of an electromagnetic field. In Eq. (3.4.41) the energy-momentum tensor $T_{\mu\nu}$ is given by Eq. (3.4.39).

The Einstein field equations (3.4.41) and the Maxwell's equations (5.28) (Chapter 5 of Carmeli and Malin's *Theory of Spinors*) constitute *the coupled Einstein-Maxwell field equations.*

In the next section an exact solution of the Einstein field equations, the Schwarzschild solution, is obtained.

3.5 The Schwarzschild Solution

In spite of the nonlinearity of the Einstein field equations, there are numerous exact solutions to these equations. Moreover, there are other solutions which are not exact but approximate. Exact solutions are usually obtained using special methods.

The simplest of all exact solutions to Einstein's field equations is that of Schwarzschild.[2] The solution is *spherically symmetric* and *static*. Such a field can be produced by a spherically symmetric distribution and motion

---

[2]Karl Schwarzschild (Born: 9 Oct 1873 in Frankfurt am Main, Germany; Died: 11 May 1916 in Potsdam, Germany)

Karl Schwarzschild's parents were Henrietta Sabel and Moses Martin Schwarzschild. The family was Jewish, with Karl's father being a well-off member of the business community in Frankfurt. Karl had five younger brothers and one sister. His family were cultured people with interests mainly in art and music. He was the first member of his family to become a scientist.

Karl attended a Jewish primary school in Frankfurt and at the age of eleven he entered the Gymnasium. At this stage he became interested in astronomy and saved his pocket money to buy himself materials such as lens from which he could construct a telescope. Karl's father was friendly with Professor J. Epstein who had his own private observatory. Professor Epstein's son Paul and Karl shared an interest in astronomy, and Karl learnt how to use a telescope and also learnt some advanced mathematics from his friend Paul Epstein. At the age of sixteen Karl wrote his first two papers on the theory of orbits of double stars, which were published in "Astronomische Nachrichten" in 1890.

At the University of Strasbourg during the years 1891-93 Schwarzschild learnt a great deal of practical astronomy, then he obtained his doctorate at the University of Munich. His dissertation, on an application of Poincaré's theory of stable configurations of rotating bodies to tidal deformation of moons and to Laplace's origin of the solar system, was supervised by Hugo von Seeliger. In October 1896 Schwarzschild became an assistant at the Von Kuffner Observatory in Vienna. He held this appointment until June 1899. At the Observatory he worked on ways to determine the apparent brightness of stars using photographic plates.

In June 1899 he became a Privatdozent at the University of Munich, having submitted his work on measuring stellar magnitudes as his habilitation thesis "Beiträge zur photographischen Photometrie der Gestirne." This work led him to make several important discoveries. He saw that the photographic magnitudes which he measured differed from the visual magnitudes which had been tabulated and ascribed the difference to different colors of the stars. For variable stars he noticed that the range of magnitude change measured by his photographic methods was much greater than the range of change in visual magnitude, and he realized that this was due to changes in surface temperature of the variable star through its cycle.

In 1900 Schwarzschild published a paper giving a lower limit for the radius of curvature of space as 2500 light years. He also worked on radiation pressure from the sun and calculated the size of the particles in comet tails. He knew that radiation pressure had to overcome gravitation, and he also knew that the particles did not scatter light. This allowed him to deduce that the diameters of the particles had to be between 0.07 and 1.5 microns.

From 1901 until 1909 he was extraordinary professor at Göttingen and also director of the Observatory there. In Göttingen he collaborated with Klein, Hilbert and Minkowski. In less than a year he had been promoted to Ordinary Professor. Being a man of "wide interests in all branches of mathematics and physics," Schwarzschild published on electrodynamics and geometrical optics; he carried out a large survey of stellar magnitudes while at the Göttingen Observatory, publishing Aktinometrie (the first part in 1910, the second in 1912). In 1906 he studied the transport of energy through a star by radiation and published an important paper on radiative equilibrium of the atmosphere of the sun.

On 22 October 1909 he married Else Posenbach, the daughter of a professor of surgery at Göttingen. They had three children, Agathe, Martin, who became a professor of astronomy at Princeton, and Alfred. After his marriage, near the end of 1909, Schwarzschild left Göttingen to take up an appointment as director of the Astrophysical Observatory in Potsdam. He had the opportunity to study photographs of the return of Halley's comet in 1910 taken by a Potsdam expedition to Tenerife. He also made major

of matter. It thus follows that the requirement of spherical symmetry alone is sufficient to yield a static solution.

The spherical symmetry of the metric means that the expression for the interval $ds = \sqrt{g_{\mu\nu}dx^\mu dx^\nu}$ must be the same for all points located at the same distance from the center. In the flat space of an expanding Universe the distance is equal to the radius vector, and the metric is given by (c is

---

contributions to spectroscopy.

In 1913 Schwarzschild was elected to the Berlin Academy. In his admission speech he gave a good indication of his attitude toward science:

"Mathematics, physics, chemistry, astronomy, march in one front. Whichever lags behind is drawn after. Whichever hastens ahead helps on the others. The closest solidarity between astronomy and the whole circle of exact science. ... from this aspect I may count it well that my interest has never been limited to the things beyond the moon, but has followed the threads which spin themselves from there to our sublunar knowledge; I have often been untrue to the heavens. That is an impulse to the universal which was strengthened unwittingly by my teacher Seeliger, and afterwards was further nourished by Felix Klein and the whole scientific circle at Göttingen. There the motto runs that mathematics, physics, and astronomy constitute one knowledge, which, like the Greek culture, is only comprehended as a perfect whole."

On the outbreak of war in August 1914 Schwarzschild volunteered for military service. In Belgium he was put in charge of a weather station, in France where he was given the task of calculating missile trajectories, and then he went to Russia.

In Russia he wrote two papers on Einstein's relativity theory and one on Planck's quantum theory. The latter paper explained that the Stark effect (splitting of the spectral lines of hydrogen by an electric field, the amount being proportional to the field strength) could be proved from the postulates of quantum theory.

Schwarzschild's relativity papers give the first exact solution of Einstein's general gravitational equations, giving an understanding of the geometry of space near a point mass. He sent the first paper to Einstein who replied: "I had not expected that one could formulate the exact solution of the problem in such a simple way".

The work presented in these two papers formed the basis for a later study of black holes, showing that bodies of sufficiently large mass would have an escape velocity exceeding the speed of light and so could not be seen. However, Schwarzschild himself makes clear that he believes that the theoretical solution is physically meaningless, so making it very clear that he did not believe in the physical reality of black holes.

In Russia he contracted a rare autoimmune blistering disease of the skin. In Schwarzschild's time there was no known treatment and, after being invalided home in March 1916, he died two months later.

During his lifetime Schwarzschild was elected to the Scientific Society of Göttingen in 1905, the Royal Astronomical Society of London on 11 June 1909, and the German Academy of Sciences in 1913. He also received posthumous honours: in particular an observatory, founded in 1960 in Tautenburg as an affiliated Institute of the German Academy of Sciences, was named after him. The dedication described him as "the greatest German astronomer of the last hundred years." The German Astronomical Society established a special lectureship in his honor in 1959 and a Karl Schwarzschild Medal. The first recipient was his son Martin Schwarzschild.

The above report on Karl Schwarzschild is based on the article by J J O'Connor and E F Robertson.

taken as equal to 1):

$$ds^2 = dt^2 - dr^2 - r^2 \left(d\theta^2 + \sin^2\theta d\phi^2\right). \qquad (3.5.1)$$

In a non-Euclidean space, such as the Riemannian space that we have in the presence of a gravitational field, there is no quantity which has all the properties of the flat space radius vector, such as that it is equal both to the distance from the center and to the length of the circumference divided by 2π. Therefore, the choice of a radius vector is here arbitrary.

When a mass with spherical symmetry is introduced at the origin, the flat space line element (3.5.1) must be modified but in a way that retains spherical symmetry. The most general spherically symmetric expression for ds^2 is

$$ds^2 =$$

$$a\left(r,t\right)dt^2 + b\left(r,t\right)dr^2 + c\left(r,t\right)drdt + d\left(r,t\right)\left(d\theta^2 + \sin^2\theta d\phi^2\right). \qquad (3.5.2)$$

Because of the arbitrariness in the choice of the coordinate system in general relativity theory, we can perform a coordinate transformation which does not destroy the spherical symmetry of ds^2. Hence we can choose new coordinates r' and t' given by some functions $r' = r'(r,t)$ and $t' = t'(r,t)$.

Making use of these transformations, we can choose the new coordinates so that the coefficient $c(r,t)$ of the mixed term $drdt$ vanishes and the coefficient $d(r,t)$ of the angular part is $-r'^2$, in the metric (3.5.2). The latter condition implies that the radius vector is now defined in such a way that the circumference of a circle whose center is at the origin of the coordinates is equal to $2\pi r$. It is convenient to express the functions $a(r,t)$ and $b(r,t)$ in exponential forms, e^ν and $-e^\lambda$, respectively, where ν and λ are functions of the new coordinates r' and t'. Consequently, the line element (3.5.2) will have the form

$$ds^2 = e^\nu dt^2 - e^\lambda dr^2 - r^2\left(d\theta^2 + \sin^2\theta d\phi^2\right), \qquad (3.5.3)$$

where, for brevity, we have dropped the primes from the new coordinates r' and t', and the speed of light in vacuum c is taken as equal to 1.

We now denote the coordinates t, r, θ, ϕ by x^0, x^1, x^2, x^3, respectively. Hence the components of the covariant metric tensor are given by:

$$g_{\mu\nu} = \begin{pmatrix} e^\nu & 0 & 0 & 0 \\ 0 & -e^\lambda & 0 & 0 \\ 0 & 0 & -r^2 & 0 \\ 0 & 0 & 0 & -r^2\sin^2\theta \end{pmatrix}, \qquad (3.5.4a)$$

whereas those of the contravariant metric tensor are:

$$g^{\mu\nu} = \begin{pmatrix} e^{-\nu} & 0 & 0 & 0 \\ 0 & -e^{-\lambda} & 0 & 0 \\ 0 & 0 & -r^{-2} & 0 \\ 0 & 0 & 0 & -r^{-2}\sin^{-2}\theta \end{pmatrix}. \tag{3.5.4b}$$

To find out the differential equations that the functions ν and λ have to satisfy, according to Einstein's field equations, we first need to calculate the Christoffel symbols associated with the metric (3.5.4). The nonvanishing components are:

$$\Gamma^0_{00} = \frac{\dot{\nu}}{2}, \qquad \Gamma^0_{10} = \frac{\nu'}{2}, \qquad \Gamma^0_{11} = \frac{\dot{\lambda}}{2}e^{\lambda-\nu}, \tag{3.5.5a}$$

$$\Gamma^1_{00} = \frac{\nu'}{2}e^{\nu-\lambda}, \qquad \Gamma^1_{10} = \frac{\dot{\lambda}}{2}, \qquad \Gamma^1_{11} = \frac{\lambda'}{2}, \tag{3.5.5b}$$

$$\Gamma^1_{22} = -re^{-\lambda}, \qquad \Gamma^1_{33} = -r\sin^2\theta\, e^{-\lambda}, \qquad \Gamma^2_{12} = \frac{1}{r}, \tag{3.5.5c}$$

$$\Gamma^2_{33} = -\sin\theta\cos\theta, \qquad \Gamma^3_{13} = \frac{1}{r}, \qquad \Gamma^3_{23} = \cot\theta, \tag{3.5.5d}$$

where dots and primes denote differentiation with respect to t and r, respectively.

With these Christoffel symbols, we compute the following expressions for the nonvanishing components of the Einstein tensor:

$$G_0{}^0 = -e^{-\lambda}\left(\frac{1}{r^2} - \frac{\lambda'}{r}\right) + \frac{1}{r^2} = \kappa T_0{}^0, \tag{3.5.6a}$$

$$G_0{}^1 = -e^{-\lambda}\frac{\dot{\lambda}}{r} = \kappa T_0{}^1, \tag{3.5.6b}$$

$$G_1{}^1 = -e^{-\lambda}\left(\frac{\nu'}{r} + \frac{1}{r^2}\right) + \frac{1}{r^2} = \kappa T_1{}^1, \tag{3.5.6c}$$

$$G_2{}^2 = -\frac{1}{2}e^{-\lambda}\left(\nu'' + \frac{\nu'^2}{2} + \frac{\nu'-\lambda'}{r} - \frac{\nu'\lambda'}{2}\right)$$

$$+ \frac{1}{2}e^{-\nu}\left(\ddot{\lambda} + \frac{\dot{\lambda}^2}{2} - \frac{\dot{\lambda}\dot{\nu}}{2}\right) = \kappa T_2{}^2, \tag{3.5.6d}$$

$$G_3{}^3 = G_2{}^2 = \kappa T_3{}^3. \tag{3.5.6e}$$

All other components vanish identically.

The gravitational field equations can now be integrated exactly for the spherical symmetric field in vacuum, i.e., outside the masses producing the field. Setting Eqs. (3.5.6) equal to zero leads to the independent equations:

$$e^{-\lambda}\left(\frac{\nu'}{r}+\frac{1}{r^2}\right)-\frac{1}{r^2}=0, \qquad (3.5.7a)$$

$$e^{-\lambda}\left(\frac{\lambda'}{r}-\frac{1}{r^2}\right)+\frac{1}{r^2}=0, \qquad (3.5.7b)$$

$$\dot{\lambda}=0. \qquad (3.5.7c)$$

From Eq. (3.5.7a) and (3.5.7b) we find $\nu'+\lambda'=0$, so that $\nu+\lambda=f(t)$, where $f(t)$ is a function of t only. If we perform now the coordinate transformation $x^0=h\left(x'^0\right)$, $x^k=x'^k$, then $g'_{00}=h^2 g_{00}$. Such a transformation amounts to adding an arbitrary function of time to the function ν, while leaving the other components of the metric unaffected. Hence we can choose the function h so that $\nu+\lambda=0$. Consequently, we see, by Eq. (3.5.7c), that both ν and λ are time-independent. In other words, the spherically symmetric gravitational field in vacuum is automatically static.

Equation (3.5.7b) can now be integrated. It gives:

$$e^{-\lambda}=e^{\nu}=1-\frac{K}{r}, \qquad (3.5.8)$$

where K is an integration constant. We see that for $r\to\infty$, $e^{-\lambda}=e^{\nu}=1$, i.e., far from the gravitational bodies, the metric reduces to that of the flat space (3.5.1). The constant K can easily be determined from the requirement that Newton's law of motion be obtained at large distances from the central mass. From the geodesic equation it follows that the radial acceleration of a small test mass at rest with respect to the central mass is (see Problem 3.9.11):

$$-\Gamma^1_{00}=-\frac{1}{2}\left(1-\frac{K}{r}\right)\frac{K}{r^2}\to-\frac{K}{2r^2}. \qquad (3.5.9)$$

Comparing this expression with the Newtonian value $-Gm/r^2$ gives $K=2Gm$, where m is the central mass and G is the Newton constant.

The constant $2Gm$, or $2Gm/c^2$ in units where c is not taken as equal to 1, is often called the *Schwarzschild radius* of the mass m. For example, the Schwarzschild radius for the Sun is 2.95 km, that for the Earth is 8.9 mm, and that for an electron is 13.5×10^{-56} cm.

We therefore obtain for the spherically symmetric metric the form:

$$g_{\mu\nu} = \begin{pmatrix} 1 - 2Gm/r & 0 & 0 & 0 \\ 0 & -(1 - 2Gm/r)^{-1} & 0 & 0 \\ 0 & 0 & -r^2 & 0 \\ 0 & 0 & 0 & -r^2 \sin^2\theta \end{pmatrix}. \quad (3.5.10)$$

It is known as the *Schwarzschild solution* and describes the most general spherically symmetric solution of the Einstein field equations in a region of space where the energy-momentum tensor $T^{\mu\nu}$ vanishes. Although $g_{\mu\nu}$ goes to the flat space metric when r goes to infinity, it was *not* necessary to require this asymptotic behavior to obtain the solution.

It is worth mentioning that all spherically symmetric solutions of the Einstein field equations in vacuum, which satisfy the boundary conditions at infinity mentioned above, are equivalent to the Schwarzschild field, i.e., their time-dependence can be eliminated by a suitable coordinate transformation. This result is due to Birkhoff (see Figures 3.5.1 and 3.5.2).

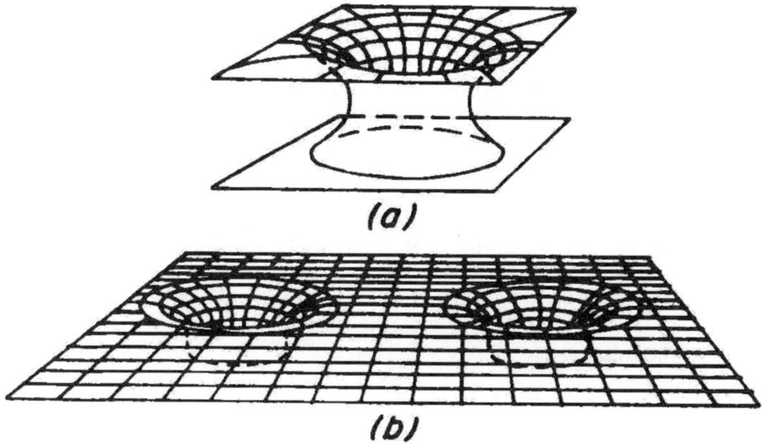

Fig. 3.5.1: Two interpretations of the three-dimensional "maximally extended Schwarzschild metric" at time $t = 0$. (a) A connection or bridge in the sense of Einstein and Rosen between two otherwise Euclidean spaces. (b) A wormhole in the sense of Wheeler connecting two regions in *one* Euclidean space, in the limiting case where these regions are extremely far apart compared to the dimensions of the throat of the wormhole.

Finally, it is convenient to introduce Cartesian coordinates by means of

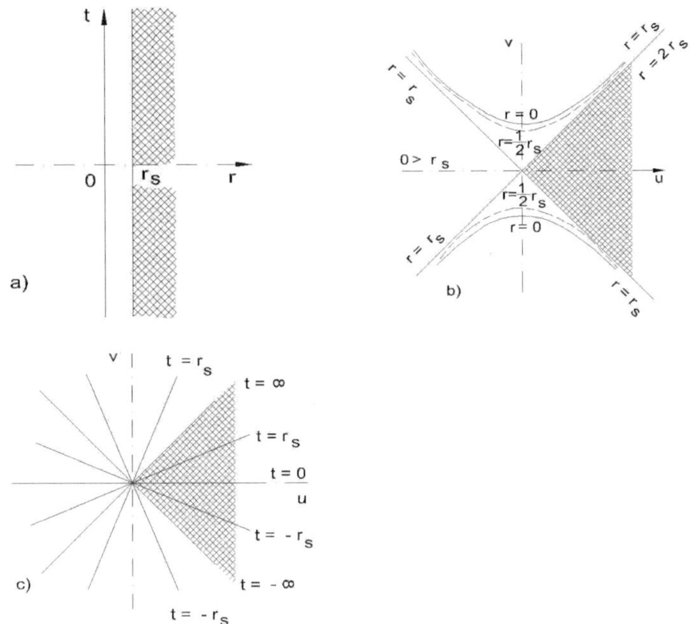

Fig. 3.5.2:   Kruskal diagram. Corresponding regions of the (r, t) and (u, v) planes. In the latter, curves of constant r are hyperbolas asymptotic to the lines $r = r_s$, while t is constant on straight lines through the origin. The exterior of the singular sphere $r > r_s$ corresponds to the region $|v| < u$ (hatched areas). The whole line $r = r_s$ in the (r, t) plane corresponds to the origin $u = v = 0$, while two one-dimensional families of ideal limit points with $r \to r_s$ and $t \to \pm\infty$ correspond to the remaining boundary points $u = |v| > 0$. In the (u, v) plane the metric is entirely regular not only in the hatched area, but in the entire area between the two branches of the hyperbola $r = 0$. This comprises two images of the exterior of the spherical singularity and two of its interior. The purely radial $(d\theta = d\phi = 0)$ null geodesics are lines inclined at $45°$. The points with $r = r_s$ have no local topological distinction, but rather a global one: if a test particle crosses $r = r_s$ into the interior (where r is timelike), it can never get back out, but must inevitably hit the irremovable singularity $r = 0$.

the coordinate transformation

$$\begin{aligned}
x^1 &= r \sin\theta \cos\phi, \\
x^2 &= r \sin\theta \sin\phi, \\
x^3 &= r \cos\theta.
\end{aligned} \tag{3.5.11}$$

In terms of these coordinates, the Schwarzschild metric (3.5.10) will then have the form

$$g_{00} = 1 - \frac{2Gm}{r},$$

$$g_{0r} = 0, \tag{3.5.12}$$

$$g_{rs} = -\delta_{rs} - \frac{2Gm/r}{1 - 2Gm/r} \frac{x^r x^s}{r^2}.$$

In the next section it is shown how the general relativity theory is verified experimentally.

3.6 Experimental Tests of General Relativity

Up to a few years ago, general relativity was verified by three tests: the gravitational redshift, the deflection of light near massive bodies and the planetary orbit effect on the planets. The first phenomenon could also be explained, in fact, without the use of the Einstein field equations. However, this picture has been changed.

3.6.1 *The gravitational redshift*

Consider the clocks at rest at two points 1 and 2. The rate of change of times at these points are then given by $ds(1) = \sqrt{g_{00}(1)}dt$ and $ds(2) = \sqrt{g_{00}(2)}dt$. The relation between the rates of identical clocks in a gravitational field is therefore given by $\sqrt{g_{00}(2)/g_{00}(1)}$. The frequency of an atom, ν_0, located at point 1, when it is seen by an observer at point 2 is, hence, given by

$$\nu = \nu_0 \sqrt{\frac{g_{00}(1)}{g_{00}(2)}}. \tag{3.6.1}$$

For a gravitational field like that of Schwarzschild, one therefore obtains for the frequency shift per unit frequency:

$$\frac{\Delta\nu}{\nu_0} = \frac{\nu - \nu_0}{\nu_0} \approx -\frac{Gm}{c^2}\left(\frac{1}{r_1} - \frac{1}{r_2}\right), \tag{3.6.2}$$

to first order in $Gm/c^2 r$. If we take r_1 to be the observed radius of the Sun and r_2 the radius of the Earth's orbit around the Sun (thus neglecting completely the Earth's gravitational field), then

$$\frac{\Delta\nu}{\nu_0} = -\frac{GM_\odot}{c^2 R_\odot} = -2.12 \times 10^{-6}.$$

This frequency shift is usually referred to as the *gravitational redshift*.

The gravitational redshift was tested for the Sun and for white dwarfs, and it was suggested that to be tested by atomic clocks. The redshift was also observed directly using the Mössbauer effect by Pound and Rebka, and by Cranshaw, Schiffer and Whitehead. The latter employed Fe^{57} and a total height difference of 12.5 metres. A redshift 0.96 ± 0.45 times the predicted value was observed by them. Pound and Rebka's result is more precise. They obtained a redshift 1.05 ± 0.10 times the predicted value.

3.6.2 *Effects on planetary motion*

One assumes that test particles move along geodesics in the gravitational field (see next section), and that planets have small masses as compared with the mass of the Sun, thus behaving like test particles. Consequently, to find the equation of motion of a planet moving in the gravitational field of the Sun one has to write the geodesic equation in the Schwarzschild field. In fact one does not need the exact solution (3.5.12) but rather its first approximation,

$$g_{00} = 1 - \frac{2Gm}{r},$$
$$g_{0r} = 0, \qquad\qquad\qquad (3.6.3)$$
$$g_{rs} = -\delta_{rs} - 2Gm\frac{x^r x^s}{r^3}.$$

In the above equations the speed of light is taken as unity.

Using the approximate metric (3.6.3) in the geodesic equation (3.1.43) gives (see Problem 3.9.15)

$$\ddot{\mathbf{x}} - Gm\nabla\frac{1}{r}$$

$$= Gm\left\{2\left(\dot{\mathbf{x}}^2\right)\nabla\frac{1}{r} - 2Gm\frac{1}{r}\nabla\frac{1}{r} - 2\left(\dot{\mathbf{x}}\cdot\nabla\frac{1}{r}\right)\dot{\mathbf{x}} + \frac{3}{r^5}\left(\mathbf{x}\cdot\dot{\mathbf{x}}\right)^2\mathbf{x}\right\}, \quad (3.6.4)$$

where we have used three-dimensional notation, and a dot denotes differentiation with respect to t. Multiplying Eq. (3.6.4), vectorially, by the radius vector $\mathbf{x}$ gives

$$\mathbf{x}\times\ddot{\mathbf{x}} = -2Gm\left(\dot{\mathbf{x}}\cdot\nabla\frac{1}{r}\right)\left(\mathbf{x}\times\dot{\mathbf{x}}\right), \qquad (3.6.5)$$

thus leading to the first integral

$$\mathbf{x}\times\dot{\mathbf{x}} = \mathbf{J}e^{-2Gm/r}, \qquad\qquad (3.6.6)$$

where $\mathbf{J}$ is a *constant* vector, the angular momentum per mass unit.

Hence the radius vector $\mathbf{x}$ moves in a plane perpendicular to the vector $\mathbf{J}$, as in Newtonian mechanics. Introducing polar coordinates r, ϕ in this plane to describe the motion of the planet, the equation of motion (3.6.4), consequently, decomposes into

$$\ddot{r} - r\dot{\phi}^2 + \frac{Gm}{r^2} = \frac{Gm}{r^2}\left\{3\dot{r}^2 - 2r^2\dot{\phi}^2 + 2\frac{Gm}{r}\right\}, \qquad (3.6.7a)$$

$$r^2\dot{\phi} = Je^{-2Gm/r}, \qquad (3.6.7b)$$

where J is the magnitude of the vector $\mathbf{J}$.

Introducing now the new variable $u = 1/r$, one can rewrite Eqs. (3.6.7) in terms of $u\,(\phi)$:

$$u'' + u - \frac{Gm}{J^2} = Gm\left(-u'^2 + 2u^2 + 2\frac{Gm}{J^2}u\right). \qquad (3.6.8)$$

Here a prime denotes a derivative with respect to the angle ϕ.

Let us try a solution of the form

$$u = b\,(1 + \epsilon\cos\alpha\phi). \qquad (3.6.9)$$

Here ϵ is the eccentricity, and α is some parameter to be determined, whose value in the usual nonrelativistic mechanics is unity. The other constant b is related to J in the nonrelativistic mechanics by $Gm/J^2 = b$. Using the above solution in Eq. (3.6.8) and equating coefficients of $\cos\alpha\phi$ gives (see Figures 3.6.1 and 3.6.2)

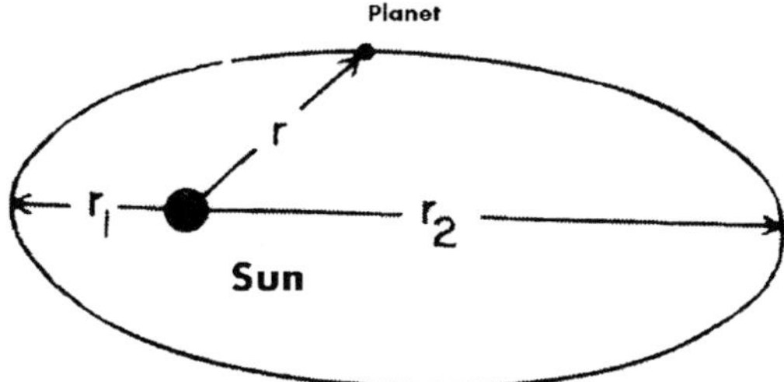

Fig. 3.6.1:   Newtonian limit of planetary motion. The motion is described by a closed ellipse if the effect of other planets is completely neglected.

$$\alpha^2 = 1 - 2Gm\left(2b + \frac{Gm}{J^2}\right). \qquad (3.6.10)$$

Substituting for Gm/J^2 its nonrelativistic value b then gives $\alpha^2 = 1 - 6Gmb$, or, to a first approximation in Gm,

$$\alpha = 1 - 3Gmb. \qquad (3.6.11)$$

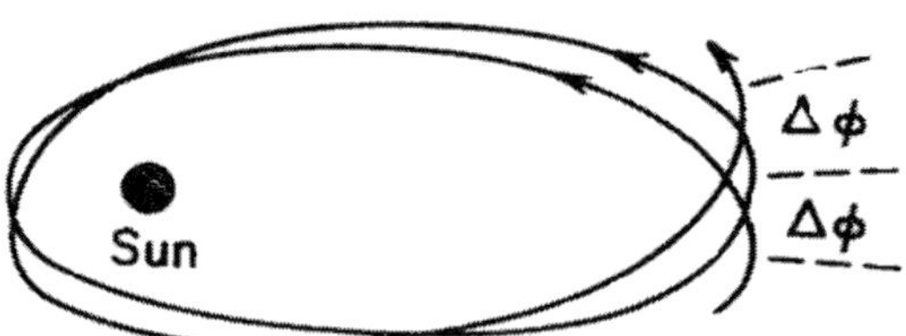

Fig. 3.6.2: Planetary elliptic orbit with perihelion advance. The effect is a general relativistic one. The advance of the perihelion is given by $\Delta\phi$ in radians per revolution, where $\Delta\phi = 6\pi GM/c^2 a(1 - \epsilon^2)$, with M being the mass of the Sun, a the semimajor axis, and ϵ the eccentricity of the orbit of the planet.

Successive perihelia occur when

$$(1 - 3Gmb)\,(2\pi + \Delta\phi) = 2\pi. \tag{3.6.12}$$

Consequently, there will be an advance in the perihelion of the orbit per revolution given by $\Delta\phi = 6\pi Gmb$, or $\Delta\phi = 6\pi Gm/a\,(1 - \epsilon^2)$ if we use the nonrelativistic value of the constant b, where a is the semimajor axis of the orbit. Reinstating now c, the speed of light, finally gives for the perihelion advance

$$\Delta\phi = \frac{6\pi Gm}{c^2 a\,(1 - \epsilon^2)}, \tag{3.6.13}$$

in radians per revolution (see Figures 3.6.1 and 3.6.2).

We list below the calculated values of $\Delta\phi$ per century for four planets:

Planet	$\Delta\phi$
Mercury	$43.03''$
Venus	$8.60''$
Earth	$3.80''$
Mars	$1.35''$

The astronomical observations for the planet Mercury give 43.11 ± 0.45 sec per century, in good agreement with the calculated value.

3.6.3 *The deflection of light*

To discuss the deflection of light in the gravitational field we must again solve the geodesic equation, but now with the null conditions $ds = 0$. Using the approximate solution (3.6.3) then gives for $g_{\mu\nu}dx^\mu dx^\nu = 0$

$$\dot{\mathbf{x}} \cdot \dot{\mathbf{x}} + \frac{2Gm}{r^3}\left(\mathbf{x} \cdot \dot{\mathbf{x}}\right)^2 = 1 - \frac{2Gm}{r}. \tag{3.6.14a}$$

Divide now both sides of Eq. (3.6.14a) by $(1 - \frac{2Gm}{r})$, and using

$$\left(1 - \frac{2Gm}{r}\right)^{-1} \approx \left(1 + \frac{2Gm}{r}\right),$$

we obtain

$$\left(1 + \frac{2Gm}{r}\right)\left[\dot{\mathbf{x}} \cdot \dot{\mathbf{x}} + \frac{2Gm}{r^3}\left(\mathbf{x} \cdot \dot{\mathbf{x}}\right)^2\right] = 1. \tag{3.6.14b}$$

Using polar coordinates r, ϕ, by means of the transformation $x = r\cos\phi$, $y = r\sin\phi$, consequently, gives to the first approximation in Gm

$$\dot{r}^2 + r^2\dot{\phi}^2 + \frac{4Gm\dot{r}^2}{r} + 2Gmr\dot{\phi}^2 = 1. \tag{3.6.15}$$

Again changing variables into $u(\phi) = 1/r$, and using Eq. (3.6.7b), gives

$$u'^2 + u^2 + 2Gmu\left(2u'^2 + u^2\right) = J^{-2}e^{4Gmu}. \tag{3.6.16}$$

Differentiation of this equation with respect to ϕ gives

$$u'' + u + Gm\left(2u'^2 + 4uu'' + 3u^2\right) = 2GmJ^{-2}, \tag{3.6.17}$$

to the first approximation in Gm.

To solve Eq. (3.6.17) we note that in the lowest approximation one has

$$u'^2 \approx J^{-2} - u^2, \tag{3.6.18}$$

$$u'' \approx -u. \tag{3.6.19}$$

Using these values in Eq. (3.6.17) gives

$$u'' + u = 3Gmu^2, \tag{3.6.20}$$

for the orbit of the light ray. In the lowest approximation u satisfies $u'' + u = 0$, whose solution is a straight line

$$\frac{1}{r} = u = \frac{\sin\phi}{R}, \tag{3.6.21}$$

where R is a constant. This shows that $r = 1/u$ has a minimum value R at $\phi = \pi/2$. Substituting into the right-hand side of Eq. (3.6.20) then gives

$$u'' + u = 3\frac{Gm}{R^2}\sin^2\phi. \tag{3.6.22}$$

The solution of this equation is

$$u = \frac{\sin\phi}{R} + \frac{Gm}{R^2}\left(1 + \cos^2\phi\right). \tag{3.6.23}$$

Introducing now Cartesian coordinates $x = r\cos\phi$ and $y = r\sin\phi$, the above equation gives

$$y = R - \frac{Gm}{R}\frac{2x^2 + y^2}{\sqrt{x^2 + y^2}}. \tag{3.6.24}$$

For large values of $|\,x\,|$ this equation becomes (see Figures 3.6.3 and 3.6.4)

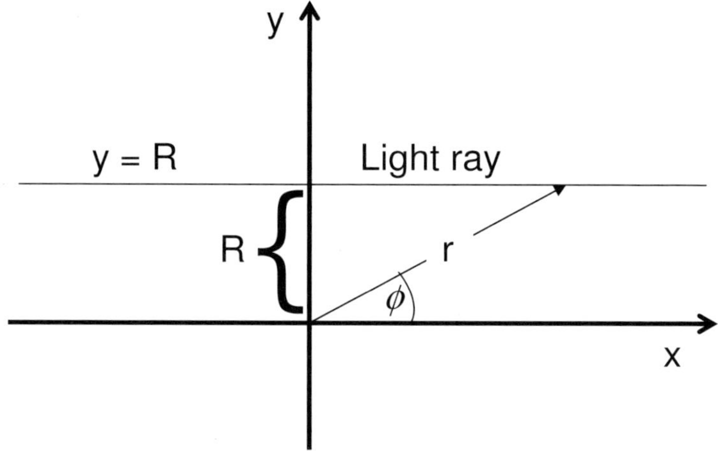

Fig. 3.6.3:   Light ray when the effect of the central body's gravitational field is completely neglected. The light ray then moves along the straight line $y = r\sin\phi = R =$ constant, namely, $u = 1/r = (1/R)\sin\phi$.

$$y \approx R - \frac{2Gm}{R}\,|\,x\,|. \tag{3.6.25}$$

Hence, asymptotically, the orbit of the light ray is a straight line in space. This result is expected, since far away from the central mass the space is flat. The angle $\Delta\phi$ between the two asymptotes is, however, equal to

$$\Delta\phi = 4\frac{Gm}{c^2 R}, \tag{3.6.26}$$

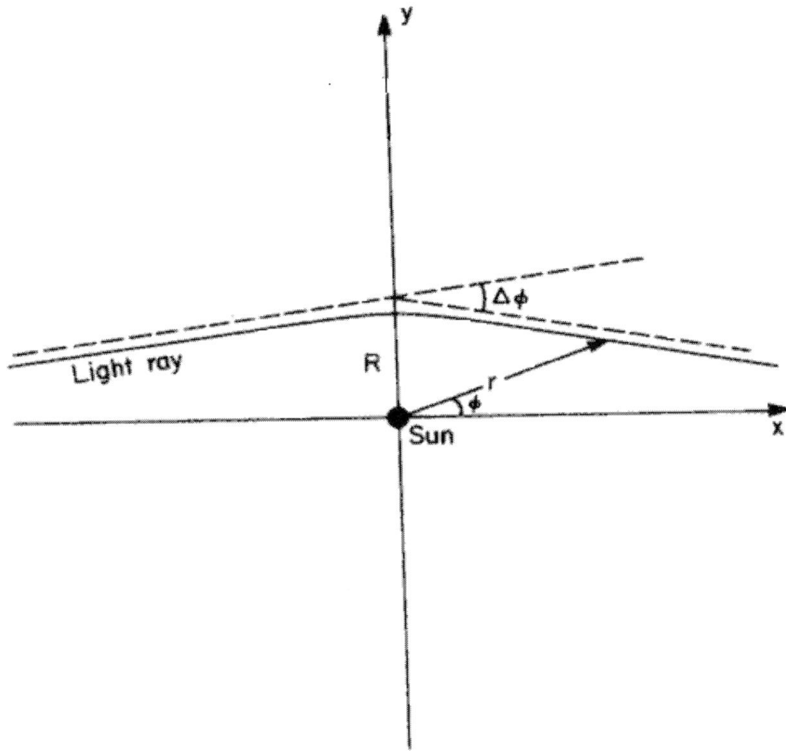

Fig. 3.6.4: Bending of a light ray in the gravitational field of a spherically symmetric body. The angle of deflection $\Delta\phi = 4GM/c^2 R$, where M is the mass of the central body and R is the closest distance of the light ray from the center of the body.

in units in which c is different from unity (see Figures 3.6.3 and 3.6.4).

The angle $\Delta\phi$ represents the *angle of deflection* of a light ray in passing through the Schwarzschild field. For a light ray just grazing the Sun Eq. (3.6.26) gives $\Delta\phi = 1.75$ sec. Observations indeed confirm this result; one of the latest results gives 1.75 ± 0.10 sec.

3.6.4 *Gravitational radiation experiments*

Weber developed methods to detect gravitational waves that Einstein's gravitational field equations predict. His experiment involved detectors at opposite ends of a 1000 km baseline. Sudden increases in the detector output were claimed by him, coincident within a resolution time of 25 msec. In 1969 he reported in *PRL* some two dozen coincident detections at the two locations in an 81-day period, and again in 1970 he reported many more, this time from the galactic center. Later his claims were discredited by others, but his imagination and determination inspired many to search for gravitational waves, a quest that has culminated in the construction of enormous optical interferometers such as the Laser Interferometer Gravitational Wave Observatory (LIGO) project.

Weber's apparatus measures the Fourier transform of the Riemann tensor. The method uses the fact that the distance η^μ between two neighboring test particles, which follow geodesics, satisfies the *geodesic deviation* equation

$$\frac{\delta^2 \eta^\mu}{\delta s^2} + R^\mu{}_{\alpha\nu\beta}\lambda^\alpha \eta^\nu \lambda^\beta = 0, \tag{3.6.27}$$

where λ^α is the tangent vector to one of the geodesics, and $\delta/\delta s = \lambda^\alpha \nabla_\alpha$ is a directional covariant derivative. Weber measured the strain of a large aluminum cylinder, having mass of the order 10^6 grams, by means of a piezoelectric crystal attached to the cylinder which transforms the mechanical movement into an electric current. The detector was developed to operate in the vicinity of 1662 Hz. A high frequency source was developed for dynamic gravitational fields and the detector was tested by doing a communication experiment with high frequency Coulomb fields.

3.6.5   *Radar experiment*

Shapiro designed a radar experiment to test general relativity by measuring the effect of solar gravity on time delays of round-trip travel times of radar pulses transmitted from the Earth toward an inner planet, i.e., Venus or Mercury. The experiment is based on the phenomenon that electromagnetic waves "slow down" in a gravitational field. Within the framework of general relativity there should be an anomalous delay of 200 microseconds in the arrival time of a radar echo from Mercury, positioned on the far side of the Sun near the limb.

For example, if we calculate the proper time τ at $r = r_2$ for a radial round-trip travel $r_2 \to r_1 \to r_2$, with $r_2 > r_1$, of a radar pulse in the Schwarzschild field, and subtract from τ the corresponding value τ_0 when the spherical mass $m = 0$, we find

$$\Delta\tau = \frac{4Gm}{c^3}\left(\ln\frac{r_2}{r_1} - \frac{r_2 - r_1}{r_2}\right) + O\left(m^2\right). \tag{3.6.28}$$

In general one finds

$$\Delta\tau \approx \frac{4Gm}{c^3}\left(\ln\frac{r_e + r_p + R}{r_e + r_p - R}\right), \tag{3.6.29}$$

where r_e is the Earth-Sun distance, r_p is the planet-Sun distance, and R is the Earth-planet distance.

Shapiro found that the retardation of radar signals are 1.02 ± 0.05 times the corresponding effect predicted by general relativity.

3.6.6 *Low-temperature experiments*

Schiff has proposed an experiment to check the equations of motion in general relativity by means of a gyroscope, which is forced to go around the Earth either in a stationary laboratory fixed to the Earth or in a satellite. The unique experiment is made possible by complete use of a low-temperature environment, and the properties of superconductors, including the use of zero magnetic fields and ultrasensitive magnetometry. Schiff has calculated, using results obtained by Papapetrou for the motion of spinning bodies in general relativity, that a perfect gyroscope subject to no torques will experience an anomalous precession with respect to the fixed stars as it travels around the Earth.

In the next section equations of motions are derived and the Einstein-Infeld-Hoffmann method is explained.

3.7 Equations of Motion

3.7.1 *The geodesic postulate*

In Section 3.1 it was assumed that the planet's motion around the Sun is described by the geodesic equation (3.1.43). The assumption that the equations of motion of a test particle, moving in gravitational field, are given by the geodesic equation is known as the *geodesic postulate* and was suggested by Einstein in his first article on the general theory of relativity.

Eleven years later Einstein and Grommer showed that the geodesic postulate need not be assumed, but that it rather follows from the gravitational field equations; this is a consequence of nonlinearity of the field equations along with the fact that they satisfy the four contracted Bianchi identities (see Section 3.1). The discovery of Einstein and Grommer is considered to be one of the most important achievements, and one of the most attractive features of the general theory of relativity. Later on Infeld and Schild showed that the equations of motion of a test particle are given by the geodesic equation in an *external gravitational* field. This result, however, does not differ from the geodesic postulate because, by definition, a test particle has no self-field.

3.7.2 *Equations of motion as a consequence of field equations*

In order to establish the relation between the Einstein field equations and the equations of motion one proceeds as follows. As we have seen in Section 3.4, it follows from the contracted Bianchi identities that the energy-momentum tensor $T^{\mu\nu}$ satisfies a generally covariant conservation law of the form given by Eq. (3.4.3). Consequently, one obtains for the energy-momentum tensor *density* $\mathcal{T}^{\mu\nu}$

$$\nabla_\nu \mathcal{T}^{\mu\nu} = \partial_\nu \mathcal{T}^{\mu\nu} + \Gamma^\mu_{\alpha\beta} \mathcal{T}^{\alpha\beta} = 0, \tag{3.7.1}$$

where $\mathcal{T}^{\mu\nu} \equiv \sqrt{-g} T^{\mu\nu}$.

For a system of N particles of finite masses, represented as singularities of the gravitational field, $\mathcal{T}^{\mu\nu}$ may be taken in the form

$$\mathcal{T}^{\mu\nu} = \sum_{A=1}^{N} m_A v_A^\mu v_A^\nu \delta_A \left(\mathbf{x} - \mathbf{z}_A \right). \tag{3.7.2}$$

Here z_A^μ are the coordinates of the Ath particle. (Capital Roman indices, $A, B, \cdots$, run from 1 to N. For these indices the summation convention will be suspended.) Also $v^\mu = \dot{z}^\mu = dz^\mu/dt$ $(v_A^0 = \dot{z}_A^0 = 1)$, and δ is the three-dimensional Dirac delta function satisfying the following conditions:

$$\delta \left(\mathbf{x} \right) = 0; \quad \text{for} \quad x \neq 0, \tag{3.7.3a}$$

$$\int \delta \left(\mathbf{x} - \mathbf{z} \right) d^3 x = 1, \tag{3.7.3b}$$

$$\int f \left(\mathbf{x} \right) \delta \left(\mathbf{x} - \mathbf{z} \right) d^3 x = f \left(\mathbf{z} \right), \tag{3.7.3c}$$

for any continuous function $f \left(\mathbf{x} \right)$ in the neighborhood of $\mathbf{z}$. In Eq. (3.7.2), m_A is a function of time which may be called the *inertial mass* of the Ath particle.

If we put the energy-momentum tensor density (3.7.2) into (3.7.1) and integrate over the three-dimensional region surrounding the first singularity, we obtain

$$\frac{dp^\mu}{dt} = \int F^\mu \delta \left(\mathbf{x} - \mathbf{z} \right) d^3 x, \tag{3.7.4}$$

where $p^\mu = mv^\mu$ and $F^\mu = -m\Gamma^\mu_{\alpha\beta} v^\alpha v^\beta$, and where we have put, for simplicity, $m = m_1$, $z^\mu = z_1^\mu$, $v^\mu = v_1^\mu$, and $\delta \left(\mathbf{x} - \mathbf{z} \right) = \delta_1 \left(\mathbf{x} - \mathbf{z}_1 \right)$.

3.7.3 *Self-action terms*

Equation (3.7.4) may be interpreted as an "exact equation of motion" of the first particle. However, since the Christoffel symbols are singular at the location of the particle, the equation contains infinite self-action terms. However, it was shown by Carmeli that these terms can be removed as follows.

Putting Eq. (3.7.2) into Eq. (3.7.1) we obtain

$$\partial_0 \left[\sum_{A=1}^{N} m_A v_A^\mu \delta_A \right] + \partial_n \left[\sum_{A=1}^{N} m_A v_A^\mu v_A^n \delta_A \right]$$

$$+ \sum_{A=1}^{N} m_A \Gamma^\mu_{\alpha\beta} v_A^\alpha v_A^\beta \delta_A = 0, \tag{3.7.5}$$

where Latin indices run from 1 to 3. The first term on the left-hand side of Eq. (3.7.1) can be written as

$$\partial_0 \left[\sum_{A=1}^{N} m_A v_A^\mu \delta_A \right] = \sum_{A=1}^{N} \partial_0 \left(m_A v_A^\mu \right) \delta_A + \sum_{A=1}^{N} m_A v_A^\mu \partial_0 \delta_A, \tag{3.7.6}$$

with

$$\partial_0 \delta_A = \partial_0 \delta_A \left(x^s - z_A^s \right) = -\partial_n \delta_A v_A^n. \tag{3.7.7}$$

Using the above results in Eq. (3.7.5), we obtain

$$\sum_{A=1}^{N} \left\{ \frac{d \left(m_A v_A^\mu \right)}{dt} + m_A \Gamma^\mu_{\alpha\beta} v_A^\alpha v_A^\beta \right\} \delta_A = 0. \tag{3.7.8}$$

Equation (3.7.8), which is identical with Eq. (3.7.1), is satisfied for any spacetime point, since otherwise the Bianchi identities or the Einstein field equations would not be satisfied.

We now examine the behavior of Eq. (3.7.8) in the infinitesimal neighborhood of the first singularity, which we assume not to contain any other singularity. In this region $\delta_B \left(\mathbf{x} - \mathbf{z}_B \right) = 0$ for $B = 2, 3, \cdots, N$. Hence Eq. (3.7.8) gives for the conservation law near the first singularity

$$\left\{ \frac{d \left(m v^\mu \right)}{dt} + m \Gamma^\mu_{\alpha\beta} v^\alpha v^\beta \right\} \delta \left(\mathbf{x} - \mathbf{z} \right) = 0. \tag{3.7.9}$$

Let us further assume that the Christoffel symbols near the first singularity can be expanded into a power series in the infinitesimal distance r, defined by $r^2 = (x^s - z^s)(x^s - z^s)$, where $z^s = z_1^s$, in the vicinity of the first particle. Then we have

$$\Gamma^{\mu}_{\alpha\beta} = {}_{-k}\Gamma^{\mu}_{\alpha\beta} + {}_{-k+1}\Gamma^{\mu}_{\alpha\beta} + \cdots + {}_0\Gamma^{\mu}_{\alpha\beta} + \cdots, \qquad (3.7.10)$$

where the indices written in subscripts on the left of a function indicate its behavior with respect to r, and k is a positive integer.

For example ${}_0\Gamma^{\mu}_{\alpha\beta}$ is the part of the Christoffel symbol which varies as r^0, i.e., is finite at the location of the first particle. When one uses spherical coordinates r, θ and ϕ, one can write

$$_{-k}\Gamma^{\mu}_{\alpha\beta} = \frac{1}{r^k} A^{\mu}_{\alpha\beta}(\theta, \phi), \qquad (3.7.11a)$$

$$_{-k+1}\Gamma^{\mu}_{\alpha\beta} = \frac{1}{r^{k-1}} B^{\mu}_{\alpha\beta}(\theta, \phi), \qquad (3.7.11b)$$

$$\cdot \quad \cdot \quad \cdot \quad \cdot \quad \cdot \quad \cdot \quad \cdot$$

$$_0\Gamma^{\mu}_{\alpha\beta} = D^{\mu}_{\alpha\beta}(\theta, \phi), \quad \text{etc.} \qquad (3.7.11c)$$

Terms like ${}_1\Gamma^{\mu}_{\alpha\beta}$, ${}_2\Gamma^{\mu}_{\alpha\beta}$, etc., however, need not be taken into account when one puts the above expansion into Eq. (3.7.9) since $r^j \delta(\mathbf{x} - \mathbf{z}) = 0$ for any positive integer j. If we denote now $m A^{\mu}_{\alpha\beta} v^{\alpha} v^{\beta}, \cdots$ by $A^{\mu}, \cdots$ we can write Eq. (3.7.9) in the form

$$\left\{ r^{-k} A^{\mu} + r^{-k+1} B^{\mu} + \cdots + r^{-1} C^{\mu} + D_1^{\mu} \right\} \delta(\mathbf{x} - \mathbf{z}) = 0, \qquad (3.7.12)$$

where we have used the notation $D_1^{\mu} = d(mv^{\mu})/dt + D^{\mu}$.

In order to get rid of terms proportional to negative powers of r in Eq. (3.7.12) we proceed as follows. Multiplying Eq. (3.7.12) by r^k and using $r^j \delta(\mathbf{x} - \mathbf{z}) = 0$ we obtain

$$A^{\mu}(\theta, \phi) \delta(\mathbf{r}) = 0, \qquad (3.7.13)$$

the integration of which over the three-dimensional region yields, using spherical coordinates,

$$\iint A^{\mu}(\theta, \phi) \sin\theta \, d\theta \, d\phi \int r^2 \delta(\mathbf{r}) = 0. \qquad (3.7.14)$$

From the property of the delta-function

$$\int \delta(\mathbf{r}) \, d^3x = \iint \sin\theta \, d\theta \, d\phi \int \delta(\mathbf{r}) \, r^2 dr = \frac{1}{4\pi}, \qquad (3.7.15)$$

one obtains $\int \delta(\mathbf{r}) \, r^2 dr = \frac{1}{4\pi}$. Hence we obtain

$$\iint A^{\mu}(\theta, \phi) \sin\theta \, d\theta \, d\phi = 0, \qquad (3.7.16)$$

independent of the value of the variable r. Thus the angular distribution of $A^\mu(\theta, \phi)$ is such that its average equals zero.

However, not only does the above equation hold, but also (s is any finite positive integer)

$$a(r) = \frac{1}{r^s} \int \int A^\mu(\theta, \phi) \sin\theta \, d\theta \, d\phi = 0, \tag{3.7.17}$$

for small values of r as well as when r tends to zero, as can be verified, for example, by using L'Hospital's theorem. It follows then that $a(r)$ is a function of r whose value is zero for any small r, including $r = 0$. Using the property of delta-function we obtain

$$\int r^2 \delta(\mathbf{r}) \, dr = \frac{1}{4\pi} f(0), \tag{3.7.18}$$

for any continuous function of r. Since $a(r)$ is certainly continuous, one obtains

$$\int r^2 \delta(\mathbf{r}) \, a(r) \, dr = 0. \tag{3.7.19}$$

Hence when one integrates Eq. (3.7.12) over the three-dimensional space, there will be no contribution from the first term.

In order to show that the second term of Eq. (3.7.12) will not contribute to the three-dimensional integration of the same equation either, we multiply it by r^{k-1}. We obtain now, after neglecting terms that do not contribute,

$$\left\{ \frac{1}{r} A^\mu(\theta, \phi) + B^\mu(\theta, \phi) \right\} \delta(\mathbf{r}) = 0. \tag{3.7.20}$$

Integration of this equation, again using spherical coordinates, shows that the first term will not contribute anything because of Eq. (3.7.19), and we are left with

$$\int \int B^\mu(\theta, \phi) \sin\theta \, d\theta \, d\phi \int r^2 \delta(\mathbf{r}) \, dr = 0. \tag{3.7.21}$$

Hence we have

$$\int \int B^\mu(\theta, \phi) \sin\theta \, d\theta \, d\phi = 0, \tag{3.7.22}$$

independent of r. From this equation one obtains another one, analogous to Eq. (3.7.19), but with B^μ instead of A^μ:

$$\int r^2 \delta(\mathbf{r}) \, b(r) \, dr = 0, \tag{3.7.23}$$

with

$$b\left(r\right) = \frac{1}{r^s} \int \int B^\mu\left(\theta, \phi\right) \sin\theta d\theta d\phi = 0. \tag{3.7.24}$$

Proceeding in this way, one verifies that the angular distribution of all functions A^μ, B^μ, etc., is such that they all satisfy equations like Eqs. (3.7.16) and (3.7.22). Hence it is clear that one obtains

$$\int D_1^\mu\left(\theta, \phi\right) \delta\left(\mathbf{r}\right) d^3x = 0, \tag{3.7.25}$$

which gives

$$\frac{dp^\mu}{dt} + mv^\alpha v^\beta \int {}_0\Gamma^\mu_{\alpha\beta}\delta\left(\mathbf{r}\right) d^3x = 0, \tag{3.7.26}$$

or equivalently

$$\dot{v}^k + v^\alpha v^\beta \int \left({}_0\Gamma^k_{\alpha\beta} - v^k{}_0\Gamma^0_{\alpha\beta}\right) \delta\left(\mathbf{r}\right) d^3x = 0. \tag{3.7.27}$$

Equation (3.7.27) is the "exact equation of motion".

3.7.4   *The Einstein-Infeld-Hoffmann method*

Now that we have found the law of motion (3.7.26), one can proceed to find the equation of motion of two finite masses, each moving in the field produced by both of them. In the following we find such an equation of motion in the case for which the particles' velocities are much smaller than the speed of light. Moreover, we will confine ourselves to an accuracy of post-Newtonian. This means the equation of motion obtained will contain the Newtonian equation as a limit, but is a first generalization of the latter. Such an equation was first obtained by Einstein, Infeld, and Hoffmann. To obtain this equation we solve the field equations and formulate the equations of motion explicitly by means of an approximation method, the Einstein-Infeld-Hoffmann (EIH) method, to be described below.

Let us assume a function ϕ developed in a power series in the parameter $\lambda = 1/c$, where c is the speed of light. One then has

$$\phi = {}_0\phi + {}_1\phi + {}_2\phi + \cdots. \tag{3.7.28}$$

The indices written as left subscripts indicate the order of λ absorbed by the ϕ's.

If a function $\phi\left(x\right)$ varies rapidly in space but slowly with x^0, then we are justified in not treating all its derivatives in the same manner. The derivatives with respect to x^0 will be of a higher order than the space derivatives. We thus write

$$\partial_0 \left({}_l\psi \right) = {}_{l+1}\psi. \tag{3.7.29}$$

That is, differentiation with respect to x^0 raises the order by one. Thus if the coordinates z^s of a particle are considered to be of order zero, $\dot{z}^s$ will be of order one, and $\ddot{z}^s$ of order two. Using now the Newtonian approximation mass $\times$ acceleration=mass $\times$ mass/(distance)^2, we see the mass is of order two. In all the power developments we take into account only even or only odd powers of $1/c$. (The expansion of the metric tensor, etc., in a power series in c^{-2} (such as $\phi = {}_0\phi + {}_2\phi + \cdots$, or $\phi = {}_1\phi + {}_3\phi + \cdots$) corresponds to the choice of the symmetric Green function, thus excluding radiation.)

Thus, because of the order with which we start m and $\dot{z}^s$, we have

$$\mathcal{T}^{00} = {}_2\,\mathcal{T}^{00} + {}_4\,\mathcal{T}^{00} + \cdots ,$$

$$\mathcal{T}^{0n} = {}_3\,\mathcal{T}^{0n} + {}_5\,\mathcal{T}^{0n} + \cdots , \tag{3.7.30}$$

$$\mathcal{T}^{mn} = {}_4\,\mathcal{T}^{mn} + {}_6\,\mathcal{T}^{mn} + \cdots .$$

As to the metric tensor, we write

$$g_{\mu\nu} = \eta_{\mu\nu} + h_{\mu\nu}, \qquad g^{\mu\nu} = \eta^{\mu\nu} + h^{\mu\nu}. \tag{3.7.31}$$

The gravitational field equations can be written as

$$\sqrt{-g}R_{\alpha\beta} = \kappa \left(\mathcal{T}_{\alpha\beta} - \frac{1}{2}g_{\alpha\beta}\mathcal{T} \right), \tag{3.7.32}$$

where $\mathcal{T} = \mathcal{T}_{\mu\nu}g^{\mu\nu}$, and $R_{\alpha\beta}$ is the Ricci tensor. From the right-hand side of the field equations it follows that R_{00} and R_{mn} (when $m = n$) start with order two, R_{mn} (when $m \neq n$) start with order four, while R_{0m} starts with order three. The lowest order expressions of the left-hand side are

$$R_{00} \approx \frac{1}{2}h_{00,ss},$$

$$R_{0m} \approx \frac{1}{2} \left(h_{0m,ss} - h_{0s,ms} - h_{ms,0s} + h_{ss,0m} \right), \tag{3.7.33}$$

$$R_{mn} \approx \frac{1}{2} \left(h_{mn,ss} - h_{ms,ns} - h_{ns,ms} - h_{00,mn} + h_{ss,mn} \right),$$

where a comma denotes a partial derivative, $\phi_{,s} = \partial_s\phi$. Hence we have

$$h_{00} = {}_2h_{00} + {}_4h_{00} + \cdots ,$$

$$h_{0m} = {}_3h_{0n} + {}_5h_{0n} + \cdots , \tag{3.7.34}$$

$$h_{mn} = {}_2h_{mn} + {}_4h_{mn} + \cdots .$$

3.7.5   *The Newtonian equation of motion*

We now find the equation of motion in the lowest (Newtonian) approximation. We do it in such a way as to make the generalization to the post-Newtonian approximation as simple as possible.

Because of Eqs. (3.7.32) and (3.7.33), the field equations of the lowest order are in h_{00},

$$\frac{1}{2}{}_2 h_{00,ss} = \kappa \left({}_2 T^{00} - \frac{1}{2}{}_2 T^{00} \right) = \frac{\kappa}{2}{}_2 T^{00} = \frac{\kappa}{2} \sum_{A=1}^{2} \mu_A \delta_A, \qquad (3.7.35)$$

where, for simplicity, we have put $\mu_A = {}_2 m_A$. Hence the equation obtained is

$$_2 h_{00,ss} = \kappa \sum_{A=1}^{2} \mu_A \delta_A. \qquad (3.7.36)$$

The solution of this equation that represents two masses is

$$_2 h_{00} = -2G \sum_{A=1}^{2} \mu_A r_A^{-1}, \qquad (3.7.37)$$

where $r_A^2 = (x^s - z_A^s)(x^s - z_A^s)$. Using ${}_2 h_{00}$ in the equation of motion (3.7.27), we obtain in the lowest (second) order for the equation of motion of the first particle

$$\ddot{z}_1^k - G \int \partial_k \left(\mu_2 r_2^{-1} \right) \delta \left(\mathbf{x} - \mathbf{z}_1 \right) d^3 x = 0. \qquad (3.7.38)$$

This gives

$$\ddot{z}_1^k = G \frac{\partial}{\partial z_1^k} \frac{\mu_2}{z}, \qquad (3.7.39)$$

where $z^2 = (z_1^s - z_2^s)(z_1^s - z_2^s)$. Equation (3.7.39) is, of course, the Newtonian equation of motion.

3.7.6   *The Einstein-Infeld-Hoffmann equation*

To find the equation of motion up to the fourth order, we must know besides ${}_2 h_{00}$ the functions ${}_4 h_{00}$, ${}_3 h_{0n}$ and ${}_2 h_{mn}$. The second and third functions are easy to find. The left-hand side of the corresponding equations is written out in Eq. (3.7.33), whereas the right-hand side is given by Eq. (3.7.32) and it is $-\kappa \sum \mu_A \dot{z}_A^m \delta_A$ for the $0m$ component, and $\frac{\kappa}{2} \delta_{mn} \sum \mu_A \delta_A$ for the mn component. Therefore, for the ${}_2 h_{mn}$ we have the equation

$$_2 h_{mn,ss} - {}_2 h_{ms,ns} - {}_2 h_{ns,ms} + {}_2 h_{ss,mn} - {}_2 h_{00,mn} = \delta_{mn}{}_2 h_{00,ss}, \qquad (3.7.40)$$

whose solution is

$$_2 h_{mn} = \delta_{mn}\, _2 h_{00}.\tag{3.7.41}$$

The equation for $_3 h_{0n}$ is

$$_3 h_{0n,ss} - _3 h_{0s,ns} - _2 h_{ns,0s} + _2 h_{ss,n0} = -2\kappa \sum \mu_A \dot{z}_A^n \delta_A.\tag{3.7.42}$$

Using the value of $_2 h_{mn}$ in terms of the $_2 h_{00}$ found above, we obtain

$$_3 h_{0n,ss} - _3 h_{0s,ns} + 2\,_2 h_{00,n0} = -2\kappa \sum_{A=1}^{2} \mu_A \dot{z}_A^n \delta_A.\tag{3.7.43}$$

The solution of this equation is

$$_3 h_{0n} = 4G \sum_{A=1}^{2} \mu_A \dot{z}_A^n r_A^{-1}.\tag{3.7.44}$$

Calculation of $_4 h_{00}$ is somewhat more complicated. The relevant part of $_4 h_{00}$, for two masses, that contributes to the equation of motion of the first particle, is

$$_4 h_{00} \approx G \left\{ \frac{2G\mu_2^2}{r_2^2} - \frac{3\mu_2 \dot{z}_2^s \dot{z}_2^s}{r_2} - \mu_2 r_{2,00} + \frac{2G\mu_1 \mu_2}{z r_2} \right\}.\tag{3.7.45}$$

Using these values for $_4 h_{00}$, $_3 h_{0n}$, and $_2 h_{mn}$ in the equation of motion (3.7.27) gives, for the two-body problem (Problem 3.9.16):

$$\ddot{z}_1^n - \mu_2 \frac{\partial\,(1/z)}{\partial z_1^n}$$

$$= \mu_2 \Big\{ \left(\dot{z}_1^s \dot{z}_1^s + \frac{3}{2} \dot{z}_2^s \dot{z}_2^s - 4\dot{z}_1^s \dot{z}_2^s - 4\frac{\mu_2}{z} - 5\frac{\mu_1}{z} \right) \frac{\partial\,(1/z)}{\partial z_1^n}$$

$$+ \left[4\dot{z}_1^s \left(\dot{z}_2^n - \dot{z}_1^n \right) + 3\dot{z}_1^n \dot{z}_2^s - 4\dot{z}_2^n \dot{z}_2^s \right] \frac{\partial\,(1/z)}{\partial z_1^s}$$

$$+ \frac{1}{2} \dot{z}_2^s \dot{z}_2^r \frac{\partial^3 z}{\partial z_1^s \partial z_1^r \partial z_1^n} \Big\}.\tag{3.7.46}$$

In Eq. (3.7.46) the Newton gravitational constant G was taken as equal to 1. The equation of motion for the second particle is obtained by replacing μ_1, μ_2, z_1, z_2 by μ_2, μ_1, z_2, z_1, respectively.

Equation (3.7.46) is known as the Einstein-Infeld-Hoffmann equation of motion, and is a generalization of the Newton equation. The essential relativistic correction may be obtained by fixing one of the particles. Writing M for μ_2, neglecting μ_1 and $\dot{z}_2^s$, and using an obvious three-dimensional vector notation, Eq. (3.7.46) simplifies to

$$\ddot{\mathbf{z}} - M\nabla \left(\frac{1}{z} \right) = M \left\{ \left(\dot{\mathbf{z}} \cdot \dot{\mathbf{z}} - \frac{4M}{z} \right) \nabla \left(\frac{1}{z} \right) - 4\dot{\mathbf{z}} \left(\dot{\mathbf{z}} \cdot \nabla \frac{1}{z} \right) \right\},\tag{3.7.47}$$

where $\mathbf{z}$ denotes the three-vector z_1^s.

In the next section decomposition of the Riemann tensor to its irreducible components is shown.

3.8   Decomposition of the Riemann Tensor

The Riemann curvature tensor $R_{\alpha\beta\gamma\delta}$ can be decomposed into its *irreducible components*. These are the Weyl conformal tensor $C_{\alpha\beta\gamma\delta}$, the tracefree Ricci tensor $S_{\alpha\beta}$, and the Ricci scalar curvature R. The tensor $S_{\alpha\beta}$ is defined by

$$S_{\alpha\beta} = R_{\alpha\beta} - \frac{1}{4} g_{\alpha\beta} R, \tag{3.8.1}$$

where $R_{\alpha\beta}$ is the ordinary Ricci tensor.

The decomposition can be written symbolically as

$$R_{\alpha\beta\gamma\delta} = C_{\alpha\beta\gamma\delta} \oplus S_{\alpha\beta} \oplus R. \tag{3.8.2}$$

No new quantities can be obtained from any of the above three irreducible components by contraction of their indices.

When written out in full, the decomposition (3.8.2) has the form:

$$R_{\rho\sigma\mu\nu} = C_{\rho\sigma\mu\nu} + \frac{1}{2} \left(g_{\rho\mu} S_{\sigma\nu} - g_{\rho\nu} S_{\sigma\mu} - g_{\sigma\mu} S_{\rho\nu} + g_{\sigma\nu} S_{\rho\mu} \right)$$

$$- \frac{1}{12} \left(g_{\rho\nu} g_{\sigma\mu} - g_{\rho\mu} g_{\sigma\nu} \right) R. \tag{3.8.3}$$

It can also be written in the form:

$$R_{\rho\sigma\mu\nu} = C_{\rho\sigma\mu\nu} + \frac{1}{2} \left(g_{\rho\mu} R_{\sigma\nu} - g_{\rho\nu} R_{\sigma\mu} - g_{\sigma\mu} R_{\rho\nu} + g_{\sigma\nu} R_{\rho\mu} \right)$$

$$+ \frac{1}{6} \left(g_{\rho\nu} g_{\sigma\mu} - g_{\rho\mu} g_{\sigma\nu} \right) R. \tag{3.8.4}$$

3.9   Problems

P 3.9.1.   Prove the transformation laws (3.1.15) and (3.1.16) of the Christoffel symbols of the first and second kinds.

Solution: Using the transformation laws for the metric tensor leads to Eqs. (3.1.15) and (3.1.16).

P 3.9.2. Prove Eq. (3.1.24).

Solution: The solution is left for the reader.

P 3.9.3. Show that the covariant derivatives of the tensors $T_{\alpha\beta}$, $T^{\alpha\beta}$ and $T^{\alpha}{}_{\beta}$ are given by

$$\nabla_\gamma T_{\alpha\beta} = \frac{\partial T_{\alpha\beta}}{\partial x^\gamma} - \Gamma^\delta_{\beta\gamma} T_{\alpha\delta} - \Gamma^\delta_{\alpha\gamma} T_{\delta\beta},$$

$$\nabla_\gamma T^{\alpha\beta} = \frac{\partial T^{\alpha\beta}}{\partial x^\gamma} + \Gamma^{\alpha}_{\delta\gamma} T^{\delta\beta} + \Gamma^{\beta}_{\delta\gamma} T^{\alpha\delta},$$

$$\nabla_\gamma T^{\alpha}_{\ \beta} = \frac{\partial T^{\alpha}_{\ \beta}}{\partial x^\gamma} + \Gamma^{\alpha}_{\delta\gamma} T^{\delta}_{\ \beta} - \Gamma^{\delta}_{\beta\gamma} T^{\alpha}_{\ \delta}.$$

From this find the general rule for covariant differentiation.

Solution: The solution is left for the reader.

P 3.9.4. Show that the covariant differentiation of the sum, difference, outer and inner products of tensors obeys the usual rules of ordinary differentiation.

Solution: The solution is left for the reader.

P 3.9.5. Generalize Eq. (3.1.30) for a tensor $T_{\mu\nu}$.

Solution: The solution is left for the reader.

P 3.9.6. If $T_{\alpha\beta}$ is the curl of a covariant vector, show that

$$\nabla_\gamma T_{\alpha\beta} + \nabla_\alpha T_{\beta\gamma} + \nabla_\beta T_{\gamma\alpha} = 0,$$

and that this is equivalent to

$$\partial_\gamma T_{\alpha\beta} + \partial_\alpha T_{\beta\gamma} + \partial_\beta T_{\gamma\alpha} = 0.$$

Solution: The solution is left for the reader.

P 3.9.7. Show that the divergence $\nabla_\mu V^\mu$ of the vector V^μ is given by

$$\nabla_\mu V^\mu = \frac{1}{\sqrt{-g}} \frac{\partial}{\partial x^\mu} \left(V^\mu \sqrt{-g} \right).$$

Also show that for a skew-symmetric tensor $F^{\alpha\beta}$ the covariant divergence is

$$\nabla_\beta F^{\alpha\beta} = \frac{1}{\sqrt{-g}} \frac{\partial}{\partial x^\beta} \left(F^{\alpha\beta} \sqrt{-g} \right).$$

Solution: The solution is left for the reader.

P 3.9.8. Find the expression for the Riemann tensor $R_{\alpha\beta\gamma\delta}$. Using it prove Eqs. (3.1.32).

Solution: The solution is left for the reader.

P 3.9.9. Show that a curve with a covariantly constant tangent vector is necessarily geodesic.

Solution: Let the curve be denoted by $x^\alpha = x^\alpha(s)$ and the tangent vector by dx^α/ds. If the tangent vector is covariantly constant, then

$$\nabla_\mu \frac{dx^\alpha}{ds} = 0, \tag{1}$$

or explicitly

$$\frac{\partial}{\partial x^\mu}\left(\frac{dx^\alpha}{ds}\right) + \Gamma^\alpha_{\mu\nu}\frac{dx^\nu}{ds} = 0. \tag{2}$$

Multiplying Eq. (2) by dx^μ/ds and using the identity

$$\frac{dx^\mu}{ds}\frac{\partial}{\partial x^\mu} = \frac{d}{ds}, \tag{3}$$

gives

$$\frac{d^2 x^\alpha}{ds^2} + \Gamma^\alpha_{\mu\nu}\frac{dx^\mu}{ds}\frac{dx^\nu}{ds} = 0. \tag{4}$$

P 3.9.10 Discuss the constancy of the weak and gravitational coupling constants.

Solution: The solution is left for the reader.

P 3.9.11 Use the geodesic equations, Eq. (3.1.43), to determine the force per unit mass on a body at rest, and show that it is given by $F^i = -c^2\Gamma^i_{00}$ where $i = 1, 2, 3$. In the weak field approximation $g^{i\alpha}$ are very close to the Lorentz metric, and for a time-independent metric $F^i = c^2\Gamma^i_{00} \approx \left(c^2/2\right)\partial_i g_{00}$. Show that in the weak field case Eq. (3.4.2) reduces to the Poisson equation (3.4.1), where $g_{00} \approx 1 + 2\phi/c^2$. Using the latter result, show that the constant κ in Eq. (3.4.2) is given by $\kappa = 8\pi G/c^4$.

Solution: The solution is left for the reader.

P 3.9.12 Prove Eqs. (3.4.10) and (3.4.11).

Solution: The solution is left for the reader.

P 3.9.13 Derive the gravitational field equations (3.4.2) using the calculus of variation by treating both $g_{\mu\nu}$ and $\Gamma^\mu_{\alpha\beta}$ as independent variants, and obtain thereby equations that determine both objects. Such a procedure is known as the Palatini formalism. The procedure is analogous to the one employed in deriving the electromagnetic field equations from a variational principle where both the field $f^{\mu\nu}$ and the potential A_μ are variants of an action principle.

Solution: The solution is left for the reader.

P 3.9.14 Find the energy-momentum tensor $T_{\mu\nu}$ for: (1) a system of neutral particles of inertial mass M (function of time); (2) the electromagnetic field; and (3) a scalar field ϕ. Show that they are given by:

$$(1)\ T^{\mu\nu} = \sum M \dot{z}^\mu \dot{z}^\nu \delta\left(\mathbf{x} - \mathbf{z}\right),$$

$$(2)\ T_{\mu\nu} = \frac{1}{4\pi} \left\{ \frac{1}{4} g_{\mu\nu} f_{\alpha\beta} f^{\alpha\beta} - f_{\mu\alpha} f_\nu{}^\alpha \right\},$$

$$(3)\ T_{\mu\nu} = \partial_\mu \phi \partial_\nu \phi - \frac{1}{2} g_{\mu\nu} \left(\nabla^\alpha \phi \nabla_\alpha \phi - m^2 \phi^2 \right).$$

Solution: The solution is left for the reader.

P 3.9.15 Use the approximate metric (3.6.3) in the geodesic equation (3.1.43) to show that the equation obtained is (3.6.4).

Solution: The solution is left for the reader.

P 3.9.16 Prove Eqs. (3.7.45) and (3.7.46).

Solution: The solution is left for the reader.

3.10 Suggested References

B. Bertotti, D. Brill and R. Krotkov, Experiments on gravitation, in: *Gravitation: An Introduction to Current Research* (L. Witten, Editor), (John Wiley, New York, 1962).

G. Birkhoff, *Relativity and Modern Physics* (Harvard University Press, Cambridge, Massachusetts, 1923).

M. Carmeli, Equations of motion without infinite self-action terms in general relativity, *Phys. Rev. B* **140**, 1441 (1965).

M. Carmeli and S. Malin, *Theory of Spinors: An Introduction* (World Scientific, Singapore, 2000).

T.F. Cranshaw, S.P. Schiffer and A.B. Whitehead, Measurement of the gravitational red shift using the Mössbauer effect in Fe^{57}, *Phys. Rev. Lett.* **4**, 163 (1960).

R.H. Dicke, Experimental relativity, in: *Relativity, Groups and Topology* (C. DeWitt *et al.*, Eds.) (Gordon and Breach, New York, 1964), p.163.

A. Einstein, *Ann. Phys.* **49**, 761 (1916); English translation in: *The Principle of Relativity* (Dover, New York, 1923).

L.P. Eisenhart, *Riemannian Geometry* (Princeton University Press, New Jersey, 1949).

E. Fermi, *Atti Accad. Naz. Lincei* **21**, 21 and 51 (1922).

L. Infeld and A. Schild, On the motion of test particles in general relativity, *Rev. Mod. Phys.* **21**, 408 (1949).

L.D. Landau and E.M. Lifshitz, *Course of Theoretical Physics: Mechanics* (Pergamon Press, 1960).
T. Levi-Civita, Math. Ann. **97**, 291 (1926).

A. Papapetrou, Spinning test-particles in general relativity. I, *Proc. R. Soc. London (A)* **209**, 248 (1951).

F.A.E. Pirani, Introduction to gravitational radiation theory, in: *Lectures on General Relativity* (1964 Brandeis Summer School) (Prentice-Hall, Englewood Cliffs, New Jersey, 1965).

R.V. Pound and G.A. Rebka, Jr., Apparent weight of photons, *Phys. Rev. Lett.* **4**, 337 (1960).

P.G. Roll, R. Krotkov and R.H. Dicke, The equivalence of inertial and passive gravitational mass, *Ann. Phys. (N.Y.)* **26**, 442 (1964).

L.I. Schiff, Motion of a gyroscope according to Einstein's theory of gravitation, *Proc. Natl. Acad. Sci.* **46**, 871 (1960).

I.I. Shapiro, Testing general relativity: Progress, problems, and prospects, *Gen. Relat. Grav.* **3**, 135 (1972).

A. Trautman, Foundations and current problems of general relativity, in: *Lectures on General Relativity* (1964 Brandeis Summer School) (Prentice-Hall, Englewood Cliffs, New Jersey, 1965).

J. Weber, Gravitational radiation experiments, in: *Relativity* (M. Carmeli, S.I. Fickler and L. Witten, Eds.), (Plenum Press, New York, 1970).
J. Weber, Gravitational-Wave-Detector Events, *Phys. Rev. Lett.* **20**, 1307 (1968).
J. Weber, Evidence for Discovery of Gravitational Radiation, *Phys. Rev. Lett.* **22**, 1320 (1969).
J. Weber, Anisotropy and Polarization in the Gravitational-Radiation Experiments, *Phys. Rev. Lett.* **25**, 180 (1970).

Chapter 4

Cosmological General Relativity

Moshe Carmeli

In this chapter we develop the theory of cosmological general relativity (CGR). It is an extension of cosmological special relativity (CSR), presented in Chapter 2, to curved Riemannian manifold. The transition from CSR to CGR, in principle, is similar, but not identical, to the transition from Einstein's special relativity (ESR) to ordinary general relativity (EGR). In the framework of CGR gravitation is described by a curved four-dimensional Riemannian spacevelocity . CGR incorporates the Hubble-Carmeli constant τ at the outset. This constant describes the Big Bang time and is the analogue of the speed of light in vacuum c in Einstein's relativity theory. The Hubble law is assumed in CGR as a fundamental law. CGR, in essence, extends Hubble's law so as to incorporate gravitation in it; it is actually a distribution theory *that relates distances and velocities between galaxies . The theory involves only measured quantities and it takes a picture of the Universe as it is at any instant of time. No cosmological constant is required in this theory. In the four-dimensional space-velocity ds expresses the proper velocity just as in ordinary relativity it expresses the proper time. The null condition ds $= 0$ describes the expansion of the Universe in this theory just as the null condition ds $= 0$ in Einstein's theory describes the propagation of light. The energy-momentum tensor in this theory is a modification of the standard one so as to adjust it properly to the cosmological theory. The spherically symmetric metric is presented and the field equations are written down and solved. The equations for the expansion of the Universe and their solutions are given. The numerical value of the constant τ is given. The Tolman metric of an expanding Universe is presented in detail. Kantowski-Sachs metric describing the expanding*

161

Universe is given in detail. This is done in the space-velocity manifold. The issue of having any spacetime converted to a spacevelocity is discussed in connection to whether or not the Schwarzschild metric can be considered as an expanding Universe in the manifold of space and velocity.

4.1   Cosmology in Spacevelocity

4.1.1   *The foundations of CGR*

It is preferable to write the laws of physics in terms of measured quantities. But this is usually impossible. See what will happen if we write the dynamics of a particle when a force is acting on it. The velocity will be proportional to the force. But there is no law like that and it is actually wrong. So we go to the acceleration. And that was the greatness of Newton. But then we replace the concept of force by a field both in Newtonian theory and in Einstein's theory. On the other hand see how thermodynamics works even though the dynamics of that theory are just the measured quantities like temperature and pressure, etc.

The foundations of any gravitational theory are based on the principle of equivalence and the principle of general covariance. These two principles lead immediately to the realization that gravitation should be described by a four-dimensional (4D) curved spacetime, in our theory spacevelocity, and that the field equations and the equations of motion should be written in a generally covariant form. Hence these principles are adopted in CGR also. In a four-dimensional Riemannian manifold we use a metric $g_{\mu\nu}$ and a line element

$$ds^2 = g_{\mu\nu}dx^\mu dx^\nu. \tag{4.1.1}$$

The difference from Einstein's general relativity is that our coordinates are: x^0 is a velocitylike coordinate (rather than a timelike coordinate), thus $x^0 = \tau v$ where τ is the Big Bang time, which is the inverse of the Hubble parameter (constant) H_0 in the zero-gravity limit, and v the velocity . The coordinate $x^0 = \tau v$ is the comparable to $x^0 = ct$ where c is the speed of light and t is the time in ordinary general relativity. The other three coordinates x^k, $k = 1, 2, 3$, are spacelike, just as in general relativity theory.

4.1.2 *The null condition ds = 0*

An immediate consequence of the above choice of coordinates is that the null condition $ds = 0$ describes the expansion of the Universe in the curved spacevelocity (generalized Hubble law with gravitation) as compared to the propagation of light in curved spacetime in general relativity. This means one solves the field equations (to be given in the sequel) for the metric tensor, then from the null condition $ds = 0$ one obtains immediately the dependence of the relative distances between the galaxies on their relative velocities.

4.1.3 *Gravitational field equations*

As usual in gravitational theories, one equates geometry to physics. The first is expressed by means of a combination of the Ricci tensor and the Ricci scalar, and follows to be naturally either the Ricci trace-free tensor or the Einstein tensor. The Ricci trace-free tensor does not fit gravitation in general. The Einstein tensor is a natural candidate. The physical part is expressed by the energy-momentum tensor, which now has a different physical meaning from that in Einstein's theory. More important, the coupling constant that relates geometry to physics is now also different.

Accordingly the field equations are

$$G_{\mu\nu} = R_{\mu\nu} - \frac{1}{2}g_{\mu\nu}R = \kappa T_{\mu\nu}, \qquad (4.1.2)$$

exactly as in Einstein's theory, with κ given by $\kappa = 8\pi k/\tau^4$, (in general relativity it is given by $8\pi G/c^4$), where k is given by $k = G\tau^2/c^2$, with G being Newton's gravitational constant. When the equations of motion are written in terms of velocity instead of time, the constant k replaces G. Using the above equations one then has

$$\kappa = \frac{8\pi G}{c^2\tau^2}. \qquad (4.1.3)$$

4.1.4 *The energy-momentum tensor*

The energy-momentum tensor $T^{\mu\nu}$ is constructed, along the lines of general relativity theory, with the speed of light being replaced by the Hubble-Carmeli constant τ. If ρ is the average mass density of the Universe, then it will be assumed that $T^{\mu\nu} = \rho u^\mu u^\nu$, where $u^\mu = dx^\mu/ds$ is the four-velocity. In general relativity theory one takes $T_0^0 = \rho$. In Newtonian gravity one has the Poisson equation $\nabla^2\phi = 4\pi G\rho$. At points where $\rho = 0$

one solves the vacuum Einstein field equations in general relativity and the Laplace equation $\nabla^2 \phi = 0$ in Newtonian gravity. In both theories a null (zero) solution is allowed as a trivial case. In cosmology, however, there exists no situation where ρ can be zero because the Universe is filled with matter.

In order to be able to have zero on the right-hand side of Eq. (4.1.2) one takes T_0^0 not equal to ρ, but equal to $\rho_{eff} = \rho - \rho_c$, where ρ_c is the critical mass density, a *constant* in CGR given by

$$\rho_c = \frac{3h^2}{8\pi G},$$

where h is Hubble's constant in empty space (in the standard model it is $\rho_c = 3H_0^2/8\pi G$). Using $h = 72.17$ km/s-Mpc we obtain

$$\rho_c = 9.77 \times 10^{-30} \text{gm/cm}^3,$$

a few hydrogen atoms per cubic meter. Accordingly one takes

$$T^{\mu\nu} = \rho_{eff} u^\mu u^\nu; \qquad \rho_{eff} = \rho - \rho_c \qquad (4.1.4)$$

for the energy-momentum tensor .

In the next sections we apply CGR to obtain the accelerating expanding Universe and we discuss related subjects.

4.1.5   *The Newtonian limit in cosmological general relativity*

We now find the Newtonian limit in cosmological general relativity theory. The Newtonian limit in general relativity theory was given in detail in Subsection 3.4.3. However, the present case is less familiar and the approximation is not that of slow motion as in general relativity. Rather, it is a cosmic time approximation that will be used here.

Remark:

There is actually no Newtonian limit in this case that is similar to the Newtonian limit in general relativity. Here we are talking about the four-dimensional space of the distances of galaxies, each with its own velocity (compare Figure 4.1.1 to a single particle moving through space with a velocity in Figure 4.1.2). And there is no known Newtonian equation of motion in this space of distances and velocities. It is actually the space of the aggregate of galaxies that are discrete and do not form a continuum. But here we consider them as a continuum space

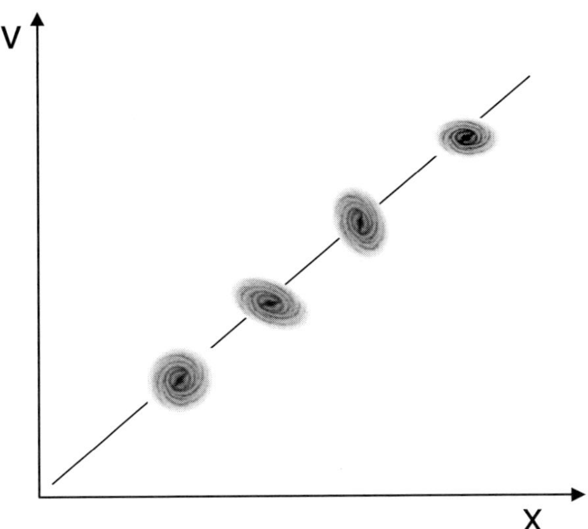

Fig. 4.1.1: Spacevelocity diagram: Galaxies at different distances from observer, each with its own velocity.

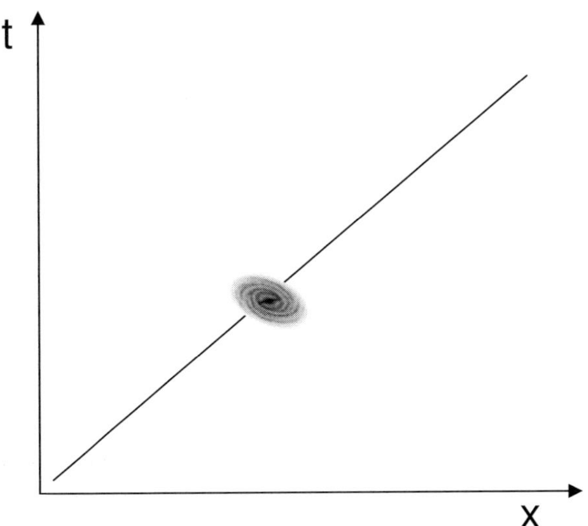

Fig. 4.1.2: Spacetime diagram: A galaxy moving in space.

in four dimensions. This is like considering gas molecules, which are also discrete, as a continuum in problems of fluid dynamics. Eventually, our limit is the Hubble law that shows a linear relationship between the distances and velocities of galaxies.

The parameter of the approximation will be t/τ, where t is the cosmic time measured backward from the present epoch (it is zero now and equal to τ at the Big Bang). It should be remembered that τ is the Big Bang time, that is the time of creation of the Universe. In other words it is the age of the Universe and it is a universal constant, $\tau = 13.56$ Gyr.

It is worthwhile comparing the new approximation in terms of the cosmic time t/τ to that used previously in general relativity of slow motion, v/c. A typical slow motion is usually identified with the motion of planets in the Solar system. If we assume that a typical velocity in the Solar system is of the order of $v = 30\text{km/s}$, then

$$\frac{v}{c} = \frac{30 \times 10^3 \times 10^2}{3 \times 10^{10}} = 10^{-4}. \tag{4.1.5}$$

What will be the cosmic time that is comparable to such a slow-motion approximation? We have $t/\tau = 10^{-4}$, or

$$t = \tau \times 10^{-4} \approx 13.6 \times 10^9 \times 10^{-4} = 1.36 \times 10^6 \text{ yr}. \tag{4.1.6}$$

If we observed galaxies moving away from us with speeds like this then Eq. (4.1.6) gives their cosmic time with respect to us.

We now proceed with the approximation that will also determine the Einstein gravitational constant for cosmological general relativity. As in the general relativity case, the geodesic equation can be considered as the equation which describes the motion of an infinitesimally small test particle moving in a gravitational field. We thus will approximate the geodesic equation to the lowest order in t/τ.

The line element

$$ds^2 = g_{\mu\nu}dx^\mu dx^\nu \tag{4.1.7}$$

can be written approximately if we notice that $dx^0 = \tau dv$, where τ is the Big Bang time. Hence the term $g_{00}dx^0dx^0 = g_{00}\tau^2 dv^2$ is one order of magnitude larger than the term $2g_{0k}dx^0dx^k = 2g_{0k}\tau dvdx^k$, where $k = 1,2,3$. The latter term, in turn, is again one order of magnitude larger than $g_{kl}dx^k dx^l$. Consequently to the lowest approximation $ds^2 \approx g_{00}dx^0dx^0$. It should be emphasized now that this approximation is valid only for cosmic times close to our time now. Remember now that $\tau \gg dx/dv$.

We have seen in Subsection 3.4.3 that the geodesic equation can be written in the form

$$\frac{d^2 x^\mu}{d\sigma^2} + \Gamma^\mu_{\alpha\beta}\frac{dx^\alpha}{d\sigma}\frac{dx^\beta}{d\sigma} = -\frac{d^2\sigma/ds^2}{(d\sigma/ds)^2}\frac{dx^\mu}{d\sigma}, \qquad (4.1.8)$$

where a new parameter σ is chosen instead of the length parameter s. We now choose the parameter σ to be $\sigma = x^0$, where $x^0 = \tau v$. Thus the latter equation can be written in the form

$$\ddot{x}^\mu + \Gamma^\mu_{\alpha\beta}\dot{x}^\alpha\dot{x}^\beta = -\frac{d^2 x^0/ds^2}{(dx^0/ds)^2}\dot{x}^\mu, \qquad (4.1.9)$$

where a dot means differentiation with respect to x^0. Using the zero component of the last equation and remembering that $\dot{x}^0 = dx^0/dx^0 = 1$ and that $\ddot{x}^0 = 0$, we obtain

$$\ddot{x}^k + \left(\Gamma^k_{\alpha\beta} - \dot{x}^k\Gamma^0_{\alpha\beta}\right)\dot{x}^\alpha\dot{x}^\beta = 0, \qquad (4.1.10)$$

where $k = 1, 2, 3$.

We have to find the lowest approximation of Eq. (4.1.10) and also of the Einstein field equations.

In Eq. (4.1.10) there appears the term $\Gamma^0_{\alpha\beta}\dot{x}^k$, and by our assumption it is much smaller than $\Gamma^k_{\alpha\beta}$. Hence the second term in the parenthesis in Eq. (4.1.10) can be neglected. Proceeding in this way, neglecting all terms with velocities in Eq. (4.1.10), then gives

$$\ddot{x}^k + \Gamma^k_{00} = 0. \qquad (4.1.11)$$

So far this is the lowest approximation of Eq. (4.1.10). We now write the Christoffel symbol in terms of the metric tensor

$$\Gamma^k_{00} = \frac{1}{2}g^{k\lambda}\left(2\frac{\partial g_{\lambda 0}}{\partial x^0} - \frac{\partial g_{00}}{\partial x^\lambda}\right)$$

$$\approx -\frac{1}{2}\eta^{k\lambda}\frac{\partial g_{00}}{\partial x^\lambda}$$

$$\approx -\frac{1}{2}\eta^{kl}\frac{\partial g_{00}}{\partial x^l}$$

$$= \frac{1}{2}\delta^{kl}\frac{\partial g_{00}}{\partial x^l}$$

$$= \frac{1}{2}\frac{\partial g_{00}}{\partial x^k}, \qquad (4.1.12)$$

where $\eta^{\mu\nu}$ is the Minkowskian metric $\eta^{\mu\nu} = (+1, -1, -1, -1)$, and $\eta^{\mu\nu} = 0$ for $\mu \neq \nu$.

Accordingly we obtain

$$\ddot{x}^k = -\frac{1}{2}\frac{\partial g_{00}}{\partial x^k}. \tag{4.1.13}$$

We now introduce the function $\psi(x)$, similarly to the Newtonian case, by

$$g_{00}(x) = 1 + \frac{2}{\tau^2}\psi(x), \tag{4.1.14}$$

and accordingly we now obtain

$$\ddot{x}^k = -\frac{1}{\tau^2}\frac{\partial \psi(x)}{\partial x^k}. \tag{4.1.15}$$

If we replace the differentiation with respect to $x^0 = \tau v$ by differentiation with respect to v we obtain

$$\frac{d^2 x^k}{dv^2} = -\frac{\partial \psi(x)}{\partial x^k}. \tag{4.1.16}$$

And we are left with finding the lowest approximation of the Einstein field equations.

We now approximate the Einstein field equations

$$R_{\mu\nu} = \kappa\left(T_{\mu\nu} - \frac{1}{2}g_{\mu\nu}T\right). \tag{4.1.17}$$

In the lowest approximation we have

$$T = T_{\mu\nu}g^{\mu\nu} \approx T_{\mu\nu}\eta^{\mu\nu} \approx T_{00}\eta^{00} = T_{00}. \tag{4.1.18}$$

The component R_{00} gives

$$R_{00} = \kappa\left(T_{00} - \frac{1}{2}g_{00}T\right)$$

$$\approx \kappa\left(T_{00} - \frac{1}{2}\eta_{00}T\right)$$

$$= \frac{1}{2}\kappa T_{00}$$

$$= \frac{1}{2}\kappa\tau^2\rho(x). \tag{4.1.19}$$

In the above equations $\rho(x)$ is the galaxy mass density.

In the lowest approximation R_{00} yields

$$R_{00} \approx \frac{\partial \Gamma^s_{00}}{\partial x^s}. \tag{4.1.20}$$

Thus we have

$$R_{00} \approx \frac{1}{\tau^2} \nabla^2 \psi\left(x\right), \tag{4.1.21}$$

where

$$\nabla^2 = \frac{\partial^2}{\partial x^1 \partial x^1} + \frac{\partial^2}{\partial x^2 \partial x^2} + \frac{\partial^2}{\partial x^3 \partial x^3}. \tag{4.1.22}$$

is the three-dimensional Laplace operator. Comparing now the two expressions obtained for R_{00} in Eqs. (4.1.19) and (4.1.21) we obtain

$$\nabla^2 \psi\left(x\right) = \frac{1}{2} \kappa \tau^4 \rho\left(x\right). \tag{4.1.23}$$

We now have a dilemma. What is the differential equation that ψ has to satisfy in the lowest approximation? In the Newtonian limit of general relativity it satisfies the Newtonian gravitational field equation, namely the Poisson equation. But here we have no Newtonian approximation for the distances and the velocities of the galaxies because, to our knowledge, there is no such a Newtonian theory. So accordingly we assume, and simplicity plays a major reason, that a Poisson equation of the form

$$\nabla^2 \psi\left(x\right) = 4\pi k \rho\left(x\right) \tag{4.1.24}$$

exists for the function $\psi\left(x\right)$. In the above equation

$$k = G\frac{\tau^2}{c^2}. \tag{4.1.25}$$

Comparing now the two expressions on the right-hand sides of Eqs. (4.1.23) and (4.1.24) gives

$$\kappa = \frac{8\pi k}{\tau^4} \tag{4.1.26}$$

for the Einstein gravitational constant in cosmological general relativity. And that is an essentially different physical result from the traditional Einstein gravitational constant, $\kappa = 8\pi G/c^4$, since the Big Bang time τ appears explicitly in the cosmological general relativity case.

Finally we find out what is the physical meaning of the equations obtained. In the lowest approximation we might take $\psi\left(x\right)$ as constant and accordingly by Eq. (4.1.16) we obtain

$$\frac{d^2 x^k}{dv^2} \approx 0.$$

Thus dx^k/dv is a constant, or

$$x^k = H_0^{-1} v. \tag{4.1.27}$$

Obviously H_0 is the Hubble constant in the absence of gravitation, that is $1/\tau$ by our previous notation. And what has been obtained can be written as

$$r = \tau v. \tag{4.1.28}$$

So we obtained not a familiar Newtonian law but rather the Hubble law. And such a limit might as well be called the Hubble limit, rather than the Newtonian limit, in cosmological general relativity.

4.1.6 *Spherically-symmetric vacuum solution of the Einstein field equations in CGR*

In Section 3.5 we derived the Schwarzschild metric, a spherically symmetric vacuum solution of the Einstein field equations. It is probably the most important solution of the Einstein field equations with its black hole interpretation. Can we find a vacuum, spherically symmetric, solution of the Einstein field equations in the space of velocities of cosmological general relativity? If yes, then will there be a Schwarzschild radius? And then what is the physical meaning of the expansion as obtained by equating the line element ds to zero? As is well known in general relativity, the null condition $ds = 0$ describes the propagation of light in the presence of gravitation. In cosmological general relativity the Universe is considered at one instant of time and there is no dependence on time. These questions will be answered in this subsection.

As in the Schwarzschild case, a spherical symmetry of the metric means that the expression for the interval $ds = \sqrt{g_{\mu\nu}dx^\mu dx^\nu}$ must be the same at all points located at the same distance from the center. In flat space of an expanding Universe their distance is equal to the radius vector, and the metric is given by:

$$ds^2 = \tau^2 dv^2 - dr^2 - r^2\left(d\theta^2 + \sin^2\theta d\phi^2\right). \tag{4.1.29}$$

In a non-Euclidean space, such as the Riemannian one we have in the presence of a gravitational field, there is no quantity which has all the properties of the flat space radius vector, such that it is equal both to the distance from the center and to the length of the circumference divided by 2π. Therefore, the choice of a radius vector is here arbitrary.

When a mass with spherical symmetry is introduced, the flat space line element (4.1.29) must be modified but in a way that retains spherical symmetry. The most general spherically symmetric expression for ds^2 is

$$ds^2 = a\left(r, v\right) dv^2 + b\left(r, v\right) dr^2 + c\left(r, v\right) drdv + d\left(r, v\right)\left(d\theta^2 \right.$$

$$\left. + \sin^2\theta d\phi^2\right). \tag{4.1.30}$$

Because of the arbitrariness in the choice of the coordinate system in general relativity theory, we can perform a coordinate transformation which does not destroy the spherical symmetry of ds^2. Hence we can choose new coordinates r' and v' given by some functions $r' = r'\left(r, v\right)$ and $v' = v'\left(r, v\right)$.

Making use of these transformations, we can choose the new coordinates so that the coefficient $c\left(r, v\right)$ of the mixed term $drdv$ vanishes and the coefficient $d\left(r, v\right)$ of the angular part to be $-r'^2$, in the metric (4.1.30). The latter condition implies that the radius vector is now defined in such a way that the circumference of a circle is equal to $2\pi r$. It is convenient to express the functions $a\left(r, v\right)$ and $b\left(r, v\right)$ in exponential forms, e^ν and $-e^\lambda$, respectively, where ν and λ are functions of the new coordinates r' and v'. Consequently, the line element (4.1.30) will have the form

$$ds^2 = e^\nu dv^2 - e^\lambda dr^2 - r^2\left(d\theta^2 + \sin^2\theta d\phi^2\right), \tag{4.1.31}$$

where, for brevity, we have dropped the primes from the new coordinates r' and v', and the Big Bang time τ is taken as equal to unity.

We now denote the coordinates v, r, θ, ϕ by x^0, x^1, x^2, x^3, respectively. Hence the components of the covariant metric tensor are given by:

$$g_{\mu\nu} = \begin{pmatrix} e^\nu & 0 & 0 & 0 \\ 0 & -e^\lambda & 0 & 0 \\ 0 & 0 & -r^2 & 0 \\ 0 & 0 & 0 & -r^2\sin^2\theta \end{pmatrix}, \tag{4.1.32a}$$

whereas those of the contravariant metric tensor are:

$$g^{\mu\nu} = \begin{pmatrix} e^{-\nu} & 0 & 0 & 0 \\ 0 & -e^{-\lambda} & 0 & 0 \\ 0 & 0 & -r^{-2} & 0 \\ 0 & 0 & 0 & -r^{-2}\sin^{-2}\theta \end{pmatrix}. \tag{4.1.32b}$$

To find out the differential equations that the functions ν and λ have to satisfy, according to the Einstein field equations, we first need to calculate the Christoffel symbols associated with the metric (4.1.32). The nonvanishing components are:

$$\Gamma^0_{00} = \frac{\dot\nu}{2}, \qquad \Gamma^0_{10} = \frac{\nu'}{2}, \qquad \Gamma^0_{11} = \frac{\dot\lambda}{2}e^{\lambda-\nu}, \tag{4.1.33a}$$

$$\Gamma^1_{00} = \frac{\nu'}{2}e^{\nu-\lambda}, \qquad \Gamma^1_{10} = \frac{\dot\lambda}{2}, \qquad \Gamma^1_{11} = \frac{\lambda'}{2}, \tag{4.1.33b}$$

$$\Gamma^1_{22} = -re^{-\lambda}, \qquad \Gamma^1_{33} = -r\sin^2\theta\, e^{-\lambda}, \qquad \Gamma^2_{12} = \frac{1}{r}, \tag{4.1.33c}$$

$$\Gamma^2_{33} = -\sin\theta\cos\theta, \qquad \Gamma^3_{13} = \frac{1}{r}, \qquad \Gamma^3_{23} = \cot\theta, \tag{4.1.33d}$$

where dots and primes denote differentiation with respect to v and r, respectively.

With these Christoffel symbols, we compute the following expressions for the nonvanishing components of the Einstein tensor:

$$G_0{}^0 = -e^{-\lambda}\left(\frac{1}{r^2} - \frac{\lambda'}{r}\right) + \frac{1}{r^2} = \kappa T_0{}^0, \tag{4.1.34a}$$

$$G_0{}^1 = -e^{-\lambda}\frac{\dot\lambda}{r} = \kappa T_0{}^1, \tag{4.1.34b}$$

$$G_1{}^1 = -e^{-\lambda}\left(\frac{\nu'}{r} + \frac{1}{r^2}\right) + \frac{1}{r^2} = \kappa T_1{}^1, \tag{4.1.34c}$$

$$G_2{}^2 = -\frac{1}{2}e^{-\lambda}\left(\nu'' + \frac{\nu'^2}{2} + \frac{\nu' - \lambda'}{r} - \frac{\nu'\lambda'}{2}\right)$$

$$+ \frac{1}{2}e^{-\nu}\left(\ddot\lambda + \frac{\dot\lambda^2}{2} - \frac{\dot\lambda\dot\nu}{2}\right) = \kappa T_2{}^2, \tag{4.1.34d}$$

$$G_3{}^3 = G_2{}^2 = \kappa T_3{}^3. \tag{4.1.34e}$$

All other components vanish identically.

The gravitational field equations can now be integrated exactly for the spherical symmetric field in vacuum, i.e., outside the masses producing the field. Setting Eqs. (4.1.34) equal to zero leads to the independent equations:

$$e^{-\lambda}\left(\frac{\nu'}{r} + \frac{1}{r^2}\right) - \frac{1}{r^2} = 0, \tag{4.1.35a}$$

$$e^{-\lambda}\left(\frac{\lambda'}{r} - \frac{1}{r^2}\right) + \frac{1}{r^2} = 0, \tag{4.1.35b}$$

$$\dot\lambda = 0. \tag{4.1.35c}$$

From Eq. (4.1.35a) and (4.1.35b) we find $\nu' + \lambda' = 0$, so that $\nu + \lambda = f(v)$, where $f(v)$ is a function of v only. If we perform now the coordinate transformation $x^0 = h(x'^0)$, $x^k = x'^k$, then $g'_{00} = \dot{h}^2 g_{00}$. Such a transformation amounts to adding an arbitrary function of velocity to the function ν , while leaving unaffected the other components of the metric. We can choose the function h so that $\nu + \lambda = 0$. Consequently, we see, by Eq. (4.1.35c), that both ν and λ are independent of velocity. In other words the spherically symmetric gravitational field in vacuum is automatically independent of the velocity of the expansion of the Universe.

Equation (4.1.35b) can now be integrated. It gives:

$$e^{-\lambda} = e^{\nu} = 1 - \frac{K}{r}, \qquad (4.1.36)$$

where K is an integration constant. We see that for $r \to \infty$, $e^{-\lambda} = e^{\nu} = 1$, i.e., far from the gravitational bodies, the metric reduces to that of the flat space (4.1.29). The constant K can easily be determined from the requirement that the law of motion be obtained at large distances from the central mass. From the geodesic equation it follows that the radial acceleration of a small test mass is (see Subsection 4.1.5):

$$-\Gamma^1_{00} = -\frac{1}{2}\left(1 - \frac{K}{r}\right)\frac{K}{r^2} \to -\frac{K}{2r^2}. \qquad (4.1.37)$$

Comparing this expression with the Newtonian value $-km/r^2$ gives $K = 2km$, where m is the central mass and $k = G\tau^2/c^2$, where G is the Newtonian constant.

The constant $2Gm$, or $2Gm/c^2$ (in units where c is not taken as equal to 1), is called the *Schwarzschild radius* of the mass m.

We therefore obtain for the spherically symmetric metric the form:

$$g_{\mu\nu} = \begin{pmatrix} 1 - \dfrac{2km}{\tau^2 r} & 0 & 0 & 0 \\ 0 & -\left(1 - \dfrac{2km}{\tau^2 r}\right)^{-1} & 0 & 0 \\ 0 & 0 & -r^2 & 0 \\ 0 & 0 & 0 & -r^2\sin^2\theta \end{pmatrix}. \qquad (4.1.38)$$

The solution is valid for all velocities, and thus at all locations (outside of the masses producing the field) of the expanding Universe. This is exactly the *Schwarzschild solution*, since $2km/\tau^2 r = 2Gm/c^2 r$, and it describes the most general spherically symmetric solution of the Einstein field equations in a region of space where the energy-momentum tensor $T^{\mu\nu}$ vanishes. Although $g_{\mu\nu}$ goes to the flat space metric when r goes to infinity, it was *not* necessary to require this asymptotic behavior to obtain the solution.

When the equations of motion are written in terms of velocity instead of time G is replaced by the constant k. Using the above equations one then has

$$\kappa = \frac{8\pi G}{c^2 \tau^2} \qquad (4.1.39)$$

It remains to find out the meaning of the solution obtained in spacevelocity as compared to the Schwarzschild solution in spacetime. In the latter case, when $ds = 0$, we have

$$c^2 \left(1 - \frac{2Gm}{c^2 r}\right) dt^2 - \frac{dr^2}{1 - \dfrac{2Gm}{c^2 r}} - r^2 d\Omega^2 = 0, \qquad (4.1.40)$$

where

$$d\Omega^2 = d\theta^2 + \sin^2\theta d\phi^2. \qquad (4.1.41)$$

For radial motion $d\Omega = 0$, and we obtain

$$\frac{dr}{dt} = c\left(1 - \frac{2Gm}{c^2 r}\right). \qquad (4.1.42)$$

This is the radial equation of the propagation of light in the Schwarzschild metric.

A similar situation exists in the spacevelocity expanding Universe case. The null condition $ds^2 = 0$ gives

$$\tau^2 \left(1 - \frac{2Gm}{c^2 r}\right) dv^2 - \frac{dr^2}{1 - \dfrac{2Gm}{c^2 r}} - r^2 d\Omega^2 = 0, \qquad (4.1.43)$$

thus we have for the case of radial expansion

$$\frac{dr}{dv} = \tau\left(1 - \frac{2Gm}{c^2 r}\right). \qquad (4.1.44)$$

Except for the constants c and τ, the right hand sides of Eqs. (4.1.42) and (4.1.44) are identical. Let us assume that the mass m, which appears in these equations, is the mass of the Sun, and r is the radius of the Sun. Then both equations involve the Schwarzschild radius of the Sun R_s divided by the radius of the Sun $R_\odot$, $R_s/R_\odot$.

Equation(4.1.44) can further be related to cosmology in an interesting way. From the Hubble expansion formula

$$v = H_0 r, \qquad (4.1.45)$$

we obtain

$$\frac{dr}{dv} = H_0^{-1}, \qquad (4.1.46)$$

since H_0 is constant at the observation time. Hence Eq. (4.1.44) gives

$$H_0^{-1} = \tau\left(1 - \frac{2Gm}{c^2 r}\right), \qquad (4.1.47)$$

or

$$H_0 = \frac{1}{\tau} \frac{1}{1 - \dfrac{R_s}{R_\odot}} = \frac{h}{1 - \dfrac{R_s}{R_\odot}}, \qquad (4.1.48)$$

where h is the Hubble constant in empty space.

We do not have to confine ourselves to the Sun and might consider huge masses along the way of the signals to the Hubble Space Telescope. In that case the Hubble constant H_0 might deviate substantially from the Hubble constant in vacuum h. We then have

$$H_0 = \frac{h}{1 - \dfrac{2Gm}{c^2 R}},$$

from which we obtain

$$\frac{2Gm}{c^2 R} = 1 - \frac{h}{H_0}.$$

For example if $h = 72.17$ km/s-Mpc, and $H_0 = 80.17$ km/s-Mpc, then we have

$$\frac{2Gm}{c^2 R} = 0.1.$$

It is interesting to know what kind of matter distribution may cause this kind of deviation.

It thus follows that the Hubble constant depends on how close to the Sun electromagnetic signals pass. It appears that there is a gravitational effect similar to that of the light bending in classical general relativity. An experiment to measure this effect will have to be designed.

In the next section we find the analogue to the Schwarzschild metric in an expanding Universe.

4.2 Spherically-Symmetric Metric

In this section we solve the gravitational field equations in the four-dimensional spacevelocity.

4.2.1 *Energy-momentum tensor with pressure*

In the previous section we wrote down the energy-momentum tensor for the Universe without pressure. In general, we need the energy-momentum tensor that includes pressure. This is given by

$$T^{\mu\nu} = (\rho_{eff} + p)\, u^\mu u^\nu - p g^{\mu\nu}, \qquad (4.2.1)$$

where p is the pressure and the speed of light in empty space c was taken as unity.

4.2.2 *The metric*

The gravitational field that is sought is assumed to be both independent of velocity and spherically symmetric, and is therefore given by

$$ds^2 = e^\nu \tau^2 dv^2 - e^\lambda dr^2 - r^2 \left(d\theta^2 + \sin^2\theta d\phi^2\right) \qquad (4.2.2)$$

where ν, λ are functions of r alone, and $u^0 = u_0^{-1} = (g_{00})^{-1/2}$, and $u^\alpha = u_\alpha = (1,0,0,0)$; other components of u^α are zero. As has been described in the last section, the Universe expansion is obtained by the null requirement, $ds = 0$. Since the Universe expands in a spherically symmetric way, one also has $d\theta = d\phi = 0$. As a result, Eq. (4.2.2) reduces to

$$e^\nu \tau^2 dv^2 - e^\lambda dr^2 = 0, \qquad (4.2.3)$$

which yields for the Universe expansion the very simple formula

$$\frac{dr}{dv} = \tau e^{(\nu-\lambda)/2}. \qquad (4.2.4)$$

Using now the Hubble expansion formula

$$v = H_0 r,$$

we have

$$\frac{dv}{dr} = H_0.$$

Hence

$$H_0 = \frac{1}{\tau} e^{(\lambda-\nu)/2}.$$

4.2.3 *The field equations*

The nonvanishing components of the mixed Einstein tensor G_μ^ν (not yet assuming independence on v) gives the following for the gravitational field equations:

$$G_0^0 = -e^{-\lambda}\left(\frac{1}{r^2} - \frac{\lambda'}{r}\right) + \frac{1}{r^2} = \kappa T_0^0, \qquad (4.2.5a)$$

$$G_0^1 = -e^{-\lambda}\frac{\dot\lambda}{r} = \kappa T_0^1, \qquad (4.2.5b)$$

$$G_1^1 = -e^{-\lambda}\left(\frac{\nu'}{r} + \frac{1}{r^2}\right) + \frac{1}{r^2} = \kappa T_1^1, \tag{4.2.5c}$$

$$G_2^2 = -\frac{1}{2}e^{-\lambda}\left(\nu'' + \frac{\nu'^2}{2} + \frac{\nu' - \lambda'}{r} - \frac{\nu'\lambda'}{2}\right) + \frac{1}{2}e^{-\nu}\left(\ddot{\lambda} + \frac{\dot{\lambda}^2}{2} - \frac{\dot{\nu}\dot{\lambda}}{2}\right)$$

$$= \kappa T_2^2, \tag{4.2.5d}$$

$$G_3^3 = G_2^2 = \kappa T_3^3. \tag{4.2.5e}$$

All other components of the Einstein tensor vanish identically, a prime denotes differentiation with respect to r, and a dot denotes differentiation with respect to v.

The above equations (now assuming independence on v) then yield

$$e^{-\lambda}\left(\frac{1}{r^2} - \frac{\lambda'}{r}\right) - \frac{1}{r^2} = -\kappa\rho_{eff}, \tag{4.2.6}$$

$$e^{-\lambda}\left(\frac{1}{r^2} + \frac{\nu'}{r}\right) - \frac{1}{r^2} = \kappa p, \tag{4.2.7}$$

$$\frac{1}{2}e^{-\lambda}\left(\nu'' + \frac{1}{2}\nu'^2 + \frac{\nu' - \lambda'}{r} - \frac{1}{2}\nu'\lambda'\right) = \kappa p. \tag{4.2.8}$$

The conservation law $\nabla_\nu T^{\mu\nu} = 0$ yields

$$p' = -\frac{1}{2}\nu'\left(p + \rho_{eff}\right). \tag{4.2.9}$$

Equation (4.2.9) is not independent of Eqs. (4.2.6)-(4.2.8) since it is a consequence of the contracted Bianchi identities. One therefore has three equations for the four unknown functions ν, λ, ρ, p. One assumes a functional dependence of ρ on r, calculate ν, λ from this knowledge, and finally calculate p.

4.2.4 *Solutions*

The solution of Eq. (4.2.6) is given by

$$e^{-\lambda} = 1 - \frac{\kappa}{4\pi}\frac{m(r)}{r}, \tag{4.2.10}$$

where

$$m(r) = 4\pi \int_0^r \rho_{eff}(r')\, r'^2 dr' \tag{4.2.11}$$

is the mass of the fluid contained in a ball of radius r. The solution given by Eq. (4.2.10) is chosen so that $g_{\mu\nu}$ is regular at $r = 0$ and goes to the Schwarzschild form

$$e^{-\lambda} = 1 - \frac{r_s}{r}, \tag{4.2.12}$$

where $r_s = 2Gm$ (divided by c^2) and $m = m(r_0)$, if $\rho_{eff}(r) = 0$ for $r > r_0$.

We now assume that ρ is a constant for $r \leq r_0$. We then obtain from Eqs. (4.2.9), (4.2.6), (4.2.7) and (4.2.10) the following:

$$e^{-\lambda} = 1 - \frac{r^2}{R^2}, \tag{4.2.13}$$

$$e^{\nu/2} = A - B\sqrt{1 - \frac{r^2}{R^2}}, \tag{4.2.14}$$

$$p = \frac{1}{\kappa R^2}\left[\frac{3B\sqrt{1 - \frac{r^2}{R^2}} - A}{A - B\sqrt{1 - \frac{r^2}{R^2}}}\right], \tag{4.2.15}$$

where A and B are constants, and

$$R^2 = \frac{3}{\kappa\rho_{eff}}. \tag{4.2.16}$$

The constants A and B can be fixed by the requirements that $p = 0$ and e^ν join smoothly the Schwarzschild field on the surface of the sphere. One obtains

$$A = \frac{3}{2}\sqrt{1 - \frac{r_0^2}{R^2}}, \qquad B = \frac{1}{2}, \tag{4.2.17}$$

$$e^{\nu/2} = \frac{3}{2}\sqrt{1 - \frac{r_0^2}{R^2}} - \frac{1}{2}\sqrt{1 - \frac{r^2}{R^2}}, \tag{4.2.18}$$

$$p = \rho\left[\frac{\sqrt{1 - \frac{r^2}{R^2}} - \sqrt{1 - \frac{r_0^2}{R^2}}}{3\sqrt{1 - \frac{r_0^2}{R^2}} - \sqrt{1 - \frac{r^2}{R^2}}}\right], \tag{4.2.19}$$

with the condition that $r_0^2 < R^2$. If one assumes that the pressure inside the fluid is finite everywhere, one obtains from Eq. (4.2.19) the more restrictive condition

$$r_0^2 < \frac{8}{9} R^2. \tag{4.2.20}$$

The spacetime-coordinate version of the solutions presented above are due to K. Schwarzschild.

4.2.5 *The Universe expansion*

Using the above results in the equation for the Universe expansion (4.2.4) we obtain

$$\frac{dr}{dv} = \tau \left[A \sqrt{1 - \frac{r^2}{R^2}} - B \left(1 - \frac{r^2}{R^2} \right) \right]. \tag{4.2.21}$$

We now confine ourselves to the linear approximation, getting

$$\frac{dr}{dv} \approx \tau \left(\tilde{A} + \tilde{B} \frac{r^2}{R^2} \right), \tag{4.2.22}$$

where $\tilde{A} = A - B$ and $\tilde{B} = B - A/2$, or

$$\frac{dv}{dr} \approx \frac{1}{\tau} \left(\tilde{A} - \tilde{B} \frac{r^2}{R^2} \right). \tag{4.2.23}$$

A simplification is also obtained if we confine to the linear approximation of A and B, hence $A = 3/2$, $B = 1/2$, thus $\tilde{A} = 1$, $\tilde{B} = -1/4$. Using now the standard notation $\Omega = \rho/\rho_c$, we obtain

$$\frac{dr}{dv} \approx \tau \left[1 + \frac{(1 - \Omega) r^2}{4c^2 \tau^2} \right] \tag{4.2.24}$$

for the equation of the expansion of the Universe. Except for the factor 4, Eq. (4.2.24) is identical to Eq. (15) of M. Carmeli, *Commun. Theor. Phys.*, 1996 and Eq. (5.10) of Behar and Carmeli, *Intern. J. Theor. Phys.*, 2000 (see Suggested References) (see Problem 4.2.1). With $A = 1$ and $B = 0$ we get the same result as in the latter.

The second term in the square bracket in the above equation represents the deviation from the standard Hubble law due to gravity. For without that term, Eq. (4.2.24) reduces to $dr/dv = \tau$, thus $r = \tau v + \text{const}$. The constant can be taken zero if one assumes, as usual, that at $r = 0$ the velocity should also vanish. Thus $r = \tau v$, or $v = H_0 r$ (since $H_0 = 1/\tau$). Accordingly, the equation of motion (4.2.24) describes the expansion of the Universe when $\Omega = 1$, namely when $\rho = \rho_c$, the equation coincides with the standard Hubble law.

4.2.6 *Problem (a)*

P 4.2.1. Use the Hubble expansion formula

$$v = H_0 r \tag{1}$$

in order to find the value of the Hubble constant H_0 from the expansion formula for the Universe Eq. (4.2.24).

Solution: From the Hubble law we obtain, since H_0 is constant,

$$dv = H_0 dr. \tag{2}$$

Therefore

$$\frac{1}{H_0} = \frac{dr}{dv}. \tag{3}$$

Using Eq. (4.2.24) we then obtain

$$\frac{1}{H_0} = \tau \left[1 + \frac{(1 - \Omega)\, r^2}{4c^2\tau^2} \right]. \tag{4}$$

For the case $\Omega = 1$ we then have

$$H_0 = \frac{1}{\tau}. \tag{5}$$

Otherwise

$$H_0 = \frac{1}{\tau} \left[1 + \frac{(1 - \Omega)\, r^2}{4c^2\tau^2} \right]^{-1}. \tag{6}$$

For $r \to 0$ we have:

$$H_0 = \frac{1}{\tau}, \tag{7}$$

and for $r \to c\tau$, we have

$$H_0 = \frac{1}{\tau} \left(1 + \frac{1 - \Omega}{4} \right). \tag{8}$$

4.2.7 *Integration of equation of motion*

The equation of motion (4.2.24) can easily be integrated exactly by the substitutions

$$\sin \chi = \frac{r}{2c\tau} \sqrt{\Omega - 1}; \quad \Omega > 1, \tag{4.2.25a}$$

$$\sinh \chi = \frac{r}{2c\tau} \sqrt{1 - \Omega}; \quad \Omega < 1. \tag{4.2.25b}$$

One then obtains, using Eqs. (4.2.24) and (4.2.25),

$$dv = \frac{2c}{\sqrt{\Omega - 1}} \frac{d\chi}{\cos \chi}; \quad \Omega > 1, \tag{4.2.26a}$$

$$dv = \frac{2c}{\sqrt{1 - \Omega}} \frac{d\chi}{\cosh \chi}; \quad \Omega < 1. \tag{4.2.26b}$$

We give below the exact solutions for the expansion of the Universe for each of the cases, $\Omega > 1$ and $\Omega < 1$. As will be seen, the case of $\Omega = 1$ can be obtained at the limit $\Omega \to 1$ from both cases.

4.2.7.1 *The case $\Omega > 1$.*

From Eq. (4.2.26a) we have

$$\int dv = \frac{2c}{\sqrt{\Omega - 1}} \int \frac{d\chi}{\cos \chi}, \qquad (4.2.27)$$

where $\sin \chi = \frac{r}{a}$, and $a = c\tau\sqrt{\Omega - 1}$. A simple calculation gives

$$\int \frac{d\chi}{\cos \chi} = \ln \left| \frac{1 + \sin \chi}{\cos \chi} \right|. \qquad (4.2.28)$$

A straightforward calculation then gives

$$v = \frac{a}{\tau} \ln \left| \frac{1 + \dfrac{r}{a}}{1 - \dfrac{r}{a}} \right| = c\sqrt{\Omega - 1}\, \ln \left| \frac{1 + \dfrac{r}{a}}{1 - \dfrac{r}{a}} \right|. \qquad (4.2.29)$$

As is seen, when $r \to 0$ then $v \to 0$, and using the L'Hospital lemma one obtains $v \to r/\tau$ as $a \to 0$ (and thus $\Omega \to 1$).

4.2.7.2 *The case $\Omega < 1$.*

From Eq. (4.2.26b) we now have

$$\int dv = \frac{2c}{\sqrt{1 - \Omega}} \int \frac{d\chi}{\cosh \chi}, \qquad (4.2.30)$$

where $\sinh \chi = \dfrac{r}{b}$, and $b = c\tau\sqrt{1 - \Omega}$. A straightforward calculation then gives

$$\int \frac{d\chi}{\cosh \chi} = \arctan e^\chi. \qquad (4.2.31)$$

We then obtain

$$\cosh \chi = \sqrt{1 + \frac{r^2}{b^2}}, \qquad (4.2.32)$$

$$e^\chi = \sinh \chi + \cosh \chi = \frac{r}{b} + \sqrt{1 + \frac{r^2}{b^2}}. \qquad (4.2.33)$$

Equations (4.2.30) and (4.2.31) now give

$$v = \frac{2c}{\sqrt{1 - \Omega}} \arctan e^\chi + K, \qquad (4.2.34)$$

where K is an integration constant which is determined by the requirement that at $r = 0$, v is zero. We obtain

$$K = -\frac{\pi c}{2\sqrt{1 - \Omega}}, \qquad (4.2.35)$$

and thus

$$v = \frac{c}{\sqrt{1-\Omega}} \left(\arctan e^{\chi} - \frac{\pi}{4} \right). \qquad (4.2.36)$$

A straightforward calculation then gives

$$v = \frac{b}{\tau} \left\{ \arctan \left(\frac{r}{b} + \sqrt{1 + \frac{r^2}{b^2}} \right) - \frac{\pi}{2} \right\}$$

$$= c\sqrt{1-\Omega} \left\{ \arctan \left(\frac{r}{b} + \sqrt{1 + \frac{r^2}{b^2}} \right) - \frac{\pi}{2} \right\}. \qquad (4.2.37)$$

As for the case $\Omega > 1$ one finds that $v \to 0$ when $r \to 0$, and again, using L'Hospital lemma, $r = \tau v$ when $b \to 0$ (and thus $\Omega \to 1$).

4.2.8 *Physical meaning*

To see the physical meaning of these solutions, however, one does not need the exact solutions. Rather, it is enough to write down the solutions in the lowest approximation in τ^{-1}. Assuming $A = 1$ and $B = 0$ in Eq. (4.2.24) one obtains by differentiating it with respect to v where $\Omega > 1$,

$$\frac{d^2 r}{dv^2} = -kr; \qquad k = \frac{\Omega - 1}{c^2}, \qquad (4.2.38)$$

the solution of which is

$$r(v) = A \sin \frac{v}{c} \alpha + B \cos \frac{v}{c} \alpha, \qquad (4.2.39)$$

where $\alpha^2 = \Omega - 1$ and A and B are constants. The latter can be determined by the initial condition $r(0) = 0 = B$ and $dr(0)/dv = \tau = A\alpha/c$, thus

$$r(v) = \frac{c\tau}{\alpha} \sin \frac{v}{c} \alpha. \qquad (4.2.40)$$

This is obviously a closed Universe, and presents a decelerating expansion.

For $\Omega < 1$ we have

$$\frac{d^2 r}{dv^2} = \frac{(1-\Omega) r}{c^2}, \qquad (4.2.41)$$

whose solution, using the same initial conditions, is

$$r(v) = \frac{c\tau}{\beta} \sinh \frac{v}{c} \beta, \qquad (4.2.42)$$

where $\beta^2 = 1 - \Omega$. This is now an open accelerating Universe.

For $\Omega = 1$ we have, of course, $r = \tau v$ (see Figure 4.2.1).

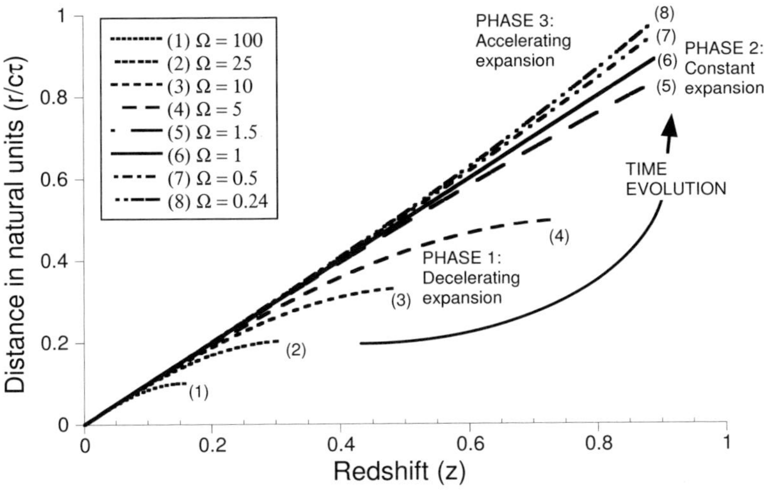

Fig. 4.2.1:   Hubble's diagram describing the tri-phase evolution of the Universe according to cosmological general relativity theory (adapted from Behar and Carmeli).

4.2.9   *Expansion at present epoch of time*

We finally determine which of the three cases of expansion is the one at present epoch of time. To this end we have to write the solutions (4.2.40) and (4.2.42) in ordinary Hubble's law form $v = H_0 r$. Expanding Eqs. (4.2.40) and (4.2.42) into power series in v/c and keeping terms up to the second order, we obtain

$$r = \tau v \left(1 - \frac{\alpha^2 v^2}{6c^2} \right) \qquad (4.2.43a)$$

$$r = \tau v \left(1 + \frac{\beta^2 v^2}{6c^2} \right) \qquad (4.2.43b)$$

for $\Omega > 1$ and $\Omega < 1$, respectively. Using now the expressions for α and β, Eqs. (4.2.43) then reduce into the single equation

$$r = \tau v \left[1 + \frac{(1 - \Omega)\, v^2}{6c^2} \right]. \qquad (4.2.44)$$

Inverting now this equation by writing it as $v = H_0 r$, we obtain in the lowest approximation

$$H_0 = h \left[1 - \frac{(1 - \Omega)\, v^2}{6c^2} \right], \qquad (4.2.45)$$

Table 4.1:   The Cosmic Times with respect to the Big Bang, the Cosmic Temperature and the Cosmic Pressure for each of the Curves in Figure 4.2.1.

Curve No[*]	Ω	Time in Units of τ	Time (Gyr)	Temperature (K)	Pressure (g/cm^2)
		DECELERATING EXPANSION			
1	100	3.1×10^{-6}	4.23×10^{-5}	1096	-4.142
2	25	9.8×10^{-5}	1.32×10^{-3}	195.0	-1.004
3	10	3.0×10^{-4}	4.07×10^{-3}	111.5	-0.377
4	5	1.2×10^{-3}	1.63×10^{-2}	58.20	-0.167
5	1.5	1.3×10^{-2}	1.76×10^{-1}	16.43	-0.021
		CONSTANT EXPANSION			
6	1	3.0×10^{-2}	4.07×10^{-1}	11.15	0
		ACCELERATING EXPANSION			
7	0.5	1.3×10^{-1}	1.76	5.538	+0.021
8	0.245	1.0	13.56	2.730	+0.032

[*]The curve number corresponds to curves in Figure 4.2.1. The calculations are made using the cosmological transformation, Eq. (2.2.5), that relates physical quantities at different cosmic times when gravity is extremely weak.

For example, we denote the temperature by θ, and the temperature at the present time by θ_0, we then have

$$\theta = \frac{\theta_0}{\sqrt{1 - \dfrac{t^2}{\tau^2}}} = \frac{\theta_0}{\sqrt{1 - \dfrac{(\tau - T)^2}{\tau^2}}} = \frac{2.73K}{\sqrt{\dfrac{2\tau T - T^2}{\tau^2}}} = \frac{2.73K}{\sqrt{\dfrac{T}{\tau}\left(2 - \dfrac{T}{\tau}\right)}},$$

where T is the time with respect to the Big Bang time.

The formula for the pressure is given by $p = c(1 - \Omega)/8\pi G\tau$. Using $c = 3 \times 10^{10}$cm/s, $\tau = 4.28 \times 10^{17}$s and $G = 6.67 \times 10^{-8}$cm^3/gs^2, we obtain

$$p = 4.184 \times 10^{-2} \left(1 - \Omega\right) g/cm^2.$$

where $h = \tau^{-1}$. To the same approximation one also obtains

$$H_0 = h\left[1 - \frac{(1 - \Omega)z^2}{6}\right] = h\left[1 - \frac{(1 - \Omega)r^2}{6c^2\tau^2}\right], \tag{4.2.46}$$

where z is the redshift parameter (see Problem 4.2.2 for an alternative way to find the value of the Hubble constant H_0).

As is seen, and it is confirmed by experiments, H_0 depends on the distance it is being measured; this fact has been emphasized by Peebles (see Peebles 1993). Accordingly the Hubble parameter (constant) H_0 has physical meaning only at the zero-distance limit (and thus at the zero gravity limit), namely when measured *locally*, in which case it becomes $h = 1/\tau$.

It follows that the measured value of H_0 depends on the "short" and the "long" distance scales (see Peebles 1993). The greater the distances over which H_0 is measured, the smaller the value obtained. It follows from Eq.

Table 4.2: Cosmological parameters in cosmological general relativity and in standard theory.

	COSMOLOGICAL RELATIVITY	STANDARD THEORY
Theory type	Spacevelocity	Spacetime
Expansion type	Tri-phase: decelerating, constant, accelerating	One phase
Inflation	Follows from theory	Assumed
Present expansion	Accelerating (predicted)	One of three possibilities
Pressure	Positive	Negative
Cosmological constant	None	Depends
$\Omega_T = \Omega + \Omega_\Lambda$	1.0	Depends
Constant-expansion occurs at	8.5Gyr ago (Gravity included)	No prediction
Constant-expansion duration	Fraction of second	Not known
Temperature at constant expansion	146K (Gravity included)	No prediction

(4.2.46) that this is possible only when $\Omega < 1$, hence when the Universe is accelerating.

4.2.10 Problem (b)

P 4.2.2. Find the value of the Hubble constant H_0 by using the Hubble expansion formula

$$v = H_0 r, \tag{1}$$

and Eq. (4.2.44).

Solution: The solution is left for the reader.

4.2.11 The value of the constant τ

To find the numerical value of the Hubble-Carmeli constant τ we use the relationship between $h = \tau^{-1}$ and H_0 given by Eq. (4.2.46).

We recall that τ is the inverse of h, the Hubble constant in the limit of zero gravity or zero distance. It can actually be called the Hubble constant in vacuum (like the speed of light in vacuum). We also recall that τ is the Big Bang time, thus it is the age of the Universe. But the available data published by Wendy Freedman from the Hubble Space Telescope do not include measurements at very short distances for H_0. So we have to refer to different methods.

Based on curve fitting Eq. (4.2.46) to all the available data at the time, Hartnett (communicated to Carmeli, 1 March 2006) obtained (see Figure 13.4 in this book) $h = 72.47 \pm 1.95$km/s-Mpc, from Tully-Fisher (TF) (the solid line), and $h = 72.17 \pm 0.84$km/s-Mpc, from SNe type Ia (the broken line) measurements. The error bars here are statistical fit standard errors, but the rms errors on the published data are ± 1.64km/s-Mpc for SNe type Ia determined and ± 13.24 for TF determined. We choose

$$h = 72.17 \pm 0.84 \pm 1.64 \text{km/s-Mpc}.$$

Therefore

$$\tau = (4.28 \pm 0.15) \times 10^{17} \text{s} = 13.56 \pm 0.48 \text{Gyr}.$$

This result fits the recently obtained measurements by NASA's WMAP, according to which initial cosmic inflation happened 13.7 ± 0.2Gyr ago.

Using Eq. (4.2.42) we get the status of the Universe at present as shown in Figure 4.2.2, where $\Omega < 1$.

In the next section it is shown how the familiar Tolman metric can be looked upon as an expanding Universe.

4.3   Tolman Metric as an Expanding Universe

4.3.1   *The Tolman metric*

In the four-dimensional spacevelocity the Tolman metric is given by

$$ds^2 = \tau^2 dv^2 - e^{\mu} dr^2 - R^2 \left(d\theta^2 + \sin^2 \theta d\phi^2\right), \qquad (4.3.1)$$

where μ and R are functions of v and r alone, and comoving coordinates $x^{\mu} = (x^0, x^1, x^2, x^3) = (\tau v, r, \theta, \phi)$ have been used. With the above choice of coordinates, the zero-component of the geodesic equation becomes an identity, and since r, θ and ϕ are constants along the geodesics, one has $dx^0 = ds$ and therefore

$$u^{\alpha} = u_{\alpha} = (1, 0, 0, 0). \qquad (4.3.2)$$

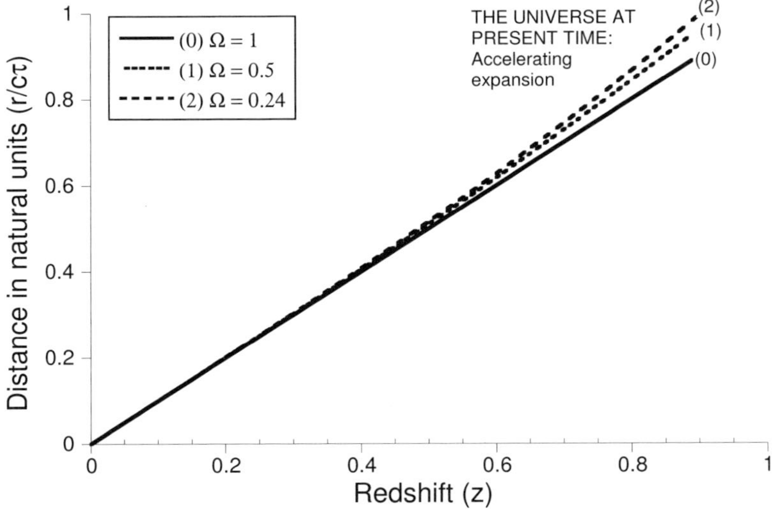

Fig. 4.2.2: Hubble's diagram of the Universe at the present phase of evolution with accelerating expansion. (Source: Behar and Carmeli)

The metric (4.3.1) shows that the area of the sphere r =constant is given by $4\pi R^2$ and that R should satisfy $R' = \partial R/\partial r > 0$. The possibility that $R' = 0$ at a point r_0 should be excluded since it would allow the lines r =constants at the neighboring points r_0 and $r_0 + dr$ to coincide at r_0, thus creating a caustic surface at which the comoving coordinates break down.

As has been shown in the previous sections, the Universe expands by the null condition $ds = 0$, and if the expansion is spherically symmetric one has $d\theta = d\phi = 0$. The metric (4.3.1) then yields

$$\tau^2 dv^2 - e^\mu dr^2 = 0, \tag{4.3.3}$$

thus

$$\frac{dr}{dv} = \tau e^{-\mu/2}. \tag{4.3.4}$$

This is the differential equation that determines the Universe expansion. In the following we solve the gravitational field equations in order to find out the function $\mu(r)$.

4.3.2   *Field equations*

The gravitational field equations (2.3), written in the form

$$R_{\mu\nu} = \kappa \left(T_{\mu\nu} - \frac{1}{2} g_{\mu\nu} T \right), \tag{4.3.5}$$

where

$$T_{\mu\nu} = \rho_{eff} u_\mu u_\nu \tag{4.3.6}$$

with $\rho_{eff} = \rho - \rho_c$ and $T = T_{\mu\nu} g^{\mu\nu}$, are now solved. Using Eq. (4.3.2) one finds that the only nonvanishing component of $T_{\mu\nu}$ is $T_{00} = \rho_{eff}$ and that $T = \rho_{eff}$.

The only nonvanishing components of the Ricci tensor are (dots and primes denote differentiation with respect to v and r, respectively):

$$R_{00} = -\frac{1}{2}\ddot{\mu} - \frac{2}{R}\ddot{R} - \frac{1}{4}\dot{\mu}^2, \tag{4.3.7a}$$

$$R_{01} = \frac{1}{R}R'\dot{\mu} - \frac{2}{R}\dot{R}', \tag{4.3.7b}$$

$$R_{11} = e^\mu \left(\frac{1}{2}\ddot{\mu} + \frac{1}{4}\dot{\mu}^2 + \frac{1}{R}\dot{\mu}\dot{R} \right) + \frac{1}{R}\left(\mu' R' - 2R'' \right), \tag{4.3.7c}$$

$$R_{22} = R\ddot{R} + \frac{1}{2}R\dot{R}\dot{\mu} + \dot{R}^2 + 1 - e^{-\mu}\left(RR'' - \frac{1}{2}RR'\mu' + R'^2 \right), \tag{4.3.7d}$$

$$R_{33} = \sin^2\theta \, R_{22}, \tag{4.3.7e}$$

whereas the Ricci scalar is given by

$$R = 2e^{-\mu}\left[\frac{2}{R}R'' + \left(\frac{R'}{R} \right)^2 - \frac{1}{R}R'\mu' \right] - \frac{2}{R}\dot{R}\dot{\mu} - 2\left(\frac{\dot{R}}{R} \right)^2$$

$$-\frac{2}{R^2} - \frac{4}{R}\ddot{R} - \ddot{\mu} - \frac{1}{2}\dot{\mu}^2. \tag{4.3.8}$$

The field equations obtained for the components 00, 01, 11, and 22 (the 33 component contributes no new information) are given by

$$-\ddot{\mu} - \frac{4}{R}\ddot{R} - \frac{1}{2}\dot{\mu}^2 = \kappa\rho_{eff} \tag{4.3.9}$$

$$2\dot{R}' - R'\dot{\mu} = 0 \tag{4.3.10}$$

$$\ddot{\mu} + \frac{1}{2}\dot{\mu}^2 + \frac{2}{R}\dot{R}\dot{\mu} + e^{-\mu}\left(\frac{2}{R}R'\mu' - \frac{4}{R}R''\right) = \kappa\rho_{eff} \tag{4.3.11}$$

$$\frac{2}{R}\ddot{R} + 2\left(\frac{\dot{R}}{R}\right)^2 + \frac{1}{R}\dot{R}\dot{\mu} + \frac{2}{R^2} + e^{-\mu}\left[\frac{1}{R}R'\mu' - 2\left(\frac{R'}{R}\right)^2 - \frac{2}{R}R''\right]$$

$$= \kappa\rho_{eff} \tag{4.3.12}$$

It is convenient to eliminate the term with the second velocity derivative of μ from the above equations. This can easily be done, and combinations of Eqs. (4.3.9)–(4.3.12) then give the following set of three independent field equations:

$$e^{\mu}\left(2R\ddot{R} + \dot{R}^2 + 1\right) - R'^2 = 0 \tag{4.3.13}$$

$$2\dot{R}' - R'\dot{\mu} = 0 \tag{4.3.14}$$

$$e^{-\mu}\left[\frac{1}{R}R'\mu' - \left(\frac{R'}{R}\right)^2 - \frac{2}{R}R''\right] + \frac{1}{R}\dot{R}\dot{\mu} + \left(\frac{\dot{R}}{R}\right)^2 + \frac{1}{R^2} = \kappa\rho_{eff} \tag{4.3.15}$$

other equations being trivial combinations of (4.3.13)–(4.3.15).

4.3.3 *Solutions*

The solution of Eq. (4.3.14) satisfying the condition $R' > 0$ is given by

$$e^{\mu} = \frac{R'^2}{1 + f(r)}, \tag{4.3.16}$$

where $f(r)$ is an arbitrary function of the coordinate r and satisfies the condition $f(r) > -1$. Substituting (4.3.16) in the other two field equations (4.3.13) and (4.3.15) then gives

$$2R\ddot{R} + \dot{R}^2 - f = 0 \tag{4.3.17}$$

$$\frac{1}{RR'}\left(2\dot{R}\dot{R}' - f'\right) + \frac{1}{R^2}\left(\dot{R}^2 - f\right) = \kappa\rho_{eff}, \tag{4.3.18}$$

respectively.

The integration of these equations is now straightforward. From Eq. (4.3.17) we obtain the first integral

$$\dot{R}^2 = f(r) + \frac{F(r)}{R}, \tag{4.3.19}$$

where $F(r)$ is another arbitrary function of r. Substituting now (4.3.19) in Eq. (4.3.18) gives

$$\frac{F'}{R^2 R'} = \kappa \rho_{eff}. \tag{4.3.20}$$

Equations (4.3.19) and (4.3.20) are now integrated for the case for which f equals zero, and Eq. (4.3.19) consequently reduces to

$$\dot{R}^2 = \frac{F(r)}{R}. \tag{4.3.21}$$

The integration of Eq. (4.3.21) gives

$$R(v,r) = \left[R^{3/2}(r) \pm \frac{3}{2} F^{1/2}(r) v \right]^{2/3}, \tag{4.3.22}$$

where

$$R(r) = R(0,r), \tag{4.3.23}$$

namely, $R(v,r)$ at $v = 0$. Differentiating Eq. (4.3.22) with respect to r and using Eq. (4.3.20), we also obtain

$$R(v,r) = (\kappa \rho_{eff})^{-2/3} \left[\frac{R^{1/2}(r) R'(r)}{F'(r)} \pm \frac{v}{2F^{1/2}(r)} \right]^{-2/3}. \tag{4.3.24}$$

Finally, from Eq. (4.3.20) we obtain

$$\frac{\partial}{\partial v} \left(\rho_{eff} R^2 R' \right) = 0, \tag{4.3.25}$$

and accordingly one has

$$\frac{dr}{dv} = \tau e^{-\mu/2} = \tau (R')^{-1/2}. \tag{4.3.26}$$

4.3.4 *The Universe expansion*

If the function f is not zero, the integration of Eq. (4.3.19) then yields for $f > 0$,

$$\tau v = \frac{1}{f} \sqrt{fR^2 + FR} - \frac{F}{\sqrt{f^3}} \operatorname{arcsinh} \sqrt{\frac{fR}{F}} + \Phi(r), \tag{4.3.27a}$$

and for $f < 0$,

$$\tau v = \frac{1}{f} \sqrt{fR^2 + FR} - \frac{F}{\sqrt{(-f)^3}} \arcsin \sqrt{-\frac{fR}{F}}$$

$$+ \Phi(r), \tag{4.3.27b}$$

where $\Phi(r)$ is an arbitrary function of r. The solutions (4.3.27) were derived in spacetime coordinates by Datt (given by Landau and Lifshitz, see Suggested References).

4.3.5 *Tolman's Universe with pressure*

The above discussion on Tolman's Universe was done without taking into account that the Universe has matter and therefore pressure. In the following we extend the discussion given above to include pressure.

The energy-momentum tensor (4.3.6) will now have to be replaced by

$$T_{\mu\nu} = \rho_{eff} u_\mu u_\nu + p\left(u_\mu u_\nu - g_{\mu\nu}\right). \tag{4.3.28}$$

Using

$$u_\alpha = u^\alpha = (1,0,0,0),$$

one can find the nonvanishing components of $T_{\mu\nu}$ to be given by:

$$T_{00} = \tau^2 \rho_{eff},$$

$$T_{11} = \frac{\tau}{c} p e^\mu,$$

$$T_{22} = \frac{\tau}{c} p R^2$$

and

$$T_{33} = \frac{\tau}{c} p R^2 \sin^2\theta,$$

and

$$T = \tau^2 \rho_{eff} - 3\frac{\tau}{c}p.$$

We proceed to write down the gravitational field equations. It will be noted that the Ricci tensor is still given by Eqs. (4.3.7) since the metric used is the same as was before. The gravitational field equations (4.3.5) therefore yield:

$$e^\mu \left(2R\ddot{R} + \dot{R}^2 + 1\right) - R'^2 = -\kappa\tau c^{-1} e^\mu R^2 p, \tag{4.3.29}$$

$$2\dot{R}' - R'\dot{\mu} = 0, \tag{4.3.30}$$

$$e^{-\mu}\left[R^{-1}R'\mu' - \left(\frac{R'}{R}\right)^2 - 2R^{-1}R''\right] + R^{-1}\dot{R}\dot{\mu} + \left(\frac{\dot{R}}{R}\right)^2 + R^{-2} = \kappa\tau^2 \rho_{eff}. \tag{4.3.31}$$

In the above equations a dot and a prime denote differentiation with respect to x^0 and r.

Equations (4.3.30) and (4.3.31) are the same as Eqs. (4.3.14) and (4.3.15), except for the factor τ^2 on the right-hand side of Eq. (4.3.31).

The solution of Eq. (4.3.30) is already known and is given by Eq. (4.3.16),

$$e^{\mu} = \frac{R'^2}{1 + f\left(r\right)}, \tag{4.3.16}$$

where $f\left(r\right)$ is an arbitrary function of the coordinate r and satisfies the condition $f\left(r\right) > -1$. Substituting the solution (4.3.16) in Eqs. (4.3.29) and (4.3.31) then yields

$$2R\ddot{R} + \dot{R}^2 - f = -\kappa c^{-1}\tau R^2 p, \tag{4.3.32}$$

$$\frac{1}{RR'}\left(2\dot{R}\dot{R}' - f'\right) + \frac{1}{R^2}\left(\dot{R}^2 - f\right) = \kappa \tau^2 \rho_{eff}. \tag{4.3.33}$$

A simple solution for the above equations is provided by

$$R(r) = r, \tag{4.3.34}$$

which satisfies the requirement $R' > 0$. It yields

$$p = \frac{c\tau}{3r}\rho_{eff}, \tag{4.3.35a}$$

$$f\left(r\right) = \frac{\kappa\tau}{c}pr^2, \tag{4.3.35b}$$

$$f' + \frac{f}{r} = -\kappa\tau^2\rho_{eff}r. \tag{4.3.35c}$$

To the solution (4.3.35b) one should add the solution of the homogeneous equation

$$f' + \frac{f}{r} = 0, \tag{4.3.35d}$$

which is easily found to be

$$-\frac{2Gm}{c^2 r}.$$

Here m is the mass of a central body located at the origin of the coordinates. Accordingly $f(r)$ will be given by

$$f\left(r\right) = \frac{\kappa\tau}{c}pr^2 - \frac{2Gm}{c^2 r}. \tag{4.3.35e}$$

Equation (4.3.16), using $R' = 1$, becomes

$$e^{\mu} = \frac{1}{1 + f\left(r\right)}. \tag{4.3.36}$$

From Eqs. (4.3.35e) and (4.3.35c) one obtains the expressions for $f(r)$ and the pressure p:

$$f(r) = \frac{(1-\Omega)\, r^2}{c^2\tau^2} - \frac{2Gm}{c^2 r},\tag{4.3.37}$$

where $\Omega = \rho/\rho_c$, and

$$p = \frac{1-\Omega}{\kappa c\tau^3} - \frac{2Gm}{c^2 r} = \frac{c}{\tau}\frac{1-\Omega}{8\pi G} - \frac{2Gm}{c^2 r}.\tag{4.3.38}$$

Finally we obtain:

$$e^{-\mu} = 1 + f(r) = 1 + \frac{\tau\kappa}{c}pr^2 - \frac{2Gm}{c^2 r}$$

$$= 1 + \frac{1-\Omega}{c^2\tau^2}r^2 - \frac{2Gm}{c^2 r}.\tag{4.3.39}$$

Since the Universe expansion is given by

$$\frac{dr}{dv} = \tau e^{-\mu/2},\tag{4.3.40}$$

we have

$$\frac{dr}{dv} = \tau\sqrt{1 + \frac{\kappa\tau}{c}pr^2 - \frac{2Gm}{c^2 r}}$$

$$= \tau\sqrt{1 + \frac{1-\Omega}{c^2\tau^2}r^2 - \frac{2Gm}{c^2 r}}.\tag{4.3.41}$$

See problem 4.3.1 for calculating the Hubble constant H_0 from Eq. (4.3.41).

With the above solution $R(r) = r$, the Tolman metric (4.3.1) will now have the simple form

$$ds^2 = \tau^2 dv^2 - e^{\mu}dr^2 - r^2\left(d\theta^2 + \sin^2\theta d\phi^2\right).\tag{4.3.42}$$

Going back to spacetime, the Tolman metric will be

$$ds^2 = c^2 dt^2 - e^{\mu}dr^2 - r^2\left(d\theta^2 + \sin^2\theta d\phi^2\right).\tag{4.3.43}$$

The next section is devoted to the Kantowski-Sachs metrics, interesting cosmological solutions of the Einstein field equations.

4.3.6 *Problem*

P 4.3.1. Calculate the Hubble constant H_0 from Eq. (4.3.41) by using the Hubble expansion formula

$$v = H_0 r.\tag{1}$$

Solution: The solution is left for the reader.

4.4   Kantowski-Sachs Metrics as Expanding Universes

4.4.1   *Introduction*

Kantowski and Sachs have obtained two new metrics that are spatially homogeneous, have shear, and have no rotation (the geometrical meaning of shear, rotation and expansion are given in Carmeli 1977). They are solutions of the Einstein field equations with an energy-momentum tensor representing a dust source

$$T^\alpha_\beta = \rho u^\alpha u_\beta, \tag{4.4.1}$$

where u^α is a timelike vector field normalized by $u^\alpha u_\alpha = 1$.

4.4.2   *Coordinate system*

They considered two cases with resulting fields that are spatially homogeneous and nonisotropic. Because of the assumed symmetries, there exist, locally, coordinates

$$x^0 = t, \quad x^1 = r, \quad x^2 = \theta, \quad x^3 = \phi. \tag{4.4.2}$$

With these coordinates one then has for the vector field

$$u^\alpha = A\delta^\alpha_0 + B\delta^\alpha_1. \tag{4.4.3}$$

One can then have two metrics:
Case 1:

$$ds^2 = dt^2 - X^2(t)dr^2 - Y^2(t)\left(d\theta^2 + \sin^2\theta d\phi^2\right). \tag{4.4.4}$$

Case 2:

$$ds^2 = dt^2 - X^2(t)dr^2 - Y^2(t)\left(d\theta^2 + \sinh^2\theta d\phi^2\right). \tag{4.4.5}$$

In the above equations and throughout this section the speed of light in vacuum c is taken as unity, and $8\pi G = 1$, where G is Newton's constant.

4.4.3   *The group generators*

There are two four-dimensional Lie algebras that are given by the operators $Z_i (i = 1, 2, 3)$ and Z_4:

Case 1:

$$[Z_1, Z_3] = Z_2, \quad [Z_3, Z_2] = Z_1, \quad [Z_2, Z_1] = Z_3, \quad [Z_4, Z_i] = 0. \tag{4.4.6}$$

Case 2:

$$[Z_1, Z_3] = Z_1, \quad [Z_3, Z_2] = Z_2, \quad [Z_2, Z_1] = 2Z_3, \quad [Z_4, Z_i] = 0. \qquad (4.4.7)$$

The operators Z_i operate on two-dimensional subspaces.

The group generators can be represented by the differential operators:

Case 1:

$$Z_1 = \frac{\partial}{\partial \phi},$$

$$Z_2 = \cos\phi \frac{\partial}{\partial\theta} - \cot\theta \sin\phi \frac{\partial}{\partial\phi},$$

$$Z_3 = \sin\phi \frac{\partial}{\partial\theta} + \cot\theta \cot\phi \frac{\partial}{\partial\phi}, \qquad (4.4.8)$$

$$Z_4 = \frac{\partial}{\partial r}.$$

Case 2:

$$Z_1 = -\cos\phi \frac{\partial}{\partial\theta} + (\coth\theta \sin\phi - 1)\frac{\partial}{\partial\phi},$$

$$Z_2 = \cos\phi \frac{\partial}{\partial\theta} - (\coth\theta \sin\phi + 1)\frac{\partial}{\partial\phi},$$

$$Z_3 = \sin\phi \frac{\partial}{\partial\theta} + \coth\theta \cos\phi \frac{\partial}{\partial\phi}, \qquad (4.4.9)$$

$$Z_4 = \frac{\partial}{\partial r}.$$

4.4.4 *The field equations*

They are given by:

Case 1:

$$2\frac{\dot X \dot Y}{XY} + \frac{1 + \dot Y^2}{Y^2} = T_0^0, \qquad (4.4.10a)$$

$$2\frac{\ddot Y}{Y} + \frac{1 + \dot Y^2}{Y^2} = T_1^1, \qquad (4.4.10b)$$

$$\frac{\ddot X}{X} + \frac{\ddot Y}{Y} + \frac{\dot X \dot Y}{XY} = T_2^2 = T_3^3, \qquad (4.4.10c)$$

$$0 = T^i_j \; (i \neq j). \tag{4.4.10d}$$

Case 2:

$$2\frac{\dot{X}\dot{Y}}{XY} - \frac{1 - \dot{Y}^2}{Y^2} = T^0_0, \tag{4.4.11a}$$

$$2\frac{\ddot{Y}}{Y} - \frac{1 - \dot{Y}^2}{Y^2} = T^1_1, \tag{4.4.11b}$$

$$\frac{\ddot{X}}{X} + \frac{\ddot{Y}}{Y} + \frac{\dot{X}\dot{Y}}{XY} = T^2_2 = T^3_3, \tag{4.4.11c}$$

$$0 = T^i_j \; (i \neq j). \tag{4.4.11d}$$

In the above equations an over-dot denotes differentiation with respect to t.

The field equations for Case 1 were given by Tolman, Bondi, and others (see Bondi 1947). However, $\partial Y / \partial r = 0$ in that case.

4.4.5   *Solutions of the field equations*

Equations (4.4.10d) and (4.4.11d) require that the vector field $u^\alpha = \delta^\alpha_0$. The remaining equations are

$$T^0_0 = \rho, \quad T^i_j = 0, \quad (i, j = 1, 2, 3). \tag{4.4.12}$$

These equations are easily solved by first solving Eq. (4.4.10b) and Eq. (4.4.11b) and then treating Y as an independent variable in Eq. (4.4.10c) and Eq. (4.4.11c). The solutions are best expressed in terms of a function $\eta(t)$:

Case 1: Closed Solution,

$$X = \epsilon + (\epsilon\eta + b)\tan\eta, \tag{4.4.13}$$

$$Y = a\cos^2\eta, \tag{4.4.14}$$

$$t - t_0 = a\left(\eta + \frac{1}{2}\sin 2\eta\right), \tag{4.4.15}$$

$$\rho = \frac{\epsilon\sec^4\eta}{a^2\left[1 + (\eta + b)\tan\eta\right]}. \tag{4.4.16}$$

In the above equations a, b and ϵ are constants satisfying

$$\epsilon = 0, 1, \quad -\infty < a < +\infty, \quad a \neq 0, \quad -\pi/2 \leq b < 1. \tag{4.4.17}$$

Case 2: Open Solution (a),

$$X = \epsilon - (\epsilon\eta + b)\tanh\eta, \tag{4.4.18}$$

$$Y = a\cosh^2\eta, \tag{4.4.19}$$

$$t - t_0 = a\left(\eta + \frac{1}{2}\sinh 2\eta\right), \tag{4.4.20}$$

$$\rho = -\frac{\epsilon\,\mathrm{sech}^4\eta}{a^2\left[1 - (\eta + b)\tanh\eta\right]}. \tag{4.4.21}$$

Open Solution (b),

$$X = \epsilon - (\epsilon\eta + b)\coth\eta, \tag{4.4.22}$$

$$Y = a\sinh^2\eta, \tag{4.4.23}$$

$$t - t_0 = a\left(\eta - \frac{1}{2}\sinh 2\eta\right), \tag{4.4.24}$$

$$\rho = -\frac{\epsilon\,\mathrm{csch}^4\eta}{a^2\left[1 - (\eta + b)\coth\eta\right]}. \tag{4.4.25}$$

For Case 2, a and ϵ have the same ranges as for Case 1, but b satisfies

$$0 \le b < \infty. \tag{4.4.26}$$

For $\epsilon = 0$, Case 2 reduces to the vacuum solutions:

Open Solution (a),

$$ds^2 = \frac{dY^2}{(1 - a/Y)} - (1 - a/Y)\,dr^2 - Y^2\left(d\theta^2 + \sinh\theta d\phi^2\right), \tag{4.4.27}$$

where $0 \le a/Y < 1$;

Open Solution (b),

$$ds^2 = \frac{dY^2}{(1 + a/Y)} - (1 + a/Y)\,dr^2 - Y^2\left(d\theta^2 + \sinh\theta d\phi^2\right), \tag{4.4.28}$$

where $0 < a/Y$.

4.4.6 *Kantowski-Sachs metrics in space-velocity manifold of cosmological general relativity*

We now represent the Kantowski-Sachs metrics in the four-dimensional spacevelocity of cosmological general relativity. This means replacing the time coordinate by the velocity coordinate (actually by τv, where τ is the Hubble-Carmeli constant). Obviously the metrics are solutions of the Einstein field equations with the same energy-momentum tensor in the new coordinates which are now:

$$x^0 = \tau v, \quad x^1 = r, \quad x^2 = \theta, \quad x^3 = \phi. \tag{4.4.29}$$

The new forms of the line elements (4.4.4) and (4.4.5) are now given by Case 1:

$$ds^2 = dv^2 - X^2(v)dr^2 - Y^2(v)\left(d\theta^2 + \sin^2\theta d\phi^2\right). \tag{4.4.30}$$

Case 2:

$$ds^2 = dv^2 - X^2(v)dr^2 - Y^2(v)\left(d\theta^2 + \sinh^2\theta d\phi^2\right). \tag{4.4.31}$$

The Hubble law is a relation between the velocities and distances of galaxies. It is usually a linear relationship, but with the addition of gravity it becomes nonlinear. We obtain two relationships from the last two line elements by putting $ds^2 = 0$ in each one of them. Accordingly we have two possibilities for the Hubble expansion:

$$dv^2 = X^2(v)dr^2 + Y^2(v)\left(d\theta^2 + \sin^2\theta d\phi^2\right), \tag{4.4.32}$$

and

$$dv^2 = X^2(v)dr^2 + Y^2(v)\left(d\theta^2 + \sinh^2\theta d\phi^2\right). \tag{4.4.33}$$

In the case of negligible shear, the terms with the angular dependence can be neglected, and we obtain for the expansion of the Universe

$$dv^2 = X^2(v)dr^2. \tag{4.4.34}$$

This is the formula for the radial expansion of the Universe when the shear factor is negligible. Accordingly we have

$$\frac{dr}{dv} = \tau X(v). \tag{4.4.35}$$

See Problem 4.4.5 to find out the value of the Hubble constant H_0 from Eq. (4.4.35) by using the Hubble expansion formula. Now $X(v)$ was already given before for the two cases:
Case 1:

$$X = \epsilon + (\epsilon\eta + b)\tan\eta, \tag{4.4.36}$$

and
Case 2:

$$X = \epsilon - (\epsilon\eta + b)\tanh\eta. \tag{4.4.37}$$

We come now to the mass density ρ appearing in Kantowski-Sachs analysis. In cosmological general relativity one uses ρ_{eff} rather than ρ, where

$$\rho_{eff} = \rho - \rho_c, \tag{4.4.38}$$

and ρ_c is the critical mass density

$$\rho_c = \frac{3}{8\pi G\tau^2}, \tag{4.4.39}$$

where G is Newton's gravitational constant $= 6.67 \times 10^{-8}\mathrm{cm}^3\mathrm{g}^{-1}\mathrm{s}^{-2}$, and $\tau = 4.28 \times 10^{17}\mathrm{s}$. A simple calculation gives

$$\rho_c \approx 10^{-29}\mathrm{g/cm}^3. \tag{4.4.40}$$

One actually uses the relative mass density $\Omega = \rho/\rho_c$ in terms of which the expansion of space is determined. Accordingly $\Omega = \rho/\rho_c \rightarrow \rho_{eff}/\rho_c$. Thus the original mass density used by Kantowski-Sachs is interpreted as ρ_{eff}. We thus have for the three cases discussed before the following:
Case 1: Closed Solution

$$\Omega = \frac{\rho}{\rho_c} = 1 + \frac{8\pi G\tau^2\epsilon\sec^4\eta}{3a^2\left[1 + (\eta + b)\tan\eta\right]}, \tag{4.4.41}$$

Case 2: Open Solution (a),

$$\Omega = \frac{\rho}{\rho_c} = 1 - \frac{8\pi G\tau^2\epsilon\operatorname{sech}^4\eta}{3a^2\left[1 - (\eta + b)\tanh\eta\right]}, \tag{4.4.42}$$

Open Solution (b),

$$\Omega = \frac{\rho}{\rho_c} = 1 - \frac{8\pi G\tau^2\epsilon\operatorname{csch}^4\eta}{3a^2\left[1 - (\eta + b)\coth\eta\right]}. \tag{4.4.43}$$

In the next section the gravitational lensing in an expanding Universe will be discussed.

4.4.7 Problems

P 4.4.1 Find the values of the function η appearing in the relative mass densities Ω for which $\Omega > 1$ (closed Universe), $\Omega = 1$ (constant expansion) and $\Omega < 1$ (open Universe).

Solution:

The solution is left for the reader.

P 4.4.2 Write the Kantowski-Sachs metrics in Cartesian coordinates,

$$x^1 = r \sin \theta \cos \phi,$$

$$x^2 = r \sin \theta \sin \phi,$$

$$x^3 = r \cos \theta.$$

Solution:
The solution is left for the reader.

P 4.4.3 Find the radial equation of motion of a star moving in the Kantowski-Sachs Universe.

Solution: Use the geodesic equation

$$\frac{d^2 x^\kappa}{ds^2} + \Gamma^\kappa_{\alpha\beta} \frac{dx^\alpha}{ds} \frac{dx^\beta}{ds} = 0. \tag{1}$$

Replace the parameter s by the velocity in Eq. (1), thus getting

$$\ddot{x}^k + \left(\Gamma^k_{\alpha\beta} - \Gamma^0_{\alpha\beta} \dot{x}^k \right) \dot{x}^\alpha \dot{x}^\beta = 0. \tag{2}$$

In the above equation a dot denotes differentiation with respect to v.

The metric is given by

$$g_{\mu\nu} = \begin{pmatrix} 1 & & & 0 \\ & -X^2 & & \\ & & -Y^2 & \\ 0 & & & -Y^2 \sin^2 \theta \end{pmatrix}, \tag{3}$$

and the contravariant components are given by

$$g^{\mu\nu} = \begin{pmatrix} 1 & & & 0 \\ & -X^{-2} & & \\ & & -Y^{-2} & \\ 0 & & & -Y^{-2} \sin^{-2} \theta \end{pmatrix}. \tag{4}$$

This is for the case of closed solution in which the coordinates are

$$x^0 = v, \quad x^1 = r, \quad x^2 = \theta, \quad x^3 = \phi. \tag{5}$$

The nonvanishing Christoffel symbols needed for the equation of motion are:

$$\Gamma^1_{01} = \frac{\dot{X}}{X}, \quad \Gamma^0_{11} = X\dot{X}, \quad \Gamma^0_{22} = Y\dot{Y}, \quad \Gamma^0_{33} = Y\dot{Y} \sin^2 \theta. \tag{6}$$

The expressions needed for the equation of motion are:

$$\Gamma^0_{\alpha\beta}\dot{x}^\alpha\dot{x}^\beta = X\dot{X}\dot{r}^2 + Y\dot{Y}\left(\dot{\theta}^2 + \sin^2\theta\dot{\phi}^2\right),\tag{7}$$

$$\Gamma^1_{\alpha\beta}\dot{x}^\alpha\dot{x}^\beta = 2\frac{\dot{X}}{X}\dot{r}.\tag{8}$$

The equation obtained is

$$\ddot{r} + \left[2\frac{\dot{X}}{X} - X\dot{X}\dot{r}^2 - Y\dot{Y}\left(\dot{\theta}^2 + \sin^2\theta\dot{\phi}^2\right)\right]\dot{r} = 0.\tag{9}$$

For the Case 2 the coordinates are

$$x^0 = v,\quad x^1 = r,\quad x^2 = \theta,\quad ,x^3 = \phi,\tag{10}$$

but the metric should be changed by taking $\sinh\theta$ instead of $\sin\theta$. The equation of motion obtained is

$$\ddot{r} + \left[2\frac{\dot{X}}{X} - X\dot{X}\dot{r}^2 - Y\dot{Y}\left(\dot{\theta}^2 + \sinh^2\theta\dot{\phi}^2\right)\right]\dot{r} = 0.\tag{11}$$

P 4.4.4 Find the equations of motion of a star in the Kantowski-Sachs Universe where the angles θ and ϕ are not constant.

Solution:

As in the previous problem, use the geodesic equation with the independent parameter v.

The nonvanishing Christoffel symbols needed, in addition to those given in the previous problem, are:

$$\Gamma^2_{02} = \frac{\dot{Y}}{Y},\quad \Gamma^2_{33} = -\sin\theta\cos\theta,\quad \Gamma^3_{03} = \frac{\dot{Y}}{Y},\quad \Gamma^3_{23} = \cot\theta.\tag{1}$$

The equations to be found are:

$$\ddot{\theta} + \left(\Gamma^2_{\alpha\beta} - \Gamma^0_{\alpha\beta}\dot{\theta}\right)\dot{x}^\alpha\dot{x}^\beta = 0,\tag{2}$$

$$\ddot{\phi} + \left(\Gamma^3_{\alpha\beta} - \Gamma^0_{\alpha\beta}\dot{\phi}\right)\dot{x}^\alpha\dot{x}^\beta = 0.\tag{3}$$

We obtain

$$\Gamma^2_{\alpha\beta}\dot{x}^\alpha\dot{x}^\beta = 2\frac{\dot{Y}}{Y}\dot{\theta} - \sin\theta\cos\theta\dot{\phi}^2,\tag{4}$$

$$\Gamma^0_{\alpha\beta}\dot{x}^\alpha\dot{x}^\beta = X\dot{X}\dot{r}^2 + Y\dot{Y}\left(\dot{\theta}^2 + \sin^2\theta\dot{\phi}^2\right),\tag{5}$$

The equation obtained for θ is:

$$\ddot{\theta} + 2\frac{\dot{Y}}{Y}\dot{\theta} - \sin\theta\cos\theta\dot{\phi}^2 - \left[X\dot{X}\dot{r}^2 + Y\dot{Y}\left(\dot{\theta}^2 + \sin^2\theta\dot{\phi}^2\right)\right]\dot{\theta} = 0. \qquad (6)$$

And for the angle ϕ we obtain

$$\Gamma^3_{\alpha\beta}\dot{x}^\alpha\dot{x}^\beta = 2\frac{\dot{Y}}{Y}\dot{\phi} + 2\cot\theta\dot{\theta}\dot{\phi}, \qquad (7)$$

and the equation of motion is:

$$\ddot{\phi} + 2\left(\frac{\dot{Y}}{Y} + \cot\theta\dot{\theta}\right)\dot{\phi} - \left[X\dot{X}\dot{r}^2 + Y\dot{Y}\left(\dot{\theta}^2 + \sin^2\theta\dot{\phi}^2\right)\right]\dot{\phi} = 0. \qquad (8)$$

P 4.4.5 Find the value of the Hubble constant H_0 by using Eq. (4.4.35) and the Hubble expansion formula

$$v = H_0 r. \qquad (1)$$

Solution: The solution is left for the reader.

4.5   Gravitational Lensing in an Expanding Universe

4.5.1   *Introduction*

The behavior of light rays in an expanding Universe is of great interest. Our Universe is expanding radially, and, unless the light ray is also moving in a radial way, the expansion of the Universe itself pushes the light ray outward.

This is like stretching a rope in a river and tie its two edges to the banks of the river. If the rope is not too tight, then the water of the river will push the rope in the direction of the flowing water. The rope will then have a shape of a curve which can be approximated to an arc of a circle, whose center is somewhere back in the middle of the river. The angle of deflection of the rope in radians will then be approximated to the length of the arc of the rope divided by the radius of the circle. And this is what is happening in our expanding Universe.

If we take the light ray to be tangent to an imaginary sphere that represents the radial expansion of the Universe, then the light ray will experience an outward push, the stronger push will be at the point where the light ray touches the imaginary sphere. The angle of deflection of the light ray will be the length of the light ray divided by the radius of the sphere. But then one should take into account the fact that the Universe is also expanding with the factor $(1 - \Omega)/c^2\tau^2$. Accordingly, there is an effect of light ray bending just as in the case of the bending being caused by a central body because of its gravitational field. In this section we will see how both a central body and the expansion of the Universe cause a lensing effect.

4.5.2 *Equation of motion of light in the Tolman expanding Universe*

In this section we find the formula for the gravitational lensing in an expanding Universe. This is a generalization of the deflection of light, well known from general relativity theory (see Chapter 3). This will be done for the Tolman metric which we already presented in the four-dimensional spacetime and which was given by Eq. (4.3.43),

$$ds^2 = c^2 dt^2 - e^\mu dr^2 - r^2 \left(d\theta^2 + \sin^2\theta d\phi^2\right), \qquad (4.5.1)$$

with μ given by

$$e^{-\mu} = 1 + \frac{\tau\kappa}{c}pr^2 - \frac{2Gm}{c^2 r}$$

$$= 1 + \frac{1-\Omega}{c^2\tau^2}r^2 - \frac{2Gm}{c^2 r}. \qquad (4.5.2)$$

We also have the equations of motion

$$\frac{dr}{dt} = c\sqrt{1 + \frac{\kappa\tau}{c}pr^2 - \frac{2Gm}{c^2 r}} \qquad (4.5.3a)$$

$$= c\sqrt{1 + \frac{1-\Omega}{c^2\tau^2}r^2 - \frac{2Gm}{c^2 r}}. \qquad (4.5.3b)$$

Light propagates as a null geodesic $ds = 0$ in spacetime, where ds is given by Eq. (4.5.1). Hence we have

$$c^2 dt^2 - e^\mu dr^2 - r^2 \left(d\theta^2 + \sin^2\theta d\phi^2\right) = 0. \qquad (4.5.4)$$

We now use the identity

$$r^2 \left(d\theta^2 + \sin^2\theta d\phi^2\right) = dx^s dx^s - dr^2 \qquad (4.5.5)$$

in Eq. (4.5.4), getting

$$dx^s dx^s - dr^2 \left(1 - e^\mu\right) = c^2 dt^2. \qquad (4.5.6)$$

Now

$$1 - e^{\mu} \approx 1 - \left(1 - \frac{1-\Omega}{c^2\tau^2}r^2 + \frac{2Gm}{c^2r}\right)$$

$$= \frac{1-\Omega}{c^2\tau^2}r^2 - \frac{2Gm}{c^2r}. \tag{4.5.7}$$

A simple calculation also yields

$$dr^2 = \frac{(x^s dx^s)^2}{r^2}, \tag{4.5.8}$$

thus we obtain

$$dx^s dx^s - \left(\frac{1-\Omega}{c^2\tau^2} - \frac{2Gm}{c^2r^3}\right)(x^s dx^s)^2 = c^2 dt^2, \tag{4.5.9}$$

or

$$\dot{x}^s \dot{x}^s - \left(\frac{1-\Omega}{c^2\tau^2} - \frac{2Gm}{c^2r^3}\right)(x^s \dot{x}^s)^2 = c^2. \tag{4.5.10}$$

The last formula describes the equation of motion of light in the Tolman expanding Universe.

Assuming that the light ray propagates in a two-dimensional plane x^1 x^2, it will be convenient to use polar coordinates r, ϕ. By the transformation

$$x^1 = r\cos\phi, \qquad x^2 = r\sin\phi, \qquad x^3 = 0, \tag{4.5.11}$$

we obtain

$$\dot{x}^s \dot{x}^s = \dot{r}^2 + r^2\dot{\phi}^2, \tag{4.5.12}$$

$$x^s \dot{x}^s = r\dot{r}. \tag{4.5.13}$$

Changing variables from r to $u(\phi) = \frac{1}{r}$, gives $r = \frac{1}{u}$, and we obtain

$$\dot{r} = -\frac{1}{u^2}u'\dot{\phi}, \tag{4.5.14}$$

where a prime denotes differentiation with respect to ϕ. Using the above results in Eq. (4.5.10), we obtain

$$\frac{1}{u^4}u'^2\dot{\phi}^2 + \frac{1}{u^2}\dot{\phi}^2 - \left(\frac{1-\Omega}{c^2\tau^2} - \frac{2Gm}{c^2r^3}\right)\frac{1}{u^6}u'^2\dot{\phi}^2 = c^2, \tag{4.5.15}$$

or

$$\frac{\dot{\phi}^2}{u^4}\left[u'^2 + u^2 - \left(\frac{1-\Omega}{c^2\tau^2} - \frac{2Gm}{c^2r^3}\right)\frac{u'^2}{u^2}\right] = c^2. \tag{4.5.16}$$

But

$$\frac{\dot{\phi}}{u^2} = r^2\dot{\phi} = Je^{-Gm/c^2 r}, \tag{4.5.17}$$

where J is the magnitude of the angular momentum per unit mass. Using Eq. (4.5.17) in Eq. (4.5.16) we obtain

$$u'^2 + u^2 - \left(\frac{1-\Omega}{c^2\tau^2} - \frac{2Gm}{c^2 r^3}\right)\frac{u'^2}{u^2} = \frac{c^2}{J^2}e^{2Gm/c^2 r}. \tag{4.5.18}$$

Differentiating this equation with respect to ϕ and dividing by $2u'$, then gives

$$u'' + u = -\frac{3Gm}{c^2}u'^2 + \left(\frac{1-\Omega}{c^2\tau^2} - \frac{2Gm}{c^2 r^3}\right)\left(\frac{u''}{u^2} - \frac{u'^2}{u^3}\right)$$

$$+ \frac{c^2}{J^2}e^{2Gm/c^2 r}\frac{Gm}{c^2}. \tag{4.5.19}$$

To the accuracy of our approximation, Eq. (4.5.19) gives

$$u'' + u = \left(\frac{1-\Omega}{c^2\tau^2} - \frac{2Gmu^3}{c^2}\right)\left(\frac{u''}{u^2} - \frac{u'^2}{u^3}\right) - \frac{3Gmu'^2}{c^2}$$

$$+ \frac{Gm}{J^2}. \tag{4.5.20}$$

4.5.2.1 *The case of non-expanding Universe*

The case of a non-expanding Universe should be obtained from our theory by going to the limit

$$\frac{1-\Omega}{c^2\tau^2} \to 0. \tag{4.5.21}$$

Equation (4.5.20) then yields

$$u'' + u = -\frac{2Gmu^3}{c^2}\left(\frac{u''}{u^2} - \frac{u'^2}{u^3}\right) - \frac{3Gmu'^2}{c^2} + \frac{Gm}{J^2}. \tag{4.5.22}$$

From Eq. (4.5.18) we have, in the lowest approximation,

$$u'^2 \approx \frac{c^2}{J^2} - u^2, \tag{4.5.23}$$

$$u'' \approx -u. \tag{4.5.24}$$

Using these approximate expressions in Eq. (4.5.22) gives, after a simple calculation,

$$u'' + u = \frac{2Gm}{c^2}\frac{c^2}{J^2} + \frac{c^2}{J^2}\frac{Gm}{c^2} - \frac{3Gm}{c^2}\left(\frac{c^2}{J^2} - u^2\right). \tag{4.5.25}$$

Thus we obtain

$$u'' + u = \frac{3Gm}{c^2}u^2,$$
(4.5.26a)

which is the standard general relativistic formula (3.6.20) obtained in general relativity theory for the orbit of light ray in a gravitational field. It yields

$$u'' + u = \frac{3Gm}{c^2R^2}\sin^2\phi$$
(4.5.26b)

upon inserting the value

$$u = \frac{1}{R}\sin\phi$$
(4.5.26c)

in its right-hand side, and gives an angle of deflection $\Delta\theta = 4Gm_\odot/c^2 r_\odot$ for the Sun.

4.5.3   *Light propagation in the lowest approximation*

We are now back with the expanding Universe. In the lowest approximation the equation of propagation of light (4.5.20) gives

$$u'' + u = 0.$$
(4.5.27)

The solution of Eq. (4.5.27) can be taken as

$$u = \frac{1}{R}\sin\phi; \qquad R = \text{const},$$
(4.5.28)

by an appropriate choice of the boundary conditions, or

$$\frac{1}{r} = \frac{1}{R}\sin\phi; \qquad R = r\sin\phi = \text{const}.$$
(4.5.29)

The light ray goes in a straight line parallel to the x axis with $y = R$ at this approximation (see Figure 3.6.3), where the assignment $y = x^1$ and $x = x^2$ has been made.

4.5.4   *The second approximation*

To get the second approximation solution we substitute the solution (4.5.28)

$$u = \frac{1}{r} = \frac{1}{R}\sin\phi,$$

on the right-hand side of Eq. (4.5.20), getting

$$u'' + u = \left(\frac{1-\Omega}{c^2\tau^2} - \frac{2Gm}{c^2r^3}\right)R\left(-\frac{1}{\sin\phi} - \frac{\cos^2\phi}{\sin^3\phi}\right)$$

$$-\frac{3Gm\cos^2\phi}{c^2R^2} + \frac{Gm}{J^2}.$$

$$(4.5.30)$$

The last expression in Eq. (4.5.30) is a constant $K = Gm/J^2$, which can be discarded by the transformation

$$u = v - K.$$

Thus the equation of light propagation is given by

$$u'' + u = -\left(\frac{1-\Omega}{c^2\tau^2} - \frac{2Gm}{c^2r^3}\right)\frac{R}{\sin^3\phi} - \frac{3Gm\cos^2\phi}{c^2R^2},$$

$$(4.5.31)$$

where we have used u instead of v for simplicity.

Thus the equation obtained is

$$u'' + u = A + \frac{B}{\sin^3\phi} + C\cos^2\phi$$

$$= (A+C) + \frac{B}{\sin^3\phi} + \frac{3Gm}{c^2R^2}\sin^2\phi,$$

$$(4.5.32)$$

where

$$A = \frac{2Gm}{c^2R^2},$$

$$(4.5.33)$$

$$B = -\frac{(1-\Omega)R}{c^2\tau^2},$$

$$(4.5.34)$$

$$C = -\frac{3Gm}{c^2R^2}.$$

$$(4.5.35)$$

We denote the solution of Eq. (4.5.32) by

$$u = u_1 + u_2 + u_3,$$

$$(4.5.36)$$

where $u_i(i = 1, 2, 3)$ denote the solutions of the equation with the corresponding three terms appearing on the right-hand side in it,

$$u_1'' + u_1 = A + C,$$

$$(4.5.37)$$

$$u_2'' + u_2 = \frac{B}{\sin^3\phi},$$

$$(4.5.38)$$

$$u_3'' + u_3 = \frac{3Gm}{c^2R^2}\sin^2\phi.$$

$$(4.5.39)$$

The solutions are

$$u_1 = -\frac{Gm}{c^2R^2},$$

$$(4.5.40)$$

$$u_2 = \frac{B}{2}\frac{1}{\sin\phi},$$

$$(4.5.41)$$

$$u_3 = \frac{Gm}{c^2R^2}\left(1 + \cos^2\phi\right).$$

$$(4.5.42)$$

One easily finds that the above solutions satisfy the differential equation (4.5.32).

Equation (4.5.39) is of a particular interest since it is identical to Eq. (3.6.22) (in which c was taken as equal to 1) familiar from general relativity theory that yields the angle of deflection $\Delta\phi = 4Gm/c^2r$ (see Figure 3.6.4).

4.5.5　*The contribution due to the expansion*

What is left is to find out the contribution due to the expansion of the Universe that comes out of the solution u_2. Adding this to Eq. (4.5.28) we get

$$u = \frac{1}{r} = \frac{1}{R}\sin\phi + \frac{B}{2}\frac{1}{\sin\phi}. \tag{4.5.43}$$

Multiplying this equation by Rr, gives

$$R = r\sin\phi + \frac{BR}{2}\frac{r}{\sin\phi}, \tag{4.5.44}$$

or

$$y = R - \frac{BR}{2}\frac{r}{\sin\phi}$$

$$= R - \frac{BR}{2}\frac{x^2 + y^2}{y}. \tag{4.5.45}$$

　　The contribution to the angle of deflection of this term is more complicated than that due to the central body, since now this angle depends on the length of the light ray. And that is not surprising since the deflection is due to the expansion and thus depends on how long the light ray is affected by the expansion. A delicate calculation then shows that the angle of deflection due to the expansion of the Universe can be given by

$$\Delta\phi_1 \approx \frac{\sqrt{R\,|\,x\,|}}{c\tau}, \tag{4.5.46}$$

and this is equal to, as expected and explained before,

$$\Delta\phi_1 = \frac{\sqrt{(1-\Omega)\,R}}{c\tau}\sqrt{|\,x\,|}. \tag{4.5.47}$$

It should be possible to measure this effect.

　　In the next chapter the properties of the gravitational field are described.

4.6　Suggested References

S. Behar and M. Carmeli, Cosmological relativity: A new theory of cosmology, *Intern. J. Theor. Phys.* **39**, 1375 (2000). (astro-ph/0008352)

H. Bondi, *M.N.* **107**, 401(1947).

M. Carmeli, *Group Theory and General Relativity* (McGraw-Hill, New York, 1977; reprinted by Imperial College Press 2000).

M. Carmeli, *Classical Fields: General Relativity and Gauge Theory* (Wiley, New York, 1982; reprinted by World Scientific, 2001).

M. Carmeli, Cosmological general relativity, *Commun. Theor. Phys.* **5**, 159 (1996).

M. Carmeli and S. Behar, Cosmological general relativity, pp. 5–26, in: *Quest for Mathematical Physics*, T.M. Karade *et al.* Editors, (New Delhi, 2000).

M. Carmeli and S. Behar, Cosmological relativity: A general relativistic theory for the accelerating Universe, Talk given at Dark Matter 2000, Los Angeles, February 2000, pp. 182–191, in: *Sources and Detection of Dark Matter/Energy in the Universe*, D. Cline, Ed. (Springer, 2001).

A. Einstein, *The Meaning of Relativity*, 5th Edition (Princeton Univ. Press, Princeton, 1955).

R. Kantowski and R.K. Sachs, Some spatially homogeneous anisotropic relativistic cosmological models, *J. Math. Phys.* **7**, 443 (1966).

L. Landau and E. Lifshitz, *The Classical Theory of Fields* (Addison-Wesley Pub. Co., Reading, Mass., 1959), Chap. 11.

P.J.E. Peebles, Status of the big bang cosmology, in: *Texas/Pascos 92: Relativistic Astrophysics and Particle Cosmology*, C.W. Akerlof and M.A. Srednicki, Editors (New York Academy of Sciences, New York, 1993), p. 84.

A.G. Riess *et al.*, *Astron. J.* **116**, 1006 (1998). [Hi-Z Supernova Team Collaboration (astro-ph/9805201)].

K. Schwarzschild, *Sitzungber. Preuss. Akad. Wiss. Berlin*, p. 424 (1916).

Chapter 5

Properties of the Gravitational Field

Moshe Carmeli

In this chapter we discuss variety of physical properties of the gravitational field both in the framework of standard general relativity theory and cosmological relativity theory. We first discuss the important law of motion of Newton and derive it from general relativity by taking the combination of the weak gravitational field and the geodesic equation. Along the same lines the Einstein gravitational constant is also determined. It follows that the weak gravitational field equations provide exactly the function that is needed in the geodesic equation, and thus Newton's law of motion is determined. We then discuss the geodesic equation in the framework of cosmology. This has not been done so far in both standard cosmological theory and in cosmological general relativity.

5.1 The Newtonian Equation of Motion

In Chapter 3 the Newtonian law of motion was derived along with its extension to post-Newtonian equations. Because of the fundamental role of Newton's law in gravitational theory we now rederive that law in a different way.

Our starting point is the geodesic equation describing the motion of a small particle in a gravitational field:

$$\frac{d^2 x^\mu}{ds^2} + \Gamma^\mu_{\alpha\beta} \frac{dx^\alpha}{ds} \frac{dx^\beta}{ds} = 0. \tag{5.1.1}$$

For the propagation of light the same geodesic equation holds but one has to add to it the null condition $ds = 0$. This will be done in conjunction

211

with the Newtonian limit of the Einstein field equations. We will see how putting them together leads to Newton's second law.

5.1.1 *The Newtonian limit of the Einstein field equations*

After having presented the Einstein gravitational field equations in Chapter 3, we now apply them for the case of a weak gravitational field. This is done also for the sake of fixing the value of Einstein's gravitational constant κ. We will also obtain the connection between the Einstein field equations and Newton's equation of gravitation. We will find that Newton's theory can be obtained as a limiting case of the Einstein equations. To obtain the Newtonian limit we proceed as follows.

5.1.2 *The Newtonian potential*

We first find out how the Newtonian potential is related to the components of the metric tensor in the lowest approximation. We have already mentioned before that the geodesic equation can be considered as the equation that describes the motion of an infinitesimally small test particle moving in a gravitational field. We use this observation in order to obtain Newton's law of motion out of the geodesic equation when the latter is approximated to its lowest order. This procedure will also single out a certain function as the one that corresponds to the Newtonian potential. After that we use the Einstein field equations in order to find out what differential equation this function satisfies.

The line element $ds^2 = g_{\mu\nu}dx^\mu dx^\nu$ can be written approximately if we notice that $dx^0 = cdt$, where c is the speed of light. Hence the term $g_{00}dx^0dx^0 = g_{00}c^2dt^2$ is one order of magnitude larger than the term $2g_{0k}dx^0dx^k = 2g_{0k}cdtdx^k$, where $k = 1, 2, 3$. The latter term, in turn, is again one order of magnitude larger than the term $g_{kl}dx^k dx^l$. Consequently, to its lowest order, $ds^2 \approx g_{00}dx^0dx^0$. It should be emphasized that this kind of approximation is valid only when the velocities of the particles producing the gravitational field are much smaller than the speed of light. This is so since our approximation is based on the assumption that $cdt \gg dx^k$, or $c \gg dx^k/dt$.

We now write the geodesic equation (5.1.1) in the alternative form

$$\frac{d^2x^\mu}{d\sigma^2} + \Gamma^\mu_{\alpha\beta}\frac{dx^\alpha}{d\sigma}\frac{dx^\beta}{d\sigma} = -\frac{d^2\sigma/ds^2}{(d\sigma/ds)^2}\frac{dx^\mu}{d\sigma}, \tag{5.1.2}$$

when one changes the parameter s into σ. We now choose the parameter $\sigma = x^0$, where x^0 is the time coordinate. The latter equation can therefore be written in the form

$$\ddot{x}^\mu + \Gamma^\mu_{\alpha\beta}\dot{x}^\alpha\dot{x}^\beta = -\frac{d^2x^0/ds^2}{\left(dx^0/ds\right)^2}\dot{x}^\mu, \tag{5.1.3}$$

where a dot indicates differentiation with respect to the coordinate x^0. The right-hand side of Eq. (5.1.3) can be written in a somewhat different form by using its zero component,

$$\ddot{x}^0 + \Gamma^0_{\alpha\beta}\dot{x}^\alpha\dot{x}^\beta = -\frac{d^2x^0/ds^2}{\left(dx^0/ds\right)^2}\dot{x}^0. \tag{5.1.4}$$

But $\dot{x} = dx^0/dx^0 = 1$, and $\ddot{x} = 0$. Hence we obtain

$$\frac{d^2x^0/ds^2}{\left(dx^0/ds\right)^2} = -\Gamma^0_{\alpha\beta}\dot{x}^\alpha\dot{x}^\beta. \tag{5.1.5}$$

Using the above result in Eq. (5.1.3), the latter can then be written in the form

$$\ddot{x}^\mu + \left(\Gamma^\mu_{\alpha\beta} - \Gamma^0_{\alpha\beta}\dot{x}^\mu\right)\dot{x}^\alpha\dot{x}^\beta = 0. \tag{5.1.6}$$

Notice that the zero component of Eq. (5.1.6) is now an identity since $\ddot{x}^0 = 0$ and $\dot{x}^0 = 1$. Consequently Eq. (5.1.6) is equivalent to the equation

$$\ddot{x}^k + \left(\Gamma^k_{\alpha\beta} - \Gamma^0_{\alpha\beta}\dot{x}^k\right)\dot{x}^\alpha\dot{x}^\beta = 0, \tag{5.1.7}$$

where $k = 1, 2, 3$.

5.1.3 *The lowest approximation*

To find the lowest approximation of Eq. (5.1.7), we notice that $\dot{x}^k = \dfrac{1}{c}\dfrac{dx^k}{dt}$. Hence $\Gamma^k_{\alpha\beta} \gg \Gamma^0_{\alpha\beta}\dot{x}^k$, and as a result the term $\Gamma^0_{\alpha\beta}\dot{x}^k$ can be neglected in Eq. (5.1.7). Moreover, since $\dot{x}^0 \gg \dot{x}^k$, all terms with velocities can be neglected. Consequently the geodesic equation (5.1.7) is reduced to the form $\ddot{x}^k + \Gamma^k_{00} \approx 0$ or

$$\ddot{x}^k \approx -\Gamma^k_{00} \tag{5.1.8}$$

in the lowest approximation.

Accordingly Γ^k_{00} acts like a Newtonian force per mass unit. In terms of the metric tensor we therefore obtain

$$\Gamma^k_{00} = \frac{1}{2}g^{k\lambda}\left(2\frac{\partial g_{\lambda 0}}{\partial x^0} - \frac{\partial g_{00}}{\partial x^\lambda}\right)$$

$$\approx -\frac{1}{2}\eta^{k\lambda}\frac{\partial g_{00}}{\partial x^{\lambda}} = \frac{1}{2}\delta^{kl}\frac{\partial g_{00}}{\partial x^{l}} = \frac{1}{2}\frac{\partial g_{00}}{\partial x^{k}}. \tag{5.1.9}$$

As a result one obtains in the lowest approximation of the geodesic equation the following:

$$\ddot{x}^{k} \approx -\frac{1}{2}\frac{\partial g_{00}}{\partial x^{k}}. \tag{5.1.10}$$

If we now write

$$g_{00}(x) = 1 + \frac{2}{c^{2}}\phi(x), \tag{5.1.11}$$

where $\phi(x)$ is a new function of the coordinates, then we obtain for the equation of motion the following:

$$\ddot{x}^{k} \approx -\frac{1}{c^{2}}\frac{\partial \phi(x)}{\partial x^{k}}. \tag{5.1.12}$$

Replacing now x^{0} by ct, we finally obtain

$$\frac{d^{2}x^{k}}{dt^{2}} \approx -\frac{\partial \phi}{\partial x^{k}}. \tag{5.1.13}$$

Consequently we see that the function $\phi(x)$ acts like a Newtonian potential. It remains to be seen that the function $\phi(x)$ indeed satisfies the Poisson equation, as is the case in the Newtonian theory of gravitation.

5.1.4   *The function $\phi(x)$*

To find out what a differential equation the function $\phi(x)$ satisfies, we now refer to the Einstein field equations

$$R_{\mu\nu} = \kappa\left(T_{\mu\nu} - \frac{1}{2}g_{\mu\nu}T\right). \tag{5.1.14}$$

It will actually be sufficient to use only the 00 component of this equation. Again, we do that in the lowest approximation and obtain

$$T = T_{\mu\nu}g^{\mu\nu} \approx T_{\mu\nu}\eta^{\mu\nu} \approx T_{00}\eta^{00} = T_{00}. \tag{5.1.15}$$

Thus we obtain

$$R_{00} = \kappa\left(T_{00} - \frac{1}{2}g_{00}T\right)$$

$$\approx \kappa\left(T_{00} - \frac{1}{2}\eta_{00}T\right) = \frac{1}{2}\kappa T_{00} = \frac{1}{2}\kappa c^{2}\rho(x), \tag{5.1.16}$$

where $\rho(x)$ is the mass density of the matter distribution that produces the gravitational field.

The approximate value of R_{00}, on the other hand, can be found from the expression of the Ricci tensor. One finds, after neglecting the nonlinear terms and the terms that are time derivatives, the following:

$$R_{00} = \frac{\partial \Gamma^\rho_{00}}{\partial x^\rho} - \frac{\partial \Gamma^\rho_{0\rho}}{\partial x^0} + \Gamma^\sigma_{00}\Gamma^\rho_{\rho\sigma} - \Gamma^\sigma_{0\rho}\Gamma^\rho_{0\sigma}$$

$$\approx \frac{\partial \Gamma^\rho_{00}}{\partial x^\rho} \approx \frac{\partial \Gamma^s_{00}}{\partial x^s}. \tag{5.1.17}$$

Using the results obtained for Γ^k_{00}, given by Eqs. (5.1.9) and (5.1.11), we find:

$$R_{00} \approx \frac{\partial \Gamma^m_{00}}{\partial x^m} \approx \frac{1}{2}\frac{\partial^2 g_{00}}{\partial x^m \partial x^m} = \frac{1}{2}\nabla^2 g_{00} \approx \frac{1}{c^2}\nabla^2 \phi. \tag{5.1.18}$$

Here ∇^2 is the three-dimensional Laplace operator,

$$\nabla^2 = \frac{\partial^2}{\partial x^1 \partial x^1} + \frac{\partial^2}{\partial x^2 \partial x^2} + \frac{\partial^2}{\partial x^3 \partial x^3}. \tag{5.1.19}$$

Equating now the two expressions given by Eqs. (5.1.16) and (5.1.18) for R_{00} then gives the desired differential equation, which the Newtonian function $\phi(x)$ has to satisfy:

$$\nabla^2 \phi(x) = \frac{1}{2}\kappa c^4 \rho(x). \tag{5.1.20}$$

Accordingly we see that this equation can be identified with Newton's equation for the gravitational potential, provided one identifies the general relativistic term $\frac{1}{2}\kappa c^4$ with the Newtonian term $4\pi G$, where G is Newton's gravitational constant. This identification then leads to the equation

$$\kappa = \frac{8\pi G}{c^4} \tag{5.1.21}$$

for Einstein's gravitational constant. Equation (5.1.20) then becomes

$$\nabla^2 \phi(x) = 4\pi G \rho(x). \tag{5.1.22}$$

In the next section the geodesic equation is investigated in cosmology.

5.2 The Geodesic Equation in Cosmology

We recall that the geodesic equation has not been utilized in cosmology either in the standard theory or in the cosmological general relativity. In this section we consider the geodesic equation in cosmological general relativity. The following is due to Julia Goldbaum.

Our coordinate system is chosen such that

$$g_{00} = 1; \qquad g_{11} = -e^{\mu}; \qquad g_{22} = -R^2;$$

$$g_{33} = -R^2 \sin^2 \theta, \tag{5.2.1a}$$

$$g^{00} = 1; \qquad g^{11} = -e^{-\mu}; \qquad g_{22} = -R^{-2};$$

$$g_{33} = -R^{-2} \sin^{-2} \theta, \tag{5.2.1b}$$

and the line element is

$$ds^2 = g_{\mu\nu}dx^{\mu}dx^{\nu} = \tau^2 dv^2 - e^{\mu}dr^2 - R^2(d\theta^2 + \sin^2\theta d\phi^2). \tag{5.2.2}$$

The geodesic equation can then be written in the form

$$\ddot{x}^k + \left(\Gamma^k_{\alpha\beta} - \Gamma^0_{\alpha\beta}\dot{x}^k\right)\dot{x}^{\alpha}\dot{x}^{\beta} = 0, \tag{5.2.3}$$

when the independent parameter is chosen as the velocity, and where $k = 1, 2, 3$, and a dot denotes differentiation with respect to $x^0 = \tau v$ (instead of ct in ordinary gravitation). Since we are seeking the radial expansion we have $\dot{x}^2 = 0$, $\dot{x}^3 = 0$. The only component of the geodesic equation left is with $k = 1$:

$$\ddot{x}^1 + \left(\Gamma^1_{\alpha\beta} - \Gamma^0_{\alpha\beta}\dot{x}^1\right)\dot{x}^{\alpha}\dot{x}^{\beta} = 0; \qquad (\alpha, \beta = 0, 1). \tag{5.2.4}$$

The nonzero Christoffel symbols are:

$$\Gamma^1_{01} = \frac{1}{2\tau}\frac{\partial\mu}{\partial v}, \qquad \Gamma^0_{11} = \frac{e^{\mu}}{2\tau}\frac{\partial\mu}{\partial v}, \qquad \Gamma^1_{11} = \frac{1}{2}\frac{\partial\mu}{\partial r}. \tag{5.2.5}$$

We denote $x^1 = r$, and the geodesic equation yields

$$\ddot{r} + \frac{\partial\mu}{\partial v}\dot{r} + \frac{1}{2}\frac{\partial\mu}{\partial r}\dot{r}^2 - \frac{e^{\mu}}{2\tau^2}\frac{\partial\mu}{\partial v}\dot{r}^3 = 0. \tag{5.2.6}$$

All of the above is relevant to the radial expansion of the Universe which should be supplemented by the null condition $ds = 0$. From the solution of the field equations, given in Subsection 4.3.5, we have

$$e^{-\mu} = 1 + \frac{(1 - \Omega)\,r^2}{c^2\tau^2}, \tag{5.2.7a}$$

or

$$e^{\mu} = \frac{1}{1 + Cr^2}, \tag{5.2.7b}$$

where

$$C = \frac{1 - \Omega}{c^2\tau^2} = \frac{1 - \dfrac{\rho}{\rho_c}}{c^2\tau^2}. \tag{5.2.8}$$

Accordingly we have

$$\frac{de^{\mu}}{dr} = -\frac{2Cr}{(1+Cr^2)^2},$$ (5.2.9)

But $\frac{\partial e^{\mu}}{\partial v} = 0$. Equation (5.2.6) thus becomes

$$\ddot{r} = \frac{Cr}{1+Cr^2}\dot{r}^2.$$ (5.2.10)

This is the equation for the radial expansion of the Universe, but it still should be supplemented by the null condition $ds = 0$.

In the next section the analogy with the Newtonian mechanics will be discussed.

5.2.1 *Problem*

P 5.2.1 Use Eqs. (4.2.40), (4.2.42) and $r = \tau v$ to show that the formula (5.2.10) is satisfied for the three cases of expansion with $\Omega > 1$, $\Omega < 1$, and $\Omega = 1$.

Solution:

Case 1: $\Omega > 1$ corresponds to the closed Universe. Its expansion is given by Eq. (4.2.40),

$$r(v) = \frac{c\tau}{\alpha}\sin\frac{v}{c}\alpha, \qquad \alpha = \sqrt{\Omega - 1}.$$ (1a)

The first derivative with respect to v is given by

$$\dot{r}(v) = \tau\cos\frac{v}{c}\alpha,$$ (1b)

while the second derivative is given by

$$\ddot{r}(v) = \frac{d}{dv}\left(\tau\cos\frac{v}{c}\alpha\right) = -\frac{\tau\alpha}{c}\sin\frac{v}{c}\alpha.$$ (1c)

The constant C in Eq. (5.2.10) can be rewritten, using Eq. (1a), as

$$C = \frac{1-\Omega}{c^2\tau^2} = -\frac{\alpha^2}{c^2\tau^2}.$$ (1d)

The numerator of the fraction in Eq. (5.2.10) is thus given by

$$Cr = -\frac{\alpha^2}{c^2\tau^2}\frac{c\tau}{\alpha}\sin\frac{v}{c}\alpha = -\frac{\alpha}{c\tau}\sin\frac{v}{c}\alpha,$$ (1e)

while the denominator is

$$1 + Cr^2 = 1 + \left(-\frac{\alpha^2}{c^2\tau^2}\right)\left(\frac{c\tau}{\alpha}\sin\frac{v}{c}\alpha\right)^2 = \cos^2\frac{v}{c}\alpha.$$ (1f)

Substituting now Eqs. (1b), (1c), (1e) and (1f) into Eq. (5.2.10), we obtain

$$-\frac{\tau\alpha}{c}\sin\frac{v}{c}\alpha = -\frac{\dfrac{\alpha}{c\tau}\sin\left(\dfrac{v}{c}\alpha\right)}{\cos^2\left(\dfrac{v}{c}\alpha\right)}\tau^2\cos^2\frac{v}{c}\alpha, \tag{1g}$$

which is a true equality.

Case 2: $\Omega < 1$ corresponds to the open Universe. Its expansion is given by Eq. (4.2.42),

$$r(v) = \frac{c\tau}{\beta}\sinh\frac{v}{c}\beta, \qquad \beta = \sqrt{1-\Omega}. \tag{2a}$$

In the following we use the properties of hyperbolic functions:

$$\cosh^2\alpha - \sinh^2\alpha = 1, \tag{2b}$$

and their derivatives given by

$$\frac{d}{dx}\cosh x = \sinh x, \qquad \frac{d}{dx}\sinh x = \cosh x. \tag{2c}$$

The first derivative of the distance with respect to velocity is given by

$$\dot{r}(v) = \frac{d}{dv}\left(\frac{c\tau}{\beta}\sinh\frac{v}{c}\beta\right) = \tau\cosh\frac{v}{c}\beta, \tag{2d}$$

while the second derivative is given by

$$\ddot{r}(v) = \frac{d}{dv}\left(\tau\cosh\frac{v}{c}\beta\right) = \frac{\tau\beta}{c}\sinh\frac{v}{c}\beta, \tag{2e}$$

The constant C in Eq. (5.2.10) can be rewritten, using Eq. (2a), as

$$C = \frac{1-\Omega}{c^2\tau^2} = \frac{\beta^2}{c^2\tau^2}. \tag{2f}$$

The numerator of the fraction in Eq. (5.2.10) is thus given by

$$Cr = \frac{\beta^2}{c^2\tau^2}\frac{c\tau}{\beta}\sinh\frac{v}{c}\beta = \frac{\beta}{c\tau}\sinh\frac{v}{c}\beta, \tag{2g}$$

while the denominator is

$$1 + Cr^2 = 1 + \frac{\beta^2}{c^2\tau^2}\left(\frac{c\tau}{\beta}\sinh\frac{v}{c}\beta\right)^2 = 1 + \sinh^2\frac{v}{c}\beta = \cosh^2\frac{v}{c}\beta. \tag{2h}$$

Substituting now Eqs. (2d), (2e), (2g) and (2h) into Eq. (5.2.10), we obtain

$$\frac{\tau\beta}{c}\sinh\frac{v}{c}\beta = \frac{\dfrac{\beta}{c\tau}\sinh\left(\dfrac{v}{c}\beta\right)}{\cosh^2\left(\dfrac{v}{c}\beta\right)}\tau^2\cosh^2\frac{v}{c}\beta, \tag{2i}$$

which is a true equality.

Case 3: $\Omega = 1$ corresponds to the Universe with constant expansion velocity. The expansion is given by

$$r(v) = \tau v. \tag{3a}$$

Its first and second derivatives with respect to v are

$$\dot{r} = \tau, \qquad \ddot{r} = 0. \tag{3b}$$

The constant C in Eq. (5.2.10) is now

$$C = \frac{1 - \Omega}{c^2 \tau^2} = 0. \tag{3c}$$

The numerator of the fraction in Eq. (5.2.10) is thus given by

$$Cr = 0, \tag{3d}$$

while the denominator is

$$1 + Cr^2 = 1. \tag{3e}$$

Substituting now Eqs. (3b), (3d) and (3e) into Eq. (5.2.10), we obtain

$$0 = 0. \tag{3f}$$

5.3 The Dynamics of the Universe Expansion: Analogy with Newtonian Mechanics

In classical mechanics one has the equation of motion of a particle determined by coordinate, velocity and acceleration. The Hubble expansion, on the other hand, is usually determined by the Hubble parameter (constant). In order to determine the Hubble constant all one needs is the distance of the galaxy and its recession velocity. The Hubble law

$$r = H_0^{-1} v, \tag{5.3.1}$$

is just a kinematical description. This is like measuring the time it takes a car to travel between New York City and Rochester and determining the velocity of the car from the measured time.

5.3.1 *"Acceleration" in cosmology*

What is needed for the Universe is a dynamical expression that includes not only velocity and distance but an "acceleration" term, too. In the following we find such an expression and we compare it to that known in classical mechanics.

The equation equivalent to Eq. (5.3.1) in classical mechanics is

$$x = vt. \tag{5.3.2}$$

The acceleration term in classical mechanics is

$$x \approx \frac{1}{2}\frac{dv}{dt}t^2 = \frac{1}{2}at^2. \tag{5.3.3}$$

Thus the expression for the dynamics of a particle in classical mechanics will have the form

$$x = x_0 + vt + \frac{1}{2}at^2, \tag{5.3.4}$$

where a is the acceleration of the particle. By analogy to Eq. (5.3.3) one can write for the Universe

$$r \approx \frac{1}{2}\frac{dH_0^{-1}}{dv}v^2 = \frac{1}{2}{''a''}v^2, \tag{5.3.5}$$

thus one should expect for the Universe to have the expression

$$r = r_0 + H_0^{-1}v + \frac{1}{2}{''a''}v^2. \tag{5.3.6}$$

The term

$$''a'' = \frac{dH_0^{-1}}{dv} \tag{5.3.7}$$

is the "acceleration" in cosmology for the Universe expansion. Equation (5.3.6) gives a general expression for the dynamics of the expanding Universe, and it is completely analogous to Eq. (5.3.4) in classical mechanics.

Using Hubble law (5.3.1) one can write the acceleration $''a''$ in terms of distances instead of velocity, since distances are usually used,

$$''a'' = \frac{dH_0^{-1}}{dv} = \frac{dH_0^{-1}}{dr}\frac{dr}{dv}. \tag{5.3.8}$$

5.3.2 The "acceleration" term explicitly: Determining the type of the Universe

We now write the "acceleration" term explicitly. From Eq. (4.3.41)

$$\frac{dr}{dv} = \tau\sqrt{1 + \kappa\tau c^{-1}pr^2} = \tau\sqrt{1 + \frac{1-\Omega}{c^2\tau^2}r^2},\tag{5.3.9}$$

we can find the second derivative of r with respect to v. We have for the "acceleration," after a simple calculation,

$$''a'' = \frac{d^2r}{dv^2} = \frac{\kappa\tau^2 c^{-1}pr}{\sqrt{1 + \kappa\tau c^{-1}pr^2}}\frac{dr}{dv} = \frac{1-\Omega}{c^2}r = \frac{dH_0^{-1}}{dv}.\tag{5.3.10}$$

Accordingly we obtain

$$\dot{H}_0 = \frac{\Omega - 1}{c^2}H_0^2 r,\tag{5.3.11}$$

where a dot denotes differentiation with respect to v, or

$$\frac{dH_0}{H_0^2} = \frac{\Omega - 1}{c^2}r\,dv.\tag{5.3.12}$$

By integration we obtain for the last equation

$$\left[\frac{1}{H_0}\right]_1^2 = \frac{1-\Omega}{c^2}\int_1^2 r\,(v)\,dv.\tag{5.3.13}$$

Thus we have

$$H_0^{-1}(2) - H_0^{-1}(1) = \frac{1-\Omega}{c^2}\int_1^2 r\,(v)\,dv.\tag{5.3.14}$$

But, according to Eq. (4.2.44), when $v \ll c$

$$r = \tau v\left[1 + \frac{(1-\Omega)\,v^2}{6c^2}\right].\tag{5.3.15}$$

Hence we obtain

$$H_0^{-1}(2) - H_0^{-1}(1) = \frac{1-\Omega}{c^2}\left[\frac{\tau}{2}\left(v_2^2 - v_1^2\right) + \frac{\tau\,(1-\Omega)}{24c^2}\left(v_2^4 - v_1^4\right)\right].\tag{5.3.16}$$

As we see from this last formula, the value of H_0 depends on the velocity (and, of course, on the distance). The velocity and distance are determined by the measurements being made at the Hubble Space Telescope. More important is the dependence on the value of Ω (i.e. greater than, less than or equal to unity).

The second term in Eq. (5.3.16) includes the term $(v_2^4 - v_1^4)/c^4$. This term is much smaller than the first. One can thus simplify Eq. (5.3.16) into:

$$H_0^{-1}(2) - H_0^{-1}(1) = \frac{\tau}{2c^2}\left(v_2^2 - v_1^2\right)(1 - \Omega). \qquad (5.3.17)$$

Assuming now that $v_2 > v_1$, it thus follows that the sign of $H_0^{-1}(2) - H_0^{-1}(1)$ is the same as that of $(1 - \Omega)$. If $\Omega > 1$ the sign will be negative, thus $H_0^{-1}(2) < H_0^{-1}(1)$, or $H_0(2) > H_0(1)$. If $\Omega = 1$ then $H_0^{-1}(2) = H_0^{-1}(1)$, meaning the Hubble constant H_0 is constant. If $\Omega < 1$ the sign will be positive, and we have $H_0(2) < H_0(1)$, the opposite of the first case.

As we see the Hubble "constant" H_0 is crucially dependent on Ω. It increases with the distance if $\Omega > 1$, does not depend on the distance if $\Omega = 1$ and it decreases with the distance if $\Omega < 1$.

In the next section Hook's law for the Universe is given.

5.4 Hook's Law of the Universe

As we have seen, the motion of the Universe is determined by the value of Ω (greater than, less than or equal to unity) or, equivalently, by the sign of the pressure. We have, by Eq. (4.3.41),

$$\frac{dr}{dv} = \tau\sqrt{1 + \frac{1 - \Omega}{c^2\tau^2}r^2} = \tau\sqrt{1 + \kappa\tau c^{-1}pr^2}, \qquad (5.4.1)$$

where p is the pressure. The second derivative of r with respect to v, may be derived by repeating the use of Eq. (5.4.1), and results in

$$\frac{d^2r}{dv^2} = \frac{\Omega - 1}{c^2\tau^2}\tau^2 r = \kappa c^{-1}\tau^3 pr. \qquad (5.4.2)$$

The last formula can be written in the form

$$\frac{d^2r}{dv^2} + kr = 0, \qquad (5.4.3)$$

where

$$k = \frac{\Omega - 1}{c^2} = -\frac{\kappa\tau^3}{c}p. \qquad (5.4.4)$$

The constant k is Hook's constant for the Universe. Depending on its sign, the Universe will contract or expand. If $\Omega > 1$, or the pressure p is negative, then k is positive and the radial motion of the Universe behaves with a sine or cosine dependence, which means the Universe is closed. When

$\Omega < 1$, or the pressure is positive, on the other hand, then k is negative and the solution of Hook's law yields hyperbolic functions like sinh or cosh, and the Universe is open.

In the next chapter the five-dimensional special relativity theory of spacetime-velocity is presented.

5.5　Suggested References

M. Carmeli, *Classical Fields: General Relativity and Gauge Theory* (John Wiley and Sons, New York, 1982; reprinted by World Scientific, 2001).

M. Carmeli, *Cosmological Relativity: The Special and General Theories for the Structure of the Universe* (World Scientific, Singapore, 2006).

Chapter 6

Cosmological Special Relativity in Five Dimensions

Moshe Carmeli

In this chapter cosmological special relativity is extended to five dimensions by adding time to the three spatial dimensions and the velocity of the Hubble expansion. As a consequence of this extension, equations of electrodynamics are considered through the extended skew-symmetric tensor, in which a new field is included along with the electric and magnetic fields. This new field is due to the Higgs interaction associated with the expansion of the Universe. It is unified with the electromagnetic interaction in the frame of cosmology. The field equations are developed in five dimensions. In addition to the well-known Maxwell equations new equations that describe the mix-up of different fields are obtained.

6.1 Introduction

We are now in a position to extend cosmological special relativity to five dimensions by adding the time to the four dimensions of space and velocity. Accordingly, the coordinates will be taken as $x^\mu=(x^0,\, x^1,\, x^2,\, x^3,\, x^4)=(ct,\, x,\, y,\, z,\, \tau v)$. Thus Greek letters take the values 0,...,4. Notice that the time coordinate is now taken as the first and the velocity coordinate as the last one, and in the rest of this section we choose units in which $c = \tau = 1$.

The line element is now given by

$$ds^2 = dt^2 - (dx^2 + dy^2 + dz^2) + dv^2, \tag{6.1.1}$$

which can be written in the simple form $ds^2 = \eta_{\mu\nu}dx^\mu dx^\nu$, where $\eta_{\mu\nu}$ is a generalized Minkowskian metric in five dimensions (5D), given by (1,-1,-1,-1,1), with signature -1.

225

The invariance in these five dimensions is written as

$$dt'^2 - (dx'^2 + dy'^2 + dz'^2) + dv'^2 = dt^2 - (dx^2 + dy^2 + dz^2) + dv^2. \quad (6.1.2)$$

A subtransformation is obtained if one takes v to be unchanged. We then have the four-dimensional Lorentz transformation

$$(t, x, y, z) \to (t', x', y', z').$$

This is suitable for observers located at inertial frames moving with constant velocities. The parameter of the transformation between two frames is V/c ($c = 1$), where V is the relative velocity. The transformation of the coordinates is the familiar Lorentz transformation given previously in Section 1.2.

For invariant t we get the cosmological transformation

$$(x, y, z, v) \to (x', y', z', v').$$

An observer here, located in a cosmic frame, makes observations at a fixed time. The parameter of the transformation between two cosmic frames is now T/τ ($\tau = 1$), where T is the relative cosmic time between them. The transformation obtained is the cosmological transformation given in Section 2.2.

And finally, for unchanged spatial coordinates x's we have a two-dimensional rotation $(t, v) \to (t', v')$, that has not been discussed before. The frame is now fixed at a point in space and the parameter of the transformation is $X/R = \tan\psi$, where ψ is the angle of rotation in the $t - v$ plane, and $R = c\tau (= 1)$.

From the above discussion one can easily find the most general transformation that includes the above three subtransformations. This can be done, for example, like in the three-dimensional successive rotations in classical mechanics. We will not go through that here even though it is not complicated.

In the next sections the generalized Maxwell equations are derived in five dimensions, and the mix-up of different fields is discussed.

6.2   Some Consequences of the Extension to Five Dimensions

The extension to five dimensions of the cosmological special relativity by adding the time coordinate raises some questions, including whether or not the equations of electrodynamics have to be changed. In this section we address ourselves to this problem.

One can introduce in these five dimensions a skew-symmetric tensor $f_{\mu\nu}$ like in electrodynamics, with μ, $\nu = 0,1,2,3,4$, in the form

$$f_{\mu\nu} = \begin{pmatrix} 0 & -E_x & -E_y & -E_z & H \\ E_x & 0 & H_z & -H_y & W_x \\ E_y & -H_z & 0 & H_x & W_y \\ E_z & H_y & -H_x & 0 & W_z \\ -H & -W_x & -W_y & -W_z & 0 \end{pmatrix}, \qquad (6.2.1)$$

where $\mathbf{E}$ and $\mathbf{H}$ are the electric and magnetic fields. For μ, $\nu = 0,1,2,3$, this is exactly the usual representation for electromagnetic field. Whereas the field $(H, \mathbf{W})$ describes a new interaction.

It is well known that the classification with respect to strengths of fields and particles leads to four types of interactions: strong, electromagnetic, weak and gravitational. This classification, however, has become mixed up since the strengths of the interactions are energy dependent. At high energy the strong interaction becomes weaker and roughly equal to the electromagnetic interactions. There are also interactions of the same strength as the weak interaction, called the Higgs interactions that involve a zero mass with spin 0 particle and a W or Z particle. Moreover, there are interactions that involve a photon and a Higgs particle, with the possibility of a W or Z particle. And so one can talk about the electromagnetic forces due to photons in a very limited sense. So why not unify the Higgs and the electromagnetic interactions at high energy? This has not been possible up to recently because there has been no framework suitable to do it. But now it is possible within the framework of cosmology.

Our interpretation for the added scalar H and the vector $\mathbf{W}$ in the skew-symmetric tensor (6.2.1) is that they are the Higgs massless particle and the W or the Z particle. In this way the Higgs interaction is unified with the electromagnetic interaction within the framework of cosmology, where the Higgs interaction is associated with the expansion of the Universe.

What is needed is to write the field equations which are assumed to be given by

$$\frac{\partial f^{\alpha\beta}}{\partial x^\beta} = 4\pi j^\alpha, \qquad (6.2.2)$$

$$\frac{\partial f_{\alpha\beta}}{\partial x^\gamma} + \frac{\partial f_{\beta\gamma}}{\partial x^\alpha} + \frac{\partial f_{\gamma\alpha}}{\partial x^\beta} = 0, \qquad (6.2.3)$$

where j^α is the current, along with the Lorentz force law

$$\frac{d^2 x^\mu}{ds^2} = \frac{e}{c^2} f^\mu_\nu \frac{dx^\nu}{ds}.$$

(6.2.4)

For $\alpha = 0, 1, 2, 3$ the components of j^α are the ordinary electric current (with $j^0 = \rho$ the charge density, j^m (m=1,2,3) $= \mathbf{j}$ the vector current density) and for $\alpha = 4$ an additional component j^4 is the velocity-component charge density which might be denoted by $\tilde{\rho}$.

The tensor $f^{\alpha\beta}$ is the contravariant tensor and is given by

$$f^{\alpha\beta} = \begin{pmatrix} 0 & E_x & E_y & E_z & H \\ -E_x & 0 & H_z & -H_y & -W_x \\ -E_y & -H_z & 0 & H_x & -W_y \\ -E_z & H_y & -H_x & 0 & -W_z \\ -H & W_x & W_y & W_z & 0 \end{pmatrix}.$$

(6.2.5)

We introduce the potential A_μ in five dimensions as in electrodynamics by

$$f_{\mu\nu} = \frac{\partial A_\mu}{\partial x^\nu} - \frac{\partial A_\nu}{\partial x^\mu},$$

(6.2.6)

with $A_\mu = (A_0, A_m, A_4) = (\phi, -\mathbf{A}, \tilde{\phi})$, where ϕ and $\mathbf{A}$ are the usual scalar and vector electromagnetic potentials, and A_4 is an additional potential related to the expansion of the Universe.

6.3   Generalized Maxwell's Equations

A straightforward calculation, using Eq. (6.2.6), gives

$$\mathbf{E} = -\nabla\phi - \frac{\partial \mathbf{A}}{\partial t},$$

(6.3.1a)

$$\mathbf{H} = \nabla \times \mathbf{A},$$

(6.3.1b)

$$\mathbf{W} = -\nabla\tilde{\phi} - \frac{\partial \mathbf{A}}{\partial v},$$

(6.3.1c)

$$H = \frac{\partial \phi}{\partial v} - \frac{\partial \tilde{\phi}}{\partial t}.$$

(6.3.1d)

From Eqs. (6.2.2) and (6.2.3) we get the rest of the generalized Maxwell's equations:

$$\nabla \cdot \mathbf{E} + \frac{\partial H}{\partial v} = 4\pi\rho,$$

(6.3.2a)

$$\nabla \times \mathbf{E} = -\frac{\partial \mathbf{H}}{\partial t}, \tag{6.3.2b}$$

$$\nabla \cdot \mathbf{H} = 0, \tag{6.3.2c}$$

$$\nabla \times \mathbf{H} = \frac{\partial \mathbf{E}}{\partial t} + \frac{\partial \mathbf{W}}{\partial v} + 4\pi \mathbf{j}, \tag{6.3.2d}$$

$$\nabla \cdot \mathbf{W} = \frac{\partial H}{\partial t} + 4\pi \tilde{\rho}, \tag{6.3.2e}$$

$$\nabla \times \mathbf{W} = \frac{\partial \mathbf{H}}{\partial v}. \tag{6.3.2f}$$

The wave equation for A_μ is then given by

$$\left(\frac{\partial^2}{\partial t^2} - \nabla^2 + \frac{\partial^2}{\partial v^2} \right) A_\mu = -j_\mu, \tag{6.3.3}$$

with the condition $\partial A^\alpha / \partial x^\alpha = 0$. There are now a delay in time and in velocity in the solutions of this equation.

6.3.1 *The mix-up*

It is important to examine the mix-up between the different fields under the transformation in five dimensions. We recall that we have three subtransformations. These are the Lorentz, the cosmological and the 2-rotation. They are given by

$$L = \frac{1}{\sqrt{1 - v^2/c^2}} \begin{pmatrix} 1 & -v/c & 0 & 0 & 0 \\ -v/c & 1 & 0 & 0 & 0 \\ 0 & 0 & 1 & 0 & 0 \\ 0 & 0 & 0 & 1 & 0 \\ 0 & 0 & 0 & 0 & 1 \end{pmatrix}, \tag{6.3.4}$$

where $y' = y, \ z' = z, \ v' = v,$

$$C = \frac{1}{\sqrt{1 - t^2/\tau^2}} \begin{pmatrix} 1 & 0 & 0 & 0 & 0 \\ 0 & 1 & 0 & 0 & -t/\tau \\ 0 & 0 & 1 & 0 & 0 \\ 0 & 0 & 0 & 1 & 0 \\ 0 & -t/\tau & 0 & 0 & 1 \end{pmatrix}, \tag{6.3.5}$$

where $t' = t$, $y' = y$, $z' = z$, and

$$R = \begin{pmatrix} \cos\psi & 0 & 0 & 0 & \sin\psi \\ 0 & 1 & 0 & 0 & 0 \\ 0 & 0 & 1 & 0 & 0 \\ 0 & 0 & 0 & 1 & 0 \\ -\sin\psi & 0 & 0 & 0 & \cos\psi \end{pmatrix}, \tag{6.3.6}$$

where $x' = x$, $y' = y$, $z' = z$, and ψ is the angle between the time and velocity axes.

To find out what are the transformed quantities, it is convenient to use the contravariant components of $f_{\mu\nu}$ which are given by Eq. (6.2.5).

One finds, for example, that under the Lorentz transformation

$$\mathbf{E}' = \mathbf{E}. \tag{6.3.7}$$

We also find that under the cosmological transformation,

$$\mathbf{W}' = \mathbf{W}. \tag{6.3.8}$$

And under the 2-rotation we find

$$\mathbf{E}' = \mathbf{E}\cos\psi + \mathbf{W}\sin\psi, \tag{6.3.9a}$$

$$\mathbf{H}' = \mathbf{H}, \tag{6.3.9b}$$

$$H' = H \tag{6.3.9c}$$

$$\mathbf{W}' = \mathbf{W}\cos\psi - \mathbf{E}\sin\psi. \tag{6.3.9d}$$

6.4 Concluding Remarks

The above cosmological special relativity corresponds to a Universe with zero curvature, i.e. $\Omega = \rho_0/\rho_c = 1$. Thus $\rho_0 = \rho_c = 3h^2/8\pi G = 9.8 \times 10^{-30}\mathrm{g/cm}^3$, a few hydrogen atoms per cubic meter, is the vacuum energy density, and ρ_0 is the present-time mean mass density. The constant h is the Hubble constant in empty space and its value is 72.17km/s-Mpc. Due to the flatness of the spacevelocity in this particular case, and only in this case (other cases are $\Omega > 1$ and $\Omega < 1$), a cosmological special relativity could have been developed since in the $\Omega > 1$ and $\Omega < 1$ cases the spacevelocity is not flat.

In a sense the theory presented here is half dynamical, since $\rho_0 \neq 0$ as opposed to ordinary special relativity, which can be considered as kinematical. It is for this reason that we could obtain results similar to those obtained from the inflationary Universe model.

In the next chapter five-dimensional cosmological general relativity as a brane world theory is presented and applied to various physical problems.

6.5 Suggested References

A.M. Anile, *Relativistic Fluids and Magneto–Fluids* (Cambridge University Press, Cambridge, New York, 1989).

W.B. Bonnor, *J. Math. Mech.* **9**, 439 (1960).

R.R. Caldwell and P.J. Steinhardt, *Phys. Rev. D* **57**, 6057 (1998).

M. Carmeli, Cosmological relativity: A special relativity for cosmology, *Found. Phys.* **25**, 1029 (1995).

M. Carmeli, Cosmological special relativity, *Found. Phys.* **26**, 413 (1996).

M. Carmeli, Space, time and velocity in cosmology, *Int. J. Theor. Phys.* **36**, 757 (1997).

M. Carmeli, *Cosmological Special Relativity: The Large-Scale Structure of Space, Time and Velocity*, Second Edition (World Scientific, Singapore, 2002).

M. Carmeli, *Cosmological Relativity: The Special and General Theories for the Structure of the Universe* (World Scientific, Singapore, 2006).

M. Cissoko, Wavefronts in a relativistic cosmic two-component fluid, *Gen. Relat. Grav.* **30**, 521 (1998).

M. de Campos, Tensorial perturbations in an accelerating Universe, *Gen. Relat. Grav.* **34**, 1393 (2002).

R. Ebert, in: *Proceedings of the Fifth Marcel Grossman Meeting on General Relativity, Part A, B,* (World Scientific, Perth, Teaneck, NJ, 1988).

G. Ferrarese, *Lezioni di Meccanica Relativistica* (Pitagora Ed., Bologna, 1985).

M. Greenhow and S. Moyo, *Philos. Trans. Roy. Soc. London Ser. A* **355**, 551 (1997).

R.T. Jantzen, P. Carini and D. Bini, The many faces of gravito-electromagnetism, *Ann. Phys.* **215**, 1 (1992).

J. Katz and D. Lyndel-Bell, *Class. Quant. Grav.* **8**, 2231 (1991).

H.P. Künzle, *Proc. Roy. Soc. Ser. A* **297**, 244 (1967).

A. Lichnerowicz, *Relativistic Hydrodynamics and Magneto–Hydrodynamics* (Benjamin, New York, 1967).

A. Lichnerowicz, Magnetohydrodynamics: Waves and shock waves in curved space-time, *Mathematical Physics Studies*, Vol. 14 (Kluwer Academic Publishers, Dordrecht, Boston, London, 1994).

A. Lifschitz, A nonlinear spectral problem with periodic coefficients occurring in magnetohydrodynamic stability theory, in: *Differential and integral operators* (Regensburg, 1995), *Oper. Theory Adv. Appl.* **102**, 97, Birkhuser, Basel (1998).

M. Manarini, *Atti Accad. Naz. Lincei. Rend. Cl. Sci. Fis. Mat. Nat.* **4**, 427 (1948).

J.M. Overduin and P.S. Wesson, *Kaluza-Klein gravity* (1998), preprint gr-qc/9805018.

I. Prigogine, J. Geheniau, E. Gunzig and P. Nardone, *Gen. Relat. Grav.* **21**, 767 (1989).

P.S. Wesson, Comments on a class of similarity solutions of Einstein equations relevant to the early Universe, *Phys. Rev. D* **34**, 3925 (1986).

P.S. Wesson, *Space, Time, Matter: Modern Kaluza-Klein Theory* (World Scientific, Singapore, 1999).

Chapter 7

Cosmological General Relativity in Five Dimensions: Brane World Theory

Moshe Carmeli

In this chapter cosmological general relativity is extended to five dimensions, the five-dimensional brane world theory. The Bianchi identities and the field equations are developed for the Universe filled up with gravity. The mass density of the Universe is obtained to be a few hydrogen atoms per cubic meter. Velocity is considered as an independent variable, rather than the coordinate time derivative. The expansion of the (spherically-symmetric) Universe is examined, and it is shown that at present time the Universe is accelerating. The Tully-Fisher law is further obtained by considering the geodesic equation in the five-dimensional general relativity theory for a bound spherically symmetric matter distribution. It is thus shown that the assumption of existence of halo dark matter is not necessary to explain the Tully-Fisher relation between the circular velocity of the star, moving around the galactic core, and the mass of the galaxy. The cosmological contribution to the redshift from celestial bodies is considered. The near universal fact of observing redshifts and not blueshifts in the cosmos indicates it is due to cosmological expansion and therefore the matter density $\Omega < 1$. Gravitational redshift from celestial bodies is generally small. Finally, the three classical general relativity tests are verified in cosmological general relativity in the weak field limit.

7.1 Introduction

In this and the following chapters we present the five-dimensional (5D) cosmological general relativity of space, time and velocity. The added extra dimension of velocity to the usual four-dimensional spacetime will be

233

evident in the sequel. Important basic issues that we face in five dimensions are also discussed.

7.1.1 *Five-dimensional manifold of space, time and velocity*

If we add the time to the cosmological flat spacevelocity line element, we obtain

$$ds^2 = c^2 dt^2 - (dx^2 + dy^2 + dz^2) + \tau^2 dv^2. \qquad (7.1.1)$$

Accordingly, we have a five-dimensional (5D) manifold of time, space and velocity. The above line element provides a group of transformations O(2,3). At v=const it yields the Minkowskian line element; at t=const it gives the cosmological line element; and at a fixed space point, $dx = dy = dz = 0$, it leads to a new two-dimensional (2D) line element

$$ds^2 = c^2 dt^2 + \tau^2 dv^2. \qquad (7.1.2)$$

The groups associated with the aforementioned line elements are, of course, O(1,3), O(3,1) and O(2), respectively. They are the Lorentz group, the cosmological group and a two-dimensional Euclidean group, respectively.

In the next section we discuss some properties of the Universe with gravitation in five dimensions. That includes the Bianchi identities, the gravitational field equations, the velocity as an independent coordinate and the energy density in cosmology. We then find the equations of motion of the expanding Universe and show that the Universe is accelerating. Afterward we discuss the important problem of halo dark matter around galaxies by finding the equations of motion of a star moving around a spherically-symmetric galaxy. The equations obtained are *not* Newtonian and instead the Tully-Fisher formula is obtained from our theory. We then show that the Universe is infinite and open, now by applying redshift analysis, using a new formula that is derived here. Finally, the concluding remarks and some mathematical conventions are presented.

7.2 Universe with Gravitation

The Universe is, of course, not flat but filled up with gravity. When gravitation is invoked, the above spaces become curved Riemanian with the line element

$$ds^2 = g_{\mu\nu} dx^\mu dx^\nu,$$

where μ, ν take the values 0, 1, 2, 3, 4. The coordinates are: $x^0 = ct$, x^1, x^2, x^3 are time and spatial coordinates, and $x^4 = \tau v$ (the role of the velocity as an independent coordinate will be discussed in the sequel). The signature is $(+ - - - +)$. The metric tensor $g_{\mu\nu}$ is symmetric, and thus we have fifteen independent components. They will be a solution of the Einstein field equations in five dimensions. A discussion on the generalization of the Einstein field equations from four to five dimensions will also be given.

7.2.1 *The Bianchi identities*

The restricted Bianchi identities are given by

$$\left(R^{\nu}_{\mu} - \frac{1}{2} \delta^{\nu}_{\mu} R \right)_{;\nu} = 0, \tag{7.2.1}$$

where $\mu,\nu=0,\ldots,4$. They are valid in five dimensions just as they are in four dimensions. In Eq. (7.2.1) R^{ν}_{μ} and R are the Ricci tensor and scalar, respectively, and a semicolon denotes covariant differentiation. As a consequence we now have five coordinate conditions that permit us to determine five coordinates. For example, one can choose $g_{00} = 1$, $g_{0k} = 0$, $g_{44} = 1$, where $k=1$, 2, 3. These are the co-moving coordinates in five dimensions that keep the clocks and the velocity-measuring instruments synchronized. We will not use these coordinates in this chapter.

7.2.2 *The gravitational field equations*

In four dimensions these are the Einstein field equations:

$$R_{\mu\nu} - \frac{1}{2} g_{\mu\nu} R = \kappa T_{\mu\nu}, \tag{7.2.2}$$

or equivalently

$$R_{\mu\nu} = \kappa \left(T_{\mu\nu} - \frac{1}{2} g_{\mu\nu} T \right), \tag{7.2.3}$$

where $T = g_{\alpha\beta} T^{\alpha\beta}$, and we have $R = -\kappa T$. In five dimensions if one chooses Eq. (7.2.2) as the field equations then Eq. (7.2.3) is *not* valid (the factor $\frac{1}{2}$ will have to be replaced by $\frac{1}{3}$, and $R = -\frac{1}{2}\kappa T$ by $-\frac{2}{3}\kappa T$), and thus there is no symmetry between R and $-\kappa T$.

7.2.3   *The velocity as an independent coordinate*

First we have to iterate what do we mean by coordinates in general and how one measures them. The time coordinate is measured by clocks as was emphasized by Einstein repeatedly. So are the spatial coordinates: they are measured by meters, as was originally done in special relativity theory by Einstein, or by use of Bondi's more modern version of k-calculus.

But what about the velocity as an independent coordinate? One might be inclined to think that if we know the spatial coordinates, then the velocities are just their time derivatives, and they are not independent coordinates. This is, indeed, the situation for a dynamical system when the coordinates are given as functions of the time. But in general the situation is different, especially in cosmology. Take, for instance, the Hubble law $v = H_0 x$. Obviously v and x are independent parameters and v is not the time derivative of x. Basically one can measure v by instruments like those used by traffic police.

7.2.4   *Effective mass density in cosmology*

To finish this introductory section we discuss the important concept of the energy density in cosmology. We use the Einstein field equations, in which the right-hand side includes the energy-momentum tensor. For fields others than gravitation, like the electromagnetic field, this is a straightforward expression that comes out as a generalization to curved spacetime of the same tensor appearing in special-relativistic electrodynamics. However, when dealing with matter, one should construct the energy-momentum tensor according to the physical situation (see, for example, Fock). Often a special expression for the mass density ρ is taken for the right-hand side of Einstein's equations, which sometimes is expressed as a δ-function.

In cosmology we also have the situation where the mass density is put on the right-hand side of the Einstein field equations. There is also the critical mass density $\rho_c = 3h^2/8\pi G$, where h is the Hubble constant in empty space and is equal to 72.17 km/s-Mpc. Hence the value of ρ_c is 9.8×10^{-30} g/cm^3, just a few hydrogen atoms per cubic meter throughout the cosmos. If the Universe average mass density ρ is equal to ρ_c, then the three spatial-geometry of the four-dimensional cosmological space is Euclidian. A deviation from this Euclidian geometry necessitates an increase or decrease of ρ from ρ_c. That is to say,

$$\rho_{eff} = \rho - \rho_c \tag{7.2.4}$$

is the active or the effective mass density that causes the three geometry not to be Euclidian. Accordingly, one should use ρ_{eff} on the right-hand side of the Einstein field equations. Indeed, we will use such a convention throughout this chapter. The subtraction of ρ_c from ρ in not significant for celestial bodies and makes no difference.

In the next section the theory of the accelerating Universe, a recent development in observational cosmology, is presented.

7.3   The Accelerating Universe

7.3.1   *Preliminaries*

In the last two sections we gave arguments to the fact that the Universe should be presented in five dimensions, even though the standard cosmological theory is obtained from Einstein's four-dimensional general relativity theory. The situation here is similar to that prevailed before the advent of ordinary special relativity. At that time the equations of electrodynamics, written in three dimensions, were well known to predict that the speed of light was constant. But that was not the end of the road. The abandoning of the concept of absolute space along with the constancy of the speed of light led to the four-dimensional notion. In cosmology now, we have to give up the notion of absolute cosmic time. Then this, with the constancy of the Big Bang time τ, leads us to a five-dimensional presentation of cosmology.

We recall that the field equations are those of Einstein in five dimensions,

$$R^\nu_\mu - \frac{1}{2}\delta^\nu_\mu R = \kappa T^\nu_\mu,$$

where Greek letters $\alpha, \beta, \cdots, \mu, \nu, \cdots = 0, 1, 2, 3, 4$. The coordinates: $x^0 = ct$; x^1, x^2 and x^3 are space-like coordinates, $r^2 = (x^1)^2 + (x^2)^2 + (x^3)^2$; $x^4 = \tau v$. The metric used is linearized and is given by (see Appendix A)

$$g_{\mu\nu} = \begin{pmatrix} 1+\phi & 0 & 0 & 0 & 0 \\ 0 & -1 & 0 & 0 & 0 \\ 0 & 0 & -1 & 0 & 0 \\ 0 & 0 & 0 & -1 & 0 \\ 0 & 0 & 0 & 0 & 1+\psi \end{pmatrix}. \tag{7.3.1}$$

We will keep only linear terms. The nonvanishing Christoffel symbols are given by (see Appendix A)

$$\Gamma^0_{0\lambda} = \frac{1}{2}\phi_{,\lambda}, \quad \Gamma^0_{44} = -\frac{1}{2}\psi_{,0}, \quad \Gamma^n_{00} = \frac{1}{2}\phi_{,n},$$

$$\Gamma^n_{44} = \frac{1}{2}\psi_{,n}, \quad \Gamma^4_{00} = -\frac{1}{2}\phi_{,4}, \quad \Gamma^4_{4\lambda} = \frac{1}{2}\psi_{,\lambda},$$

where $n = 1, 2, 3$ and a comma denotes partial differentiation. The components of the Ricci tensor and the Ricci scalar are given by (Appendix A)

$$R^0_0 = \frac{1}{2}\left(\nabla^2\phi - \phi_{,44} - \psi_{,00}\right), \tag{7.3.2a}$$

$$R^n_0 = \frac{1}{2}\psi_{,0n}, \quad R^0_n = -\frac{1}{2}\psi_{,0n}, \quad R^4_0 = R^0_4 = 0, \tag{7.3.2b}$$

$$R^n_m = \frac{1}{2}\left(\phi_{,mn} + \psi_{,mn}\right), \tag{7.3.2c}$$

$$R^4_n = -\frac{1}{2}\phi_{,n4}, \quad R^n_4 = \frac{1}{2}\phi_{,n4}. \tag{7.3.2d}$$

$$R^4_4 = \frac{1}{2}\left(\nabla^2\psi - \phi_{,44} - \psi_{,00}\right), \tag{7.3.2e}$$

$$R = \nabla^2\phi + \nabla^2\psi - \phi_{,44} - \psi_{,00}. \tag{7.3.3}$$

In the above equations ∇^2 is the ordinary three-dimensional Laplace operator.

7.3.2   *Expanding Universe*

The line element in five dimensions is given by

$$ds^2 = (1 + \phi)dt^2 - dr^2 + (1 + \psi)dv^2, \tag{7.3.4}$$

where $dr^2 = (dx^1)^2 + (dx^2)^2 + (dx^3)^2$, and where c and τ are taken, for brevity, as equal to 1. For an expanding Universe one has $ds = 0$. The line element (7.3.4) represents a spherically symmetric Universe.

The expansion of the Universe (the Hubble expansion) is recorded at a definite instant of time and thus $dt = 0$. Accordingly, taking into account $d\theta = d\phi = 0$, Eq. (7.3.4) gives the following equation for the expansion of the Universe at a certain moment,

$$-dr^2 + (1 + \psi)dv^2 = 0, \tag{7.3.5}$$

and thus

$$\left(\frac{dr}{dv}\right)^2 = 1 + \psi. \tag{7.3.6}$$

To find ψ we solve the Einstein field equation (noting that $T_0^0 = g_{0\alpha}T^{\alpha 0} \approx T^{00} = \rho(dx^0/ds)^2 \approx c^2\rho$, or $T_0^0 \approx \rho$ in units with $c = 1$):

$$R_0^0 - \frac{1}{2}\delta_0^0 R = 8\pi G\rho_{eff} = 8\pi G\left(\rho - \rho_c\right), \tag{7.3.7}$$

where $\rho_c = 3h^2/8\pi G$, where h is the Hubble constant in empty space.

A simple calculation then yields

$$\nabla^2\psi = 6(1 - \Omega), \tag{7.3.8}$$

where $\Omega = \rho/\rho_c$.

The solution of the field equation (7.3.8) is given by

$$\psi = (1 - \Omega)r^2 + \psi_0, \tag{7.3.9}$$

where the first part on the right-hand side is a solution for the non-homogeneous Eq. (7.3.8), and ψ_0 represents a solution to its homogeneous part, i.e. $\nabla^2\psi_0 = 0$. A solution for ψ_0 can be obtained as an infinite series in powers of r. The only term that is left is of the form $\psi_0 = -K_2/r$, where K_2 is a constant whose value can easily be shown to be the Schwarzschild radius, $K_2 = 2GM$. We therefore have

$$\psi = (1 - \Omega)r^2 - \frac{2GM}{r}. \tag{7.3.10}$$

The Universe expansion is therefore given by

$$\left(\frac{dr}{dv}\right)^2 = 1 + (1 - \Omega)\,r^2 - \frac{2GM}{r}. \tag{7.3.11}$$

For large r, or where there is no central mass M, as is the case with the Universe, the last term on the right-hand side can be neglected, and therefore

$$\left(\frac{dr}{dv}\right)^2 = 1 + (1 - \Omega)r^2, \tag{7.3.12a}$$

or

$$\frac{dr}{dv} = \sqrt{1 + (1 - \Omega)\,r^2}. \tag{7.3.12b}$$

Inserting now the constants c and τ, we finally obtain for the expansion of the Universe

$$\frac{dr}{dv} = \tau\sqrt{1 + (1 - \Omega)\frac{r^2}{c^2\tau^2}}. \tag{7.3.13}$$

The right-hand side of Eq. (7.3.13) represents the deviation from constant expansion due to gravity. For without this term, Eq. (7.3.13) reduces to $dr/dv = \tau$, thus $r = \tau v + \text{const}$. The constant can be taken as zero if one assumes, as usual, that at $r = 0$ the velocity should also vanish. Accordingly we have $r = \tau v$ or $v = hr$, where h is the Hubble constant in empty space. Hence when $\Omega = 1$, that is when $\rho = \rho_c$, it follows that we have a constant expansion.

Using the Hubble expansion formula $v = H_0 r$, for small v, at an instant of the observation time results in

$$dv = H_0 dr,$$

and thus

$$H_0 = \frac{dv}{dr} = \frac{h}{\sqrt{1 + \dfrac{(1 - \Omega)\, r^2}{c^2 \tau^2}}} \tag{7.3.14}$$

using Eq. (7.3.13). The last equation is only valid for small v.

7.3.3 *Decelerating, constant and accelerating expansions*

The equation of motion (7.3.13) can be integrated exactly (see Appendix B). We have three cases:

7.3.3.1 *Case 1*

For the $\Omega > 1$ case

$$r(v) = \frac{c\tau}{\alpha} \sin \frac{v}{c}\alpha; \qquad \alpha = \sqrt{\Omega - 1}. \tag{7.3.15a}$$

This is obviously a decelerating expansion.

7.3.3.2 *Case 2*

For $\Omega < 1$,

$$r(v) = \frac{c\tau}{\beta} \sinh \frac{v}{c}\beta; \qquad \beta = \sqrt{1 - \Omega}. \tag{7.3.15b}$$

This is now an accelerating expansion.

7.3.3.3 *Case 3*

For $\Omega = 1$ we have, from Eq. (7.3.13),

$$\frac{d^2 r}{dv^2} = 0, \tag{7.3.15c}$$

whose solution is, of course,

$$r(v) = \tau v, \qquad (7.3.16)$$

and this is a constant expansion. It will be noted that the last solution can also be obtained directly from the previous two cases for $\Omega > 1$ and $\Omega < 1$ by going to the limit $v \to 0$, using L'Hospital's lemma, showing that our solutions are consistent.

It has been shown that the constant expansion is just a transition stage between the decelerating and the accelerating expansions as the Universe evolves toward its present situation. This occurred 8.5 Gyr ago, at which time the cosmic radiation temperature was 143K.

7.3.4   *The accelerating Universe*

In order to decide which of the three cases is the appropriate one at the present time, it will be convenient to write the solutions in the ordinary Hubble law form $v = H_0 r$. Expanding Eqs. (7.3.15a,b) and keeping the appropriate terms (where $v \ll c$) then yields

$$r = \tau v \left(1 - \frac{\alpha^2 v^2}{6c^2} \right), \qquad (7.3.17)$$

$$r = \tau v \left(1 + \frac{\beta^2 v^2}{6c^2} \right), \qquad (7.3.18)$$

for the $\Omega > 1$ and $\Omega < 1$ cases, respectively. Using now the expressions for α and β, then both of the last equations can be reduced into the single equation

$$r = \tau v \left[1 + (1 - \Omega) \frac{v^2}{6c^2} \right]. \qquad (7.3.19)$$

Inverting now this equation by writing it in the form $v = H_0 r$, we obtain in the lowest approximation for H_0,

$$H_0 = h \left[1 - (1 - \Omega) \frac{v^2}{6c^2} \right], \qquad (7.3.20)$$

where $h = 1/\tau$. Using $v \approx r/\tau$, or $z \approx v/c$, we also obtain

$$H_0 = h \left[1 - (1 - \Omega) \frac{r^2}{6c^2 \tau^2} \right] = h \left[1 - (1 - \Omega) \frac{z^2}{6} \right], \qquad (7.3.21)$$

which is only valid for small redshifts z.

The above equations show that H_0 depends on the distance, or equivalently, on the redshift. Consequently H_0 has meaning only in the limits

$r \to 0$ and $z \to 0$, namely when measured *locally*, in which case it becomes the constant h. This is similar to the situation with respect to the speed of light when measured globally in the presence of gravitational field as the ratio between distance and time, the result usually depends on these parameters. Only in the limit one obtains the constant speed of light in vacuum ($c \approx 3 \times 10^{10}$cm/s).

As is seen from the above discussion, H_0 is intimately related to the sign of the factor $(1 - \Omega)$. If measurements of H_0 indicate that it increases with the redshift parameter z, then the sign of $(1 - \Omega)$ is negative, namely $\Omega > 1$. If, however, H_0 decreases when z increases, then the sign of $(1 - \Omega)$ is positive, i.e. $\Omega < 1$. If H_0 is independent of redshift then it indicates that $\Omega = 1$.

In recent years different measurements were obtained for H_0, with the so-called "short" and "long" distance scales, in which higher H_0 values were obtained over short distances and the lower H_0 values over long distances. Indications are that the greater the distances of measurement, the smaller the value of H_0. If one takes these experimental results seriously, then that is possible only for the case in which $\Omega < 1$, namely when the Universe is in an accelerating expansion phase, and the Universe is thus open. We will see in Section 7.6 that the same result is obtained via a new cosmological redshift formula.

In the next section the Tully-Fisher formula, an important indicator in cosmology, is discussed and presented, and the possibility of nonexistence of halo dark matter is presented.

7.4 The Tully-Fisher Formula: Nonexistence of Halo Dark Matter

In this section we derive the equations of motion of a star moving around a spherically symmetric galaxy and show that the Tully-Fisher formula is obtained from the five-dimensional cosmological general relativity theory. The calculation is lengthy but straightforward. The equations of motion will first be of general nature and only afterward specialized to the motion of a star around the field of a galaxy. The equations obtained are *not* Newtonian. The Tully-Fisher formula was obtained previously using combined two representations of Einstein's general relativity: the standard spacetime theory and a spacevelocity version of it. However, the present derivation is a straightforward result from the unification of space, time and velocity.

Our notation in this section is as follows: $\alpha, \beta, \gamma, \cdots = 0, \cdots, 4$; $a, b, c, d, \cdots = 0, \cdots, 3$; $p, q, r, s, \cdots = 1, \cdots, 4$; and $k, l, m, n, \cdots = 1, 2, 3$. The coordinates are: $x^0 = ct$ (timelike), $x^k = x^1, x^2, x^3$ (spacelike), and $x^4 = \tau v$ (velocitylike).

7.4.1 *The geodesic equation*

As usual, the equations of motion are obtained in general relativity theory from the covariant conservation law of the energy-momentum tensor (which is a consequence of the restricted Bianchi identities), and the result, as is well known, is the geodesic equation that describes the motion of a spherically symmetric test particle. In our five-dimensional cosmological theory we have five equations of motion. They are given by

$$\frac{d^2 x^\mu}{ds^2} + \Gamma^\mu_{\alpha\beta} \frac{dx^\alpha}{ds} \frac{dx^\beta}{ds} = 0. \tag{7.4.1}$$

We now change the independent parameter s into an arbitrary new parameter σ, then the geodesic equation becomes

$$\frac{d^2 x^\mu}{d\sigma^2} + \Gamma^\mu_{\alpha\beta} \frac{dx^\alpha}{d\sigma} \frac{dx^\beta}{d\sigma} = -\frac{d^2\sigma/ds^2}{(d\sigma/ds)^2} \frac{dx^\mu}{d\sigma}. \tag{7.4.2}$$

The parameter σ will be taken once as $\sigma = x^0$ (the time coordinate) and then $\sigma = x^4$ (the velocity coordinate). We obtain, for the first case,

$$\frac{d^2 x^p}{(dx^0)^2} + \left(\Gamma^p_{\alpha\beta} - \Gamma^0_{\alpha\beta} \frac{dx^p}{dx^0}\right) \frac{dx^\alpha}{dx^0} \frac{dx^\beta}{dx^0} = 0, \tag{7.4.3}$$

where $p = 1, 2, 3, 4$.

In exactly the same way we parametrize the geodesic equation now with respect to the velocity by choosing the parameter $\sigma = \tau v$. The result is

$$\frac{d^2 x^a}{(dx^4)^2} + \left(\Gamma^a_{\alpha\beta} - \Gamma^4_{\alpha\beta} \frac{dx^a}{dx^4}\right) \frac{dx^\alpha}{dx^4} \frac{dx^\beta}{dx^4} = 0, \tag{7.4.4}$$

where $a = 0, 1, 2, 3$.

The equation of motion (7.4.3) will be expanded in terms of the parameter v/c, assuming $v \ll c$, whereas Eq. (7.4.4) will be expanded with respect to t/τ, where t is a characteristic cosmic time, and $t \ll \tau$. We then can use the Einstein-Infeld-Hoffmann (EIH) method that is well known in general relativity in obtaining the equations of motion.

We obtain, in the first approximation,

$$\frac{d^2 x^p}{dt^2} + \Gamma^p_{\alpha\beta} \frac{dx^\alpha}{dt} \frac{dx^\beta}{dt} = 0, \tag{7.4.5}$$

$$\frac{d^2 x^a}{dv^2} + \Gamma^a_{\alpha\beta}\frac{dx^\alpha}{dv}\frac{dx^\beta}{dv} = 0. \qquad (7.4.6)$$

To find the lowest approximation of Eq. (7.4.5), since $dx^0/dt \gg dx^q/dt$, all terms with indices that are not zero-zero can be neglected. Consequently, Eq. (7.4.5) is reduced to the form

$$\frac{d^2 x^p}{dt^2} \approx -\Gamma^p_{00}, \qquad (7.4.7)$$

in the lowest approximation.

7.4.2   *The equations of motion*

Accordingly Γ^p_{00} acts like a Newtonian force per mass unit. In terms of the metric tensor we therefore obtain, since $\Gamma^p_{00} = -\frac{1}{2}\eta^{pq}\phi_{,q}$,

$$\frac{d^2 x^p}{dt^2} \approx -\frac{1}{2}\eta^{pq}\frac{\partial \phi}{\partial q}, \qquad (7.4.8)$$

where $\phi = g_{00} - 1$. We now decompose this equation into the spatial $(p = 1, 2, 3)$ and the velocity $(p = 4)$ parts, getting

$$\frac{d^2 x^k}{dt^2} = -\frac{1}{2}\frac{\partial \phi}{\partial x^k}, \qquad (7.4.9a)$$

$$\frac{d^2 v}{dt^2} = 0. \qquad (7.4.9b)$$

Using exactly the same method, Eq. (7.4.6) yields

$$\frac{d^2 x^k}{dv^2} = -\frac{1}{2}\frac{\partial \psi}{\partial x^k}, \qquad (7.4.10a)$$

$$\frac{d^2 t}{dv^2} = 0, \qquad (7.4.10b)$$

where $\psi = g_{44} - 1$. In the above equations $k = 1, 2, 3$. Equation (7.4.9a) is exactly the law of motion with the function ϕ being twice the Newtonian potential. The other three equations, Eq. (7.4.9b) and Eqs. (7.4.10a,b), are not Newtonian and are obtained only in the present theory. It remains to find out the functions ϕ and ψ.

7.4.2.1   *The functions ϕ and ψ*

To find the function ϕ, we solve the Einstein field equation (noting that $T_4^4 = g_{4\alpha}T^{\alpha 4} \approx T^{44} = \rho(dx^4/ds)^2 \approx \tau^2\rho$, and thus $T_4^4 \approx \rho$ in units in which $\tau = 1$):

$$R_4^4 - \frac{1}{2}\delta_4^4 R = 8\pi G \rho_{eff} = 8\pi G(\rho - \rho_c). \tag{7.4.11}$$

A straightforward calculation then gives

$$\nabla^2\phi = 6\left(1 - \Omega\right), \tag{7.4.12}$$

whose solution is given by

$$\phi = \left(1 - \Omega\right)r^2 + \phi_0,$$

where ϕ_0 is a solution of the homogeneous equation $\nabla^2\phi_0 = 0$. One then easily finds that $\phi_0 = -K_1/r$, where $K_1 = 2GM$, and thus

$$\phi = \left(1 - \Omega\right)r^2 - \frac{2GM}{r}. \tag{7.4.13}$$

In the same way the function ψ can be found (see Section 7.3),

$$\psi = \left(1 - \Omega\right)r^2 + \psi_0,$$

with

$$\nabla^2\psi_0 = 0,$$
$$\psi_0 = -\frac{K_2}{r},$$

and

$$K_2 = 2GM.$$

Therefore

$$\psi = \left(1 - \Omega\right)r^2 - \frac{2GM}{r}. \tag{7.4.14}$$

(When c and τ are inserted, then $K_1 = 2GM/c^2$ and $K_2 = 2GM\tau^2/c^2$.) For the purpose of obtaining equations of motion one can neglect the terms $(1 - \Omega)r^2$, actually $(1 - \Omega)r^2/c^2\tau^2$, in the solutions for ϕ and ψ. One then obtains

$$g_{00} \approx 1 - \frac{2GM}{c^2 r}, \qquad g_{44} \approx 1 - \frac{2GM\tau^2}{c^2 r}. \tag{7.4.15}$$

The equations of motion, consequently, have the forms, when inserting the constants c and τ,

$$\frac{d^2 x^k}{dt^2} = GM\left(\frac{1}{r}\right)_{,k}, \tag{7.4.16a}$$

$$\frac{d^2 x^k}{dv^2} = kM\left(\frac{1}{r}\right)_{,k}, \tag{7.4.16b}$$

where $k = G\tau^2/c^2$. It remains to integrate Eqs. (7.4.9b) and (7.4.10b). One finds that $v = a_0 t$, where a_0 is a constant which can be taken as equal to $a_0 = c/\tau \approx cH_0$. Accordingly, we see that the particle experiences an acceleration $a_0 = c/\tau \approx cH_0$.

7.4.2.2   *Non-Newtonian equations*

Equation (7.4.16a) is Newtonian, but Eq. (7.4.16b) is not. The integration of the latter is identical to that familiar in classical Newtonian mechanics, but there is an essential difference which should be emphasized. In Newtonian equations of motion one deals with a path of motion in the 3-space. In our theory we do not have the same situation. Rather, the paths here indicate locations of particles in the sense of the Hubble distribution, which now takes a different physical meaning. With that in mind we proceed as follows.

7.4.2.3   *First integrals*

Equation (7.4.16b) yields the first integral

$$\left(\frac{dr}{dv}\right)^2 = \frac{kM}{r},\tag{7.4.17a}$$

where v is the circular velocity of the particles, and spherical coordinates (r, θ, φ) of the matter distribution centered at $r = 0$ are assumed. Integrating Eq. (7.4.17a) we get

$$r = \left(\frac{3}{2}\right)^{2/3}(kM)^{1/3}v^{2/3}.\tag{7.4.17b}$$

Equation (7.4.17a) is analogous to the Newtonian case

$$\left(\frac{dr}{dt}\right)^2 = v^2 = \frac{GM}{r}.\tag{7.4.17c}$$

Comparing Eqs. (7.4.17a) and (7.4.17c), we obtain

$$\frac{ds}{dv} = \frac{\tau}{c}\frac{ds}{dt}.\tag{7.4.18}$$

Thus

$$\frac{dv}{dt} = \frac{c}{\tau}.\tag{7.4.19}$$

Accordingly, as we have mentioned before, the particle experiences an acceleration $a_0 = c/\tau \approx cH_0$.

7.4.3   **The Tully-Fisher law**

The particles described by the Eqs (7.4.17b) and (7.4.17c) undergo circular motion. Both equations are simultaneously valid so we eliminate r between them and obtain

$$v_c^4 = \frac{2}{3}GMa_0 \approx \frac{2}{3}GMcH_0.\tag{7.4.20}$$

This is the Tully-Fisher term. As is well known, astronomical observations show that for disk galaxies the fourth power of the circular velocity of stars moving around the core of the galaxy, v_c^4, is proportional to the total luminosity $\mathcal{L}$ of the galaxy to an accuracy of more than two orders of magnitude in $\mathcal{L}$, namely $v_c^4 \propto \mathcal{L}$. Since $\mathcal{L}$ is proportional to the mass M of the galaxy, one obtains $v_c^4 \propto M$. This is the Tully-Fisher law. There is no dependence on the distance of the star from the center of the galaxy as Newton's law $v_c^2 = GM/r$ requires for circular motion. In order to rectify this deviation from Newton's laws, astronomers assume the existence of halos around the galaxy, which are filled with dark matter appropriately distributed to satisfy the Tully-Fisher law for each particular situation.

In conclusion, it appears that there is no necessity to assume the existence of halo dark matter around galaxies. Rather, the result can be described in terms of the properties of *spacetimevelocity*. More thorough analyses are presented in chapter 10 for spiral galaxies and in Appendix C for elliptical galaxies.

In the next section the problem of cosmological redshift analysis is presented, and some interesting formulas result.

7.5   Cosmological Redshift Analysis

7.5.1   *The redshift formula*

In this section we derive a general formula for the redshift, in which the term $(1 - \Omega)$ appears explicitly. Since there are enough data of redshift measurements, this allows one to determine the sign of $(1 - \Omega)$: positive, zero or negative. Our conclusion is that $(1 - \Omega)$ cannot be negative or zero. This means that the Universe is infinite, and expands forever, a result favored by some cosmologists. To this end we proceed as follows.

Having the metric tensor from Section 7.4, we may now find the redshift of light emitted in the cosmos. As usual, at two points 1 and 2 we have for the wave lengths and frequencies:

$$\frac{\lambda_2}{\lambda_1} = \frac{\nu_1}{\nu_2} = \frac{ds\,(2)}{ds\,(1)} = \sqrt{\frac{g_{00}\,(2)}{g_{00}\,(1)}}. \tag{7.5.1}$$

This of course assumes that the g_{00} component derived in Section 7.4 is valid for the Universe as a whole. This is discussed in further detail later in Section 14.3.

Using now the solution for $g_{00} = 1 + \phi$, with ϕ given by Eq. (7.4.13), in Eq. (7.5.1), we obtain

$$\frac{\lambda_2}{\lambda_1} = \sqrt{\frac{1 + \dfrac{r_2^2}{a^2} - \dfrac{R_s}{r_2}}{1 + \dfrac{r_1^2}{a^2} - \dfrac{R_s}{r_1}}}. \tag{7.5.2}$$

In Eq. (7.5.2) $R_s = 2GM/c^2$ relevant to a compact object of mass M and $a = c\tau/\sqrt{|1 - \Omega|}$, which is a scale radius for the Universe.

For a sun-like body with radius R located at the coordinates origin, and an observer at a distance r from the center of the body, we then have $r_2 = r$ and $r_1 = R$, thus

$$\frac{\lambda_2}{\lambda_1} = \sqrt{\frac{1 + \dfrac{r^2}{a^2} - \dfrac{R_s}{r}}{1 + \dfrac{R^2}{a^2} - \dfrac{R_s}{R}}}. \tag{7.5.3}$$

So this equation adds a cosmological contribution to the redshift from a compact object.

7.5.2   *Particular cases*

Since $R \ll r$ and $R_s < R$ is usually the case we can write, to a good approximation,

$$\frac{\lambda_2}{\lambda_1} = \sqrt{\frac{1 + \dfrac{r^2}{a^2}}{1 - \dfrac{R_s}{R}}}. \tag{7.5.4}$$

The term r^2/a^2 in Eq. (7.5.4) is a pure cosmological one, whereas R_s/R is the standard general relativistic term. For $R \gg R_s$ we then have

$$\frac{\lambda_2}{\lambda_1} = \sqrt{1 + \frac{r^2}{a^2}} = \sqrt{1 + \frac{(1 - \Omega)\, r^2}{c^2 \tau^2}} \tag{7.5.5}$$

for the pure cosmological contribution to the redshift. If, furthermore, $r \ll a$ we then have

$$\frac{\lambda_2}{\lambda_1} = 1 + \frac{r^2}{2a^2} = 1 + \frac{(1 - \Omega)\, r^2}{2c^2 \tau^2} \tag{7.5.6}$$

to the lowest approximation in r^2/a^2, and thus

$$z = \frac{\lambda_2}{\lambda_1} - 1 = \frac{r^2}{2a^2} = \frac{(1 - \Omega)\, r^2}{2c^2 \tau^2}. \tag{7.5.7}$$

When the contribution of the cosmological term r^2/a^2 is negligible, we have

$$\frac{\lambda_2}{\lambda_1} = \frac{1}{\sqrt{1 - \dfrac{R_s}{R}}}. \tag{7.5.8}$$

The redshift could then be very large if R, the radius of the emitting body, is just a bit larger than the Schwarzschild radius R_s. For example, if $R_s/R = 0.96$ then the redshift is $z = 4$. For a typical sun like ours, $R_s \ll R$ and we can expand the right hand side of Eq. (7.5.8), getting

$$\frac{\lambda_2}{\lambda_1} = 1 + \frac{R_s}{2R}, \tag{7.5.9}$$

thus

$$z = \frac{R_s}{2R} = \frac{Gm}{c^2 R}, \tag{7.5.10}$$

the standard general relativistic result.

From Eqs. (7.5.5)–(7.5.7) it is clear that Ω cannot be larger than unity else otherwise z will be negative, which means a blueshift, and as is well known only redshifted galaxies are observed except for a few anomalous one in the local group, and that is due to local motion. If $\Omega = 1$ then $z = 0$, and for $\Omega < 1$ we have $z > 0$. The case of $\Omega = 1$ is also implausible, since the light from stars we see is redshifted more than the redshift due to the gravity of the body emitting the radiation. This is evident, for example, from our sun, whose emitted light is redshifted by only $z = 2.12 \times 10^{-6}$.

7.5.3   *Conclusions*

One can conclude that the theory of cosmological general relativity predicts that the Universe is open. As is well known, the standard FRW model does not relate the cosmological redshift to the kind of Universe.

In the next section we verify the three classical general relativity experiments in the framework of the five-dimensional cosmological theory.

7.6   Verification of the Classical General Relativity Tests in the Five-Dimensional Cosmology

7.6.1   *Comparison with general relativity*

7.6.1.1   *The Schwatzschild metric in five dimensions*

We first find the cosmological-generalization of the Schwarzschild spherically-symmetric metric in the five-dimensional cosmology. It will be useful to change variables from the classical Schwarzschild metric to new variables as follows:

$$\sin^2 \chi = r_s/r, \qquad dr = -2r_s \sin^{-3} \chi \cos \chi d\chi, \tag{7.6.1}$$

where $r_s = 2GM/c^2$ is the Schwarzschild radius. We also change the time coordinate $cdt = r_s d\eta$, thus η is a time parameter. The classical Schwarzschild solution will thus have the following form in the coordinate system η, χ, θ, ϕ:

$$ds^2 = r_s^2 \left[\cos^2 \chi d\eta^2 - 4\sin^{-6} \chi d\chi^2 - \sin^{-4} \chi \left(d\theta^2 + \sin^2 \theta d\phi^2\right)\right]. \tag{7.6.2}$$

So far this is just the classical spherically symmetric solution of the Einstein field equations in four dimensions, though written in new variables. The non-zero Christoffel symbols are given by

$$\Gamma^0_{01} = -\tan \chi, \quad \Gamma^1_{00} = -\frac{1}{4}\sin^7 \chi \cos \chi,$$

$$\Gamma^1_{11} = -3\cot \chi, \quad \Gamma^1_{22} = \frac{1}{2}\sin \chi \cos \chi,$$

$$\Gamma^1_{33} = \frac{1}{2}\sin \chi \cos \chi \sin^2 \theta, \quad \Gamma^2_{12} = -2\cot \chi, \tag{7.6.3}$$

$$\Gamma^2_{33} = -\sin \theta \cos \theta, \quad \Gamma^3_{13} = -2\cot \chi, \quad \Gamma^3_{23} = \cot \theta.$$

It is very lengthy, but one can verify that all components of the Ricci tensor $R_{\alpha\beta}$ are equal to zero identically.

We now extend this solution to five-dimensional cosmology. In order to conform with the standard notation, the zero component will be chosen as the time parameter, followed by the three space-like coordinates and then the fourth coordinate representing the velocity τdv. We will make one more change by choosing $\tau dv = r_s du$, thus u is the velocity parameter. The simplest way to have a cosmological solution of the Einstein field equation is using the so-called co-moving coordinates in which ds^2 is given by:

$$r_s^2 \left[\cos^2 \chi d\eta^2 - 4\sin^{-6} \chi d\chi^2 - \sin^{-4} \chi \left(d\theta^2 + \sin^2 \theta d\phi^2\right) + du^2\right]. \quad (7.6.4)$$

The coordinates are now $x^0 = \eta$, $x^1 = \chi$, $x^2 = \theta$, $x^3 = \phi$, and $x^4 = u$, and r_s is now assumed to be a function of the velocity u, $r_s = r_s(u)$, to be determined by the Einstein field equations in five dimensions. Accordingly we have the following form for the metric:

$$g_{\mu\nu} = r_s^2 \begin{pmatrix} \cos^2 \chi & & & & 0 \\ & -4\sin^{-6} \chi & & & \\ & & -\sin^{-4} \chi & & \\ & & & -\sin^{-4} \chi \sin^2 \theta & \\ 0 & & & & 1 \end{pmatrix},$$

$$(7.6.5a)$$

$$\sqrt{-g} = 2r_s^5 \sin^{-7} \chi \cos \chi \sin \theta. \qquad (7.6.5b)$$

The non-zero Christoffel symbols are given by

$$\Gamma_{01}^0 = -\sin \chi \cos^{-1} \chi, \qquad \Gamma_{04}^0 = \dot{r}_s r_s^{-1},$$

$$\Gamma_{00}^1 = -\frac{1}{4} \sin^7 \chi \cos \chi, \qquad \Gamma_{11}^1 = -3 \sin^{-1} \chi \cos \chi,$$

$$\Gamma_{14}^1 = \dot{r}_s r_s^{-1}, \qquad \Gamma_{22}^1 = \frac{1}{2} \sin \chi \cos \chi, \qquad \Gamma_{33}^1 = \frac{1}{2} \sin \chi \cos \chi \sin^2 \theta,$$

$$\Gamma_{12}^2 = -2\sin^{-1} \chi \cos \chi, \qquad \Gamma_{24}^2 = \dot{r}_s r_s^{-1}, \qquad \Gamma_{33}^2 = -\sin \theta \cos \theta, \qquad (7.6.6)$$

$$\Gamma_{13}^3 = -2\sin^{-1} \chi \cos \chi, \qquad \Gamma_{23}^3 = \sin^{-1} \theta \cos \theta, \qquad \Gamma_{34}^3 = \dot{r}_s r_s^{-1},$$

$$\Gamma_{00}^4 = -\dot{r}_s r_s^{-1} \cos^2 \chi, \quad \Gamma_{11}^4 = 4\dot{r}_s r_s^{-1} \sin^{-6} \chi, \quad \Gamma_{22}^4 = \dot{r}_s r_s^{-1} \sin^{-4} \chi,$$

$$\Gamma_{33}^4 = \dot{r}_s r_s^{-1} \sin^{-4} \chi \sin^2 \theta, \qquad \Gamma_{44}^4 = \dot{r}_s r_s^{-1},$$

where the dots denote derivatives with respect to the velocity parameter u.

The Ricci tensor components, after a lengthy but straightforward calculation, are given by:

$$R_{00} = -\left(\ddot{r}_s r_s^{-1} + 2\dot{r}_s^2 r_s^{-2}\right) \cos^2 \chi,$$

$$R_{11} = 4\left(\ddot{r}_s r_s^{-1} + 2\dot{r}_s^2 r_s^{-2}\right) \sin^{-6} \chi,$$

$$R_{22} = \left(\ddot{r}_s r_s^{-1} + 2\dot{r}_s^{\,2} r_s^{-2}\right) \sin^{-4}\chi, \tag{7.6.7}$$

$$R_{33} = \left(\ddot{r}_s r_s^{-1} + 2\dot{r}_s^{\,2} r_s^{-2}\right) \sin^{-4}\chi \sin^2\theta,$$

$$R_{44} = -4\left(\ddot{r}_s r_s^{-1} - \dot{r}_s^{\,2} r_s^{-2}\right).$$

All other components are identically zero.

We are interested in vacuum solution of the Einstein field equations for the spherically symmetric metric (Schwarzschild to five-dimensional cosmology), the right-hand sides of the above equations should be taken zeros. (See, however, Problem 7.6.1 for a nonvacuum solution.) A simple calculation then shows that $\dot{r}_s = 0$, $\ddot{r}_s = 0$. Accordingly the cosmological Schwarzschild metric in five dimensions is given by Eq. (7.6.5a) with a constant $r_s = 2GM/c^2$. The metric (7.6.5a) can then be written, using the coordinate transformations (7.6.1), as

$$g_{\mu\nu} = \begin{pmatrix} 1 - \dfrac{r_s}{r} & & & & 0 \\ & -\left(1 - \dfrac{r_s}{r}\right)^{-1} & & & \\ & & -r^2 & & \\ & & & -r^2 \sin^2\theta & \\ 0 & & & & 1 \end{pmatrix}, \tag{7.6.8}$$

where the coordinates are now $x^0 = ct$, $x^1 = r$, $x^2 = \theta$, $x^3 = \phi$, and $x^4 = \tau v$.

We are now in a position to compare the present theory with general relativity. By that we mean the verification of the three classical tests of general relativity in five-dimensional cosmological general relativity. This will be done in almost identical way to that used for verifying these tests in general relativity given in Chapter 3, but now use has to be made of the cosmological Schwarzschild metric in five dimensions.

7.6.2   *Problem*

P 7.6.1. In deriving the cosmological Schwarzschild metric in five dimensions (vacuum solution), we put zero for the right hand side of Eqs. (7.6.7). Find a nonvacuum solution to Eqs. (7.6.7) by taking an energy-momentum tensor that is not zero but describes a fluid. Do this first without assuming a pressure in the energy-momentum tensor, and then with pressure.

Solution: The solution is left for the reader.

7.6.3 *The gravitational redshift in five dimensions*

We start with the simplest experiment, that of the gravitational redshift. This experiment is not considered as one of the proofs of general relativity (it can be derived from conservation laws and the Newtonian theory).

Consider two clocks at rest at two points denoted by 1 and 2. The propagation of light is determined by ds at each point. Since at these points all spatial infinitesimal displacements and change in velocities vanish, one has $ds^2 = g_{00}c^2 dt^2$. Hence at the two points we have

$$ds\left(1\right) = \sqrt{g_{00}\left(1\right)}cdt, \tag{7.6.9a}$$

$$ds\left(2\right) = \sqrt{g_{00}\left(2\right)}cdt \tag{7.6.9b}$$

for the proper time.

The ratio of the rates of similar clocks, located at different places in a gravitational field, is therefore given by

$$\frac{ds\left(2\right)}{ds\left(1\right)} = \sqrt{\frac{g_{00}\left(2\right)}{g_{00}\left(1\right)}}. \tag{7.6.10}$$

The frequency ν_0 of an atom located at point 1, when measured by an observer located at point 2, is therefore given by

$$\nu = \nu_0 \sqrt{\frac{g_{00}\left(1\right)}{g_{00}\left(2\right)}}. \tag{7.6.11}$$

If the gravitational field is produced by a spherically symmetric mass distribution, then we may use the generalized Schwarzschild metric in five dimensions to calculate the above ratio at the two points. In this case $g_{00} = 1 - 2GM/c^2 r$, and therefore

$$\sqrt{\frac{g_{00}\left(1\right)}{g_{00}\left(2\right)}} \approx 1 + \frac{GM}{c^2}\left(\frac{1}{r_2} - \frac{1}{r_1}\right)$$

to first order in $GM/c^2 r$.

We thus obtain

$$\frac{\Delta\nu}{\nu_0} = \frac{\nu - \nu_0}{\nu_0} \approx -\frac{GM}{c^2}\left(\frac{1}{r_1} - \frac{1}{r_2}\right)$$

for the frequency shift per unit frequency. Taking now r_1 to be the observed radius of the Sun and r_2 the radius of the Earth's orbit around the Sun, then we find that

$$\frac{\Delta\nu}{\nu_0} \approx -\frac{GM_\odot}{c^2 R_\odot}, \tag{7.6.12}$$

where $M_\odot$ and $R_\odot$ are the mass and the radius of the Sun.

Accordingly we obtain $\Delta\nu/\nu_0 \approx -2.12 \times 10^{-6}$ for the frequency shift per unit frequency of the light emitted from the Sun. The calculation made above amounts to neglecting completely the Earth's gravitational field. The above result is the standard gravitational redshift (also known as the gravitational time dilation).

7.6.4 *Motion in a centrally symmetric gravitational field in cosmological five dimensions*

We assume that small test particles move along geodesics in the gravitational field. We also assume that planets have small masses as compared with the mass of the Sun, to the extent that they can be considered as test particles moving in the gravitational field of the Sun. As a result of these assumptions, the geodesic equation in the five-dimensional cosmological Schwarzschild field will be taken to describe the equation of motion of a planet moving in the gravitational field of the Sun. In fact, we do not need the exact solution of the cosmological Schwarzschild metric (7.6.8), but just its first approximation.

7.6.4.1 *Approximate solution of the five-dimensional Schwarzschild metric*

We obtain in the first approximation the following expressions for the components of the metric tensor:

$$g_{00} = 1 - \frac{r_s}{r}, \qquad g_{0m} = 0, \qquad g_{04} = 0,$$

$$g_{mn} = -\delta_{mn} - r_s\frac{x^m x^n}{r^3}, \qquad g_{m4} = 0, \qquad g_{44} = 1. \tag{7.6.13a}$$

The contravariant components of the metric tensor are consequently given, in the same approximation, by

$$g^{00} = 1 + \frac{r_s}{r}, \qquad g^{0m} = 0, \qquad g^{04} = 0,$$

$$g^{mn} = -\delta^{mn} + r_s\frac{x^m x^n}{r^3}, \qquad g^{m4} = 0, \qquad g^{44} = 1. \tag{7.6.13b}$$

We may indeed verify that the relation $g_{\mu\lambda}g^{\lambda\nu} = \delta^\nu_\mu$ between the contravariant and covariant components of the above approximate metric tensor is satisfied to orders of magnitude of the square of r_s/r. A straightforward calculation then gives the following expressions for the Christoffel symbols:

$$\Gamma^0_{0n} = -\frac{r_s}{2}\frac{\partial}{\partial x^n}\left(\frac{1}{r}\right),$$

$$\Gamma^k_{00} = -\frac{r_s}{2}\left(1 - \frac{r_s}{r}\right)\frac{\partial}{\partial x^k}\left(\frac{1}{r}\right),$$

$$\Gamma^k_{mn} = r_s\frac{x^k}{r^3}\delta_{mn} - \frac{3}{2}r_s\frac{x^k x^m x^n}{r^5}.$$

$$(7.6.14)$$

All other components vanish.

7.6.4.2 *The geodesic equation*

We now use these expressions for the Christoffel symbols in the geodesic equation

$$\ddot{x}^k + \left(\Gamma^k_{\alpha\beta} - \Gamma^0_{\alpha\beta}\dot{x}^k\right)\dot{x}^\alpha\dot{x}^\beta = 0, \qquad (7.6.15)$$

where a dot denotes differentiation with respect to the time coordinate x^0. We obtain

$$\Gamma^0_{\alpha\beta}\dot{x}^\alpha\dot{x}^\beta = \Gamma^0_{00} + 2\Gamma^0_{0n}\dot{x}^n + 2\Gamma^0_{04}\dot{x}^4 + \Gamma^0_{mn}\dot{x}^m\dot{x}^n + 2\Gamma^0_{m4}\dot{x}^m\dot{x}^4 + \Gamma^0_{44}\dot{x}^4\dot{x}^4$$

$$= -r_s\dot{x}^n\frac{\partial}{\partial x^n}\left(\frac{1}{r}\right), \qquad (7.6.16a)$$

$$\Gamma^k_{\alpha\beta}\dot{x}^\alpha\dot{x}^\beta = \Gamma^k_{00} + 2\Gamma^k_{0l}\dot{x}^l + 2\Gamma^k_{04}\dot{x}^4 + \Gamma^k_{mn}\dot{x}^m\dot{x}^n + 2\Gamma^k_{m4}\dot{x}^m\dot{x}^4 + \Gamma^k_{44}\dot{x}^4\dot{x}^4$$

$$= -\frac{r_s}{2}\frac{\partial}{\partial x^k}\left(\frac{1}{r}\right)$$

$$+ r_s\left[\frac{r_s}{2r}\frac{\partial}{\partial x^k}\left(\frac{1}{r}\right) - (\dot{x}^s\dot{x}^s)\frac{\partial}{\partial x^k}\left(\frac{1}{r}\right) - \frac{3}{2r^5}\left(x^s\dot{x}^s\right)^2 x^k\right]. \qquad (7.6.16b)$$

Consequently from the geodesic equation (7.6.15) we obtain the following equation of motion for the planet:

$$\ddot{x}^k - \frac{r_s}{2}\frac{\partial}{\partial x^k}\left(\frac{1}{r}\right)$$

$$= r_s\left[(\dot{x}^s\dot{x}^s) - \frac{r_s}{2r}\right]\frac{\partial}{\partial x^k}\left(\frac{1}{r}\right)$$

$$- r_s\left[\dot{x}^n\frac{\partial}{\partial x^n}\left(\frac{1}{r}\right)\dot{x}^k - \frac{3}{2r^5}\left(x^s\dot{x}^s\right)^2 x^k\right]. \qquad (7.6.17)$$

Replacing now the derivatives with respect to x^0 by those with respect to $t(\equiv x^0/c)$ in the latter equation, we obtain

$$\ddot{\mathbf{x}} - GM\nabla\frac{1}{r} = r_s$$

$$\times\left[(\dot{\mathbf{x}}^2)\nabla\left(\frac{1}{r}\right) - \frac{GM}{r}\nabla\left(\frac{1}{r}\right) - \left(\dot{\mathbf{x}}\cdot\nabla\frac{1}{r}\right)\dot{\mathbf{x}} + \frac{3}{2r^5}(\mathbf{x}\cdot\dot{\mathbf{x}})^2\mathbf{x}\right], \qquad (7.6.18)$$

where use has been made of the three-dimensional notation.

7.6.4.3 *Post-Newtonian equations of motion*

Hence the equation of motion of the planet differs from the Newtonian one since the left-hand side of Eq. (7.6.18) is proportional to terms of order of magnitude r_s instead of vanishing identically. This correction leads to a fundamental effect, namely, to a systematically secular change in the perihelion of the orbit of the planet.

To integrate the equation of motion (7.6.18) we multiply it vectorially by the radius vector $\mathbf{x}$. We obtain

$$\mathbf{x} \times \ddot{\mathbf{x}} = -r_s \left(\dot{\mathbf{x}} \cdot \nabla \frac{1}{r} \right) (\mathbf{x} \times \dot{\mathbf{x}}). \tag{7.6.19}$$

All other terms in Eq. (7.6.18) are proportional to the radius vector $\mathbf{x}$ and thus contribute nothing. Equation (7.6.19) may be integrated to yield the first integral

$$\mathbf{x} \times \dot{\mathbf{x}} = \mathbf{J} e^{-r_s/r}. \tag{7.6.20}$$

Here $\mathbf{J}$ is a constant vector, the *angular momentum* per unit mass of the planet. One can easily check that the first integral (7.6.20) indeed leads back to Eq. (7.6.19) by taking the time derivatives of both sides of Eq. (7.6.20).

From Eq. (7.6.20) we see that the radius vector $\mathbf{x}$ moves in a plane perpendicular to the constant angular momentum vector $\mathbf{J}$, thus the planet moves in a plane similar to the case in Newtonian mechanics. If we now introduce in this plane coordinates r and ϕ to describe the motion of the planet, the equation of motion (7.6.18) consequently decomposes into two equations. Introducing now the new variable $u = 1/r$, we can then rewrite the equations in terms of $u(\phi)$, using

$$\dot{r} = -\frac{u'}{u^2}\dot{\phi},$$

$$\ddot{r} = \frac{2u'^2}{u^3}\dot{\phi}^2 - \frac{u''}{u^2}\dot{\phi}^2 - \frac{u'}{u^2}\ddot{\phi},$$

where a prime denotes differentiation with respect to the angle ϕ. We subsequently obtain

$$\ddot{\phi} = 2\frac{u'}{u}\dot{\phi}^2 - \frac{2GM}{c^2}u'\dot{\phi}^2.$$

A straightforward calculation then gives, using the expression for $\ddot{\phi}$,

$$u'' + u - GM\left(\frac{u^2}{\dot{\phi}}\right)^2 = \frac{GM}{c^2}\left[2u^2 - u'^2 - 2GMu\left(\frac{u^2}{\dot{\phi}}\right)^2\right]. \tag{7.6.21}$$

The latter equation can be further simplified if we use the first integral

$$r^2 \dot{\phi} = J e^{-2GM/c^2 r}.$$

We obtain

$$\frac{u^2}{\dot{\phi}} = \frac{1}{J} e^{2GMu/c^2},$$

$$\left(\frac{u^2}{\dot{\phi}} \right)^2 = \frac{1}{J^2} e^{4GMu/c^2} \approx \frac{1}{J^2} \left(1 + \frac{4GM}{c^2} u \right).$$

Hence, to an accuracy of $1/c^2$, Eq. (7.6.21) gives

$$u'' + u - \frac{GM}{J^2} = \frac{GM}{c^2} \left(2u^2 - u'^2 + 2\frac{GM}{J^2} u \right). \qquad (7.6.22)$$

7.6.4.4 *The Newtonian limit*

Equation (7.6.22) can be used to determine the motion of the planet. The Newtonian equation of motion that corresponds to Eq. (7.6.22) is one whose left-hand side is identical to the above equation, but is equal to zero rather than to the terms on the right-hand side. This fact can easily be seen if one lets GM/c^2 go to zero in Eq. (7.6.22). Therefore in the Newtonian limit we have

$$u'' + u - \frac{GM}{J^2} \approx 0, \qquad (7.6.23)$$

whose solution can be written as

$$u \approx u_0 \left(1 + \epsilon \cos \phi \right). \qquad (7.6.24)$$

Here u_0 is a constant, and ϵ is the eccentricity of the ellipse, $\epsilon = (1 - b^2/a^2)^{1/2}$, where a and b are the semimajor and semiminor axes of the ellipse. Using the solution (7.6.24) in the Newtonian limit of the equation of motion (7.6.23) then determines the value of the constant u_0, as $u_0 = GM/J^2$.

7.6.4.5 *Motion beyond Newton*

To solve the equation of motion (7.6.22), we therefore assume a solution of the form

$$u = u_0 \left(1 + \epsilon \cos \alpha\phi \right), \qquad (7.6.25)$$

where α is some parameter to be determined, whose value in the usual nonrelativistic mechanics is unity. The appearance of the parameter $\alpha \neq 1$

in our solution is an indication that the motion of the planet will no longer be a closed ellipse.

Using the above solution in Eq. (7.6.22), and equating coefficients of $\cos \alpha\phi$, then gives

$$\alpha^2 = 1 - \frac{2GM}{c^2}\left(2u_0 + \frac{GM}{J^2}\right).$$

If we substitute for GM/J^2 in the above equation its nonrelativistic value u_0, then the error will be of a higher order. Hence the latter equation can be written as

$$\alpha^2 = 1 - \frac{6GM}{c^2}u_0,$$

or

$$\alpha \approx 1 - \frac{3GM}{c^2}u_0. \tag{7.6.26}$$

Successive perihelia occur at two angles ϕ_1 and ϕ_2 when $\alpha\phi_2 - \alpha\phi_1 = 2\pi$. Since the parameter α is smaller than unity, we have $\phi_2 - \phi_1 = 2\pi/\alpha > 2\pi$. Hence we can write $\phi_2 - \phi_1 = 2\pi + \Delta\phi$, with $\Delta\phi > 0$, or

$$\alpha\left(\phi_2 - \phi_1\right) = \alpha\left(2\pi + \Delta\phi\right) = \left(1 - \frac{3GM}{c^2}u_0\right)\left(2\pi + \Delta\phi\right) = 2\pi. \tag{7.6.27}$$

As a result there will be an *advance* in the perihelion of the planet orbit per revolution given by Eq. (7.6.27) or, to the first order, by

$$\Delta\phi = 6\pi\frac{GMu_0}{c^2}. \tag{7.6.28}$$

The constant u_0 can also be expressed in terms of the eccentricity, using the Newtonian approximation. Denoting the radial distances of the orbit, which correspond to the angles $\phi_2 = 0$ and $\phi_1 = \pi$, by r_2 and r_1, respectively, we have from Eq. (7.6.24),

$$\frac{1}{r_2} = u_0\left(1 + \epsilon\right), \qquad \frac{1}{r_1} = u_0\left(1 - \epsilon\right).$$

Hence since $r_1 + r_2 = 2a$, we obtain

$$2a = r_1 + r_2 = \frac{2}{u_0\left(1 - \epsilon^2\right)},$$

where a is the semimajor axis of the orbit, and therefore

$$u_0 = \frac{1}{a\left(1 - \epsilon^2\right)}.$$

Using this value for u_0 in Eq. (7.6.28) for $\Delta\phi$, we obtain for the perihelion advance the expression

$$\Delta\phi = \frac{6\pi GM}{c^2 a\left(1 - \epsilon^2\right)} \tag{7.6.29}$$

in radians per revolution. This is the standard general relativistic formula for the advance of the perihelion.

In the next subsection we discuss the deflection of a light ray moving in a gravitational field.

7.6.5 *The deflection of light in a gravitational field within the five-dimensional theory*

To discuss the effect of gravitation on the propagation of light signals, we may use the geodesic equation, along with the null condition $ds = 0$ at a fixed velocity. A light signal propagating in the gravitational field of the Sun, for instance, will thus be described by the null geodesics in the 5D cosmological Schwarzschild field at $dv = 0$.

Using the approximate solution for the five-dimensional cosmological Schwarzschild metric, given by Eq. (7.6.13a), we obtain

$$g_{\mu\nu}dx^\mu dx^\nu = \left(1 - \frac{2GM}{c^2 r}\right) c^2 dt^2$$

$$- \left[dx^s dx^s + \frac{2GM}{c^2} \frac{(x^s dx^s)^2}{r^3}\right] = 0. \tag{7.6.30}$$

Hence we have, after a simple calculation, to the first approximation in GM/c^2, the following equation of motion for the propagation of light in a gravitational field:

$$\left(1 + \frac{2GM}{c^2 r}\right) \left[(\dot{x}^s \dot{x}^s) + \frac{2GM}{c^2} \frac{(x^s \dot{x}^s)^2}{r^3}\right] = c^2, \tag{7.6.31}$$

where a dot denotes differentiation with respect to the time coordinate $t(\equiv x^0/c)$.

7.6.5.1 *Equation of motion*

Just as in the case of planetary motion (see previous subsection), the motion here also takes place in a plane. Hence in this plane we may introduce the polar coordinates r and ϕ. The equation of motion (7.6.31) then yields, to the first approximation in GM/c^2, the following equation in the polar coordinates:

$$\left(\dot{r}^2 + r^2 \dot{\phi}^2\right) + \frac{4GM}{c^2} \frac{\dot{r}^2}{r} + \frac{2GM}{c^2} r\dot{\phi}^2 = c^2. \tag{7.6.32}$$

Changing now variables from r to $u(\phi) \equiv 1/r$, we obtain

$$\left[u'^2 + u^2 + \frac{2GMu}{c^2}\left(2u'^2 + u^2\right)\right] \left(\frac{\dot{\phi}}{u^2}\right)^2 = c^2, \tag{7.6.33}$$

where a prime denotes differentiation with respect to the angle ϕ.

Moreover we may use the first integral of motion,

$$r^2 \dot\phi = J e^{-2GM/c^2 r},$$ (7.6.34)

in Eq. (7.6.33), thus getting

$$u'^2 + u^2 + \frac{2GMu}{c^2}\left(2u'^2 + u^2\right) = \left(\frac{c}{J}\right)^2 e^{4GMu/c^2}.$$ (7.6.35)

Differentiation of this equation with respect to ϕ then gives

$$u'' + u + \frac{GM}{c^2}\left(2u'^2 + 4uu'' + 3u^2\right) = \frac{2GM}{J^2},$$ (7.6.36)

where terms have been kept to the first approximation in GM/c^2 only.

To solve Eq. (7.6.36), we notice that, in the lowest approximation, we have, from Eq. (7.6.35),

$$u'^2 \approx \left(\frac{c}{J}\right)^2 - u^2,$$ (7.6.37)

$$u'' \approx -u.$$ (7.6.38)

Hence using these approximate expressions in Eq. (7.6.36) gives

$$u'' + u = \frac{3GM}{c^2}u^2$$ (7.6.39)

for the equation of motion of the orbit of the light ray propagating in a spherically symmetric gravitational field.

7.6.5.2   *Lowest approximation*

In the lowest approximation, namely, when the gravitational field of the central body is completely neglected, the right-hand side of Eq. (7.6.39) can be taken as zero, and therefore u satisfies the equation $u'' + u = 0$. The solution of this equation is a straight line given by

$$u = \frac{1}{R}\sin\phi,$$ (7.6.40)

where R is a constant. This equation for the straight line shows that $r \equiv 1/u$ has a minimum value R at the angle $\phi = \pi/2$. If we denote $y = r\sin\phi$, the straight line (7.6.40) can then be described by

$$y = r\sin\phi = R = \text{constant}.$$ (7.6.41)

7.6.5.3 *Next approximation*

We now use the approximate value for u, Eq. (7.6.40), in the right-hand side of Eq. (7.6.39), since the error introduced in doing so is of higher order. We therefore obtain the following for the equation of the light ray orbit:

$$u'' + u = \frac{3GM}{c^2 R^2} \sin^2 \phi. \tag{7.6.42}$$

The solution of this equation is then given by

$$u = \frac{1}{R} \sin \phi + \frac{GM}{c^2 R^2} \left(1 + \cos^2 \phi\right). \tag{7.6.43}$$

Introducing now the Cartesian coordinates $x = r \cos \phi$ and $y = r \sin \phi$, the above solution can be written as

$$y = R - \frac{GM}{c^2 R} \frac{2x^2 + y^2}{(x^2 + y^2)^{1/2}}. \tag{7.6.44}$$

We thus see that for large values of $|x|$ the above solution asymptotically approaches the following expression:

$$y \approx R - \frac{2GM}{c^2 R} |x|. \tag{7.6.45}$$

As seen from Eq. (7.6.45), asymptotically, the orbit of the light ray is described by two straight lines in the spacetime. These straight lines make angles with respect to the x axis given by $\tan \phi = \pm \left(2GM/c^2 R\right)$. The angle of deflection $\Delta \phi$ between the two asymptotes is therefore given by

$$\Delta \phi = \frac{4GM}{c^2 R}. \tag{7.6.46}$$

This is the angle of *deflection* of a light ray in passing through the gravitational field of a central body, described by the cosmological Schwarzschild metric. For a light ray just grazing the Sun, Eq. (7.6.46) gives the value

$$\Delta \phi = \frac{4GM_\odot}{c^2 R_\odot} = 1.75 \text{ seconds.}$$

This is the standard general-relativistic formula. Observations indeed confirm this result. One of the latest measurements gives 1.75 ± 0.10 seconds. It is worth mentioning that only general relativity theory and the present theory predict the correct factor of the deflection of light in the gravitational field.

In the next chapter the particle production mechanism is considered by means of five-dimensional cosmological special and general relativity theories.

7.7   Suggested References

S. Behar and M. Carmeli, *Int. J. Theor. Phys.* **39**, 1375 (2000), astro-ph/0008352.

H. Bondi, Brandeis Summer School 1955.

M. Carmeli, *Phys. Rev.* **138**, B1003 (1965).

M. Carmeli, *Classical Fields: General Relativity and Gauge Theory* (John Wiley, New York, 1982); reprinted by World Scientific Publishing Company (2001).

M. Carmeli, Cosmological general relativity, *Commun. Theor. Phys.* **5**, 159 (1996).

M. Carmeli, Is galaxy dark matter a property of spacetime? *Int. J. Theor. Phys.* **37**, 2621-2625 (1998).

M. Carmeli, Derivation of the Tully-Fisher law: Doubts about the necessity and existence of halo dark matter, *Int. J. Theor. Phys.* **39**, 1397 (2000), astro-ph/9907244.

M. Carmeli, *Cosmological Special Relativity: The Large-Scale Structure of Space, Time and Velocity*, Second Edition (World Scientific, River Edge, NJ. and Singapore, 2002).

M. Carmeli, Accelerating Universe: theory versus experiment (2002); astro-ph/0205396.

M. Carmeli, The line elements in the Hubble expansion, in: *Gravitation and Cosmology*, Eds. A. Lobo *et al.* (Universitat de Barcelona, 2003); astro-ph/0211043.

A. Einstein, *Autobiographical Notes*, Ed. P.A. Schilpp (Open Court Pub. Co., La Salle and Chicago, 1979).

F. Eisenhauer, R. Schödel, R. Genzel, T. Ott, M. Tecza, R. Abuter, A. Eckart and T. Alexander, A geometric determination of the distance to the galactic center, *Astrophys. J.* **597**, L121-L124 (2003).

V. Fock, *The Theory of Space, Time and Gravitation* (Pergamon Press, Oxford, 1959).

L. Landau and E. Lifshitz, *The Classical Theory of Fields* (Addisson-Wesley Publishing Company, Reading, Massachusetts, 1959).

P.J.E. Peebles, Status of the big bang cosmology, p. 84, in: *Texas/ Pascos 92: Relativistic Astrophysics and Particle Cosmology*, Eds. C.W. Akerlof and M.A. Srednicki, Vol. 688 (The New York Academy of Sciences, New York, 1993).

B.C. Whitemore, Rotation curves of spiral galaxies in clusters, in: *Galactic Models*, J.R. Buchler, S.T. Gottesman, J.H. Hunter. Jr., Eds., (New York Academy Sciences, New York, 1990).

Chapter 8

Particle Production in Five-Dimensional Cosmological Relativity

Gianluca Gemelli[1]

In this chapter we consider the five-dimensional (5D) extension of cosmological special and general relativity. In this framework a single 5D tensor conservation law can represent the equations of hydrodynamics, unifying the continuity equation and the conservation of the stress-energy tensor. This results in additional terms, which can be interpreted as representing matter creation (or annihilation). Thus, in principle, this picture permits the interpretation of the particle production phenomena as a cosmological effect. The following is based on Gemelli 2006 and 2007 (see references).

8.1   Introduction

The differential system of special relativistic hydrodynamics is a set of five scalar equations: four come from the stress-energy tensor conservation $\partial_\alpha T^{\alpha\beta} = 0$; the fifth is the continuity equation $\partial_\alpha (rU^\alpha) = 0$, where r is the baryon number and U the proper velocity, which corresponds to number density conservation (α, $\beta = 0$, 1, 2, 3).

It is then reasonable to imagine their unification into a single 5D tensor conservation law: $\partial_A T^{AB} = 0$ (A, $B = 0$, 1, 2, 3, 4). This however lets some additional terms, due to the fifth dimension, arise (see Section 8.3). In fact these terms turn the energy-momentum and the number-density conservation laws into balance laws; this seems to suggest that the fifth dimension plays a role in problems, such as relativistic inflationary cosmology, where the baryon number is not conserved. And, rather unexpectedly, this holds

---

[1]L. S. "B. Pascal", Via P. Nenni n. 48, 00040 Pomezia, Roma, Italy; Email: gianluca.gemelli@poste.it

265

in flat Minkowski spacetime, i.e. in the limit of negligible gravitation.

This is what we are going to do here: consider hydrodynamics in 5D cosmological special relativity and then in 5D cosmological general relativity, and show that this leads in a natural way to particle production (or annihilation) effects.

The picture in fact does not change significantly when adding gravitation to the scheme and moving to a 5D curved manifold, with the presence of a 5D self-gravitating cosmological fluid. This is not the case, of course, for the link between dynamics (including particle production) and the geometry of the space-time, which in the former case is absent, while in the latter case is governed by the 5D Einstein equations. In any case stress-energy is conserved in 5D, and particle production seems interpretable as a collateral effect of the presence of the fifth dimension (velocity, or time, depending on the point of view), or, rather, of our non-perceiving of it.

Relativistic inflationary cosmology often considers particle production phenomena (see e.g. Cissoko 1998). In fact cosmological particle production accounts for negative pressure, which arises in the modeling of the accelerating Universe in the standard model (see e.g. de Campos 2002). The cosmological scenario with particle production is sometimes called open system cosmology (see e.g. Prigogine et al. 1989). It will be interesting to see how such a scenario fits in a natural way in the framework of 5D cosmological relativity. However the effect here seems to be independent of the sign of the pressure of the cosmological fluid.

We recall that the famous Kaluza-Klein unified theory of gravity and electromagnetism is the prototype 5D extension of general relativity. One of the aims of this theory is however to show how 5D vacuum turns into 4D stress-energy; rather, here instead we have that 5D matter conservation turns into 4D particle creation. For an overview on modern Kaluza-Klein theory, see e.g. Wesson 1999.

8.2 Relativistic Hydrodynamics

Let $\mathcal{N}_4$ denote the space-time of General Relativity, with signature $+---$. Let Greek indices run from 0 to 3. Units are chosen such that the speed of light in empty space $c \equiv 1$.

A relativistic continuum system is characterized by a world tube $\Omega \subset \mathcal{N}_4$, generated by the set of the worldlines of its constituting elementary particles, with tangent unit timelike vector field U^α. The world tube Ω

is the support of a stress-energy tensor $T^{\alpha\beta}$, which, as a consequence of the Einstein gravitational equations, must satisfy the conservation laws $\nabla_\alpha T^{\alpha\beta} = 0$.

Let us follow Lichnerowicz 1967 and 1994 and recall the main features of relativistic hydrodynamics. For a perfect fluid the energy-momentum tensor is a function of the dynamical and thermodynamical variables:

$$T^{\alpha\beta} = (\rho + p)U^\alpha U^\beta - pg^{\alpha\beta}, \tag{8.2.1}$$

where g is the spacetime metric, $\rho \geq 0$ is the proper energy density and $p \geq 0$ the proper pressure. We suppose $\rho + p > 0$. We moreover set

$$\rho = r(1 + \mathcal{E}), \tag{8.2.2}$$

where $r \geq 0$ is the matter density (baryon number) and $\mathcal{E} \geq 0$ the internal energy. Let us introduce the proper temperature T and the specific entropy S. We also introduce the variable $f = (\rho + p)/r$, which is called fluid index, and it is equivalently defined by $f = i + 1$, where $i = p/r + \mathcal{E}$ is the specific enthalpy. The thermodynamical variables are linked by the thermodynamic differential principle:

$$d\rho = f dr + rT dS. \tag{8.2.3}$$

It is convenient to adopt p and S as the fundamental thermodynamical variables, and to introduce a generic equation of state of the form: $r = r(p, S)$. Thus from Eq. (8.2.3) we have two independent relations: $T = (\partial \mathcal{E}/\partial S) - (p/r^2)(\partial r/dS)$, and $T(\partial S/\partial p) = (\partial \mathcal{E}/\partial p) - (p/r^2)(\partial r/\partial p)$. We therefore have 5 independent variables: p, S (or any other pair of thermodynamical variables) and the three independent components of U^α ($U^\alpha U_\alpha = 1$).

The differential system of relativistic hydrodynamics is

$$\nabla_\alpha(rU^\alpha) = 0, \tag{8.2.4a}$$

$$\nabla_\alpha T^{\alpha\beta} = 0. \tag{8.2.4b}$$

Equation (8.2.4a) is the continuity equation, which means number density conservation. Therefore matter creation is neglected within this scheme. Equation (8.2.4b) is the ordinary conservation equation for the stress-energy tensor of the fluid, which follows from the Einstein equations and the Bianchi identities. As for (8.2.4a), instead, it is not a consequence of the Einstein equations, but a specific axiom. The total number of scalar equations in Eq. (8.2.4) is 5.

An equivalent formulation of Eqs. (8.2.4) is obtained by replacing Eq. (8.2.4a) with $U^\alpha \partial_\alpha S = 0$. However formulation (8.2.4) is preferable for its conservative form.

In the particular case of special relativity, the Minkowskian spacetime $\mathcal{M}_4$ replaces the generic space-time pseudo-Riemannian manifold $\mathcal{N}_4$, and the Minkowskian metric η replaces the generic pseudo-Lorentzian metric g. The signature is the same, and the functional form of the stress-energy tensor in terms of the hydrodynamical variables is again (8.2.1). Also the thermodynamical principle (8.2.3) is unchanged. The continuity equation and the stress-energy conservation equations still hold, both axiomatically, but partial derivatives ∂_α replace covariant derivatives in Eqs. (8.2.4) resulting in

$$\partial_\alpha (r U^\alpha) = 0, \tag{8.2.5a}$$
$$\partial_\alpha T^{\alpha\beta} = 0. \tag{8.2.5b}$$

8.3   Five-Dimensional Relativity

Let us now introduce a generic (non flat) 5D manifold $\mathcal{N}_5$ in general coordinates, i.e. 5D general relativity. The signature of cosmological relativity is $+ - - - +$, where the fifth dimension is velocity and the first is time. However here we prefer to leave for the moment the possibility for the signature to be $+ - - - -$ as well; to this aim we will introduce into the equations a scalar ϵ which can assume the values $+1$ or -1. Let capital Latin indices run from 0 to 4. Note that in Gemelli 2006 and 2007 the signature is opposite than in this book.

Let now ξ be a regular field of unit vectors ($\xi^A \xi_A = \epsilon$) with support on $\mathcal{N}_5$ (although this assumption can be relaxed to some domain, subset of $\mathcal{N}_5$), so that the tangent space at any point of $\mathcal{N}_5$ is the sum of the direction of ξ and of a local neighborhood $\Omega_4 \subset \mathcal{N}_4$ as orthogonal complement. Thus ξ represents, in a sense, the direction of the fifth dimension.

Any tensor index can be projected by means of the projector orthogonal to ξ: $\delta_A^{\ B} - \epsilon \xi_A \xi^B$ and that parallel to ξ: $\epsilon \xi_A \xi^B$. Thus any tensor can be split into the sum of its pure 4D space-time component (orthogonal to ξ), its pure fifth-dimension component (parallel to ξ), and some mixed components (depending on the order of the tensor). Under this point of view, this is a problem similar to that of splitting the space-time as the sum of space and time (see e.g. Jantzen *et al.* 1992).

Clearly, the definition of the 4D spacetime is local, unless one additionally supposes that the vector field ξ globally admits a family of orthogonal 4D leaf-manifolds. In this case such leaves represent the global branes $\mathcal{N}_4$ at any value of the fifth dimension parameter, and one can consider orthonormal coordinates, i.e. such that the 5D line element is

$$g_{AB}dx^A dx^B = g_{\alpha\beta}dx^\alpha dx^\beta + \epsilon\,(d\xi)^2\,, \tag{8.3.1}$$

where, with a slight abuse of notation, we have denoted $\xi = x^4$ (in the cosmological relativity case we have $\epsilon = 1$ and $\xi = v$), and where we suppose $g_{\alpha\beta}$ to be of signature $+---$. We also have $\xi^A = \delta_4{}^A$, $\xi_B = \epsilon\delta_B{}^4$ and $\xi^A \xi_A = \epsilon$. In the general case orthogonal coordinates would not exist globally.

Let us consider a generic 5D symmetric stress-energy 2-tensor $\mathcal{T}^{AB}$, representing the mass-energy-momentum contents of the 5D spacetime. We don't prescribe, for the moment, a given form for this tensor.

The (unique) splitting of $\mathcal{T}^{AB}$ along the ξ-direction and the orthogonal complement of $\mathcal{N}_5$ (which is $\mathcal{N}_4$) is

$$\mathcal{T}^{AB} = T^{AB} + P^A\xi^B + P^B\xi^A + E\xi^A\xi^B, \tag{8.3.2}$$

where T^{AB} and P^A are orthogonal to ξ, and T^{AB} is symmetric (it is the ordinary energy-momentum tensor).

If we postulate, in a natural way, the 5D Einstein equations: $\mathcal{G}_{AB} = \kappa\mathcal{T}_{AB}$, where $\mathcal{G}_{AB}$ is the 5D Einstein tensor, constructed in the usual way by means of the 5D curvature, we consequently have the conservation of the stress-energy source tensor $\mathcal{T}^{AB}$, i.e.

$$\nabla_A \mathcal{T}^{AB} = 0. \tag{8.3.3}$$

We will see in the next section that (8.3.3) can be considered as unifying, in the 5D spacetime, Eqs. (8.2.4a) and (8.2.4b). An additional scalar law which plays the role of the continuity equation in fact arises from the splitting of (8.3.3) according to (8.3.2). Thus the complete set of relativistic hydrodynamics follows from the Einstein equations, while in 4D one has to add the continuity equation to the scheme. This means that 5D relativity is a somewhat natural framework for relativistic hydrodynamics. The price to pay is that conservation equations will be turned into balance laws, which can also be considered as a useful enrichment of the scheme.

Let us take a look at the particular case of a 5-dimensional flat manifold $\mathcal{M}_5$. In this case we even have global orthonormal coordinates, i.e.

$$g_{AB}dx^A dx^B = dt^2 - \left(dx^2 + dy^2 + dz^2\right) + \epsilon d\xi^2, \tag{8.3.4}$$

where we have denoted $t = x^0$, $x = x^1$, $y = x^2$, $z = x^3$ and $\xi = x^4$ (in the cosmological relativity case we have $\epsilon = 1$ and $\xi = v$). The branes at any different value of the fifth dimension parameter ξ are in this case all coincident with the usual 4D Minkowski spacetime. Here in practice we set both the speed of light in vacuo $c = 1$ and the Hubble-Carmeli constant (analogous to c for the fifth dimension) $\tau = 1/h = 1$, where h is the Hubble constant in empty space (see Carmeli 2002).

We can then generalize special relativistic dynamics by adopting a conservation law analogous to (8.3.3), with partial derivatives in place of covariant derivatives

$$\partial_A T^{AB} = 0. \tag{8.3.5}$$

The splitting of Eq. (8.3.5) along ξ and its orthogonal complement is trivial:

$$\partial_A P^A + E' = 0 \tag{8.3.6a}$$

$$\partial_A T^{AB} + P^{B\prime} = 0, \tag{8.3.6b}$$

where we have denoted partial derivatives with respect to ξ by a prime.

In a curved 5D spacetime the splitting of Eq. (8.3.3) is in general rather complicated, but things are simpler if we adopt coordinates such that Eq. (8.3.1) holds (locally or globally). In this case the evolution equations are very similar to Eqs. (8.3.6). We have, in fact, the following system, equivalent to Eq. (8.3.3):

$$\nabla_\alpha P^\alpha + E' = 0, \tag{8.3.7a}$$

$$\nabla_\alpha T^{\alpha\beta} + P^{\beta\prime} + (1/2)g^{\beta\alpha}(g_{\gamma\alpha})' P^\gamma = 0. \tag{8.3.7b}$$

Both Eqs. (8.3.7a) and (8.3.7b) are consequences of the 5D Einstein equations.

8.4   Particle Production

Now let us compare the 4D differential system (8.2.5), governing special relativistic hydrodynamics, with the 5D system (8.3.6). We see that (8.3.6) is a good candidate to represent a generalization of its 4D analogue; the natural identifications to have a match are

$$P^\alpha = rU^\alpha, \quad T^{\alpha\beta} = (\rho + p)U^\alpha U^\beta - pg_4^{\alpha\beta}, \tag{8.4.1}$$

where $g_4{}^{AB} = g^{AB} - \epsilon\xi^A\xi^B$ is the 4D metric. If such identification is made, we have, in particular, from Eq. (8.3.6a):

$$\partial_\alpha(rU^\alpha) + E' = 0. \tag{8.4.2}$$

Such an equation is the replacement of the continuity equation, but it also includes a source term $-E'$. From the time component of Eq. (8.3.6b) we get,

$$(\rho + p)\partial_\alpha U^\alpha + \dot\rho + r' = 0, \tag{8.4.3}$$

where a dot denotes the time derivative $U^\alpha\partial_\alpha$.

Now from Eq. (8.4.3) we eliminate the divergence of U and obtain

$$-f(E' + \dot r) + \dot\rho + r' = 0. \tag{8.4.4}$$

Then, from (8.4.4) and the thermodynamical principle (8.2.3), we obtain the following relation for E':

$$E' = f^{-1}(rT\dot S + r'). \tag{8.4.5}$$

We obtain exactly the same expression also in the general relativistic case. Starting from a comparison of Eqs. (8.2.4) with (8.3.7), we again get (8.4.1) and proceed in a similar way. In this case we have

$$\nabla_\alpha(rU^\alpha) + E' = 0, \tag{8.4.6}$$

where Eq. (8.4.5) is unchanged, since contributions from covariant derivatives are dropped by the splitting, or are contained in the term $\nabla_\alpha U^\alpha$ and then canceled. This means, in a sense, that *5D Einstein equations directly prescribe 4D particle non-conservation*. In fact they provide a source for it, at least if they are considered in the presence of a 5D stress-energy tensor which can be given a 4D hydrodynamical interpretation as in Eqs. (8.4.1).

Is it matter creation we are talking about, or annihilation? It is clear that the sign of E' cannot be evaluated without additional assumptions, however, if we take $\dot S < 0$, as usual, from Eq. (8.4.5) we have at least a negative component. In particular, if the term r' should be negligible, we would have $E' < 0$, and thus $-E'$ would be a source for particle production.

This is a rather interesting result. In recent times cosmology has begun to be increasingly regarded as an experimental science. Recent measurements on magnitude and redshift of supernova seem to indicate that we live in an accelerating Universe (see e.g Riess *et al.* 1998 and Perlmutter *et al.* 1999). It is thought, in the standard model, that such acceleration is due to some kind of repulsive gravitational force, and that a cosmological perfect fluid with negative pressure can account for it. Negative pressure can be introduced in the energy-momentum of a perfect fluid by taking into account the balance of the positive thermodynamic pressure with a negative pressure scalar due to particle creation. In order to take into account particle production one introduces by hand a source term at right hand side of the matter density conservation equation (see e.g. Cissoko 1998), which then reads as a balance law. Here, instead, Eq. (8.4.6) follows from the Einstein equations, and in the flat case Eq. (8.4.2) is also a consequence of Eq. (8.3.5), i.e. of the 5D model. It is worth saying that, in our framework, even energy is not conserved, as from Eq. (8.3.6b) or Eq. (8.3.7b).

It seems that effects due to the fifth dimension are not negligible when particle production phenomena are present; conversely, such kind of phenomena can in principle be interpreted as effects due to the fifth dimension. Moreover, in the 5D general relativistic case, the 5D Einstein equations suggest that particle production phenomena could influence the geometry of the spacetime.

8.5   5D Hydrodynamics

We have not assumed any prescribed form for the 5D stress-energy yet. Then, let us define, by analogy with the case of perfect hydrodynamics

$$\mathcal{T}^{AB} = (M + Q)V^A V^B - Qg^{AB}. \tag{8.5.1}$$

Here the "thermodynamical" fields M and Q are supposed to be defined and regular in a 5D "world tube" generated by a geometrical congruence of lines tangent to V. Let s be a privileged parameter along one such line, which we denote by ℓ, with parametric equations $x^A = X^A(s)$; we thus have, along ℓ: $M = M(s)$, $Q = Q(s)$, and

$$V^A = \frac{dx^A}{ds}. \tag{8.5.2}$$

Now let us introduce the following splitting,

$$V^A = W^A + \mu\xi^A, \tag{8.5.3}$$

where, for the moment, μ is a free parameter. Also, at the moment, we avoid prescribing any sign to the norm of V.

We have $V^\alpha = W^\alpha = dx^\alpha/ds$ and $V^\xi = \mu = d\xi/ds$.

Let us denote by a star the derivative with respect to s (and, as before, by a prime the derivative with respect to ξ), so that we have

$$(\;)^\star = V^A \partial_A = W^\alpha \partial_\alpha + \mu(\;)'. \tag{8.5.4}$$

We also have: $V^A = (X^A)^\star$, $\xi^\star = \mu$. Now let us compare Eqs. (8.3.2), (8.4.1) and (8.5.1); we must have:

$$\mu(M + Q)W^\alpha = ru^\alpha, \qquad (8.5.5a)$$

$$(M + Q)W^\alpha W^\beta + Qg^{\alpha\beta} = (\rho + p)u^\alpha u^\beta - pg^{\alpha\beta}. \qquad (8.5.5b)$$

One thus necessarily finds that

$$Q = p, \qquad (8.5.6a)$$

$$M = r\mu^{-2}f^{-1} - p, \qquad (8.5.6b)$$

and

$$W^\alpha = \mu f U^\alpha. \qquad (8.5.6c)$$

Consequently we have

$$E = rf^{-1} - \epsilon p. \qquad (8.5.7)$$

We also have

$$(V \cdot V) = V^A V_A = \mu^2 \left[\epsilon + f^2\right]. \qquad (8.5.8)$$

Therefore for $\epsilon = 1$, i.e. if the signature is that of Carmeli's Cosmological Relativity, we necessarily have $V \cdot V > 0$, i.e. V timelike. If $\epsilon = -1$, instead, V could in principle be timelike, spacelike or lightlike.

Note that in the cited literature there are different signs in Eqs. (8.5.7) and (8.5.8), due to the signature adopted.

For the sake of brevity we denote by a dot, as usual, derivative with respect to proper time, i.e. $(\ )^{\cdot} = U^\alpha \partial_\alpha$. From Eq. (8.5.4) we then have

$$(\ )^\star = \mu[f(\ )^{\cdot} + (\ )']. \qquad (8.5.9)$$

Now let us apply Eq. (8.5.9) to ξ and compare with $\xi^\star = \mu$; we are led to the following identity:

$$f\dot{\xi} = 0. \qquad (8.5.10)$$

We discard for the moment the singular situation $f = 0$ (otherwise $\rho + p = 0$) and conclude from Eq. (8.5.10) that $\dot{\xi} = 0$.

8.6 The Isentropic Case

From comparison between Eqs. (8.4.5) and (8.5.7) we obtain,

$$rf^{-1}T(\dot{S} + f^{-1}S') = -(\epsilon + f^{-2})p',\qquad(8.6.1)$$

or equivalently, from Eq. (8.5.9),

$$r^2 f^{-2}\mu^{-1}TS^\star = -(\epsilon + f^{-2})p'.\qquad(8.6.2)$$

A first consequence of (8.6.1) is that if all fields are independent on ξ, like in ordinary 4D hydrodynamics, we have $\dot{S} = 0$, which in fact is a well known consequence of the original systems (8.2.4) and (8.2.5). Conversely, it is easy to check that if the flow is isentropic then particle production is absent. In fact if $dS = 0$ we have,

$$(\epsilon + f^{-2})p' = 0,$$

and therefore there are two possible situations: $p' = 0$ or $\epsilon = -f^{-2}$. Now, if the fluid is isentropic the equation of state must be of the kind: $p = p(r)$ (see e.g. Lichnerowicz 1967 and 1994), thus from Eq. (8.2.3) if $p' = 0$ we have $r' = \rho' = 0$ and consequently $E' = 0$. On the other hand, if $\epsilon = -f^{-2}$, then we must have $\epsilon = -1$ (in other words this cannot happen in Carmeli's cosmological relativity) and $f^2 = 1$, and consequently $r^2 = (\rho + p)^2$. We then have from Eq. (8.2.3),

$$rdr = (\rho + p)(d\rho + dp).$$

Now since $dS = 0$ we have

$$d\rho = r^{-1}(\rho + p)dr,$$

so that in the end we have

$$(\rho + p)dp = 0.\qquad(8.6.3)$$

Thus, excluding the singular case $\rho + p = 0$ we conclude $dp = 0$ and consequently $dr = d\rho = 0$, which implies $dE = 0$ and thus again $E' = 0$.

Thus in any case the source of particle production vanishes if $dS = 0$. We remark that such a result holds both in the flat case and in that of curved spacetime.

8.7 Simulation of Friedmann Cosmology in Flat Spacetime

We have described above a particle production process which is mainly a consequence of the extra dimension, with the presence of gravity being practically ineffectual. But to what extent the dynamic of the mass-energy contents of the Universe can be described in special relativistic terms? In this section we try and answer this question by obtaining a set of evolution equations for the Universe expansion. The signature of the fifth dimension, which we have not specified above, will be instead directly involved in such Friedmann-like equations, and will determine, in a sense, the type of universe. The following is based on Gemelli 2007, with some generalizations and corrections of mistakes. Moreover, note that the signature of the metric used in (Gemelli 2007) is the opposite to that used here.

Friedmann equations are (Carmeli 2002 pp. 168-169 and Wesson 1999 p. 15):

$$\left(\frac{\dot{R}}{R}\right)^2 = \frac{1}{3}\left(\kappa\rho_F + \Lambda\right) - \frac{k}{R^2}, \tag{8.7.1a}$$

$$\frac{\ddot{R}}{R} = -\frac{\kappa}{6}\left(\rho_F + 3p_F\right) + \frac{\Lambda}{3}, \tag{8.7.1b}$$

where R is the scale factor, κ is the gravitational constant, Λ is the cosmological constant, $k = +1, 0, -1$ and ρ_F and p_F are density and pressure of the cosmological matter-distribution of curved spacetime. We have had to introduce the suffix $|_F$ to distinguish the fields from analogous fields introduced above: simulating the Friedmann equations will in fact mean obtaining them by a suitable redefinition of those fields. In (8.7.1) dots mean derivatives with respect to the time coordinate.

The fact that Eqs. (8.7.1) can be formally obtained in a 5D special relativistic framework, which we are going to see, is not a completely surprising fact, since the full 5D hydrodynamical system gives us some parameters to play with, yet it is certainly non trivial, since Friedmann equations are Einstein's gravitational equations, while we are working in a flat-spacetime, with no gravitation.

The significant idea then is that 5D special relativistic hydrodynamics can simulate, at least to some extent, general relativistic cosmology.

Let us consider the original 5D system (8.3.5). In relativistic hydrodynamics the variable $\mathcal{T} = f/r$ has the characteristics of a volume, and is, in fact, called dynamical volume (see Lichnerowicz 1994, p. 99). Let

$\Phi = \mu^{-2}\mathcal{T}^{-1}$. From Eqs. (8.5.6a) and (8.5.6b) we then have

$$M + Q = \Phi. \tag{8.7.2}$$

It is a useful idea, on physical terms, to imagine that the dynamical volume should be proportional to the cube of a parameter R, representing a "typical length", and that therefore our variable Φ should be proportional to R^{-3}; we will introduce this hypothesis later on.

From Eqs. (8.3.5) and (8.5.1) we then obtained the following general form of the 5D system:

$$\Phi^\star V^B + \Phi(V^B)^\star + \Phi\partial_A V^A V^B - \partial^B p = 0. \tag{8.7.3}$$

Now we will consider some useful consequences of the system (8.7.3).

By multiplying Eq. (8.7.3) by X_B we get

$$\Phi X_B(V^B)^\star + (\Phi^\star + \Phi\partial_A V^A)X_B V^B - X^B\partial_B p = 0. \tag{8.7.4}$$

By multiplying Eq. (8.7.3) by V_B we get

$$\Phi V_B(V^B)^\star + (\Phi^\star + \Phi\partial_A V^A)V_B V^B - p^\star = 0. \tag{8.7.5}$$

Finally, by taking the 5D divergence of Eq. (8.7.3), i.e. in practice by multiplying it by ∂_B, we get

$$\Phi^{\star\star} + 2\Phi^\star\partial_A V^A + (V^B)^\star\partial_B\Phi + \Phi[\partial_B(V^B)^\star + (\partial_B V^B)^\star]$$

$$+\Phi(\partial_A V^A)^2 - \partial_A\partial^A p = 0. \tag{8.7.6}$$

Now, since $V^B = (X^B)^\star$, we have

$$X_B V^B = (X^B X_B)^\star/2, \tag{8.7.7a}$$

$$X_B(V^B)^\star = (X^B X_B)^{\star\star}/2 - V_B V^B, \tag{8.7.7b}$$

$$V_B(V^B)^\star = (V_B V^B)^\star/2. \tag{8.7.7c}$$

Moreover, since $\Phi = \Phi(s)$, we write

$$(V^B)^\star\partial_B\Phi = \Phi^\star\partial_A V^A.$$

In fact,

$$\Phi^\star(V^B)^\star\frac{ds}{dX^B} = \Phi^\star\frac{(V^B)^\star}{(X^B)^\star},$$

and we also have,

$$\frac{(V^B)^\star}{(X^B)^\star} = \frac{dV^B}{ds}\frac{ds}{dX^B} = \frac{dV^B}{dX^B}.$$

Therefore, denoting, for the sake of brevity: $X^2 = -X_B X^B$, $\partial_X = X^B \partial_B$, $\Delta = \partial_A \partial^A$ and $\nabla V = \partial_A V^A$, then Eqs. (8.7.4)-(8.7.6) take the following form:

$$\frac{1}{2}\Phi[(-X^2)^{\star\star} - 2(V \cdot V)] + \frac{1}{2}(\Phi^\star + \Phi \nabla V)(-X^2)^\star - \partial_X p = 0, \quad (8.7.8a)$$

$$\Phi^\star(V \cdot V) + \frac{1}{2}\Phi(V \cdot V)^\star + \Phi \nabla V(V \cdot V) - p^\star = 0, \quad (8.7.8b)$$

$$\Phi^{\star\star} + 3\Phi^\star \nabla V + \Phi[\partial_A(V^A)^\star + \nabla V^\star] + \Phi \nabla(V \cdot V) - \Delta p = 0. \quad (8.7.8c)$$

Note that with our use of the symbol X^2 we implicitly assume, for the sake of simplicity, $X_A X^A < 0$, which is restrictive, since $X^A X_A = t^2 - x_i x^i + \epsilon \xi^2$ in general could be non negative. However we may be considering $t = 0$ (present time) and "large distances" in a sense, such that X can be considered spacelike.

Now let us introduce the typical length parameter R, in a crude and simple way, i.e. by taking $X^2 = R^2$ and $\Phi = R^{-3}$. This R should not be confused with the Ricci scalar of general relativity. It also should not be taken as implying the existence of a physical boundary. From Eqs. (8.7.8a) - (8.7.8c) we then have respectively:

$$\Phi\left[(R^\star)^2 + RR^{\star\star} + (V \cdot V)\right] + (-3\Phi R^\star/R + \Phi \nabla V)RR^\star + \partial_X p = 0, \quad (8.7.9a)$$

$$-3\Phi(V \cdot V)R^\star/R + (1/2)\Phi(V \cdot V)^\star + \Phi \nabla V(V \cdot V) - p^\star = 0, \quad (8.7.9b)$$

$$-\Phi\left[3R^{\star\star}/R - 12(R^\star/R)^2 + 9\nabla V R^\star/R\right] + \Phi[(\nabla V)^\star + \nabla(V^\star)]$$

$$+\Phi \nabla(V \cdot V) - \Delta p = 0. \quad (8.7.9c)$$

We now need to introduce some estimates for the pressure gradient. We suppose the cosmological fluid to have a quasi-isotropic and slow-varying pressure, i.e. such that $dp \propto R^{-2}$. This generalizes the hypothesis of Gemelli 2007, where all terms depending on the derivatives of the pressure were simply dropped. Now, if $dp \propto R^{-2}$ we also have: $\partial_X p \propto R^{-1}$ (since $|X| = R$) and $\Delta p \propto d^2 p \propto R^{-3}$. As for p^* we see from Eq. (8.7.7b) that in a sense $V^\star \propto R^{-1}$, so we are led to assume $p^\star \propto R^{-3}$.

Now from Eq. (8.7.9b) we have

$$\nabla V = 3\frac{R^\star}{R} - \alpha, \quad (8.7.10)$$

where we have denoted

$$\alpha = \frac{1}{2}\frac{(V \cdot V)^\star}{(V \cdot V)} - R^3 p^\star. \qquad (8.7.11)$$

Replacing ∇V by (8.7.10) in Eqs. (8.7.9a) and (8.7.9b) we have:

$$\frac{R^{\star\star}}{R} + \left(\frac{R^\star}{R}\right)^2 - \alpha\frac{R^\star}{R} + \frac{(V \cdot V)}{R^2} + R\partial_X p = 0, \qquad (8.7.12a)$$

$$\frac{R^{\star\star}}{R} + 2\left(\frac{R^\star}{R}\right)^2 - \alpha\frac{R^\star}{R} - \frac{1}{3}(q + \alpha^2), \qquad (8.7.12b)$$

where we have denoted

$$q = (\nabla V)^\star + \nabla(V^\star) - R^3 \Delta p. \qquad (8.7.12c)$$

Taking Eq. (8.7.12b) minus Eq. (8.7.12a), and denoting $\beta = q + 3R\partial_X p$, we then have

$$\left(\frac{R^\star}{R}\right)^2 = \frac{1}{3}(\beta + \alpha^2) + \frac{(V \cdot V)}{R^2}. \qquad (8.7.13)$$

For our purposes, and under our hypothesis on dp, the fields α, β and q can be regarded as constants.

We recognize in Eq. (8.7.13) the same structure of the Friedmann equation (8.7.1a). The correspondence is only formal, since we have to somehow identify the time derivative with the star derivative, and the scale factor of Friedmann cosmology (which comes from the metric of the curved 4D spacetime) with our "typical length". Yet such correspondence is significant. In particular, Eq. (8.7.13) reduces to Eq. (8.7.1a) if

$$\kappa\rho_F + \Lambda = \beta + \alpha^2, \qquad (8.7.14a)$$

$$k = -(V \cdot V). \qquad (8.7.14b)$$

We know from Eq. (8.5.8) that the case $\epsilon = 1$ (Carmelian relativity) means $V \cdot V > 0$. Thus from Eq. (8.7.14b) we see that our model in connection with Carmeli's cosmological relativity actually predicts $k = -1$. It directly gives $k < 0$, but an exact match with the value -1 may be obtained by proper choice of our free parameter μ; see Eq. (8.5.8).

It is clear that in general the sign of k is determined by that of $V \cdot V$: we have $k = 0$ if $V \cdot V = 0$, $k = -1$ if $V \cdot V > 0$ and $k = 1$ if $V \cdot V < 0$. Therefore if instead $\epsilon = -1$ then our model fits with all the three possible values of k, and does not select one.

Now replacing $R^\star/R$ from Eq. (8.7.13) in Eq. (8.7.12a), we get

$$\frac{R^{\star\star}}{R} = -\frac{2\beta + \alpha^2 - q}{3} - 2\frac{(V \cdot V)}{R^2}$$

$$+\alpha\nu\sqrt{\frac{1}{3}(\beta + \alpha^2) + \frac{(V \cdot V)}{R^2}}, \tag{8.7.15}$$

where ν can assume the values $+1$ or -1 (depending on the sign assigned to $R^\star/R$, which, by Eq. (8.7.13), is unknown). By power series expansion in terms of R^{-1} we have

$$\sqrt{\frac{(\beta + \alpha^2)}{3} + \frac{(V \cdot V)}{R^2}}$$

$$= \frac{\sqrt{3(\beta + \alpha^2)}}{3} + \sqrt{\frac{3}{\beta + \alpha^2}}\frac{(V \cdot V)}{2R^2} + O(R^{-4}). \tag{8.7.16}$$

Thus, dropping terms of higher orders, from Eq. (8.7.14) we get

$$\frac{R^{\star\star}}{R} = \frac{\alpha\nu\sqrt{3\beta + 3\alpha^2} - 2\beta - \alpha^2 + q}{3}$$

$$- \left(2 - \frac{\alpha\nu}{2}\sqrt{\frac{3}{\beta + \alpha^2}}\right)\frac{(V \cdot V)}{R^2}. \tag{8.7.17}$$

Note that, besides the different signature of the metric, Eq. (53) in Gemelli 2007, analogous to Eq. (8.7.17) above, misses a factor 2 and some of the constants, which were erroneously forgotten.

Equation (8.7.17) appears to introduce a correction term of order R^{-2} to our analogue to Friedmann equation (8.7.1b). However we can still match the terms of order zero if:

$$\frac{\kappa}{6}(\rho_F + 3p_F) - \frac{\Lambda}{3} = \frac{1}{3}\left[(2\beta + \alpha^2 - q) - \nu\alpha\sqrt{3\beta + 3\alpha^2}\right]. \tag{8.7.18}$$

A solution to system (8.7.14) - (8.7.18) is the following:

$$\kappa\rho_F = \beta + \alpha^2 - \Lambda, \tag{8.7.19a}$$

$$\kappa p_F = \Lambda + \frac{1}{3}\left[3\beta + \alpha^2 - 2q - 2\sqrt{3\beta + 3\alpha^2}\right]. \tag{8.7.19b}$$

The possible additional condition: $\beta = -(13/16)\alpha^2$ lets the R^{-2} correction vanish, so that both Friedmann equations (8.7.1) are formally recovered, and leads to the following values for ρ_F and p_F:

$$\kappa\rho_F = \frac{3}{16}\alpha^2 - \Lambda, \tag{8.7.20a}$$

$$\kappa p_F = \frac{1}{16}\alpha^2 + \frac{2}{3}(\beta - q) - \frac{1}{2}\nu\alpha|\alpha| + \Lambda. \qquad (8.7.20b)$$

We have obtained a formal recovering of the Friedmann equations, with the possible correction displayed by Eq. (8.7.17). We also have obtained the condition $k = -1$ from compatibility with Carmeli's cosmological relativity. Such are in the writer's opinion, somewhat interesting results. However it is doubtful if such results can be considered compatible with the particle production mechanism considered in Sections 8.4 and 8.6 and how, since density and pressure involved in the two cases are different fields.

In the next chapter the properties of gravitational waves are examined in the framework of the five-dimensional brane world theory.

8.8 Suggested References

M. Carmeli, *Cosmological Special Relativity: The Large-Scale Structure of Space, Time and Velocity*, Second Edition (World Scientific, Singapore, 2002).

M. Carmeli, *Cosmological Relativity: The Special and General Theories for the Structure of the Universe* (World Scientific, Singapore, 2006).

M. Cissoko, Wavefronts in a relativistic cosmic two-component fluid, *Gen. Relat. Grav.* **30**, 521 (1998).

M. de Campos, Tensorial perturbations in an accelerating Universe, *Gen. Relat. Grav.* **34**, 1393 (2002).

G. Gemelli, Particle production in 5-dimensional cosmological relativity, *Int. J. Theor. Phys.* **45**, 2226 (2006).

G. Gemelli, Hydrodynamics in 5-dimensional cosmological special relativity, *Int. J. Theor. Phys.* **46**, 1431 (2007).

R.T. Jantzen, P. Carini and D. Bini, The many faces of gravito-electromagnetism, *Ann. Phys.* **215**, 1 (1992).

A. Lichnerowicz, *Relativistic Hydrodynamics and Magneto–Hydrodynamics* (Benjamin, New York, 1967).

A. Lichnerowicz, *Magnetohydrodynamics: Waves and shock waves in curved space-time,* Mathematical Physics Studies, Vol. 14 (Kluwer Academic Publishers, Dordrecht, Boston, London, 1994).

I. Prigogine, J. Geheniau, E. Gunzig and P. Nardone, Thermodynamics and cosmology, *Gen. Relat. Grav.* **21**, 767 (1989).

A. G. Riess et al. Observational evidence from supernovae for an accelerating universe and a cosmological constant *Astron. J.* **116**, 1009, (1998).

S. Perlmutter et al. Measurements of Ω and Λ from 42 high-redshift supernovae *Astrophys. J.* **517**, 565, (1999).

P.S. Wesson, *Space, Time, Matter: Modern Kaluza-Klein Theory* (World Scientific, Singapore, 1999).

Chapter 9

Properties of Gravitational Waves in an Expanding Universe

John Hartnett[1] & Michael Tobar[2]

We have seen in the previous chapters that the 5D cosmological general relativity theory reproduces all the results that have been successfully tested for Einstein's 4D theory. However the theory, because of its fifth dimension, namely the velocity of the expanding Universe, predicts something different for the propagation of gravity waves on cosmological distance scales. This analysis indicates that gravitational radiation may not propagate as an unattenuated wave where effects of the Hubble expansion are felt. In such cases the energy does not travel over very large length scales but is evanescent and dissipated into the surrounding space as heat. The following is based on Hartnett and Tobar (see Hartnett and Tobar 2006). This chapter uses the linearized gravitational field equations while general relativity is highly nonlinear. To take this further the next step would be to use the exact theory of gravitation to handle this problem.

9.1 Introduction

In recent decades, the search for gravity waves has intensified with large high powered laser-based interferometric detectors coming on line. See LIGO and TAMA for example. These detectors have already reached sensitivities that should enable them to "see" well beyond the local galactic Group. On the other hand, the Hulse-Taylor binary ring-down energy

---

[1]School of Physics, the University of Western Australia, Crawley 6009 WA, Australia; Email: john@physics.uwa.edu.au

[2]School of Physics, the University of Western Australia, Crawley 6009 WA, Australia; Email: mike@physics.uwa.edu.au

budget is a precise test of general relativity and a clear indication of the existence of gravitational radiation, and it seems that the first direct detection is just a matter of time. That analysis, however, did not involve an expanding Universe.

In standard general relativity the expanding Universe has no impact on the properties of gravitational waves, except for the well known effect of redshift. However, in Carmeli cosmology the expansion of the Universe (or redshift of the gravitational wave) manifests as a fifth dimension and in this chapter we calculate the effect and how this might impact on a possible direct detection.

9.1.1 *Cosmological general relativity — A brief review*

We use the Carmelian 5D *spacetimevelocity* Universe with two timelike and three spacelike coordinates in the metric. The signature is then $(+ - - - +)$. The Universe is represented by a 5-dimensional Riemannian manifold with a metric $g_{\mu\nu}$ and a line element $ds^2 = g_{\mu\nu}dx^\mu dx^\nu$. This differs from general relativity in that here the $x^4 = \tau v$ coordinate is more correctly *velocitylike* instead of *timelike* as is the case of $x^0 = ct$, where c is the speed of light, a universal constant and t is the time coordinate. In this theory $x^4 = \tau v$, where τ is also a universal constant, the Hubble-Carmeli time constant. The other three coordinates $x^k, k = 1, 2, 3$, are spatial and *spacelike*, as in general relativity.

It has been shown that all the results predicted by general relativity and experimentally verified are also predicted by CGR. However Carmeli discussed the one consequence that was not exactly reproduced, and that was gravity waves in 5 dimensions. The new metric resulted in a redshift dependence with a more general wave equation incorporating 5 dimensions $(ct, x^1, x^2, x^3, \tau v)$.

9.1.2 *Linearized gravitational field equations*

As is the usual practice in a weak gravitational field we write the metric in the form

$$g_{\mu\nu} = \eta_{\mu\nu} + h_{\mu\nu}, \qquad (9.1.1)$$

where $\eta_{\mu\nu}$ is the Minkowskian metric, extended here from the usual 4D Minkowskian metric to 5D with signature $(+ - - - +)$. Here $\eta_{\mu\nu}$ is perturbed due to gravity sources with $h_{\mu\nu} \ll 1$.

A useful tool is to define the trace-reversed $h_{\mu\nu}$ as

$$\bar{h}_{\mu\nu} = h_{\mu\nu} - \frac{1}{2}\eta_{\mu\nu}h, \qquad (9.1.2)$$

where $h = \eta^{\alpha\beta}h_{\alpha\beta}$ is the trace of $h_{\mu\nu}$. Consequently

$$h_{\mu\nu} = \bar{h}_{\mu\nu} - \frac{1}{2}\eta_{\mu\nu}\bar{h}, \qquad (9.1.3)$$

where $\bar{h} = \eta^{\alpha\beta}\bar{h}_{\alpha\beta}$ and $\bar{h} = -h$.

Then the linearized Einstein field equations to first order in $\bar{h}_{\mu\nu}$ yield

$$\Box\bar{h}_{\mu\nu} = -2\kappa T_{\mu\nu}, \qquad \text{plus } \eta^{\alpha\beta}\bar{h}_{\mu\alpha,\beta} = 0, \qquad (9.1.4)$$

where $\Box$ is the D'Alembertian operator in 5D and may be expressed as

$$\Box = \left(\frac{1}{c^2}\frac{\partial^2}{\partial t^2} - \nabla^2 + \frac{1}{\tau^2}\frac{\partial^2}{\partial v^2}\right). \qquad (9.1.5)$$

For conservation of energy and momentum, excluding gravity, it follows from Eq. (9.1.4) that

$$\eta^{\alpha\beta}T_{\mu\alpha,\beta} = 0. \qquad (9.1.6)$$

From Eqs. (9.1.4) and (9.1.5) it is clear that (9.1.4) is a generalized wave equation that reduces to

$$\Box\bar{h}_{\mu\nu} = 0 \qquad (9.1.7)$$

in vacuum.

So the gravitational waves depend not only on space and time but also on the expansion velocity of the source in the Hubble flow. So v represents the velocity of the expansion of the space through which the wave passes.

The solution of Eq. (9.1.4) is the sum of the solutions of the homogeneous equation (9.1.7), and a particular solution. The following is the special time independent retarded solution in the absence of source-less radiation. The contravariant form is

$$\bar{h}^{\mu\nu} = -2\kappa \int \frac{T^{\mu\nu}d^3x'}{|\mathbf{x} - \mathbf{x}'|}, \qquad (9.1.8)$$

where the source mass is located at $\mathbf{x}'$ and the potential measured at $\mathbf{x}$. To evaluate the integral in Eq. (9.1.8), provided the measurement point determined by the vector $\mathbf{x}$ is far away from the source, a Taylor expansion of $1/|\mathbf{x} - \mathbf{x}'|$ about $\mathbf{x}' = 0$ is taken retaining only the first two terms,

$$\frac{1}{|\mathbf{x} - \mathbf{x}'|} \approx \frac{1}{r} + \frac{x^k x'^k}{r^3}, \qquad (9.1.9)$$

where $r^2 = x^k x^k$.

The integral in Eq. (9.1.8) is then written as

$$\bar{h}^{\mu\nu} = -\frac{GM}{r} - \frac{G}{2}\epsilon^{kln}S^n\frac{x^k}{r^3}, \tag{9.1.10}$$

where the dipole term has been eliminated by choosing the origin of the coordinates to coincide with the center of the source mass. The first term in Eq. (9.1.10) involves the integral

$$\int T^{00}(\mathbf{x}')d^3x' = M$$

identified with the source mass and the second term

$$\int x'^k T^{l0}(\mathbf{x}')d^3x' = 1/2\epsilon^{kln}S^n$$

where S^n is the spin angular momentum of the system. Here $k, l, n = 1, 2, 3$ for the spatial coordinates.

9.2 Wave Equation in Curved *Spacevelocity*

Now considering the time dependence again we can write Eq. (9.1.7) as

$$\left(\frac{1}{c^2}\frac{\partial^2}{\partial t^2} - \nabla^2\right)\bar{h}^{\mu\nu} = -\frac{1}{c^2\tau^2}\frac{\partial^2}{\partial z^2}\bar{h}^{\mu\nu}, \tag{9.2.1}$$

where the substitution $v/c \to z$ has been made. Here z is the redshift of the wave, and the substitution is valid where $z =< 0.1$, which is approximately $400\ Mpc$. We assume it is approximately valid beyond that. Now the solution to Eq. (9.2.1) is the sum of the solution to the homogeneous equation (9.2.1), which is the usual gravity wave solution in general relativity, and a particular solution of Eq. (9.2.1), which has a redshift-dependent source term.

In fact, in CGR, because the Hubble law is assumed *a priori*, the expansion velocity v (or gravitational wave redshift z), is not independent of r and depends on the matter density of the Universe. In fact, in the case of CGR, Eq. (9.2.1) can be written as

$$\left(\frac{1}{c^2}\frac{\partial^2}{\partial t^2} - \nabla^2 + \frac{1}{c^2\tau^2}\frac{\partial^2}{\partial z^2}\right)\bar{h}^{\mu\nu} = 0, \tag{9.2.2}$$

with

$$\frac{\partial^2}{\partial z^2} = \left\{\left(\frac{\partial r}{\partial z}\right)^2\frac{\partial^2}{\partial r^2} + \frac{\partial^2 r}{\partial z^2}\frac{\partial}{\partial r}\right\} \tag{9.2.3}$$

when the chain rule is applied.

9.2.1 *Plane wave solution*

Let us look for a plane wave solution of the form

$$\bar{h}^{\mu\nu} = \varepsilon^{\mu\nu}\cos k_\alpha x^\alpha, \qquad (9.2.4)$$

where the 3-space co-ordinates are (x^1, x^2, x^3). Here x^1 and x^2 are orthogonal to the direction of propagation x^3 from source to detector. Here $\varepsilon^{\mu\nu}$ is a constant tensor and k_α is a constant vector. Therefore we look for a wave propagating in the r direction which would have $k^\alpha = (\omega/c, 0, 0, k_r)$.

This means we can retain only the r derivative in ∇^2 and effectively re-write Eq. (9.2.2) in spherical co-ordinates as

$$\left(\frac{1}{c^2}\frac{\partial^2}{\partial t^2} - \frac{\partial^2}{\partial r^2} - \frac{2}{r}\frac{\partial}{\partial r} + \frac{1}{c^2\tau^2}\frac{\partial^2}{\partial z^2}\right)\bar{h}^{\mu\nu} = 0. \qquad (9.2.5)$$

Equations (9.2.2) and (9.2.5) are only valid where the Hubble law applies. When it doesn't apply, that is, where $\partial r/\partial z = 0$ in Eq. (9.2.3), Eq. (9.2.2) becomes the normal wave equation for gravity waves in source free regions.

However where the Hubble law is applicable, in flat (i.e. $\Omega = 1$) *spacevelocity* $\partial r/\partial z = c\tau$, which is the Hubble law in the zero distance/zero gravity limit. In the general curved *spacevelocity*, the form of the derivative is given by

$$\frac{1}{c^2\tau^2}\left(\frac{\partial r}{\partial z}\right)^2 = 1 + (1-\Omega)\frac{r^2}{c^2\tau^2}, \qquad (9.2.6)$$

where $\Omega = \rho/\rho_c$ is the mass/energy density at some epoch expressed as a fraction of the 'critical' density, $\rho_c = 3/8\pi G\tau^2$.

Substituting Eq. (9.2.6) into Eq. (9.2.5) with Eq. (9.2.3) we get

$$\left(\frac{1}{c^2}\frac{\partial^2}{\partial t^2} - \frac{2}{r}\frac{\partial}{\partial r} + \frac{1-\Omega}{c^2\tau^2}\left\{r^2\frac{\partial^2}{\partial r^2} + r\frac{\partial}{\partial r}\right\}\right)\bar{h}^{\mu\nu} = 0, \qquad (9.2.7)$$

which now only has dependence on r and t coordinates not z. This is a new equation and depends on the surrounding matter density Ω.

When the gravity wave is very far from the source and when $r \gg c\tau/\sqrt{1-\Omega}$ the second term of Eq. (9.2.7) is much smaller than the term in curly brackets, we assume the second term negligible. Therefore Eq. (9.2.7) can be approximated for large r as

$$\left(\frac{1}{c^2}\frac{\partial^2}{\partial t^2} + \frac{1-\Omega}{c^2\tau^2}\left\{r^2\frac{\partial^2}{\partial r^2} + r\frac{\partial}{\partial r}\right\}\right)\bar{h}^{\mu\nu} = 0. \qquad (9.2.8)$$

9.2.2 *Solutions of the field equations*

The solution of Eq. (9.2.8) can be obtained by separation of variables assuming a solution of the form $\bar{h}^{\mu\nu} \propto R(r)e^{i(\omega t + k_z)}$. Substituting the latter into Eq. (9.2.8) yields

$$\frac{\omega^2}{c^2} + \frac{\Omega - 1}{c^2 \tau^2} = 0, \tag{9.2.9}$$

with $k_z \approx 0$ and $R(r) = a_1 r^{-1}$ for $r \gg c\tau/\sqrt{1 - \Omega}$. More generally $R(r) = \sum a_n r^{-n}$ is a polynomial expression with an index $n > 0$. Furthermore by taking a hint from the solution of the usual heterogeneous equation shown in Eq. (9.1.10) $R(r)$ is determined as

$$R(r) = -\frac{GM}{r} - \mathcal{O}(\frac{1}{r})^3. \tag{9.2.10}$$

Equation (9.2.9) is a resonance condition. For this solution, which spans the whole extent of the Universe, the Universe acts like a resonant mode with a characteristic scale radius

$$R_\Omega = \sqrt{|R_\Omega^2|} = \sqrt{\left|\frac{c^2}{\omega^2}\right|} = \frac{c\tau}{\sqrt{|1 - \Omega|}}, \tag{9.2.11}$$

and resonance frequency

$$\omega = \frac{\sqrt{1 - \Omega}}{\tau}. \tag{9.2.12}$$

For values of $\tau = 4.28 \times 10^{17}$ s and $\Omega = 0.02$ the scale radius is $R_\Omega \approx 4.19\,Gpc$ and the characteristic frequency is $\omega/2\pi \approx 3.68 \times 10^{-19}$ Hz.

When $r \approx< c\tau/\sqrt{|1 - \Omega|}$ the second and fourth terms of Eq. (9.2.7) are much smaller than the third, and hence can be neglected. Therefore Eq. (9.2.7) can be approximated as

$$\left(\frac{1}{c^2}\frac{\partial^2}{\partial t^2} + \frac{1 - \Omega}{c^2 \tau^2} r^2 \frac{\partial^2}{\partial r^2}\right)\bar{h}^{\mu\nu} = 0. \tag{9.2.13}$$

Now for a spherically symmetric expanding Universe Carmeli obtains the relation between redshift and distance of the emitting source,

$$\frac{r}{c\tau} = \frac{\sinh(\varsigma\sqrt{1 - \Omega})}{\sqrt{1 - \Omega}}, \tag{9.2.14}$$

where $\varsigma = ((1 + z)^2 - 1)/((1 + z)^2 + 1)$. Assuming Eq. (9.2.14) also holds for gravity waves Eq. (9.2.8) becomes

$$\left(\frac{1}{c^2}\frac{\partial^2}{\partial t^2} + \sinh^2(\varsigma\sqrt{1 - \Omega})\frac{\partial^2}{\partial r^2}\right)\bar{h}^{\mu\nu} = 0. \tag{9.2.15}$$

This is now a wave equation with a solution of the form $\bar{h}^{\mu\nu} \propto e^{i(k_r r + \omega t)}$, assuming the $\sinh^2$ term is approximately constant on the scale of interest. Then it results in the following dispersion relation

$$k_r^2 \approx -\frac{\omega^2/c^2}{\sinh^2(\varsigma\sqrt{1-\Omega})}, \qquad (9.2.16)$$

which can be approximated for $z \ll 1$ as

$$k_r^2 \approx \frac{\omega^2/c^2}{(\Omega-1)z^2}. \qquad (9.2.17)$$

It would be more accurate to solve Eq. (9.2.15) numerically, but at least for $z \ll 1$ this approach is reasonably valid.

When $\Omega > 1$ the wave number is real and approximately $k_r = \omega/(cz\sqrt{\Omega-1})$ and hence gravity waves propagate yet are dependent on redshift. When $\Omega < 1$ the wave number is imaginary and the amplitude is attenuated with a decay constant $\kappa = ik_r = \omega/(cz\sqrt{1-\Omega})$.

In Figure 9.2.1 we plot the decay constant κ normalized by ω/c, using k_r from Eq. (9.2.16). From the Figure it is apparent that the approximation of Eq. (9.2.17) is good for $z < 0.1$.

9.2.3 *Phase and group velocities*

It follows from Eq. (9.2.16) that the phase and group velocities are determined from

$$\frac{\partial\omega}{\partial k_r} = \frac{\omega}{k_r} = c\sin(\varsigma\sqrt{\Omega-1}). \qquad (9.2.18)$$

where the identity $i\sin\theta = \sinh(i\theta)$ has been used. It is valid where $\Omega \geq 1$. The latter can be approximated for $z \ll 1$ as $cz\sqrt{\Omega-1}$. In the general cosmos where $\Omega > 1$ gravity wave propagate with the velocity $c\sin(\varsigma\sqrt{\Omega-1}) \to c$ where $v \to c$. In the cosmos where $\Omega < 1$ we have evanescent decay.

The dependence in Eq. (9.2.6) is not the same within a bound galaxy of stars and gas as it is for the large scale structure of the expanding Universe, which considers only the center of mass motion of galaxies within it. This is because within a galaxy (or cluster) the full effect of the Hubble expansion is not felt with respect to the center of mass, of which the gravitational radiation must travel with respect to, and here we show it results in the same solution as standard General Relativity.

For spherically symmetric distribution of matter in a galaxy with the region of interest far from the central potential of a fixed mass, in the disk

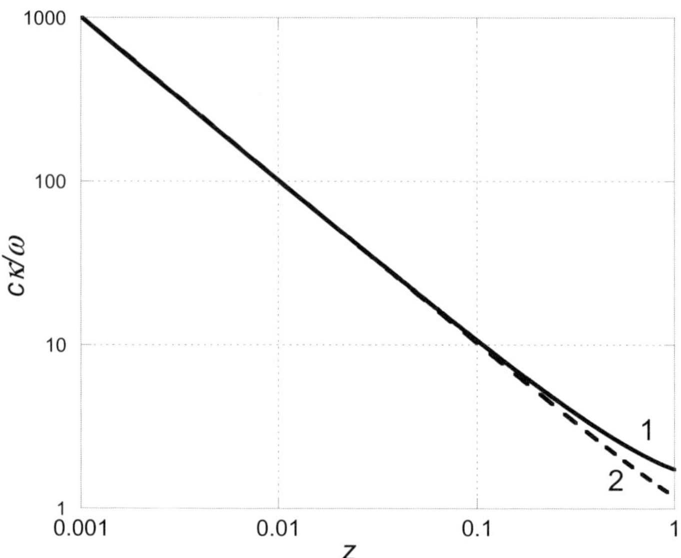

Fig. 9.2.1:   The decay constant $c\kappa/\omega$ as function of redshift, z shown from Eq. (9.2.16) as the solid curve 1 and from the approximated Eq. (9.2.17) as the broken curve 2. $|c\kappa/\omega|$ asymptotes to unity in the limit as $v \to c$ or when $z \to \infty$. The value of Ω changes from $\Omega < 1$ to $\Omega > 1$ at some value of z but the sinh function is continuous changing into a sine function as the argument of the sinh function changes from real to imaginary. (Source: Hartnett & Tobar, 2006).

region, from Eq. (7.4.10a) and the first term of Eq. (7.4.14) it may be derived that

$$\frac{1}{c^2\tau^2}\left(\frac{\partial r}{\partial z}\right)^2 = (\Omega - 1)\frac{r^2}{c^2\tau^2}, \qquad (9.2.19)$$

where $z = v/c$ and Ω is the mass density, with the origin of coordinates coinciding with the origin of the spherically symmetric gravitational potential.

Using Eq. (9.2.19) in Eq. (9.2.5) with Eq. (9.2.3) results in a wave equation

$$\left(\frac{1}{c^2}\frac{\partial^2}{\partial t^2} - \frac{\partial^2}{\partial r^2} - \frac{2}{r}\frac{\partial}{\partial r} + \frac{\Omega - 1}{c^2\tau^2}\left\{r^2\frac{\partial^2}{\partial r^2} + r\frac{\partial}{\partial r}\right\}\right)\bar{h}^{\mu\nu} = 0, \qquad (9.2.20)$$

where we can neglect the terms in the curly brackets because they are insignificant on the scale of a galaxy. Also after neglecting the third term for distant sources, we get the normal gravity wave equation of GR, that is,

$$\left(\frac{1}{c^2}\frac{\partial^2}{\partial t^2} - \frac{\partial^2}{\partial r^2}\right)\bar{h}^{\mu\nu} = 0. \tag{9.2.21}$$

9.3 Density Scales in the Universe

On the local scale the Hubble law $v = r/\tau$ does not apply or is so insignificant as to be negligible. On that scale therefore effects of *spacevelocity* are negligible, or in other words, $dv \to 0$. That is the realm where CGR reduces to the usual special and general relativity theory. Gravity waves propagate as is usually expected according to Eq. (9.2.21).

On the cosmological scale we expect the Hubble law to be very significant and hence that is the realm where *spacevelocity* is operative. On that scale we can now also consider Eq. (9.2.13) to represent a genuine modification to the usual 4D *spacetime* equation found in general relativity textbooks. The effect of *spacevelocity* is contained in the modified D'Alembertian operator. This means close to the source gravitational energy is emitted in the usual fashion as described by Eq. (9.1.10). Over cosmological length scales, however, CGR predicts the gravitational waves from distant galaxies will be fully attenuated by the time they reach the Earth. Thus, projects such as the Large-scale Cryogenic Gravitational-wave Telescope (LCGT) in Japan could be very important to test CGR theory. For example this project will detect gravity-waves from coalescing neutron-star binary systems 200 Mpc away at a SNR of 10. In contrast the TAMA field of view is 1 Mpc and LIGO is 20 Mpc. Thus, if the LGCT field of view includes a much larger volume of the Universe than LIGO or TAMA, and a smaller event rate than predicted by GR, this could be evidence for CGR.

The best estimates of the local baryonic matter density puts it around $\Omega_m \approx 0.04$ for the present epoch. As has been shown, using the Carmelian theory it is not necessary to assume any dark matter in the cosmos, therefore the matter density between galaxies should be $\Omega < 1$ even out to a redshift $z = 2$. Equation (9.2.6) is valid for large z therefore the analysis applies.

Using the averaged matter density of the Universe at the current epoch $\Omega = 0.02$, which is approximately related by $\Omega = \Omega(1+z)^3$ as a function of redshift for $z \leq 1$, and the form of Eq. (9.2.16) the decay constant (κ)

is shown as $c\kappa/\omega$ in Figure 9.2.1 as a function of redshift, z. The value of $c\kappa/\omega$ in the figure is limited by the unknown form of $\Omega(z)$ for $z \gg 1$. However, over the epochs shown, CGR predicts that gravitational waves do not propagate at scales beyond galaxies (and clusters).

Gravity waves that are generated within a galaxy quickly decay in the void between. Gravitational radiation therefore leaks into the surrounding space according to Eq. (9.2.13) with attenuated amplitudes when Ω drops below unity. According to CGR, gravity waves will not propagate far in an expanding Universe. Therefore we would expect to see no stochastic gravity wave background spectrum. Instead we conjecture that the energy is deposited into space as heat. As a result they may contribute to the CMB blackbody temperature.

9.3.1   *The case of binary pulsar*

The binary pulsar PSR B1913+16, discovered in 1974 by Russell Hulse and Joseph Taylor, for which they won the 1993 Nobel prize, consists of two neutron stars closely orbiting their common center of mass. One of them is a pulsar with a rotational period of 59 ms and extremely stable compared to other pulsars. The two neutron stars slowly spiral toward their common center of mass radiating energy. The orbit period however is declining by about 7.5×10^{-5} seconds per year on an orbit period of 7.75 hours. This change, believed to result from the system emitting energy in the form of gravitational waves, has been a very precise and successful test of general relativity.

On the scale of the Galaxy it is expected that there are still some small effects of the Hubble law. These effects modify the dynamics of the motion of tracer gases in the outlying regions of the Galaxy but in regard to gravity waves in galaxies they propagate unhindered, as in normal GR. Therefore it is expected that the energy dissipated by the Hulse-Taylor binary does travel as gravity waves within the Galaxy, but will be attenuated outside the region of the Galaxy where the mass density Ω drops below unity.

9.4   Conclusion

We have derived the wave propagation equation for gravitational radiation in an expanding Universe where an additional constraint has been placed on the nature of space itself. This is the introduction into the metric, the fundamental assumption of the expansion of space according to the Hubble law. It is then found that no unattenuated gravity wave propagation may be possible in regimes where the Hubble expansion has effect. Then, depending on the density of matter, the propagation constant of any gravitational wave is either real or imaginary. If imaginary it represents an evanescent wave, which we conjecture means the energy is dissipated into the surrounding space as heat. When gravity waves are eventually detected, a test of this theory would be the detection of gravity waves from within the Galaxy but not from extra-galactic sources.

In the next chapter the spiral galaxy rotation curves are examined in the framework of the five-dimensional brane world theory.

9.5 Problems

P 9.5.1. Prove the Landau-Lifshitz famous formula for the gravitational radiation emitted by a system of masses in the linear approximation of general relativity. Use Eq. (9.1.1) but neglect the last term in the operator with the dependence on the velocity. Show that

$$\frac{dE}{dt} = -\frac{G}{45c^5} \left(\frac{d^3 D_{kl}}{dt^3} \right)^2, \tag{1}$$

where t is the time and G is Newton's gravitational constant. Here D_{kl} is the quadrupole moment of the mass system,

$$D_{kl} = \int \rho \left(3x^k x^l - \delta^{kl} \mathbf{x} \cdot \mathbf{x} \right) d^3 x, \tag{2}$$

where ρ is the mass density (see Landau and Lifshitz 1959).

Solution: The solution is left for the reader.

P 9.5.2. Generalize the result of Problem 9.5.1 so as to include the dependence on the redshift. You will have to use the full Eq. (9.1.1). Use the fact that the time and the velocity appear in that equation with the same sign, so you replace ct by $ct + \tau v$, or t can be replaced by $t + \tau v/c$. For nonrelativistic velocities one can replace v/c by z. Thus we have for the new time parameter

$$\tilde{t} = t + \tau z. \tag{1}$$

Hence one can replace the time derivative in the formula for the gravitational radiation by using the generalized time $\tilde{t}$,

$$\frac{dE}{d\tilde{t}} = -\frac{G}{45c^5} \left(\frac{d^3 D_{kl}}{d\tilde{t}^3} \right)^2. \tag{2}$$

But

$$\frac{d}{d\tilde{t}} = \frac{d}{dt} + \frac{1}{\tau}\frac{d}{dz}, \tag{3}$$

thus Eq. (2) yields

$$\frac{dE}{dt} + \frac{1}{\tau}\frac{dE}{dz} = -\frac{G}{45c^5}\left(\frac{d^3 D_{kl}}{d\tilde{t}^3}\right)^2. \tag{4}$$

Solution: The solution is left for the reader.

P 9.5.3. Generalize the formula (4) in Problem 9.5.2 to the case where the redshift parameter z is relativistic ($z \geq 1$). Prove the relativistic Doppler formula

$$\frac{v}{c} = \frac{(1+z)^2 - 1}{(1+z)^2 + 1}. \tag{1}$$

Thus the new time parameter $\tilde{t}$ will now be

$$\tilde{t} = t + \tau\frac{(1+z)^2 - 1}{(1+z)^2 + 1}. \tag{2}$$

Solution: The solution is left for the reader.

P 9.5.4. Discuss the 5D electromagnetic wave equation in an expanding Universe.

Solution: The solution is left for the reader.

9.6   Suggested References

S. Behar and M. Carmeli, *Int. J. Theor. Phys.* **39**, 1375 (2000), astro-ph/0008352.

M. Carmeli, Is galaxy dark matter a property of spacetime? *Int. J. Theor. Phys.* **37**, 2621 (1998).

M. Carmeli, *Cosmological Special Relativity: The Large-Scale Structure of Space, Time and Velocity*, Second Edition (World Scientific, River Edge, NJ. and Singapore, 2002).

M. Carmeli, Accelerating Universe: Theory versus experiment (2002), astro-ph/0205396.

M. Carmeli, The line elements in the Hubble expansion, in: *Gravitation and Cosmology*, Eds. A. Lobo *et al.* (Universitat de Barcelona, 2003); astro-ph/0211043.

M. Fukugita, C.J. Hogan and P.J.E. Peebles, *Astrophys. J.* **503**, 518 (1998).

J.G. Hartnett, Carmeli's accelerating Universe is spatially flat without dark matter, *Int. J. Theor. Phys.* **44**, 485 (2005); gr-qc/0407083.

J.G. Hartnett, The Carmeli metric correctly describes spiral galaxy rotation curves, *Int. J. Theor. Phys.* **44**, 359 (2005); gr-qc/0407082.

J.G. Hartnett, The distance modulus determined from Carmeli's cosmology fits the accelerating Universe data of the high-redshift type Ia supernovae without dark matter, *Found. Phys.* **36**(6), 839-861 (2006); astro-ph/0501526.

J.G. Hartnett and M.E. Tobar, Properties of gravitational waves in cosmological general relativity, *Int. J. Theor. Phys.* **45**, 2213 (2006); gr-qc/0603067.

R.A. Hulse and J.H. Taylor, Discovery of a pulsar in a binary system, *Astrophys. J.* **195**, L51 (1975).

Kuroda *et al.*, *Int. J. Mod. Phys. D* **8**, 557 (1999).

LIGO, http://www.ligo.caltech.edu/

H.C. Ohanian, *Gravitation and Spacetime* (New York London, W.W. Norton, 1976).

F.J. Oliveira, Quantised intrinsic redshift in cosmological general relativity, gr-gc/0508094 (2005).

F.J. Oliveira, Exact solution of a linear wave equation in cosmological general relativity, *Int. J. Mod. Phys. D* **15**(11), 1963-1967 (2006); gr-qc/0509115.

TAMA, http://tamago.mtk.nao.ac.jp/

J.H. Taylor, L.A. Fowler and J.M. Weisberg, Measurements of general relativistic effects in the binary pulsar PSR1913+16, *Nature* **277**, 437 (1979).

Chapter 10

Spiral Galaxy Rotation Curves in the Brane World Theory in Five Dimensions

John Hartnett

Equations of motion, in cylindrical coordinates, for the observed rotation of gases within the gravitational potential of spiral galaxies are derived from Carmeli's 5-Dimensional Cosmological General Relativity theory. A Tully-Fisher type relation results, and rotation curves are reproduced without the need for non-baryonic halo dark matter. Two acceleration regimes are found that are separated by a critical acceleration $\approx 4.75 \times 10^{-10}$ m/s^2. For accelerations larger than the critical value the Newtonian force law applies, but for accelerations less than the critical value the Carmelian regime applies. In the Newtonian regime the accelerations fall off as r^{-2}, but in the Carmelian regime the accelerations fall off as r^{-1}. This is new physics, but it fits Milgrom's phenomenological MOND theory. The following is based on the work of Hartnett (Hartnett 2006).

10.1 Introduction

The rotation curves, highlighted by the circular motion of stars, or more accurately characterized by the spectroscopic detection of the motion of neutral hydrogen and other gases in the disk regions of spiral galaxies, have caused concern for astronomers for many decades. Newton's law of gravitation predicts much lower orbital speeds than those measured in the disk regions of spiral galaxies.

The most luminous galaxies show slightly declining rotation curves (orbital speed vs. radial position from nucleus) in the regions outside the star bearing disk, coming down from a broad maximum in the disk. Intermediate mass galaxies have mostly nearly flat rotation speeds along the disk

297

radius. Lower luminosity galaxies usually have monotonically increasing orbital velocities across the disk.

The traditional solution has been to invoke halo 'dark matter' (Begeman *et al.*, 1991) that surrounds the galaxy, but is transparent to all forms of electromagnetic radiation. In fact, astronomers have traditionally resorted to 'dark matter' whenever known laws of physics were unable to explain the observed dynamics. For example the advance of the perihelion of the planet Mercury was at first considered by astronomers as due to dark matter, but Einstein's theory of general relativity explained the phenomenon without the need of dark matter.

In 1983 Milgrom introduced his MOND (Milgrom 1983a,b,c), an empirical approach, which attempts to modify Newtonian dynamics in the region of very low acceleration. Newton's law describes a force proportional to r^{-2}, where r is the radial position from the center of the matter distribution, but Milgrom finds that a r^{-1} law fits the data very well (Begeman *et al.*, 1991).

Carmeli (2000, 2002) approached the problem from a different perspective. Using the theory described in this book Carmeli provided a theoretical derivation of the Tully-Fisher law (Carmeli, 1998), described in Section 7.4.

Following Carmeli's lead, Hartnett (2005) showed that the same line of reasoning leads to plausible galaxy rotation curves. The latter used a density model for spiral galaxies, that assumed that most of the mass of the galaxy was in the nuclear bulge and that the density in the disk region was constant. However by perturbing the density, one could also get the variation in rotation speeds as typically observed. That paper also used spherical coordinates and a hyperbolic density distribution of disk matter, conditions which are not appropriate for exponential-density-model galaxies.

In this chapter we take the analysis further, and in a more rigorous way, model the gravitational potential and the resulting forces determining how test particles move in the disks of spiral galaxies using cylindrical coordinates and an exponential density distribution. Two acceleration regimes are found to exist. In one, normal Newtonian gravitation applies. In that regime the effect of the Hubble expansion is not observed or is extremely weak. It is as if the particles' accelerations are so great that they slip across the expanding space. In the other, new physics is needed. There, the Carmelian theory provides it. In this regime the accelerations of particles are so weak that their motions are dominated by the Hubble expansion and as a result particles move under the combined effect of both the Newtonian force and a post-Newtonian contribution.

10.2 Gravitational Potential

In the weak gravitational limit, where Newtonian gravitation applies, it is sufficient to assume the Carmelian metric with non-zero elements $g_{00} = 1 + 2\phi/c^2$, $g_{44} = 1 + 2\psi/\tau^2$, $g_{kk} = -1$, $(k = 1, 2, 3)$ in the lowest approximations in both $1/c$ and $1/\tau$. The potential functions ϕ and ψ are determined by Einstein's field equations and from their respective Poisson equations,

$$\nabla^2\phi = 4\pi G\rho, \tag{10.2.1}$$

$$\nabla^2\psi = \frac{4\pi G\rho}{a_0^2}, \tag{10.2.2}$$

where ρ is the mass density and a_0 a universal characteristic acceleration $a_0 = c/\tau$. As usual c is the speed of light in vacuo.

Normally in Carmelian theory $\rho_{eff} = \rho - \rho_c$ is used instead of matter density ρ, but because in a galaxy $\rho \gg \rho_c$, the critical mass density ρ_c can be neglected in this chapter.

A comparison of ϕ and ψ in Eqs. (10.2.1) and (10.2.2) leads to $\psi = \phi/a_0^2$ within an arbitrary additive constant. Since both potentials are defined with respect to the same coordinate system, in reality, we only need deal with one potential function, the gravitational potential ϕ.

In cylindrical coordinates (r, θ, z) the potential ϕ that satisfies Eq. (10.2.1) can be found from Toomre (1963),

$$\phi(r) = -2\pi G \int_0^\infty J_0(kr)dk \int_0^\infty \rho(r')J_0(kr')r'dr', \tag{10.2.3}$$

where $J_0(kr)$ is the zeroth order Bessel function and k is the z coordinate scale factor $(k = 1/b)$. It is also assumed that the density function can be modeled as a delta function of the vertical coordinate z. Therefore the density $\rho(r, z) = \rho(r)\rho(z) = \rho(r)\delta(z)$ with no θ dependence. To correctly model the effect of the spiral arms a θ dependence may be needed, but for our model it is assumed independent. The requirement on the z dependence is satisfied with density functions of the form

$$\rho(z) = \frac{1}{2b}\text{sech}\left(\frac{z}{b}\right)^2 \quad \text{or} \quad e^{-|z|/b}. \tag{10.2.4}$$

Here, provided the scale length b is much smaller than the limit of the actual matter distribution in the z direction then the integral over all z yields a contribution to the mass of unity. This is the thin disk approximation which seems to be fairly applicable over both disk and galactic bulge.

The integral over dk in (10.2.3) is the surface density which may be calculated once the form of the density ρ is known. Following from observation we choose an exponential function of the form

$$\rho(r) = \frac{M}{2\pi a^2} e^{-r/a} \tag{10.2.5}$$

for the radial dependence, where a is a radial scale length and M is the mass of the galaxy.

10.3   Equations of Motion

The Hubble law describes the expansion of the cosmos and the matter embedded in it. Therefore the line element for any two points in this new *space-time-velocity* is $ds^2 = g_{00}c^2 dt^2 + g_{kk}(dx^k)^2 + g_{44}\tau^2 dv^2$. Here $k = 1, 2, 3$. The relative separation in 3 spatial coordinates $r^2 = (x^1)^2 + (x^2)^2 + (x^3)^2$ and the relative velocity between points connected by ds is v. The Hubble-Carmeli constant, τ, is a constant for all observers.

The equations of motion to lowest approximation in $1/c$ are reproduced here,

$$\frac{d^2 x^k}{dt^2} = -\frac{1}{2}\frac{\partial \phi}{\partial x^k}. \tag{10.3.1}$$

This is the usual equation derived from general relativity but now in 5 dimensions derived in Eq. (8.4.9a). And the second is a new equation derived in Eq. (8.4.10a)

$$\frac{d^2 x^k}{dv^2} = -\frac{1}{2}\frac{\partial \psi}{\partial x^k}. \tag{10.3.2}$$

10.3.1   *Newtonian*

It follows from Eqs. (10.3.1), (10.2.5) and (10.2.3), and the usual form of the circular motion equation

$$\frac{v^2}{r} = \frac{d\phi}{dr}, \tag{10.3.3}$$

that

$$v^2 = \frac{GMr^2}{2a^3}\Pi, \tag{10.3.4}$$

where G denotes the gravitational constant and

$$\Pi = I_0\left(\frac{r}{2a}\right) K_0\left(\frac{r}{2a}\right) - I_1\left(\frac{r}{2a}\right) K_1\left(\frac{r}{2a}\right),\qquad(10.3.5)$$

where I and K are standard zeroth and first order Bessel functions.

Equation (10.3.4) is the usual Newtonian result for the speed of circular motion in a cylindrical gravitational potential. This equation has been plotted in curve 3 of Figures 10.3.2(a)–10.5.5(a) as a function of radial position from the center of a galaxy in kiloparsecs (kpc) where $kpc \approx 3\times10^{19}$ m. Best fits were determined with M and a as free parameters. Throughout this chapter M is expressed in solar mass units $M_\odot \approx 2 \times 10^{30}$ kg.

10.3.2 *Carmelian*

Using $\psi = \phi/a_0^2$ in Eq. (10.3.2) results in a new equation

$$v = a_0 \int_0^r \frac{dr}{\sqrt{-\phi}},\qquad(10.3.6)$$

where we have integrated and solved for v as a function of r. Using the potential ϕ, determined from (10.2.5) and (10.2.3), in (10.3.6), results in

$$v = \frac{2}{3}a_0 \frac{r^{3/2}}{\sqrt{GM}},\qquad(10.3.7)$$

which describes the expansion of space within a galaxy.

It must be realized that the only direction in the cylindrical coordinates of a galaxy that is free to expand in the Hubble flow is the azimuthal (Hartnett, 2005). Therefore (10.3.7) describes the velocity component in that direction. Carmeli applied this line of reasoning (Carmeli, 1998, 2000).

To establish the combined result of the two equations of motion (10.3.4) and (10.3.7), the simultaneous speed of test particles must be determined by the elimination of r between the two equations. The physical meaning can be understood in terms of particles that simultaneously satisfy both (10.3.1) and (10.3.2).

The usual Newtonian expression (10.3.1) describes motion under the central potential but assumes that spatial coordinates are fixed. Whereas the new equation (10.7) describes the expansion of space itself within a galaxy. Therefore we must find the combined (simultaneous) effect of Eqs. (10.3.4) and (10.3.7). The result is a post-Newtonian equation,

$$v^{2/3} = \frac{(GM)^{5/3}}{(\frac{2}{3}a_0)^{4/3}2a^3}\Pi,\qquad(10.3.8)$$

derived from Eq. (10.3.4) where the following substitution

$$r \to \left(\frac{GMv^2}{(\frac{2}{3}a_0)^2} \right)^{1/3} ,$$

has been made from Eq. (10.3.7). The resulting equation, hereafter

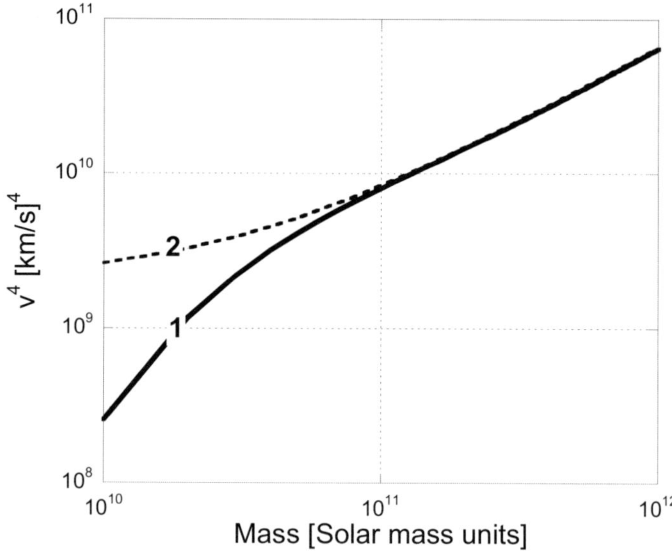

Fig. 10.3.1: Tully-Fisher law plotted on logarithmic axes. Curve 1 (solid line) represents the fourth order dependence of the rotational speeds of tracer gases in galaxies determined from the Carmelian equation (10.3.8). The masses are expressed in solar mass units of $M_\odot = 2 \times 10^{30} kg$. Curve 2 (broken line) represents the straight line $v^4 = 2 \times 10^9 + 0.064M$ (Mass). (Source: Hartnett, 2006)

referred to as Carmelian, cannot be solved analytically. However using the Mathematica software package Eq. (10.3.8) can be solved numerically.

The result is plotted in curve 1 of Figure 10.3.1 where it has been assumed that $a = 1$ *kpc* and it is compared with the straight line $v^4 = 2 \times 10^9 + 0.064M$ (curve 2). For large M the small offset can be neglected. This result indicates that the fourth order dependence on rotational speed (v) is directly proportional to mass (M) for large masses.

Assuming that the masses of the galaxies studied are directly proportional to their luminosity, this dependence then becomes the Tully-Fisher relationship. This extends the work of Carmeli (2000), and derives the underlying theoretical framework upon which the Tully-Fisher law is founded.

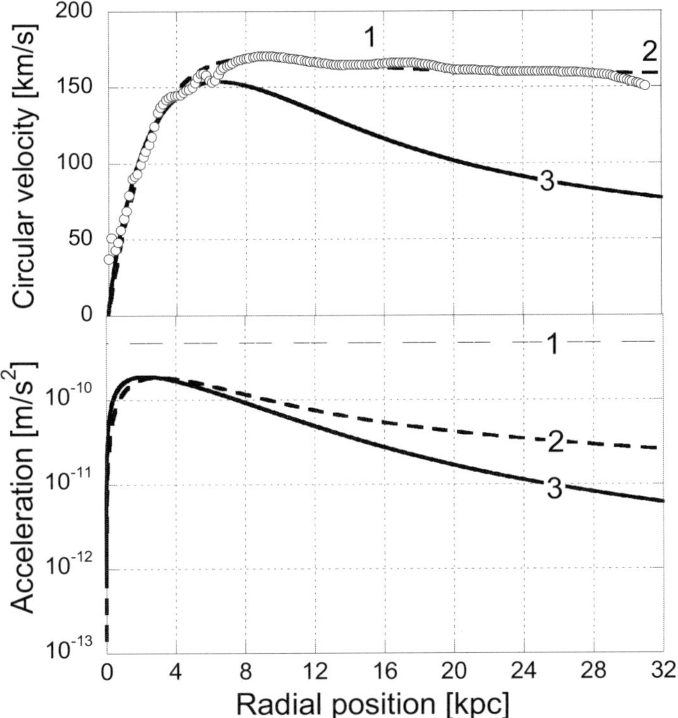

Fig. 10.3.2:   (a) Above: The rotational speeds of tracer gases in NGC 3198 (SBc barred spiral)(circles - curve 1). Theoretical curve fits from the Carmelian equation (10.3.11) (broken curve 2) and from the Newtonian equation (10.3.4) (curve 3). (b) Below: The critical acceleration $\frac{2}{3}a_0$ (curve 1). The rotational accelerations determined from the Carmelian (curve 2) and the Newtonian (curve 3) equations with their respective values of a and M. (Source: Hartnett, 2006)

10.3.3   *Rotation curves*

In Refs. Carmeli (1998, 2000) and Hartnett (2005), using spherical co-ordinates, it was found that in the limit of large r and where all the matter was interior to the position of a test particle, such a particle is also subject to an additional circular motion described by Eq. (10.3.7). Apparently this is the result of the expansion of space itself within the galaxy but in an azimuthal direction to the usual center of coordinates of the galaxy. In this section also the same result Eq. (10.3.7) was obtained but in this case derived using cylindrical coordinates.

In Section 7.4 Carmeli determined a Tully-Fisher type relation using the

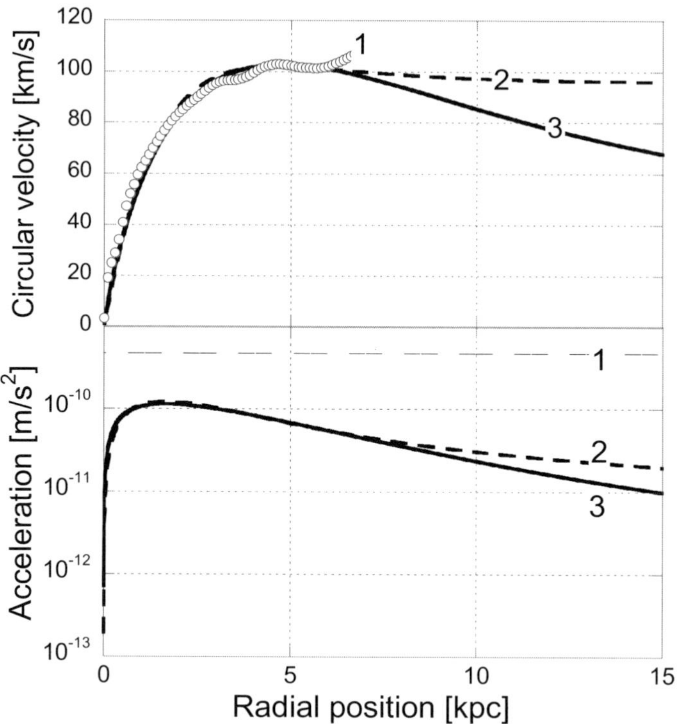

Fig. 10.3.3:   (a) Above: The rotational speeds of tracer gases in NGC 0598 (Sc spiral) (circles - curve 1). Theoretical curve fits from the Carmelian equation (10.3.11) (curve 2) and from the Newtonian equation (10.3.4) (curve 3). (b) Below: The critical acceleration $\frac{2}{3}a_0$ (curve 1). The corresponding rotational accelerations determined from the Carmelian (curve 2) and the Newtonian (curve 3) equations. (Source: Hartnett, 2006)

Newtonian circular velocity equation expressed in spherical coordinates,

$$v^2 = \frac{GM}{r}, \qquad (10.3.9)$$

where it is assumed that test particles orbit at radius r outside of a fixed mass M. Then by eliminating r between Eqs. (10.3.9) and (10.3.7) we get the result. This is achieved by taking the 3/2 power of (10.3.9) and multiplying it with (10.3.7) yielding

$$v^4 = GM\frac{2}{3}a_0. \qquad (10.3.10)$$

So by applying the same approach with Eq. (10.3.4) (raising it to the 3/2 power) and multiplying it with Eq. (10.3.7) we can derive an equation describing the rotation curves in galaxies. The result is

$$v^4 = GM\frac{2}{3}a_0 \left\{ \left(\frac{r}{2a}\right)^{9/2} 8\ \Pi^{3/2} \right\}, \qquad (10.3.11)$$

remembering Π is a function of $r/2a$. It is easily confirmed that as $r \to \infty$,

$$\left(\frac{r}{2a}\right)^{9/2} 8\ \Pi^{3/2} \to 1,$$

which is the radial position (or r) dependent part of Eq. (10.3.11). Hence (10.3.11) then recovers the form of the Tully-Fisher relation (10.3.10).

By taking the 4th root of (10.3.11) we get an expression for the circular velocity of test particles as a function of their radial position r. That result has been plotted in curve 2 of Figures 10.3.2(a)–10.5.5(a) for each galaxy with a and M determined as fit parameters. The resulting curves have the characteristic flat shape for large radial position r. At small values of r the rotation speeds determined from the Newtonian equation (curve 3) dominate as seen in Figures 10.4.1(a)–10.5.5(a).

10.4 Accelerations

The acceleration $\frac{2}{3}a_0$ in Eq. (10.3.7) can be considered to be a critical acceleration. Therefore when we compare the accelerations derived from the Newtonian equation (10.3.4) and the Carmelian equation (10.3.11) with this critical acceleration we notice two regimes develop. See Figures 10.3.2(b)–10.5.5(b).

For example, Figure 10.4.1(b) is very instructive. There the straight line (curve 1) is the critical acceleration $\frac{2}{3}a_0$, curve 2 represents the acceleration derived from the Carmelian equation (10.3.11) and curve 3 represents the acceleration derived from the Newtonian equation (10.3.4). For the values, determined from the fits, of the mass (M) and the radial scale length (a), which determine how the matter density varies as a function of radial position r, the curves 2 and 3 cross each other very close to the critical acceleration. The significance is that for accelerations less than the critical acceleration the Carmelian force applies and for accelerations greater than the critical acceleration the Newtonian force applies. Note also that the Newtonian curve 3 has a r^{-2} dependence and the Carmelian curve 2 has a r^{-1} dependence above 10 kpc. When curves of the form r^{-x} were fitted to the functions used in Figure 10.4.1(b) between 15 and 20 kpc, the coefficients x were determined to be $x = 2.003$ and $x = 1.025$ respectively.

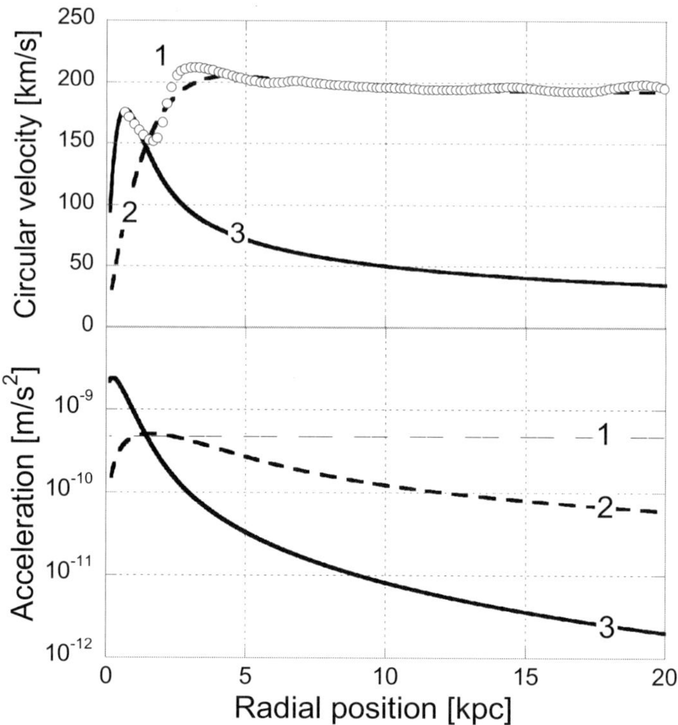

Fig. 10.4.1:   (a) Above: The rotational speeds of tracer gases in NGC 2903 (Sc spiral) (circles - curve 1). Theoretical curve fits from the Carmelian equation (10.3.11) (curve 2) and from the Newtonian equation (10.3.4) (curve 3). (b) Below: The critical acceleration $\frac{2}{3}a_0$ (curve 1). The corresponding rotational accelerations determined from the Carmelian (curve 2) and the Newtonian (curve 3) equations. (Source: Hartnett, 2006)

From Eq. (10.3.4) the gravitational acceleration (v^2/r) can be calculated in the limit of $r \to \infty$, outside most of the matter of the galaxy. As expected for the Newtonian model it tends to GM/r^2. And similarly from Eq. (10.3.11) the gravitational acceleration (v^2/r) can also be calculated in the limit of large r, for the Carmelian model. In this case it is evident from Eq. (10.3.10) that it must tend to $\sqrt{GM\frac{2}{3}a_0}/r$. In this regime the accelerations are very weak. This is very significant as alternative theories of gravity have been suggested (for example, Milgrom, 1983a,b,c) where the force of gravity falls away as r^{-1} for small accelerations. However for small r, that is, where $r \to 0$, close to the origin of the central gravitational potential, the effect of the Carmelian force law becomes extremely small and is many orders of magnitude smaller than that for the Newtonian force

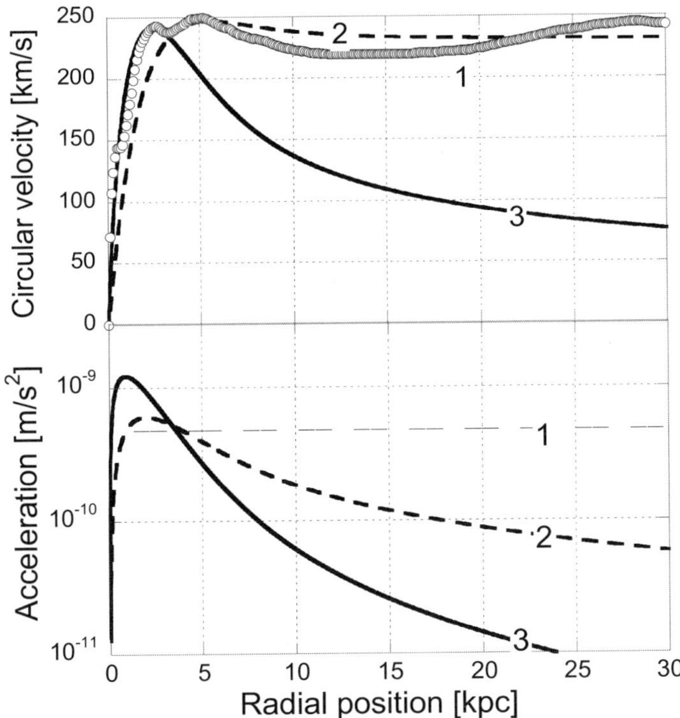

Fig. 10.4.2: (a) Above: The rotational speeds of tracer gases in NGC 7331 (Sbc spiral) (circles - curve 1). Theoretical curve fits from the Carmelian equation (10.3.11) (curve 2) and from the Newtonian equation (10.3.4) (curve 3). (b) Below: The critical acceleration $\frac{2}{3}a_0$ (curve 1). The corresponding rotational accelerations determined from the Carmelian (curve 2) and the Newtonian (curve 3) equations. (Source: Hartnett, 2006)

law.

10.5 Sample of Galaxy Rotation Curves

A sample of 9 galaxy fits are shown in Figures 10.3.2–10.5.5. The top (a) figures show the rotation curve fits and the bottom (b) figures show the resulting acceleration regimes. In each figure, the measured rotational speeds of tracer gases in the chosen spiral galaxy is shown as a function of radial position (curve 1). Measured data are taken from Sofue *et al.* (1999). Theoretical curves from the Carmelian equation (10.3.11) (curve 2) and from the Newtonian equation (10.3.4) (curve 3) are fitted over the range of r which best fit the data by allowing a and M to be free parameters.

The accelerations in the bottom (b) figures are derived from the Carmelian (curve 2) and the Newtonian (curve 3) equations respectively with values of a and M derived from the fits in the (a) figures. These are compared with (curve 1) the critical acceleration $\frac{2}{3}a_0 \approx 4.75 \times 10^{-10}$ m/s^2 determined elsewhere from $\tau \approx 4.28 \times 10^{17}$ s, which is the reciprocal of the Hubble parameter at zero distance $h = 72.17$ km/s-Mpc.

10.5.1 *Extragalactic spirals*

In the following, we discuss individual galaxy curve fits, starting with Figure 10.3.2 showing the barred spiral NGC 3198. In each case, because of the possibility of different acceleration regimes, both Carmelian and Newtonian curve fit were attempted. In Figure 10.3.2(a) the Carmelian fit is shown by the broken curve 2 to be the only good fit. The scale radius $a = 1.85$ kpc and $M = 0.984 \times 10^{10} M_\odot$ determined from the fit. Curve 3 shows the best Newtonian fit with $a = 2.99$ kpc and $M = 4.2 \times 10^{10} M_\odot$ but it doesn't fit well at high values of r. The Newtonian fit results in a mass at least 4 times greater than that from the Carmelian fit. The scale radius determined from luminous matter for this galaxy is $a = 2.5$ kpc which is closer to the Carmelian curve determination.

In the Figure 10.3.2(b) curve 2 shows the acceleration using the Carmelian determined values of the scale radius a and mass M. Curve 3 shows the acceleration for Newtonian fit determined values. Clearly curve 2 is dominant and is always less than the critical acceleration $\frac{2}{3}a_0$. In this model, when the accelerations are less than the critical value, the Carmelian force applies.

Next the data for NGC 0598, a Sc spiral galaxy, shown in Figure 10.3.3(a), fits both a Carmelian and a Newtonian curve. From the Carmelian model the scale radius $a = 1.85$ kpc and $M = 0.13 \times 10^{10} M_\odot$ determined from the fit. Clearly the Carmelian curve 2 is the better fit over the Newtonian curve 3 for the following reasons. Firstly, the data (circles - curve 1) continue to rise or at least are not falling at the extremity of the available measured range. The Newtonian curve indicates it should fall. Secondly, from Figure 10.3.3(b) the accelerations are much less than the critical value $\frac{2}{3}a_0$ and hence in this regime the Carmelian force law applies.

From the Newtonian model a scale radius of $a = 2.22$ kpc and $M = 1.42 \times 10^{10} M_\odot$ were determined but the fit doesn't conform to the model. Nevertheless the Newtonian fit results in a mass 10 times greater than the fit for the Carmelian model.

Table 10.1:   Important galaxy data where valid curve fits are found.

Figure	Galaxy	Type	Scale radius a published	Newton a from fit	Carmeli a from fit	Newton M from fit	Carmeli M from fit
10.3.2	NGC 3198	SBc	2.5	-	1.85	-	0.984
10.3.3	NGC 0598	Sc	2.7	-	1.01	-	0.13
10.4.1	NGC 2903	Sc	1.9	0.31	0.98	0.54	2.12
10.4.2	NGC 7331	Sbc	4.7	1.15	1.2	4.0	4.45
10.5.1	NGC 2841	Sb	2.3	2.26	2.30	8.12	9.00
10.5.2	IC 0342	Sc		0.74	1.05	1.28	1.81
10.5.3	NGC 1097	SBb		0.62	2.12	4.80	9.74
10.5.4	NGC 2590	Sb	2.1	0.5	2.73	2.90	6.96
10.5.5	MW Galaxy	Sb		0.12	1.09	0.45	2.31

Figure	Galaxy	Newton M_{10}	Carmeli M_{10}	Ratio	R	M_R
10.3.2	NGC 3198	6.554	0.956	6.85	-	-
10.3.3	NGC 0598	-	-	-	-	-
10.4.1	NGC 2903	8.810	2.120	4.16	1.39	0.54
10.4.2	NGC 7331	11.496	4.440	2.59	3.38	3.16
10.5.1	NGC 2841	17.721	8.377	2.11	2.11	5.09
10.5.2	IC 0342	8.200	1.809	4.53	2.08	1.00
10.5.3	NGC 1097	22.997	9.242	2.49	3.94	4.74
10.5.4	NGC 2590	17.213	6.127	2.81	3.73	2.89
10.5.5	MW Galaxy	9.302	2.308	4.03	1.22	0.45

Mass M_R calculated at $r = R$. Mass M_{10} calculated at $r = 10$ *kpc*. a & R in *kpc*, M, M_R & M_{10} in $10^{10} M_\odot$ units.

Figures 10.4.1 and 10.4.2 show both Newtonian and Carmelian models fit the rotation curve data for the galaxies NGC 2903 and NGC 7331. In the high acceleration regime a Newtonian fit is applicable and when the acceleration drops below $\frac{2}{3}a_0$ the Carmelian applies. In Table 10.1 the best fit determined values of a and M for each galaxy are listed. They are compared with published values of a and masses determined from different methods.

In these and all subsequent top (a) figures, the fits for curves 2 and 3, respectively, apply only for accelerations less than and greater than the critical acceleration. In the rotation curve fits, in the top (a) figures, the circular velocities are not added but each apply over their respective regimes. This means the masses determined from the Newtonian fits must be less than those from the Carmelian fits because the Newtonian determined mass must lay within the radius R where the two curves cross. The Newtonian fits occur in the stronger acceleration regimes ($> \frac{2}{3}a_0$) close to the galactic center as indicated by all of the bottom (b) figures.

Table 10.1 lists the radial position (R) where the Newtonian and Carmelian regimes meet, the mass (M_R) for all the mass where $r < R$ and the total mass M shown in Col. 8 (Carmeli M from fit), which is the Carmelian model determined mass. As expected the values of $M_R \leq M$ determined from the Newtonian regime.

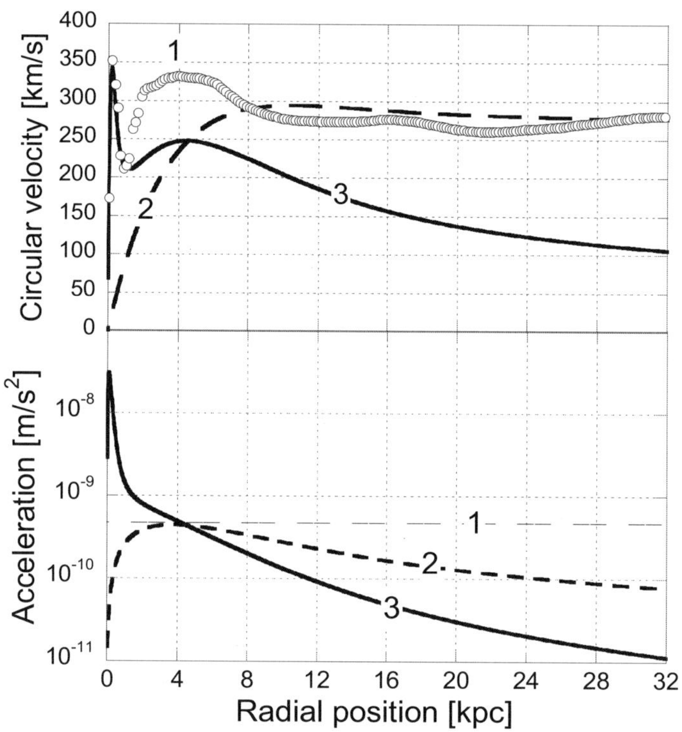

Fig. 10.5.1: (a) Above: The rotational speeds in NGC 2841 (Sb spiral) (circles - curve 1). Theoretical curve fits from the Carmelian equation (10.3.11) (curve 2) and from the Newtonian equation (10.3.4) (curve 3). (b) Below: The critical acceleration $\frac{2}{3}a_0$ (curve 1). The corresponding rotational accelerations determined from the Carmelian (curve 2) and the Newtonian (curve 3) equations. (Source: Hartnett, 2006)

Figures 10.5.1 and 10.5.2, respectively, show NGC 2841 and IC 0342, which have been modeled with two central components with different scale radii. Both seem to have a dense mass concentration toward their centers. The scale radii for these inner most concentrations are $a = 0.09$ kpc and $a = 0.05$ kpc for NGC 2841 and IC 0342 respectively. The combined rotation curve for IC 0342 is a much better fit to the data using the two-

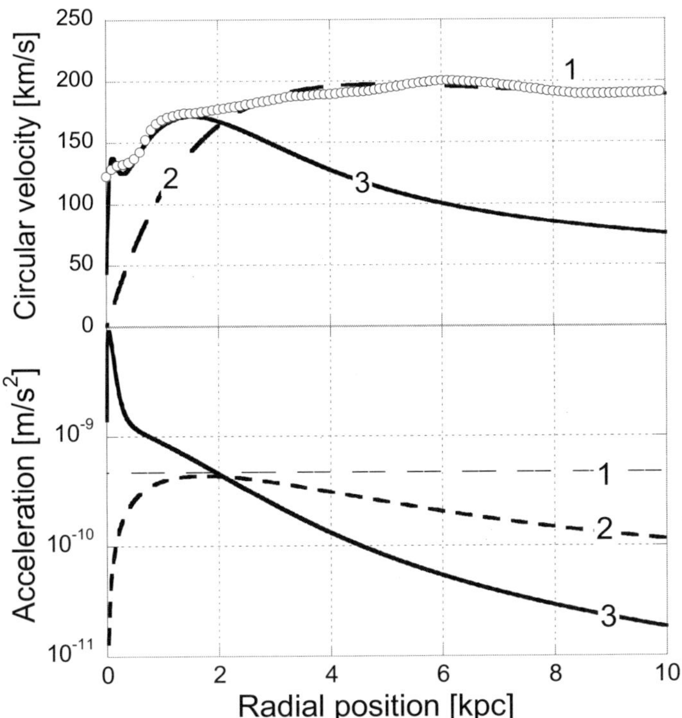

Fig. 10.5.2:   (a) Above: The rotational speeds in IC 0342 (Sc spiral)(circles - curve 1). Theoretical curve fits from the Carmelian equation (10.3.11) (curve 2) and from the Newtonian equation (10.3.4) (curve 3). (b) Below: The critical acceleration $\frac{2}{3}a_0$ (curve 1). The corresponding rotational accelerations determined from the Carmelian (curve 2) and the Newtonian (curve 3) equations. (Source: Hartnett, 2006)

acceleration-regime model than that of the former, NGC 2841. Nevertheless the theory works well in these type of galaxies also.

Figures 10.5.3 and 10.5.4, respectively, show NGC 1097 and NGC 2590 modeled with only one central dense Newtonian component and they fit the model very well. Deviations in most cases are believed can be attributed to the fact that no azimuthal dependence has been added to the model, nor does it allow for anything but constant scale radii over regions of the galaxy where the fits apply. That is clearly unphysical but seems to be a reasonable approximation.

Table 10.1 also shows, in Col. 9, the mass (M_{10}) determined at $r = 10$

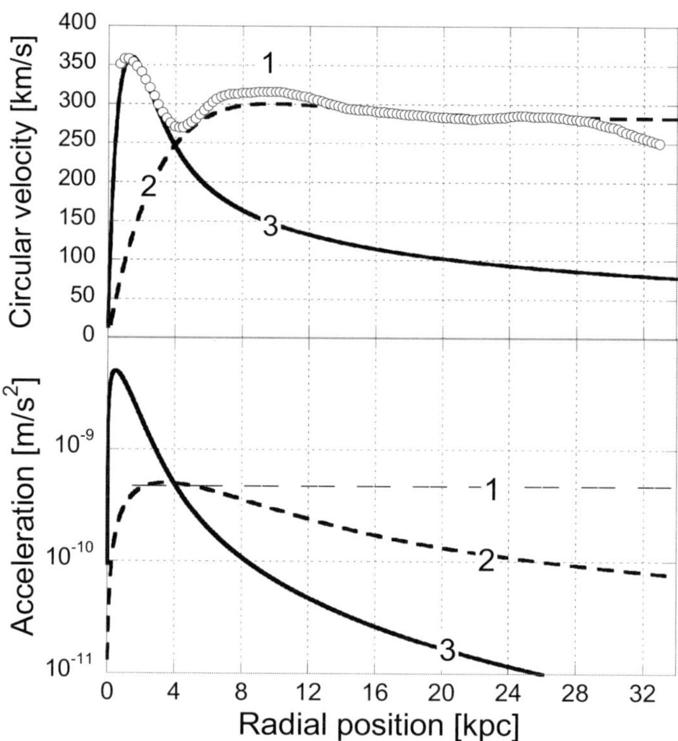

Fig. 10.5.3: (a) Above: The rotational speeds of tracer gases in NGC 1097 (SBb barred spiral) (circles - curve 1). Theoretical curve fits from the Carmelian equation (10.3.11) (curve 2) and from the Newtonian equation (10.3.4) (curve 3). (b) Below: The critical acceleration $\frac{2}{3}a_0$ (curve 1). The corresponding rotational accelerations determined from the Carmelian (curve 2) and the Newtonian (curve 3) equations. (Source: Hartnett, 2006)

kpc using the Newtonian formula

$$M = \frac{v^2 r}{G} \tag{10.5.1}$$

assuming that most of the mass is internal to $r = 10$ kpc and the measured rotation speeds. This calculation is compared with the mass (M_{10}), in Col. 10, derived from the Carmelian equation with $r = 10$ kpc. The ratios of these masses are shown in Col. 11. It indicates that using standard Newtonian/Keplerian dynamics seriously over estimates galaxy masses by between 2 and 7 times. These values are consistent with the needed mass levels of halo dark matter to achieve the correct rotation curves.

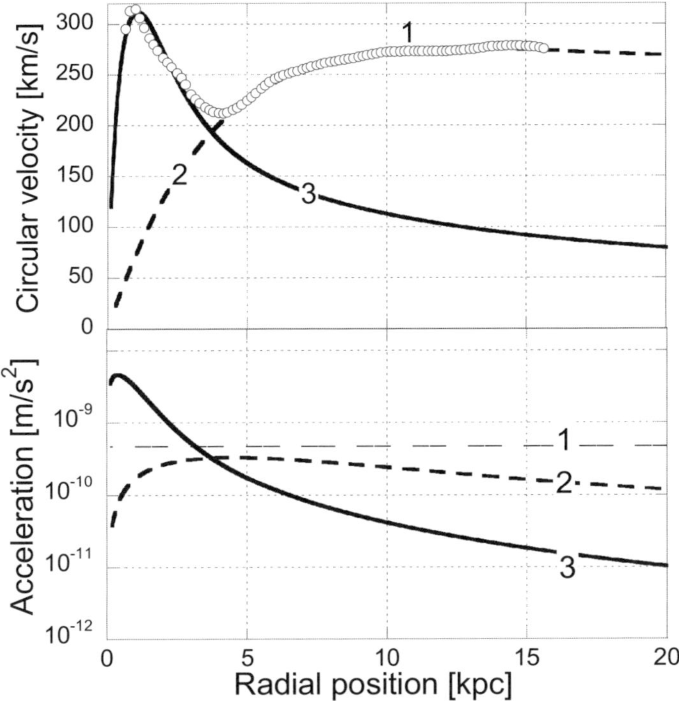

Fig. 10.5.4: (a) Above: The rotational speeds of tracer gases in NGC 2590 (Sb spiral) (circles - curve 1). Theoretical curve fits from the Carmelian equation (10.3.11) (curve 2) and from the Newtonian equation (10.3.4) (curve 3). (b) Below: The critical acceleration $\frac{2}{3}a_0$ (curve 1). The corresponding rotational accelerations determined from the Carmelian (curve 2) and the Newtonian (curve 3) equations. (Source: Hartnett, 2006)

10.5.2 *The Galaxy*

Considering the (Milky Way) Galaxy the same analysis has been applied to the data from Honma and Sofue (1997). See the rotation curve in Figure 10.5.5(a). Other observers don't record the central peak indicative of a large central mass concentration. However the compact radio source Sagittarius A* at the Galactic center is widely believed to be associated with the supermassive black hole with a mass of about $3.59 \pm 0.59 \times 10^6 M_\odot$ (Eisenhauer *et al.*, 2003). Then there is the vast concentration of matter in the Galactic bulge. From the Newtonian curve fit we'd expect that within 1.2 kpc of the center there is a mass of about $M = 4.5 \times 10^9 M_\odot$.

The Carmelian curve fit (curve 2 in Figure 10.5.5(a)) over the range 3 to 10 kpc is an excellent fit. The acceleration regimes, consistent with the

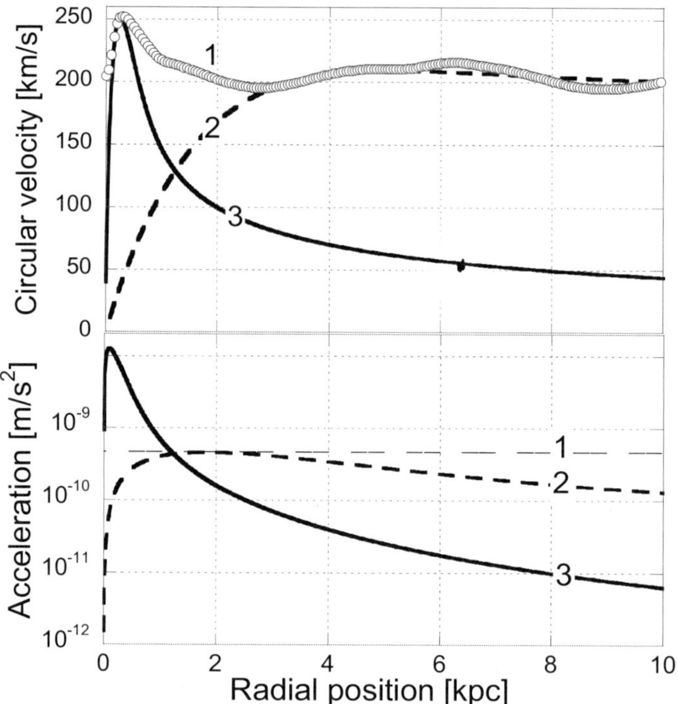

Fig. 10.5.5: (a) Above: The rotational speeds of tracer gases in the Galaxy (Milky Way Sb spiral). Theoretical curve from the Carmelian equation (10.3.11) (curve 2) and from the Newtonian equation (10.3.4) (curve 3) Measured data shown in curve 1. (b) Below: The critical acceleration $\frac{2}{3}a_0$ (curve 1). The corresponding rotational accelerations determined from the Carmelian (curve 2) and the Newtonian (curve 3) equations respectively. (Source: Hartnett, 2006)

curve fits (curves 2 and 3) in Figure 10.5.5(a), are shown in Figure 10.5.5(b). Except for the intervening region the fits agree well with the theory. The discrepancy could be due to an unmodeled higher concentration of mass in the central bulge region. Deviations in the spiral arms from a smooth exponential density decay are consistent with the oscillations about curve 2 in Figure 10.5.5(a).

The distance of our solar system from the Galactic center has been estimated at 9.95 kpc (Honma and Sofue, 1997) and more recently as $7.94\pm$ 0.42 kpc (Eisenhauer *et al.*, 2003). Likewise the enclosed mass may be calculated using both Eqs. (10.3.11) and (10.5.1). When one uses the normal Newtonian/Keplerian calculation (10.5.1) and $v = 200.78$ km/s the speed of the solar system orbiting the Galactic center, it results in an estimate of the enclosed mass of $M_{10} = 9.3 \times 10^{10} M_\odot$ as compared with $M_{10} = 2.3 \times 10^{10} M_\odot$ from the Carmelian calculation (10.3.11), which is 4 times smaller. For a distance 7.94 kpc the Keplerian calculation of the enclosed mass is $M_{10} = 7.6 \times 10^{10} M_\odot$, which is 3.3 times greater than the Carmelian calculation for the enclosed mass at that distance.

10.6   Conclusion

Cosmological General Relativity provides a solution to the rotation curve anomaly in the outer regions of spiral galaxies. Equations of motion have been derived from Carmeli's 5D brane world theory using a gravitational potential in cylindrical coordinates. A Tully-Fisher type relation results, and the rotation curves for spiral galaxies are accurately reproduced without the need for non-baryonic halo dark matter.

Two acceleration regimes are discovered that are separated by a critical acceleration $\frac{2}{3}a_0$. For accelerations larger than the critical value the Newtonian force law applies, but for accelerations less than the critical value the Carmelian regime applies. In the Newtonian regime the accelerations fall off as r^{-2} as expected, but in the Carmelian regime the accelerations fall off as r^{-1}. This is new physics but it fits what is suggested by Milgrom's phenomenological MOND theory. Thus a theoretical basis is found, whereas until now, no theory has been found for Milgrom's MOND. This theory also provides an understanding of the connection between the two regimes.

In the next chapter the Friedmann Universe and the FRW metric will be discussed in details.

10.7   Suggested References

K.G. Begeman, A.H. Broeils *et al.*, Extended rotation curves of spiral galaxies: dark haloes and modified dynamics, *Mon. Not. R. Astr. Soc.* **249**, 523-537 (1991).

M. Carmeli, *Classical Fields: General Relativity and Gauge Theory* (John Wiley, New York, 1982); reprinted by World Scientific Publishing Company (2001).

M. Carmeli, Cosmological general relativity, *Commun. Theor. Phys.* **5**, 159 (1996).

M. Carmeli, Is galaxy dark matter a property of spacetime? *Int. J. Theor. Phys.* **37**, 2621 (1998).

M. Carmeli, Derivation of the Tully-Fisher Law: Doubts About the Necessity and Existence of Halo Dark Matter, *Int. J. Theor. Phys.* **39**, 1397 (2000), astro-ph/9907244.

M. Carmeli, *Cosmological Special Relativity: The Large-Scale Structure of Space, Time and Velocity*, Second Edition (World Scientific, River Edge, NJ. and Singapore, 2002).

F. Eisenhauer, R. Schödel, R. Genzel, T. Ott, M. Tecza, R. Abuter, A. Eckart and T. Alexander, A geometric determination of the distance to the galactic center, *Astrophys. J.* **597**, L121-L124 (2003).

Peebles, *Astrophys. J.* **503**, 518 (1998).

J.G. Hartnett, Carmeli's accelerating Universe is spatially flat without dark matter, *Int. J. Theor. Phys.* **44**, 485 (2005); gr-qc/0407083.

J.G. Hartnett, The Carmeli metric correctly describes spiral galaxy rotation curves, *Int. J. Theor. Phys.* **44**, 359 (2005); gr-qc/0407082.

J.G. Hartnett, Spiral galaxy rotation curves determined from Carmelian general relativity, *Int. J. Theor. Phys.* **45**, 2118 (2006); astro-ph/0511756.

M. Honma and Y. Sofue, Rotation curve of the Galaxy, *Publ. Astron. Soc. Jpn.* **49**, 453 (1997).

M. Milgrom, A modification of the Newtonian dynamics - Implications for galaxies, *Astrophys. J.* **270**, 371 (1983a).

M. Milgrom, A Modification of the Newtonian Dynamics - Implications for Galaxy Systems, *Astrophys. J.* **270**, 384 (1983b).

M. Milgrom, A modification of the Newtonian dynamics as a possible alternative to the hidden mass hypothesis, *Astrophys. J.* **270**, 365 (1983c).

Y. Sofue, Y. Tutui, H. Honma, A. Tomita, T. Takamiya, J. Koda and Y. Takeda, Central rotation curves of spiral galaxies, *Astrophy. J.* **523**, 136-146 (1999).

A. Toomre, On the distribution of matter within highly flattened galaxies, *Astrophys. J.* **138**, 385-392 (1963).

B.C. Whitemore, Rotation curves of spiral galaxies in clusters, in: *Galactic Models*, J.R. Buchler, S.T. Gottesman, J.H. Hunter. Jr., Eds., (New York Academy Sciences, New York, 1990).

Chapter 11

The Friedmann Universe: FRW Metric

Moshe Carmeli

We are now in a position to present the "standard" theory of cosmology that is based on the original Einstein general relativity theory. That means in particular the introduction of the Friedmann Universe. For that one needs to introduce preliminary concepts, such as the deceleration parameter, *an important parameter in this approach. We then determine the geometry of the spatial three-dimensional spaces which represent our three-dimensional space, which is assumed to be homogeneous and isotropic. A classification of these three-dimensional spaces into those with positive, negative or zero curvature is then given. Solutions of the Einstein field equations are then presented and related to the three-dimensional subspaces. From that the theories of Friedmann and others follow.*

11.1 Introduction

In this chapter we present the Friedmann theory of classification of the three-dimensional subspaces of the Einstein four-dimensional Riemannian space of general relativity theory. This is one of the cases where mathematics preceded physics in predicting major results in physics.

In 1922, Alexander Alexandrovich Friedmann [1], professor in the Physics

---

[1] Alexander Alexandrovich Friedmann (Born 16 June 1888 in St Petersburg, Russia; died 16 September 1925 in Leningrad now St Petersburg again, Russia). Friedmann's father was a ballet dancer and his mother a pianist. However, the parents divorced when Alexander was nine years old. The church sided with the father who got custody of Alexander. Alexander stayed with his father who soon remarried. Alexander entered the Second St Petersburg Gymnasium in August 1897. Soon Alexander became one of the top two pupils in his class, the other outstanding pupil was Yakov Tamarkin, also

319

an extraordinary mathematician, and the two boys were close friends, almost always together during their years at school and university.

In 1905 Friedmann and Tamarkin wrote a paper on Bernoulli numbers and submitted it to Hilbert for publication. The paper was accepted and appeared in print in 1906. Friedmann graduated from school in 1906 and entered the University of St Petersburg. Friedmann was also influenced by Ehrenfest who moved to St Petersburg in 1906 and set up a modern physics seminar which was attended also by the two young mathematicians Friedmann and Tamarkin. While Friedmann was an undergraduate at St Petersburg his father died. Friedmann began studying for his Master's Degree in 1911 and became involved with a circle formed to study mathematical analysis and mechanics; other members of the circle included Tamarkin, Smirnov, Petelin, Shokhat and Besicovitch. Friedmann lectured on Clebsch's work on elasticity and other topics including Goursat's books. Friedmann also lectured at the Mining Institute and at the Railway Engineering Institute and became interested in aeronautics. By 1913 Friedmann had completed the necessary examinations for the Master's Degree and was subsequently appointed to a position in the Aerological Observatory in Pavlovsk, a suburb of St Petersburg.

In 1914 Friedmann went to Leipzig to study with Vilhelm Bjerknes, the leading theoretical meteorologist at the time. When Austria gave Serbia an ultimatum after the June 1914 assassination of Archduke Francis Ferdinand, Russia supported Serbia and Germany came to the support of Austria. World War I broke out on 1 August 1914, and Friedmann soon asked permission to join the volunteer aviation detachment and began flying aircraft and was soon involved in bombing raids, but continued to study mathematics with Steklov by letter. And in a letter to Steklov of 5 February 1915 Friedmann writes: "My life is fairly even, except such accidents as a shrapnel explosion twenty feet away, the explosion of an Austrian bomb within half a foot, which turned out almost happily, and falling down on my face and head, which resulted in a ruptured upper lip and headaches. But one gets used to all this, of course, particularly seeing things all around which are a thousand times more awful." In a second letter Friedmann wrote: "I have recently had a chance to verify my ideas during a flight over Przemysl; the bombs turned out to be falling almost the way the theory predicts. To have conclusive proof of the theory I'm going to fly again in a few days."

Friedmann was awarded the George Cross for bravery with his flights over Przemysl. After that Friedmann was sent to Kiev where he gave lectures on aeronautics for pilots and was appointed Head of the Central Aeronautical Station in Kiev. On 13 April 1918 Friedmann was elected an extraordinary professor in the Department of Mathematics and Physics at the University of Perm. At Perm Friedmann established an Institute of Mechanics and became a member of the editorial board of the newly founded Physico-Mathematical Society of Perm University. Perm became under evacuation including the University. Some time later the White Army occupied Perm. They controlled the town until August 1919 when the Red Army took control again of the city. In the spring of 1920 Friedmann returned to St Petersburg (now named Petrograd) to take a post at the Geophysical Observatory and began teaching mathematics and mechanics at Petrograd University, became a professor in the Physics and Mathematics Faculty of the Petrograd Polytechnic Institute, worked in the Department of Applied Aeronautics at Petrograd Institute of Railway Engineering, worked at the Naval Academy and undertook research at the Atomic Commission at the Optical Institute. In 1922, nine years after completing the examinations for his Master's Degree, Friedmann submitted his Master's dissertation, on the Hydrodynamics of a Compressible Fluid.

After returning to Petrograd Friedmann took interest in Einstein's theory of relativity. Although the theory was published in 1915 it was not known in Russia due to War World

I and the Civil War. By late 1920 Friedmann wrote to Ehrenfest telling him about his work on relativity and that he had developed formulas for a world with one spatial dimension which are more general than the Lorentz transformations.

Friedmann sent the article "On the curvature of Space" to "Zeitschrift für Physik" and it was received by the journal on 29 June 1922. In the paper Friedman showed that the radius of curvature of the Universe can be either an increasing or a periodic function of time. Friedmann, writing about the results of the paper in a book a little later, describes the results as follows:

"The stationary type of Universe comprises only two cases which were considered by Einstein and de Sitter. The variable type of Universe represents a great variety of cases; there can be cases of this type when the world's radius of curvature... is constantly increasing in time; cases are also possible when the radius of curvature changes periodically..."

Einstein quickly responded to Friedmann's article. His reply was received by "Zeitschrift für Physik" about on 18 September 1922:

"The results concerning the non-stationary world, contained in [Friedmann's] work, appear to me suspicious. In reality it turns out that the solution given in it does not satisfy the field equations."

On 6 December Friedmann wrote to Einstein:

"Considering that the possible existence of a non-stationary world has a certain interest, I will allow myself to present to you here the calculations I have made ... for verification and critical assessment. [The calculations are given] ... Should you find the calculations presented in my letter correct, please be so kind as to inform the editors of the "Zeitschrift für Physik" about it; perhaps in this case you will publish a correction to your statement or provide an opportunity for a portion of this letter to be published."

However by the time the letter reached Berlin, Einstein had already left on a trip to Japan. He did not return to Berlin until March but he still did not seem to have read Friedmann's letter. Only when Krutkov, a colleague of Friedmann's from Petrograd, met Einstein at Ehrenfest's house in Leiden in May 1923 and told him of the details contained in Friedmann's letter did Einstein admit his error. He wrote immediately to Zeitschrift für Physik :

"In my previous note I criticized [Friedmann's work "On the curvature of Space"]. However, my criticism, as I became convinced by Friedmann's letter communicated to me by Mr Krutkov, was based on an error in my calculations. I consider that Mr Friedmann's results are correct and shed new light."

In July 1923 Friedmann left Petrograd to visit Germany and Norway. In Germany he visited Berlin, Hamburg, Potsdam and Göttingen. In Norway he visited Christiania (Oslo). He discussed meteorology, aeronautics and mechanics. In Göttingen he talked to Prandtl and Hilbert, talking to Hilbert about his work in relativity. The following year, 1924, Friedmann was traveling again, this time to the First International Congress for Applied Mathematics held at Delft. He wrote about the congress:

"Everything went well at the congress, the attitude towards the Russians was wonderful; in particular, I was included among the members of the committee for convening the next conference. ... Courant from Göttingen got interested in Tamarkin's work. Blumenthal, Kármán and Levi-Civita got interested in my and my colleagues work."

In July 1925 Friedmann made a record-breaking ascent in a balloon to 7400 meters to make meteorological and medical observations. He returned to Leningrad (Petrograd had been renamed Leningrad in 1924). Near the end of August 1925 Friedmann began to feel unwell. He was diagnosed as having typhoid and taken to hospital where he died two weeks later.

and Mathematics Faculty of the Petrograd Polytechnic Institute, Russia, used his skills in mathematics to make profound progress in cosmology. He used the fact that in four-dimensional Riemannian spaces one can find three-dimensional spatial subspaces, which have different curvatures, positive, negative or zero. By linking these subspaces to the solutions of the Einstein field equations he obtained expanding or contracting Universes. That was before the discovery of Hubble (see references) that the Universe was indeed expanding. Friedmann wrote to Einstein about his finding and Einstein thought that Friedmann's calculation was wrong. But Friedmann showed Einstein that his calculation was correct. And Einstein did not like the idea that our Universe is expanding, so he added a "cosmological constant" to his field equations. By choosing this additional constant, negative or positive, such a term has the effect of stopping the expansion.

11.1.1 *Some preliminary concepts*

We are now in the territory of Einstein's general relativity theory and, as expected, physical quantities will be time-dependent.

Our starting point is the Hubble expansion formula at the present time which has been presented and used throughout the previous chapters,

$$v = H_0 r, \tag{11.1.1}$$

where H_0 is the Hubble constant at the present time. This formula will now be extended so as to assume its validity at any time, and accordingly will have the form

$$v = Hr, \tag{11.1.2}$$

where H is assumed to be the Hubble "constant" at time t (to distinguish it from the Hubble constant H_0 at the present time). By differentiating Eq.

In the book of E.A. Tropp, V Ya Frenkel and A D Chernin, entitled "Alexander A. Friedmann : The man who made the universe expand" (Cambridge, 1993), Friedmann's contributions are summed up as follows:

"... Friedmann is seen as a profound, independent-minded, and daring thinker who destroys scientific prejudices, myths and dogmas; his intellect sees what others do not see, and will not see what others believe to be obvious but for which there are no grounds in reality. He rejects the centuries-old tradition which chose, prior to any experience, to consider the Universe eternal and eternally immutable. He accomplishes a genuine revolution in science. As Copernicus made the Earth go round the Sun, so Friedmann made the Universe expand."

The above description of Friedmann is based on article by J J O'Connor and E F Robertson.

(11.1.2) with respect to t we obtain

$$\frac{dv}{dt} = Hv + \frac{dH}{dt} r$$

$$= \left(H^2 + \frac{dH}{dt} \right) r. \tag{11.1.3}$$

One then usually defines the quantity called the *deceleration parameter* q by

$$q = -\left(1 + \frac{1}{H^2} \frac{dH}{dt} \right). \tag{11.1.4}$$

Accordingly Eq. (11.1.3) can be written as

$$\frac{dv}{dt} = -qH^2 r. \tag{11.1.5}$$

One can then show that q is related to the density of matter in the Universe ρ by means of

$$q = \frac{4\pi G}{3H^2} \rho, \tag{11.1.6}$$

and in particular

$$q_0 = \frac{4\pi G}{3H_0^2} \rho_0, \tag{11.1.7}$$

where the subscript 0 indicates the present time, and G is Newton's gravitational constant.

In the next section the geometry of three-dimensional isotropic and homogeneous spaces with different curvature is discussed.

11.2 The Geometry of the Three-Dimensional Homogeneous and Isotropic Space

Observations of our three-dimensional space show that, on the very largest scales, it is homogeneous and isotropic. That means, if we observe at any angle we should see the same thing. Also if we move from one point to another we will also see the same thing. Of course our space is actually not that smooth at all, but to the first approximation we believe that it is.

Our interest is in the three-dimensional space surrounding us. It is actually a part of the four-dimensional spacetime described by general relativity theory. If the metric in the four-dimensional spacetime has a signature $(+ - - -)$, then the three-dimensional part will have the minus signs. It is convenient to associate this part of spacetime with a metric γ_{kl}, where $k, l = 1, 2, 3$, to distinguish it from the part of the Riemannian metric $g_{\mu\nu}$, $\mu, \nu = 0, 1, 2, 3$, so actually $\gamma_{kl} = -g_{kl}$.

The geometry of the three-dimensional space has very interesting properties, and it has its own curvature tensor in three dimensions which we denote by $S^i{}_{jkl}$, just as the Riemann tensor $R^\alpha{}_{\beta\gamma\delta}$ is the curvature of the four-dimensional Riemannian spacetime. The line element in the three-dimensional space will be given by

$$dl^2 = \gamma_{kl} dx^k dx^l. \tag{11.2.1}$$

Because of the isotropy of the space one finds that its curvature can be given by

$$S^i{}_{jkl} = K \left(\delta^i_l \gamma_{kj} - \delta^i_k \gamma_{lj} \right), \tag{11.2.2}$$

(see Problem 11.2.1). The isotropy at all points implies that K is a constant, namely it implies homogeneity. One furthermore finds that

$$S_{ij} = -2K\gamma_{ij}, \qquad S_{ij} = S^k{}_{ikj}, \tag{11.2.3}$$

and

$$S = -6K. \tag{11.2.4}$$

(See Problem 11.2.2.)

We thus see that the curvature properties of an isotropic space are determined by the constant K. According to this constant we have three possibilities for the spatial metric. These are (1) space with positive curvature; (2) space with negative curvature; and (3) space with zero curvature, depending on whether K is positive, negative or zero.

11.2.1 *Choice of coordinate system*

The coordinate system used in the four-dimensional spacetime will be co-moving, which means

$$g_{00} = 1, \qquad g_{0k} = 0, \qquad k = 1, 2, 3, \tag{11.2.5}$$

and thus

$$ds^2 = (dx^0)^2 + g_{kl} dx^k dx^l$$

$$= (dx^0)^2 - \gamma_{kl} dx^k dx^l. \tag{11.2.6}$$

Such a coordinate system satisfies and fits the fact that the three-dimensional space is homogeneous and isotropic.

11.2.2 *Space with constant positive curvature*

We want to obtain a three-dimensional space of uniform curvature. To this end we take the surface of a four-dimensional hypersphere that is known to be isotropic in a fictitious four-dimensional space. The equation of the surface of a four-dimensional hypersphere is given by

$$(x^1)^2 + (x^2)^2 + (x^3)^2 + (x^4)^2 = a^2, \tag{11.2.7}$$

where a is the radius of the sphere. The infinitesimal distance on the surface of the sphere is given by

$$dl^2 = (dx^1)^2 + (dx^2)^2 + (dx^3)^2 + (dx^4)^2, \tag{11.2.8}$$

where x^4 is a fictitious coordinate that should be eliminated by using Eq. (11.2.7).

The result is

$$dl^2 = (dx^1)^2 + (dx^2)^2 + (dx^3)^2 + \frac{\left(x^1 dx^1 + x^2 dx^2 + x^3 dx^3\right)^2}{a^2 - \left(x^1\right)^2 - \left(x^2\right)^2 - \left(x^3\right)^2}. \tag{11.2.9}$$

One then finds that

$$K = \frac{1}{a^2}. \tag{11.2.10}$$

(See Problem 11.2.3.)

To study the geometry of the space with positive curvature, we use spherical coordinates r, θ, ϕ,

$$\begin{aligned}
x^1 &= r \sin\theta \cos\phi \\
x^2 &= r \sin\theta \sin\phi \\
x^3 &= r \cos\theta.
\end{aligned} \tag{11.2.11}$$

In these coordinates the line element (11.2.9) can then be expressed as

$$dl^2 = \frac{dr^2}{1 - \dfrac{r^2}{a^2}} + r^2 d\theta^2 + r^2 \sin^2\theta d\phi^2. \tag{11.2.12}$$

It should be emphasized that the rectangular coordinates x^k and the radial coordinate r are actually periodic coordinates. Furthermore, the ratio between the circumference and the radius of a circle in this space is not equal to 2π (see Problem 11.2.4).

It follows that the three-sphere is a *closed space*; it has a finite volume but has no boundaries.

One can also introduce a new angular coordinate χ by means of

$$r = a \sin\chi, \quad 0 < \chi < \pi. \tag{11.2.13}$$

In terms of the coordinate χ one then has

$$dl^2 = a^2 \left[d\chi^2 + \sin^2\chi \left(d\theta^2 + \sin^2\theta d\phi^2\right)\right]. \tag{11.2.14}$$

One then finds that the radial distance from the origin is given by $a\chi$.

11.2.3 *Space with constant negative curvature*

For such a space $(K < 0)$ we see, by Eq. (11.2.10), that a should be imaginary. Thus all the formulas for a negative-curvature space can be obtained from that of a positive-curvature space by replacing a by ia. It thus follows that the geometry of a negative-curvature space is obtained mathematically as the geometry in a four-dimensional pseudosphere with imaginary radius. The value of K will now be negative,

$$K = -\frac{1}{a^2}. \tag{11.2.15}$$

In spherical coordinates we thus have for the line element,

$$dl^2 = \frac{dr^2}{1 + r^2/a^2} + r^2 d\theta^2 + r^2 \sin^2\theta d\phi^2, \tag{11.2.16}$$

where $0 \leq r < \infty$. The ratio of the circumference to the radius of a circle is now larger than $1/2\pi$. Also the ratio of the area of a sphere to the square of its radius is larger than $1/4\pi$. The space of negative curvature is open and infinite. One can then introduce a dimensionless coordinate χ by,

$$r = a \sinh\chi, \tag{11.2.17}$$

and thus we obtain

$$dl^2 = a^2 \left[d\chi^2 + \sinh^2\chi \left(d\theta^2 + \sin^2\theta d\phi^2 \right) \right]. \tag{11.2.18}$$

In the next section Friedmann model is considered for Universes with different curvatures.

11.2.4 *Space with zero curvature*

In this case $(K = 0)$ we obtain a three-dimensional flat, Euclidean, space. The fact that the three-dimensional subspace of the Einstein four-dimensional spacetime is Euclidean does not violate general relativity theory. Such a space has very important role in cosmological theory. The space in this case is open and infinite.

We are now in a position to go ahead to present the Friedmann Universe. This will be done in the next section.

11.2.5 *Problems*

P 11.2.1. Prove Eq. (11.2.2).

Solution: The solution is left for the reader.

P 11.2.2. Prove Eqs. (11.2.3) and (11.2.4).

Solution: The solution is left for the reader.

P 11.2.3. Prove Eq. (11.2.10).

Solution: Use the approximate formula

$$\gamma_{kl} = \delta_{kl} + \frac{x^k x^l}{a^2} \tag{1}$$

in the vicinity of the origin.

P 11.2.4. Find the ratio of a circumference of a circle in the positive curvature space to the radius.

Solution: Take the circle defined by $r = b = $ constant around the origin. This circle has a radius that is stretched between $r = 0$ and $r = b$ and is given by

$$R = \int_0^b \frac{dr}{\sqrt{1 - \frac{r^2}{a^2}}} = \frac{a}{\sin \frac{b}{a}}. \tag{1}$$

The circumference of the circle is obtained, if we choose the circle to be in the plane $\theta = \frac{\pi}{2}$,

$$C = \int_0^{2\pi} b \sin \theta d\phi = 2\pi b. \tag{2}$$

Thus the ratio of the circumference to the radius is smaller than $1/2\pi$, which is characteristic result of spaces with positive curvature.

P 11.2.5. Find the ratio of the area of a sphere to the square of its radius in the positive curvature space and show that it is smaller than $1/4\pi$.

Solution: The solution is left for the reader.

11.3 The Friedmann Model

In the last two sections we have laid the foundations for developing of what is called the Friedmann model. The concept of decelerating parameter (q) was introduced, which has a measured value in the range $0.33 > q > -1.27$ determined by Gunn and Oke. We have also classified our three-dimensional homogeneous and isotropic space into spaces with positive, negative and zero curvatures. This latter classification was made mathematically, but in the following we will relate them to the four-dimensional solutions of the Einstein field equations. To this end we proceed as follows.

We will use for the co-moving coordinate system the one whose spatial part is χ, θ, ϕ as it is the most appropriate system that keeps distances between galaxies as the Universe expands. Each of the three cases will have to be dealt separately.

11.3.1　*Space with positive curvature*

The spacetime line element is given by

$$ds^2 = c^2 dt^2 - dl^2$$

$$= c^2 dt^2 - a^2\left(t\right)\left[d\chi^2 + \sin^2\chi\left(d\theta^2 + \sin^2\theta d\phi^2\right)\right]. \tag{11.3.1}$$

It is customary to replace the time coordinate t by a time parameter η, leaving other coordinates unchanged, defined by

$$cdt = ad\eta. \tag{11.3.2}$$

Thus the line element will have the form

$$ds^2 = a^2\left(\eta\right)\left[d\eta^2 - d\chi^2 - \sin^2\chi\left(d\theta^2 + \sin^2\theta d\phi^2\right)\right]. \tag{11.3.3}$$

The covariant components of the metric tensor will accordingly have the form

$$g_{\mu\nu} = \begin{pmatrix} a^2 & 0 & 0 & 0 \\ 0 & -a^2 & 0 & 0 \\ 0 & 0 & -a^2\sin^2\chi & 0 \\ 0 & 0 & 0 & -a^2\sin^2\chi\sin^2\theta \end{pmatrix}. \tag{11.3.4}$$

The Christoffel symbols needed are given by

$$\Gamma^0_{00} = \frac{\dot{a}}{a}, \qquad \Gamma^0_{kl} = -\frac{\dot{a}}{a^3}g_{kl}, \qquad \Gamma^k_{0l} = \frac{\dot{a}}{a}\delta^k_l,$$

$$\Gamma^0_{k0} = 0, \qquad \Gamma^k_{00} = 0, \tag{11.3.5}$$

where k, $l = 1, 2, 3$, and a dot denotes a derivative with respect to η. The Einstein tensor, after a simple calculation, is then given by

$$G_0{}^0 = R_0{}^0 - \frac{1}{2}R = \frac{3}{a^4}\left(\dot{a}^2 + a^2\right). \tag{11.3.6}$$

The energy-momentum tensor that was used by Friedmann is that of a gas without pressure and is given by

$$T_\mu{}^\nu = \rho u_\mu u^\nu, \tag{11.3.7}$$

where ρ is the mass density and $u^{\mu} = dx^{\mu}/ds$. One then finds that $u^0 = d\eta/ds = 1/a$ and $u_0 = a$ (see Problem 11.3.1). Thus T_0^0 is just the mass density ρ. The mass density is given by the total mass of the Universe M divided by the volume of the closed Universe, $2\pi^2 a^3$,

$$\rho(\eta) = \frac{M}{2\pi^2 a^3}. \tag{11.3.8}$$

The Einstein field equations can then be shown to be given by (see Problem 11.3.2)

$$\frac{3}{a^4}\left(\dot{a}^2 + a^2\right) = 8\pi G\rho = \frac{4GM}{\pi a^3}, \tag{11.3.9}$$

when the speed of light in vacuum is taken as unity. Equation (11.3.9) describes the evolution of the closed Universe in terms of the time parameter η.

The solution of Eq. (11.3.9) is given by

$$a(\eta) = \tilde{a}\left(1 - \cos\eta\right), \tag{11.3.10}$$

where

$$\tilde{a} = \frac{2GM}{3\pi} \tag{11.3.11}$$

is an integration constant. We still have to find the function a in terms of the time coordinate t instead of the time parameter η. This can easily be done by integrating the equation

$$cdt = ad\eta = \tilde{a}\left(1 - \cos\eta\right)d\eta. \tag{11.3.12}$$

We obtain

$$ct = \tilde{a}\left(\eta - \sin\eta\right). \tag{11.3.13}$$

Equations (11.3.10) and (11.3.13) give parametric representation for $a(t)$. The result is plotted in Figure 11.3.1, which is a cycloid. The diagram shows a plot of a as a function of the time coordinate t. As is seen, at $t = 0$, $\pm 2\pi\tilde{a}$, $\pm 4\pi\tilde{a}$, ..., $a(t)$ vanishes. The constant $\tilde{a}$ is the maximum value of a and can be taken as

$$\tilde{a} = c\tau, \tag{11.3.14}$$

where τ is the Big Bang time $= (13.56 \pm 0.48)$ Gyr and c is the speed of light in vacuum. Thus we obtain for the mass of the Universe

$$M = \frac{3\pi c^3 \tau}{2G} = \frac{3 \times 3.14 \times (3 \times 10^8)^3 \times 4.28 \times 10^{17}}{2 \times 6.67 \times 10^{-11}}$$

$$= 8.3 \times 10^{53} \text{kg}. \tag{11.3.15}$$

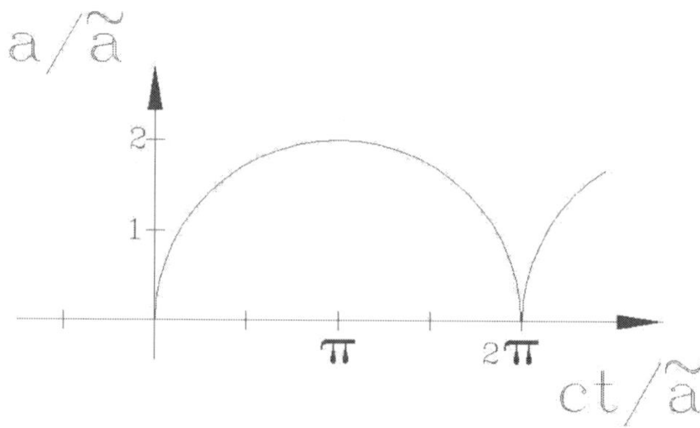

Fig. 11.3.1:   The Friedmann closed model.

11.3.2   *Remark 1*

Although the Friedmann Universe is just a mathematical model, historically it started the cosmological theory in a realistic fashion. One cannot take this model for the closed Universe very seriously these days, since we have seen repeatedly in previous chapters that the Universe is open and not closed. The idea that the Universe is closed and cycles through maxima and minima (collapses to zero) does not seem to fit the observations. It seems to be more physically correct to talk about galaxies, their receding velocities, and their distances from us, since after all the Universe is built out of galaxies, clusters of galaxies, gases, etc. And this kind of presentation of the Universe is exactly what has been done in the previous chapters. An important question remains unanswered: How is a related to the distances of galaxies?

11.3.3   *Space with negative curvature*

The case of an open isotropic Universe is obtained in exactly the same form as for the closed Universe. In this case we obtain

$$\frac{3}{a^4}\left(\dot{a}^2 - a^2\right) = 8\pi G\rho = \frac{4GM}{\pi a^3},\qquad(11.3.16)$$

with $\rho = M/2\pi^2 a^3$. But now M cannot be said to be the mass of the Universe, since the Universe in this case is infinite.

Equation (11.3.16) describes the open Friedmann model, and its solution is given by

$$a\left(\eta\right) = \tilde{a}\left(\cosh\eta - 1\right),\qquad(11.3.17)$$

with the coordinate t given by

$$ct = \tilde{a}\left(\sinh\eta - \eta\right). \qquad (11.3.18)$$

11.3.4 *Remark 2*

Although the Universe now is open as is expected, still the physical meaning of a remains obscure, and it is not clear how to relate it to distances of galaxies.

11.3.5 *Space with zero curvature*

In this case the Universe has Euclidean three-dimensional geometry, and the line element can be written as

$$ds^2 = a^2\left(\eta\right)\left[d\eta^2 - dr^2 - r^2\left(d\theta^2 + \sin^2\theta d\phi^2\right)\right]. \qquad (11.3.19)$$

The metric tensor is accordingly given by

$$g_{\mu\nu} = \begin{pmatrix} a^2 & 0 & 0 & 0 \\ 0 & -a^2 & 0 & 0 \\ 0 & 0 & -a^2 r^2 & 0 \\ 0 & 0 & 0 & -a^2 r^2 \sin^2\theta \end{pmatrix}. \qquad (11.3.20)$$

The equation for $a\left(\eta\right)$ is given by

$$\frac{3\left(da/dt\right)^2}{a^2} = 8\pi G\rho = \frac{4GM}{\pi a^3}. \qquad (11.3.21)$$

This equation has the form

$$\left(\frac{da}{dt}\right)^2 = \frac{4GM}{3\pi a}, \qquad (11.3.22)$$

whose solution is given by

$$a\left(t\right) = \left(\frac{3GM}{\pi}\right)^{1/3} t^{2/3}. \qquad (11.3.23)$$

In the next section light propagation as a null geodesics is considered.

11.3.6 *Problems*

P 11.3.1. Show that

$$u^0 = \frac{d\eta}{ds} = \frac{1}{a}, \qquad u_0 = a. \qquad (1)$$

Solution: The solution is left for the reader.

P 11.3.2. Show that the Einstein gravitational field equations for the closed Universe are given by Eq. (11.3.9).

Solution: The solution is left for the reader.

11.4   Propagation of Light in the Friedmann Model

Light propagates as a null geodesic with $ds = 0$. For a light signal that emanates at the origin of the coordinate system, the propagation will then be radially outward with $d\theta = d\phi = 0$. Thus for both the closed and the open models we obtain

$$a(\eta)\left(d\eta^2 - d\chi^2\right) = 0, \tag{11.4.1}$$

and therefore

$$\chi = \eta - \eta_0, \tag{11.4.2}$$

where η_0 is the time parameter, which characterizes time when the light signal was sent out. For two light signals that start at η_0 and $\eta_0 + \Delta\eta_0$, we then have

$$\chi = \eta - \eta_0, \qquad \chi_1 = \eta - (\eta_0 + \Delta\eta_0). \tag{11.4.3}$$

The two signals will arrive at a point χ with a difference of the time parameters

$$\Delta\eta\left(\chi\right) = \Delta\eta_0. \tag{11.4.4}$$

That is to say, their $\Delta\eta$ is independent of χ and one has to use the time coordinate t instead. We then have $c\Delta t = a\Delta\eta$. Then $c\Delta t/a$ is a constant. We then have $\Delta t = 1/\nu$, where ν is the frequency, and

$$a\nu = \text{constant}. \tag{11.4.5}$$

When the Universe is expanding, a increases and ν decreases. At a point χ, a wave that was emitted with frequency $\nu(0)$ at the point $\chi = 0$ will have the frequency given by

$$\nu(\chi) = \frac{a\left(\eta_0\right)}{a\left(\eta\right)}\nu(0) = \frac{a\left(\eta - \chi\right)}{a\left(\eta\right)}\nu\left(0\right). \tag{11.4.6}$$

The above formula can be approximated to

$$\nu\left(\chi\right) = \left[1 - \frac{\dot{a}\left(\eta\right)}{a\left(\eta\right)}\chi\right]\nu(0), \tag{11.4.7}$$

and therefore

$$\frac{\nu(\chi) - \nu(0)}{\nu(0)} = -\frac{\dot{a}(\eta)}{a(\eta)}\chi.$$
(11.4.8)

If λ is the wavelength, $\lambda = c/\nu$, and we obtain

$$\frac{\lambda(\chi) - \lambda(0)}{\lambda(0)} = \frac{\dot{a}(\eta)}{a(\eta)}\chi = \frac{\dot{a}}{a^2}l,$$
(11.4.9)

where l is the distance between the emitter and the receiver at the time parameter η. From Eq. (11.4.9) one can have the Hubble constant in terms of a and its derivative $\dot{a}$,

$$H = \frac{\dot{a}}{a^2} = \frac{1}{a^2}\frac{da}{dt}.$$
(11.4.10)

In the next section we summarize some of the formulas obtained so far by giving the Friedmann-Robertson-Walker (FRW) metric.

11.4.1 *Problems*

P 11.4.1. Show that the boundary of the visible Universe is given by the surface

$$r = a\sin\eta.$$
(1)

Solution: The solution is left for the reader.

P 11.4.2. Show that the volume of the visible part of the Universe is given by

$$2\pi a^3 (\eta - \sin\eta\cos\eta).$$
(1)

Show also that the boundary of the Universe recedes faster than the galaxies themselves, and therefore more of the Universe will be included inside the boundary.

Solution: The solution is left for the reader.

P 11.4.3. Find the intensity I of light arriving at an observation point from a source located at a distance χ. Show that for the open Friedmann model it is given by

$$I = \text{constant}\frac{a^2(\eta - \chi)}{a^4(\eta)\sinh^2\chi},$$
(1)

and for the closed model it is given by

$$I = \text{constant}\frac{a^2(\eta - \chi)}{a^4(\eta)\sin^2\chi}.$$
(2)

(See Landau and Lifshitz, p. 374).

Solution: The solution is left for the reader.

P 11.4.4. Show that a finite mass particle moving in the Friedmann model satisfies

$$\frac{va}{\sqrt{1 - \dfrac{v^2}{c^2}}} = \text{constant.} \tag{1}$$

(See Landau and Lifshitz, p. 375).

Solution: The solution is left for the reader.

P 11.4.5. Find the first two terms in the expansion of the apparent bright mass of a galaxy as a function of its redshift. (See Landau and Lifshitz, p. 375).

Solution: The solution is left for the reader.

P 11.4.6. Find the leading terms in the expansion of the number of galaxies contained inside a sphere as a function of the redshift at the boundary of the sphere. (See Landau and Lifshitz, p. 375).

Solution: The solution is left for the reader.

11.5   FRW Metric

In this section we give the familiar Friedmann-Robertson-Walker (FRW) metric.

Using Eq. (11.4.10) in Eqs. (11.3.9), (11.3.16) and (11.3.21), we can write one formula that includes all the three cases. This is given by

$$H^2 + \frac{kc}{a^2} = \frac{8\pi G\rho}{3}. \tag{11.5.1}$$

In the above equation $k = 1$, -1, 0 for the cases of positive, negative and zero curvature, respectively. Equation (11.5.1) does not include a cosmological constant Λ, but if we include such a constant, we would get

$$H^2 + \frac{kc}{a^2} = \frac{1}{3}\left(8\pi G\rho + \Lambda\right). \tag{11.5.2}$$

Accordingly we obtain

$$\Omega > 1 - \frac{\Lambda}{3H^2}, \tag{11.5.3a}$$

$$\Omega < 1 - \frac{\Lambda}{3H^2}, \tag{11.5.3b}$$

$$\Omega = 1 - \frac{\Lambda}{3H^2}, \tag{11.5.3c}$$

for the cases of positive, negative and zero curvature. In the above formulas $\Omega = \rho/\rho_c$, where ρ is the mass density and ρ_c is the critical mass density of the Universe,

$$\rho_c = \frac{3H^2}{8\pi G}, \tag{11.5.4}$$

in the standard model.

One then obtains

$$q = \frac{1}{2}\Omega - \frac{\Lambda}{3H^2}. \tag{11.5.5}$$

Eliminating Λ then gives

$$\Omega > \frac{2}{3}(1+q), \tag{11.5.6a}$$

$$\Omega < \frac{2}{3}(1+q), \tag{11.5.6b}$$

$$\Omega = \frac{2}{3}(1+q), \tag{11.5.6c}$$

where q is the deceleration parameter. Of course in the Friedmann models $\Lambda = 0$.

One then obtains

$$\Omega > 1, \tag{11.5.7a}$$

$$\Omega < 1, \tag{11.5.7b}$$

$$\Omega = 1, \tag{11.5.7c}$$

for the cases of positive, negative, and zero curvature, respectively.

Equation (11.5.2) is known as the Friedmann cosmological equation (which is actually an Einstein field equation). The FRW metric is then given by

$$ds^2 = c^2 dt^2 - a^2(t)\left\{ \frac{dr^2}{1-kr^2} + r^2\left(d\theta^2 + \sin^2\theta d\phi^2\right)\right\}. \tag{11.5.8}$$

11.5.1 *Remarks on the critical mass density ρ_c*

The critical mass density of the Universe ρ_c in the standard model (including this chapter) is given by Eq. (11.5.4), and at the present time by

$$\rho_c = \frac{3H_0^2}{8\pi G},\tag{11.5.9}$$

where H_0 is the Hubble constant at the present time. It should be emphasized that this is not the same critical mass density that has been used throughout this book (except in this chapter). Previously ρ_c was given by

$$\rho_c = \frac{3h^2}{8\pi G},\tag{11.5.10}$$

where h is the Hubble constant with no gravity (such as for short distances or $\Omega = 1$). The value of $h =72.17$ km/s-Mpc.

In the next chapter cosmological relativity is considered along with FRW theory.

11.6 Suggested References

A. Einstein, *Zeitschrift für Physik*, 18 September 1922 (Rejection of the article of Friedmann "On the curvature of Space").

A. Einstein, *Zeitschrift für Physik*, 1923 (Acceptance by Einstein of Friedmann's calculation without error as presented to him by Krutkov in Friedmann's letter to Einstein).

A. Friedmann, *Z. Phys.* **10**, 377 (1922).

A. Friedmann, *Z. Phys.* **11**, 326 (1924).

J.E. Gunn and J.B. Oke, *Astrophys. J.* **195**, 255 (1975).

E.P. Hubble, *Proc. Nat. Acad. Sci.* **15**, 168 (1927).

E.P. Hubble, *The Realm of the Nebulae* (Yale University Press, New Haven, 1936); reprinted by Dover Publications, Inc., New York, 1958.

L. Landau and E. Lifshitz, *The Classical Theory of Fields*, 4-th Edition (Addison-Wesley Publishing Company, Inc., Reading, Massachusetts, 1975).

H.C. Ohanian, *Gravitation and Spacetime* (W. W. Norton & Company, Inc., New York, 1976).

Chapter 12

CGR versus FRW

Moshe Carmeli

In this chapter we discuss the cosmic time which is shown to be a relative notion in cosmology. We then present the line elements that express the Hubble expansion. The coordinates here are the spatial coordinates x, y, z, and the velocity coordinate v, which are actually what astronomers use in their measurements. Two such line elements are presented: the first is the empty space (no matter exists), and the second including matter in the Universe. These line elements are the comparable to the standard Minkowskian and Friedmann-Robertson-Walker line elements in ordinary general relativity theory. The following is based on Carmeli 2003.

12.1 The Cosmic Time as a Relative Quantity

In the standard cosmological theory one uses the Einstein concepts of space and time as were originally introduced for the special theory of relativity and later in the general relativity theory. According to this approach all physical quantities are described in terms of the continuum spatial coordinates and the time. Using general relativity theory great progress has been made in understanding the evolution of the Universe.

Accordingly in the standard cosmological model one has the Friedmann-Robertson-Walker (FRW) line elements (see Chapter 11)

$$ds^2 = dt^2 - a^2\left(t\right)\left\{\frac{dr^2}{1 - kr^2} + r^2\left(d\theta^2 + \sin^2\theta d\phi^2\right)\right\}, \qquad (12.1.1)$$

where k is a constant, $k = -1$, open Universe, $k = 1$, closed Universe, $k = 0$, flat Universe. Here $a\left(t\right)$ is a scale function, and as follows from the

337

Einstein field equations, it is related by

$$\left(\frac{\dot{a}}{a}\right)^2 = \frac{8\pi}{3}G\rho - \frac{k}{a^2}, \qquad (12.1.2)$$

$$\ddot{a} = -\frac{4\pi}{3}G\left(\rho + 3p\right)a, \qquad (12.1.3)$$

$$\frac{d}{dt}\left(a^3\rho\right) = -p\frac{d}{dt}\left(a^3\right), \qquad (12.1.4)$$

where ρ is the mass density, p is the pressure, G is Newton's constant, and the overdot denotes derivative with respect to t. Notice that the above equations are not independent.

It is well known that both Einstein's theories are based on the fact that light propagates at a constant speed. However, the Universe also expands at a constant rate when gravity is negligible. This fact is not taken into account in Einstein's theories. Moreover, cosmologists usually measure spatial distances and redshifts of faraway galaxies in the Hubble expansion. In recent years this fact has been taken into account in the development of new theories in terms of distance and velocity (redshift). While in Einstein's special relativity the propagation of light plays the major role, in the new theory it is the expansion of the Universe that takes that role. It is the concept of cosmic time that becomes crucial in these recent theories. In the standard theory the cosmic time is considered to be an absolute concept. Thus we talk about the Big Bang time with respect to us here on Earth as an absolute quantity.

Consider, for example, another galaxy that has, let us say, a relative cosmic time with respect to us of 1 billion years. Now one may ask what will be the Big Bang time with respect to that galaxy. Will it be the Big Bang time with respect to us minus 1 billion years? A hypothetical observer on that galaxy might look at our galaxy and think that ours is also as far away from him. Does that mean, with respect to him, our galaxy is closer to the Big Bang time by 1 billion years? Or some other epoch? All this leads to the conclusion that there is no absolute cosmic time. Rather, it is a relative concept.

12.1.1   *The line element in empty space*

Based on this assumption, and using the Hubble expansion $r = \tau v$, where r is the distance from us to a galaxy, v is the recession velocity of the galaxy and τ is the Hubble-Carmeli time (a universal constant equals to approximately 13.56 Gyr - it is the standard Hubble time in the limit of zero distance and zero gravity - and it is a constant in this epoch of time).

The Hubble expansion can then be written as

$$\tau^2 v^2 - \left(x^2 + y^2 + z^2\right) = 0. \tag{12.1.5}$$

The quantity τv can alternatively be written in terms of redshift. For nonrelativistic and relativistic velocities the relationship between velocity and the redshift parameter are given respectively by $z = v/c$ and $z = \left[(1 + v/c)/(1 - v/c)\right]^{1/2} - 1$, $v/c = \left[(1 + z)^2 - 1\right]/\left[(1 + z)^2 + 1\right]$. Accordingly we have a line element

$$ds^2 = \tau^2 dv^2 - \left(dx^2 + dy^2 + dz^2\right). \tag{12.1.6}$$

It is equal to zero for the Hubble expansion, but is otherwise not vanishing for cosmic times smaller than τ. It is similar to the Minkowskian metric

$$ds^2 = c^2 dt^2 - \left(dx^2 + dy^2 + dz^2\right), \tag{12.1.7}$$

which vanishes for light propagation, but is otherwise different from zero for particles of finite mass.

If we now assume that the laws of physics are valid at all cosmic times and τ is a universal constant which has the same value at all cosmic times, then we can develop a new special relativity just like Einstein's original special relativity theory: the validity of the laws of physics at all cosmic times replaces the special relativistic assumption in Einstein's theory of their validity in all inertial coordinate systems, whereas the constancy of τ in all cosmic times replaces the constancy of the speed of light in all inertial systems in ordinary special relativity.

In this way one also obtains a cosmological transformation that relates distances and velocities (redshifts) at different cosmic times, just as in ordinary special relativity we have the Lorentz transformation that relates spatial coordinates and time at different velocities. We now will have

$$x' = \frac{x - \tau v}{\sqrt{1 - \dfrac{t^2}{\tau^2}}}, \qquad v' = \frac{v - \dfrac{xt}{\tau^2}}{\sqrt{1 - \dfrac{t^2}{\tau^2}}}, \tag{12.1.8}$$

for the case $y' = y$, $z' = z$. In the above transformation t is the relative cosmic time and is measured backward ($t = 0$ now and $t = \tau$ at the Big Bang). As seen from the above transformation, τ is the maximum cosmic time that is allowed in nature and as such can be considered as the age of Universe.

For example, if we denote the temperature at a cosmic time t by T, and the temperature at the present time by T_0 ($= 2.73$ K), we then have $T = T_0/\sqrt{1 - t^2/\tau^2}$. For instance, at $t/\tau = 1/2$ we get $T = 2 \times 2.73/\sqrt{3} = 3.15$ K. (This result does not take into account gravity that needs a correction by a factor of 13, and thus the temperature at $t = \tau/2$ is 40.95 K.) Since we always use forward rather than backward times, we can write the transformation (12.1.8) in terms of such a time $t' = \tau - t$. The resulting transformation will have the form

$$x' = \frac{x - (\tau - t)\,v}{\sqrt{\dfrac{t}{\tau}\left(2 - \dfrac{t}{\tau}\right)}}, \qquad v' = \frac{v - \dfrac{x\,(\tau - t)}{\tau^2}}{\sqrt{\dfrac{t}{\tau}\left(2 - \dfrac{t}{\tau}\right)}}, \qquad (12.1.9)$$

where primes have been dropped for brevity, $0 \le t \le \tau$, $t = 0$ at the Big Bang, $t = \tau$, now.

The above introduction gives a brief review of a new special relativity (cosmological special relativity, for more details see Chapter 2). Obviously the Universe is not empty and contains gravity , therefore one has to go to a Riemannian space with the Einstein gravitational field equations in terms of space and redshift (velocity). This was done in Chapter 4.

12.1.2 *The line element with gravity*

We obtained for the pressure

$$p = \frac{1 - \Omega}{\kappa c \tau^3} = \frac{c}{\tau}\frac{1 - \Omega}{8\pi G} = 4.544\,(1 - \Omega) \times 10^{-2} \text{g/cm}^2, \qquad (12.1.10)$$

where $\Omega = \Omega(z)$ and

$$e^{-\mu} = 1 + f(r) = 1 + \frac{\tau \kappa p}{c}r^2 = 1 + \frac{1 - \Omega_m}{c^2\tau^2}r^2. \qquad (12.1.11)$$

Accordingly, the 4D specevelocity line element of the Universe is given by

$$ds^2 = \tau^2 dv^2 - \frac{dr^2}{1 + \dfrac{1 - \Omega_m}{c^2\tau^2}r^2} - r^2\left(d\theta^2 + \sin^2\theta d\phi^2\right), \qquad (12.1.12)$$

or,

$$ds^2 = \tau^2 dv^2 - \frac{dr^2}{1 + \dfrac{\kappa \tau p}{c}r^2} - r^2\left(d\theta^2 + \sin^2\theta d\phi^2\right). \qquad (12.1.13)$$

This line element is the comparable to the FRW line element in the standard theory.

It will be recalled that the Universe expansion is determined by $dr/dv = \tau e^{-\mu/2}$. The only thing that is left to be determined is the signs of $(1 - \Omega_m)$ or of the pressure p at the current epoch. The parameter Ω_m is the current matter density of the Universe. Thus we have

$$\frac{dr}{dv} = \tau\sqrt{1 + \frac{\kappa\tau p}{c}r^2} = \tau\sqrt{1 + \frac{1 - \Omega}{c^2\tau^2}r^2}. \qquad (12.1.14)$$

Equation (12.1.14) can now be integrated and solved for a particular matter density distribution $\Omega(z)$ in the Universe.

In the next two chapters the new theory is tested against astronomical observations and compared to the standard FRW theory.

12.2 Suggested References

M. Carmeli, *Cosmological Special Relativity,* Second Edition (World Scientific, Singapore, 2002a).

M. Carmeli, Accelerating Universe: Theory versus experiment, astro-ph/0205396 (2002b).

M. Carmeli, The line element in the Hubble expansion, in: *Gravitation and Cosmology,* A. Lobo *et al.* Eds., (Universitat de Barcelona, 2003), pp. 113-130 (invited lecture in Proceedings of the Spanish Relativity Meeting, Menorca, Spain, 22-24 September 2002); astro-ph/0211043.

M. Carmeli, J.G. Hartnett and F.J. Oliveira, The cosmic time in terms of the redshift, *Found. Phys. Lett.* **19**(3), 277-283 (2006); gr-qc/0506079.

Chapter 13

Testing CGR against High Redshift Observations

John Hartnett & Firmin Oliveira[1]

Several key relations are derived for Cosmological General Relativity which are used in standard observational cosmology. These include the luminosity distance, angular size, surface brightness and matter density. Luminosity distance and matter density relations are used to fit type Ia supernova (SNe Ia) data, giving consistent, well behaved fits over a broad range of redshifts $0.1 < z < 2$. The best fit to the data for the local matter density parameter is $\Omega_m = 0.0401 \pm 0.0199$. Because Ω_m is within the baryonic budget there is no need for any dark matter to account for the SNe Ia redshift luminosity data. From this local density it is determined that the redshift where the Universe expansion transitions from deceleration to acceleration is $z_t = 1.095\,^{+0.264}_{-0.155}$. Because the fitted data covers the range of the predicted transition redshift z_t, there is no need for any dark energy to account for the expansion rate transition. We conclude that the expansion is now accelerating and that the transition from a closed to an open Universe occurred about 8.54 Gyr ago. The following is based on Hartnett and Oliveira 2007b.

13.1   Introduction

Carmeli's cosmology, also referred to as Cosmological General Relativity (CGR), is a space-velocity theory of the expanding Universe. It is a description of the Universe at a particular fixed epoch of cosmic time t. In CGR time is measured from the present back toward the beginning. The

---

[1] Joint Astronomy Center, Hilo, Hawai'i 96720, USA; Email: firmin@jach.hawaii.edu

343

theory assumes the Hubble law as fundamental. The observables are the coordinates of Hubble; proper distance and velocity of the expansion of the Universe. In practice, not velocity but redshift is used. CGR incorporates this basic law into a general 4D Riemannian geometrical theory satisfying the Einstein field equations.

In order to compare the theoretical predicted redshift distance modulus relation of CGR with the distance modulii derived from type Ia supernova data, firstly, luminosity distance must be determined in this theory. Secondly, we need to model correctly the variation of matter density with redshift of the Universe. In the following, we determine a few key relations that are used in the subsequent analysis. Then we compare the theoretical distance modulii with those measured, resulting in a good fit without the need to assume the existence of dark energy or dark matter.

13.2   Luminosity Distance

Suppose L is the total energy emitted per unit time by a source galaxy at the epoch t (that is, in the rest frame of the galaxy) to be received by an observer at the present time $t = 0$. Therefore we can write

$$dL = LI(\lambda)d\lambda, \qquad (13.2.1)$$

where I is its (normalized) intensity distribution – a function of wavelength λ. In CGR, times at cosmological distances add according to a relativistic addition law (Carmeli 2002a, Sec. 2.15.4, p. 23) when referred to the observer at $t = 0$. Hence instead of the time interval Δt, we get

$$\Delta t \rightarrow \frac{t + \Delta t}{1 + \dfrac{t\Delta t}{\tau^2}} - t = \Delta t \left\{ 1 - \frac{t^2}{\tau^2} \right\}, \qquad (13.2.2)$$

where $\tau \approx H_0^{-1} = 13.56 \pm 0.48$ Gyr is the Hubble-Carmeli time constant. From this it can be shown (Hartnett and Oliveira 2007a) that the luminosity L_0 of a source at the present time is related to the luminosity L of an identical source which emitted at time t by

$$L_0 = L \left\{ 1 - \frac{t^2}{\tau^2} \right\}. \qquad (13.2.3)$$

For the source at distance r, redshift z, emission wavelength $\lambda_0/(1+z)$ and the luminosity (13.2.3), it is straight forward to show (Narlikar 2002) that the observed flux integrated over all wavelengths is

$$\mathcal{F}_{bol} = \frac{L_{bol}}{(1+z)^2} \left\{ 1 - \frac{t^2}{\tau^2} \right\} \frac{1}{4\pi r^2} = \frac{L_{bol}}{4\pi \mathcal{D}_L^2}, \qquad (13.2.4)$$

where L_{bol} is the absolute bolometric luminosity of the source galaxy. Therefore the luminosity distance $\mathcal{D}_L$ in CGR is expressed as

$$\mathcal{D}_L = \frac{r(1+z)}{\sqrt{1 - \dfrac{t^2}{\tau^2}}}. \qquad (13.2.5)$$

It is clear that this expression for the luminosity distance in CGR when compared to that in the FRW theory (see Chapter 11) has the extra factor $(1 - t^2/\tau^2)^{-1/2}$. Hence we expect the luminosity distance to be greater in CGR than in FRW theory.

13.3 Angular Size

The line element in CGR (Carmeli 2002b)

$$ds^2 = \tau^2 dv^2 - \frac{dr^2}{1 + (1 - \Omega)\dfrac{r^2}{c^2\tau^2}} - r^2(d\theta^2 + sin^2\theta d\phi^2), \qquad (13.3.1)$$

represents a spherically symmetric isotropic Universe. See Carmeli 2002a and 2002b for details. The expansion is observed at a definite time and therefore $dt = 0$ and hence doesn't appear in Eq. (13.3.1). Carmeli solved Eq. (13.3.1) with the null condition $ds = 0$ and isotropy ($d\theta = d\phi = 0$) from which it follows that the proper distance r in spherically symmetric coordinates can be written as

$$\frac{r}{c\tau} = \frac{\sinh\left(\beta\sqrt{1 - \Omega}\right)}{\sqrt{1 - \Omega}}, \qquad (13.3.2)$$

where $\beta = t/\tau = v/c$ and Ω is the matter density, a function of redshift z. Also β can be written as a function of redshift

$$\beta = \frac{(1+z)^2 - 1}{(1+z)^2 + 1}. \qquad (13.3.3)$$

Now in CGR there is no scale factor like in the FRW theory but we can similarly define an expansion factor as $(1+z)^{-1}$. If we then make the substitution for the matter density $\Omega = \Omega_m(1+z)^3$, where Ω_m is the matter density at the current epoch, the proper distance (13.3.2) can be rewritten as a function of $(1+z)$,

$$r = \frac{c\tau \sinh\left(\beta\sqrt{1 - \Omega_m(1+z)^3}\right)}{\sqrt{1 - \Omega_m(1+z)^3}}. \qquad (13.3.4)$$

For a proper comparison with FRW theory we must use the FRW equivalent of $r/(1+z)$, which is the Hubble distance D_1 when the light we observe left the galaxy at redshift z and is given by

$$D_1 = \frac{2c}{H_0(1+z)}\left\{1 - \frac{1}{\sqrt{1+z}}\right\}, \qquad (13.3.5)$$

where a deceleration parameter $q_0 = 1/2$ (see Section 11.1) has been used. The angular size of the source galaxy in FRW theory is

$$\Delta\theta = \frac{d}{D_1}, \qquad (13.3.6)$$

where d is the actual diameter of the source galaxy and the angular distance D_1 is taken from Eq. (13.3.5).

In CGR the angular distance $\mathcal{D}_A$ is defined identically with Eq. (13.3.6)

$$\Delta\theta = \frac{d}{\mathcal{D}_A}, \qquad (13.3.7)$$

where the functional form for $\mathcal{D}_A$ is determined by its relationship to the luminosity distance $\mathcal{D}_L$. To show how $\mathcal{D}_L$ and $\mathcal{D}_A$ are related we look at the flux F_θ from a distant source of extent d which subtends an angle $\Delta\theta$ on the sky (Wright)

$$F_\theta = \Delta\theta^2 \sigma T_o^4, \qquad (13.3.8)$$

where σ is the Stephan-Boltzmann constant and T_o is the observed temperature of the source. Equating fluxes from Eqs. (13.2.4) and (13.3.8), substituting for $\Delta\theta$ from Eq. (13.3.7) and substituting $L_{bol} = 4\pi d^2 \sigma T_e^4$ with T_e the source temperature we get

$$\frac{T_e^4}{\mathcal{D}_L^2} = \frac{T_o^4}{\mathcal{D}_A^2}. \qquad (13.3.9)$$

Since for a blackbody at temperature T the radiation with average wavelength λ has energy $hc/\lambda = kT$ where k is Boltzmann's constant and since the wavelength varies with redshift as $(1+z)$ this implies $T_o = T_e/(1+z)$. We assume that this holds even for a galaxy source which may not be a perfect blackbody. Then Eq. (13.3.9) simplifies to

$$\mathcal{D}_L = \mathcal{D}_A(1+z)^2. \qquad (13.3.10)$$

This relation is the same as that for FRW theory. Hence the angular size of a source galaxy in CGR can be found

$$\Delta\theta = \frac{d}{\mathcal{D}_A} = \frac{d(1+z)}{r}\sqrt{1 - \frac{t^2}{\tau^2}}, \qquad (13.3.11)$$

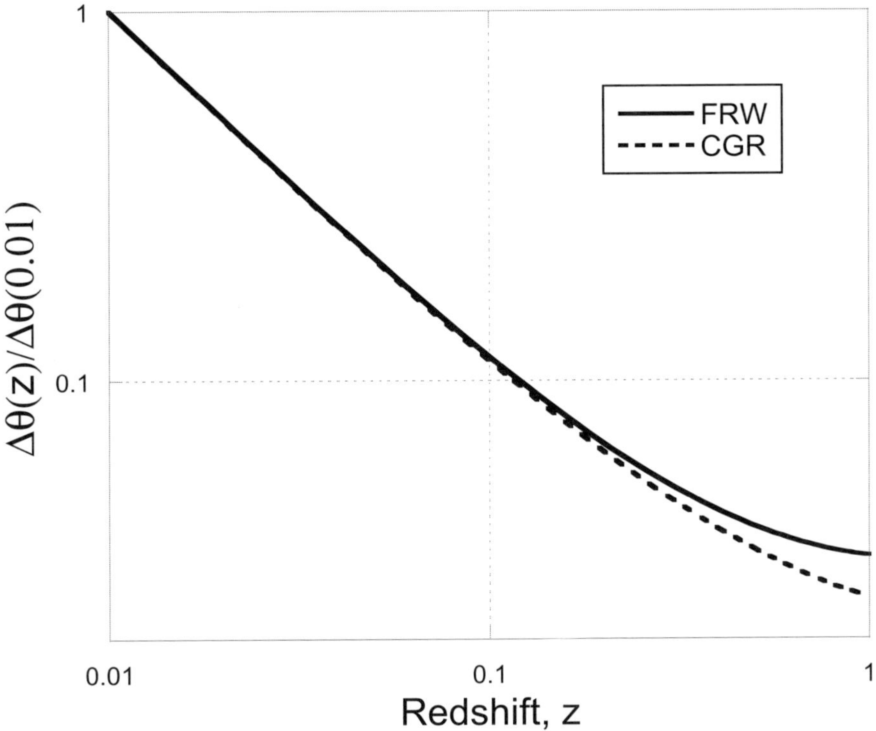

Fig. 13.3.1: Angular size shown as a function of redshift for both the FRW model (solid line) with a deceleration parameter $q_0 = 1/2$ or $\Omega_m = 1$ and the CGR model with $\Omega_m = 0.04$ (broken line). (Source: Hartnett & Oliveira, 2007a)

where Eqs. (13.2.5) and (13.3.10) have been used.

Substituting Eq. (13.3.4) in Eq. (13.3.11) produces gravitational effects on the angular size that can be called lensing. We have plotted in Figure 13.3.1 the dependence of angular size $\Delta\theta$ on redshift z for CGR using Eq. (13.3.4) in Eq. (13.3.11) but instead with the density function $\Omega(z)$ determined by Oliveira and Hartnett (2006). That density expression replaces the simple form in Eq. (13.3.4) and better characterizes the Universe at high redshifts.

In order to compare theories independently of the constants d, c and $\tau \approx H_0^{-1}$, we plot $\Delta\theta(z)/\Delta\theta(0.01)$ for both FRW and CGR theories. It is quite clear from Figure 13.3.1 that for redshifts $z \leq 0.2$ the two models are in reasonable agreement but in general $\Delta\theta_{FRW} \neq \Delta\theta_{CGR}$. For $z > 0.2$ the details depend heavily on the parameters of the models chosen.

13.4   Surface Brightness

To determine the effect of redshift variation on apparent surface brightness B of a source we need to calculate the observed flux $\mathcal{F}_{bol}$ per unit solid angle Θ,

$$B = \frac{\mathcal{F}_{bol}}{\Theta}, \qquad (13.4.1)$$

where for a source diameter of d and source angular distance $\mathcal{D}_A$, the solid angle Θ is given by

$$\Theta = \pi \left(\frac{d}{2\mathcal{D}_A} \right)^2. \qquad (13.4.2)$$

It follows from Eqs. (13.2.4), (13.3.10), (13.4.1) and (13.4.2) that the apparent surface brightness

$$B = \frac{\mathcal{F}_{bol}}{(\pi/4)\, d^2/\mathcal{D}_A^2} = \frac{L_{bol}}{\pi^2 d^2} \frac{1}{(1+z)^4}, \qquad (13.4.3)$$

which is the same as the usual FRW expression, the same $(1+z)^{-4}$ dependence Tolman (1930) produces using standard cosmology.

13.5   Matter Density of the Universe

As observers at this current epoch we observe sources in the Universe at different past epochs. Therefore to properly model the evolution of the Universe we need to know the matter density as a function of redshift $\Omega(z)$. In this chapter and Chapter 14 we offer different models for $\Omega(z)$. In this chapter we proceed as follows.

In terms of the spacevelocity expansion history, the Universe at time t has a total relativistic mass M and a total volume V. The expansion is assumed to be symmetric so that the volume V is spherical. The average matter density ρ is

$$\rho = \frac{M}{V}. \qquad (13.5.1)$$

The total relativistic mass of matter M in Cosmological Special Relativity (Carmeli 2002a) at cosmic time t is

$$M = \frac{M_0}{\sqrt{1 - t^2/\tau^2}}, \qquad (13.5.2)$$

where M_0 is the mass of the Universe at the present epoch $t = 0$.

The volume is taken to be that of a sphere

$$V = \frac{4\pi}{3} R^3,$$
(13.5.3)

where R is the radius of the portion of the Universe that just contains the mass M. In CGR, the distance r is measured from the observer at the present epoch to the source rather than the other way, e.g. as is done in the Friedmann theory of cosmology. We assume that higher density corresponds to higher velocity and that the volume decreases as velocity increases. The radius R of the Universe is therefore taken to be

$$R = c\tau - r,$$
(13.5.4)

where the redshift-distance relation r is given by Eq. (13.3.2).

R is defined this way so that for $v = 0$, $R(r = 0) = c\tau$ is the radius of the sphere of the Universe that just contains the mass of matter M_0. We define the average matter density parameter

$$\Omega = \frac{\rho}{\rho_c},$$
(13.5.5)

where $\rho_c = 3/8\pi G\tau^2$ is the critical density. An overall constraint is that, for $\Omega \geq 0$,

$$1 + \frac{(1 - \Omega)}{c^2\tau^2} r^2 > 0.$$
(13.5.6)

From Eqs. (13.5.1) - (13.5.5) the function for Ω is

$$\Omega = \frac{\Omega_m}{\sqrt{1 - \beta^2} \left[1 - \sinh\left(\beta\sqrt{1 - \Omega} \right) / \sqrt{1 - \Omega} \right]^3},$$
(13.5.7)

where

$$\Omega_m = \frac{\rho_m}{\rho_c},$$
(13.5.8)

$$\rho_m = \frac{M_0}{(4\pi/3)\,(c\tau)^3},$$
(13.5.9)

where ρ_m is the average matter density at the current epoch.

In the first order approximation where $\beta \ll 1$, $z \approx \beta$. Since $\sinh(x) \approx x$ for small x, Eq. (13.5.7) can be written

$$\Omega \approx \frac{\Omega_m \left(1 + (1/2)\beta^2 \right)}{(1 - \beta)^3} \approx \Omega_m \left(1 + z \right)^3.$$
(13.5.10)

In the Friedmann-Robertson-Walker cosmologies, the matter density parameter $\Omega = \Omega_m(1 + z)^3$ for all z in a dust dominated spatially flat Universe, but this is not the case in the present theory where the density varies more strongly than $(1 + z)^3$. This will produce significant results in the data analysis.

The derived relation Eq. (13.5.7) for Ω is transcendental. For fits to data it is more convenient to have a regular function, hence we use a second order approximation for Ω, which is briefly described in Section 13.11.

13.6 Expansion Transition Redshift z_t

In CGR the expansion has three basic phases: decelerating, coasting and finally accelerating, corresponding to density $\Omega > 1$, $\Omega = 1$, and $\Omega < 1$, respectively (Carmeli 2002a, pp. 125-127). What is the expected velocity and redshift of the transition from deceleration to acceleration? This phase shift occurs during the zero acceleration or coasting phase when $\Omega = 1$. Taking Eq. (13.5.7) to the limit $\Omega \to 1$, since $\sinh(x) \approx x$ for small x, yields

$$\lim_{\Omega \to 1} \Omega = 1 = \frac{\Omega_m}{\sqrt{1 - \beta_t^2}\,(1 - \beta_t)^3}, \qquad (13.6.1)$$

which simplifies to

$$(1 - \beta_t)^3 \sqrt{1 - \beta_t^2} = \Omega_m. \qquad (13.6.2)$$

Solving Eq. (13.6.2) for β_t, the predicted redshift z_t of the expansion transition is obtained from Eq. (13.3.3).

13.7 Comparison with High-Z Type Ia Supernovae Data

The redshift-distance relation in CGR is given by Eq. (13.3.2) and Ω is evaluated from Eq. (13.5.7).

In order to compare the redshift distance relation with the high redshift SNe Ia data from Riess *et al.* (2004) and Astier *et al.* (2006), the proper distance is converted to magnitude as follows,

$$m(z) = \mathcal{M} + 5\log\left[\mathcal{D}_L\left(z; \Omega\right)\right], \qquad (13.7.1)$$

where $\mathcal{D}_L$ is the dimensionless "Hubble constant free" luminosity distance (13.2.5). Refer Perlmutter *et al.* (1997) and Riess *et al.* (1998). Here

$$\mathcal{M} = 5\log\left(\frac{c\tau}{\text{Mpc}}\right) + 25 + M_B + a. \qquad (13.7.2)$$

The units of $c\tau$ are Mpc. The constant 25 results from the luminosity distance expressed in Mpc. However, $\mathcal{M}$ in Eq. (13.7.1) represents a scale offset for the distance modulus (m-M_B). It is sufficient to treat it as a single constant chosen from the fit. In practice we use a, a small free parameter, to optimize the fits. From Eq. (13.2.5), with $\beta = t/\tau$ the luminosity distance is given by

$$\mathcal{D}_L(z; \Omega_m) = \frac{r\,(1 + z)}{c\tau\sqrt{1 - \beta^2}}, \qquad (13.7.3)$$

using Eq. (13.3.2), hence r in units of $c\tau$. $\mathcal{D}_L$ is only a function of Ω_m and z.

The parameter $\mathcal{M}$ incorporates the various parameters that are independent of the redshift, z. The parameter M_B is the absolute magnitude of the supernova at the peak of its light-curve and the parameter a allows for any uncompensated extinction or offset in the mean of absolute magnitudes or an arbitrary zero point. The absolute magnitude then acts as a "standard candle" from which the luminosity and hence distance can be estimated.

The value of M_B need not be known, neither any other component in $\mathcal{M}$, as $\mathcal{M}$ has the effect of merely shifting the fit curve (13.7.3) along the magnitude axis.

However by choosing the value of the Hubble-Carmeli constant $\tau = 4.28 \times 10^{17}\, s = 13.58$ Gyr, which is the reciprocal of the chosen value of the Hubble constant in the gravity free limit $h = 72.17 \pm 0.84$ (statistical) km s^{-1}Mpc^{-1} (see Subsection 13.8.1) $\mathcal{M} = 43.09 + M_B + a$.

We use two SNe Ia data sets for the curved fitting analysis. The data are drawn from Table 5 of Riess *et al.* (2004), the Supernova Cosmology Project, and Tables 8 and 9 of Astier *et al.* (2006), the Supernova Legacy Survey (SNLS). Also we combined the data sets of Riess *et al.* (2004) and Astier *et al.* (2006) and found the best statistical fit to all those data.

This is shown in Figure 13.7.1 along with the curve where $\Omega_m = 0.263$, which is the value that Astier *et al.* (2006) quote for the average matter density at the current epoch. Lastly, we take the residuals between the combined the data set of Riess *et al.* (2004) and Astier *et al.* (2006) and the best fit curve of Figure 13.7.1. This is shown in Figure 13.7.2, along with the curve that represents $\Omega_m = 0.263$.

13.7.1 *Quality of curve fits*

In order to quantify the goodness of the least squares fitting we have used the χ^2 parameter which measures the goodness of the fit between the data and the theoretical curve assuming the two fit parameters a and Ω_m. Hence χ^2 is calculated from

$$\chi^2 = \sum_{i=1}^{N} \frac{1}{\sigma_i^2} \left[(m - M)(z)_i - (m - M)(z_{obs})_i \right]^2, \tag{13.7.4}$$

where N are the number of data; $(m - M)(z)$ are determined from Eq. (13.7.1) with fit values of a and Ω_m; $(m - M)(z_{obs})$ are the observed distance

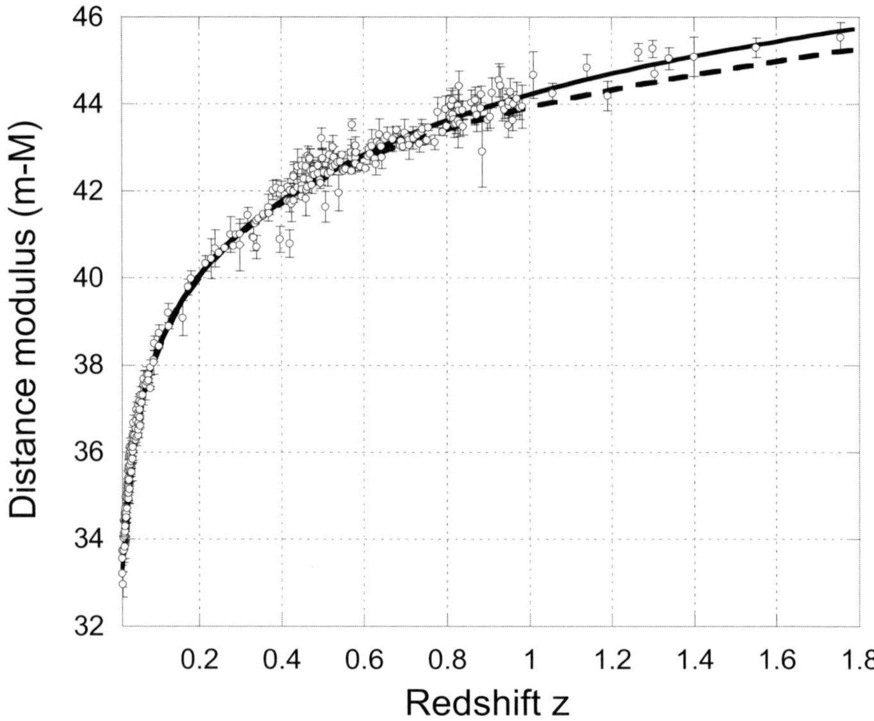

Fig. 13.7.1:   The combined data sets of Riess *et al.* (2004) and Astier *et al.* (2006). The solid line represents the statistically best fit curve with $a = 0.2284$ and $\Omega_m = 0.0401$ and the broken line represents the curve with $a = 0.2284$ and $\Omega_m = 0.263$. (Source: Oliveira & Hartnett, 2006)

modulus data at measured redshifts z_{obs}; σ_i are the published magnitude errors. The values of χ^2/N ($\approx \chi^2_{d.o.f}$) are shown in Table 13.1, calculated using published errors on the distance modulus data. In each case the best fit value of a is found for each value of Ω_m.

The published errors for Astier *et al.* (2006) data are quite small in relation to their deviation from the fitted curve, and are on average at least 0.14 magnitudes smaller than errors in the Riess *et al.* (2004) data. It appears that Astier *et al.* (2006) have underestimated the real errors in their data. Therefore to reduce the bias introduced by this data 0.14 magnitudes have been added to the error data of Astier *et al.* (2006) before being used to calculate χ^2/N values.

Table 13.1 lists the χ^2/N parameters determined for three values of Ω_m, as well as the best fit values of Ω_m determined using the Mathematica

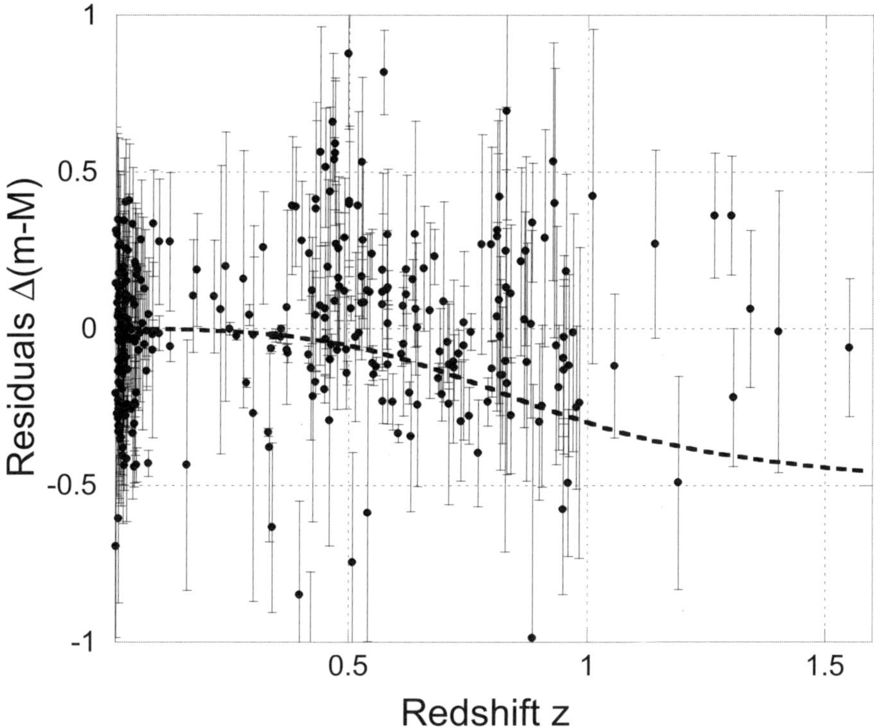

Fig. 13.7.2: Residuals vs redshift (on linear scale): the differences between the best fit curve with $\Omega_m = 0.0401$ and $a = 0.2284$ and the data of Figure 13.7.1. The mean of the residuals is 8.04×10^{-5} when all errors are assumed equal and -0.0769 when weighted by published errors. The broken line represents the curve where $\Omega_m = 0.263$. (Source: Oliveira & Hartnett, 2006)

software package. The latter are indicated by the word 'best' in the table. In the latter case the best fits are only statistically determined and hence also the standard error. In all instances the best fit value was determined for the parameter a.

From the combined data set of Riess *et al.* (2004) and Astier *et al.* (2006) the best statistical fit resulted in a value of $\Omega_m = 0.0401 \pm 0.0199$, which is consistent with the result obtained by averaging the values of Ω_m obtained from the individual data sets.

Looking at the χ^2/N values the minimum regions in each set overlap where $\Omega_m = 0.042$. This is then the region of the most probable value. This is consistent with a value of $\Omega_m = 0.0401 \pm 0.0199$ as determined from the combined data set shown in Figure 13.7.1. Therefore no exotic dark matter

Table 13.1:   Curve fit parameters.

Data set	N	a	$\sigma(stat)$	Ω_m	$\sigma(stat)$	χ^2/N	χ^2/N ($\sigma_i = 1$)
Riess *et al.*	185	0.257		0.021		1.34188	0.087620
		0.268		0.042		1.32523	0.086440
best		0.278	0.025	0.0631	0.0303	1.32152	0.086158
Astier *et al.*	117	0.158		0.021		0.71846	0.049175
best		0.161	0.043	0.0279	0.0430	0.72045	0.049428
		0.168		0.042		0.72567	0.050037
		0.177		0.063		0.73564	0.051121
Riess + Astier	302	0.219		0.021		1.19153	0.075039
best		0.228	0.018	0.0401	0.0199	1.18802	0.074726
		0.229		0.042		1.18815	0.074728
		0.239		0.063		1.19285	0.075010
		0.304		0.263		1.35816	0.086165

need be assumed as this value is within the limits of the locally measured baryonic matter budget $0.007 < \Omega_m < 0.041$ (Fukugita *et al.* 1998) where a Hubble constant of 70 km s^{-1} Mpc^{-1} was assumed.

Previously Hartnett (2006), which used some of the same data but with a different density model, the χ^2/N parameters appear to be much smaller and therefore represent better quality fits than in the former. However this is not actually the case, as a software algorithm was used in Hartnett (2006) that didn't properly calculate χ^2. The problem with the analysis was that the errors for all data were set to unity, that is, $\sigma_i = 1$. Oliveira and Hartnett (2006) calculated the correct χ^2/N parameters using Eq. (13.7.4) and published errors. Here we have done the same but increased the error by 0.14 magnitudes on the Astier *et al.* (2006) data for better comparison. Hence Table 13.1 shows the χ^2/N values where σ_i are forced to unity. The resulting $\chi^2/N(\sigma_i = 1)$ are extremely good even compared to the 185 data of Riess *et al.* fitted to in Fig. 1 of Hartnett (2006) where $\chi^2/N(\sigma_i = 1) = 0.2036$ was calculated.

The improvement has resulted from the additional factor $(1-t^2/\tau^2)^{-1/2}$ in the luminosity distance and a little from the refinement of the density model $\Omega(z)$. If we exclude the new density model and use $\Omega = \Omega_m(1 + z)^3$ where $\Omega_m = 0.04$ instead, we get $\chi^2/N(\sigma_i = 1) = 0.075986$ for the best fit to the combined data set requiring $a = 0.2152$. This indicates the improvement over Hartnett (2006) is more the result of the additional factor in the luminosity distance than the better density model.

Looking at the curve fits of Figure 13.7.1 where the distance modulus vs redshift curves with both $\Omega_m = 0.0401$ and $\Omega_m = 0.263$ are shown, it is quite clear that using the Carmeli theory a Universe with $\Omega_m = 0.263$ is ruled out and hence also the need for any dark matter. This is even more obvious from the residuals shown in Figure 13.7.2. There the fit with $\Omega_m = 0.0401$ is drawn along the $\Delta(m - M) = 0$ axis and the fit with $\Omega_m = 0.263$ is shown as a broken line. The highest redshift data clearly rules out such high matter density in the Universe.

The best fit result of this chapter, $\Omega_m = 0.0401 \pm 0.0199$, with a density function that is valid for all z over the range of observations, is also consistent with the result obtained by Hartnett (2006) $\Omega_m = 0.021 \pm 0.042$ but here the $1\,\sigma$ errors are significantly reduced.

With the best fit $\Omega_m = 0.0401$, the predicted expansion transition redshift from Eq. (13.6.2) is

$$z_t = 1.095\,^{+0.264}_{-0.155}. \tag{13.7.5}$$

This is about a factor of 2 greater than the fitted value reported by Riess *et al.* (2004) of $z_t = 0.46 \pm 0.13$, which was from a best fit to the differenced distance modulus data, a second order effect. They used a luminosity distance relation assuming a flat Euclidean space (i.e., $\Omega_{total} = 1$) and fit the difference data with the deceleration parameter $q(z) = (dH^{-1}(z)/dt) - 1$.

In the present theory, the transition redshift z_t is inherently where the density parameter $\Omega(z_t) = 1$. Thus, the transition is determined simultaneously with the initial fit of $\mathcal{D}_L$ to the data.

Moreover, Ω_m has been determined as a 'Hubble constant free' parameter because it comes from $\mathcal{D}_L(z; \Omega_m)$, which is evaluated from fits using Eq. (13.7.3). The latter is independent of the Hubble constant or more precisely in this theory τ the Hubble-Carmeli time constant. Therefore Ω_m should be compared with Ω_b and not with $\Omega_b h^2$, where h is the Hubble constant as a fraction of 100 km s^{-1} Mpc^{-1} and not to be confused with $h = 1/\tau$ used in CGR.

Nevertheless the value of $\Omega_b h^2 = 0.024$ from Spergel (2006) and $h = 0.7217$ (assuming a value of $\tau^{-1} = 72.17$ km s^{-1}Mpc^{-1}) implies $\Omega_b = 0.043$, which is in good agreement with the results of this work. Yet caution must be advised as the problem of the analysis of the WMAP data has not yet been attempted within the framework of CGR.

13.8   Values of Some Key Parameters

13.8.1   *Hubble constant*

Using the small redshift limit of Eq. (13.3.2) and the Hubble law at small redshift ($v = H_0 r$) it has been shown (Carmeli 2002a, pp. 170-172) that the Hubble parameter H_0 varies with redshift. If it applies at the low redshift limit it follows from the theory that at high redshift we can write

$$H_0 = h \frac{\beta\sqrt{1-\Omega}}{\sinh(\beta\sqrt{1-\Omega})}. \tag{13.8.1}$$

Therefore H_0 in this model is redshift dependent, not constant and $H_0 \leq h$. Only $h = \tau^{-1}$ is truly independent of redshift and constant. The condition where $H_0 = h$ only occurs at $z = 0$ and where $\Omega \to 0$.

By plotting H_0 values determined as a function of redshift, using Eq. (13.8.1), it is possible to get an independent determination of h, albeit the noise in the data is very large. This is shown in Figure 13.8.1 with values calculated by two methods with the exception of one point at $z = 0.333$. See figure caption for details. The data, even though very scattered, do indicate a trending down of H_0 with redshift.

Separate curve fits from Eq. (13.8.1), with h as a free parameter, have been applied to the two data sets, Tully-Fisher (TF) (the solid line) and SNe type Ia (the broken line) measurements. The former resulted in $h = 72.47 \pm 1.95$ (statistical) ± 13.24 (rms) km s^{-1}Mpc^{-1} and from the latter $h = 72.17 \pm 0.84$ (statistical) ± 1.64 (rms) km s^{-1}Mpc^{-1}. The rms errors are those derived from the published errors, the statistical errors are those due to the fit to the data alone. The SNe Ia determined value is more tightly constrained but falls within the TF determined value.

13.8.2   *Mass of the Universe*

It is easily shown from Eqs. (13.5.8) and (13.5.9) that

$$\Omega_m = \frac{R_s}{R_0}, \tag{13.8.2}$$

where $R_s = 2 G M_0 / c^2$ is the Schwarzschild radius if the present Universe rest mass M_0 is imagined to be concentrated at a point, and $R_0 = c\tau$ is the present radius of the Universe. From this we get the present Universe rest mass

$$M_0 = \Omega_m \frac{c^3 \tau}{2 G}, \tag{13.8.3}$$

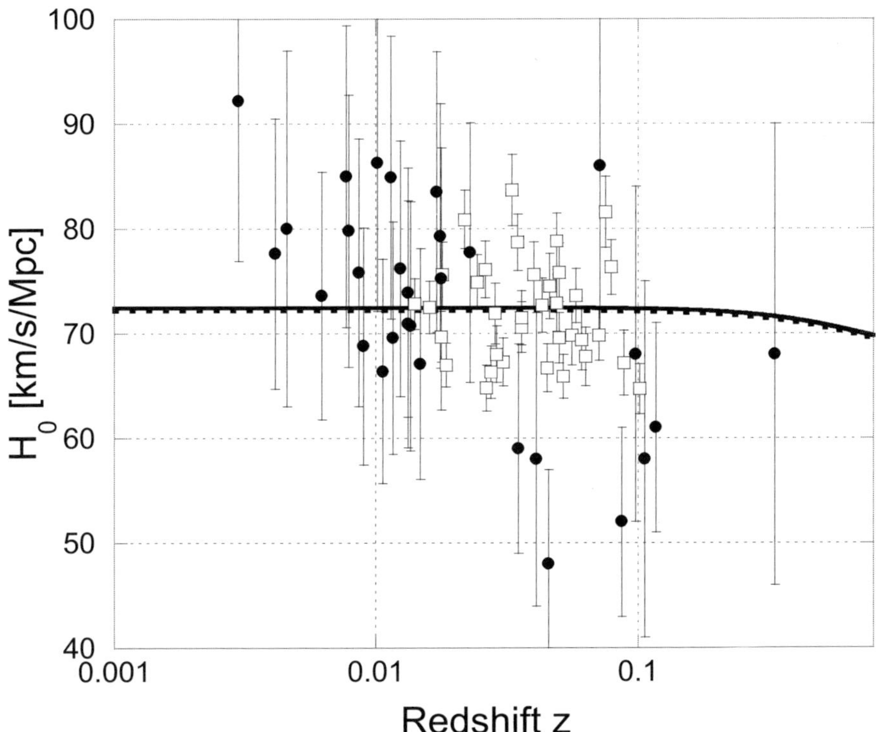

Fig. 13.8.1: Hubble constant H_0 as a function of redshift, z. The filled circles are determined from Tully-Fisher measurements taken from Freedman *et al.*, 1994, Table 5 of Tutui *et al.*, 2001 and Table 7 of Freedman *et al.*, 2001, except the point at $z = 0.333$ is from Sunyaev-Zel'dovich effect taken from Fig. 4 of Tutui *et al.*, 2001. The open squares are determined from the SN Ia measurements and taken from Table 6 of Freedman *et al.*, 2001 and Table 5 of Riess *et al.*, 2004. The errors are those quoted in the sources from which the data was taken. (Source: Oliveira & Hartnett, 2006)

which, with $\Omega_m = 0.0401 \pm 0.0199$ gives

$$M_0 = (1.74 \pm 0.86) \times 10^{21} M_\odot \,. \tag{13.8.4}$$

Likewise, the average matter density (13.5.8)

$$\rho_m = \Omega_m \rho_c = (3.92 \pm 1.94) \times 10^{-31} \mathrm{gm\,cm}^{-3} \,. \tag{13.8.5}$$

13.8.3   *Time of transition from deceleration to acceleration*

From Carmeli's cosmological special relativity (Carmeli *et al.*, 2006) we get a relation for the cosmic time in terms of the redshift. In particular, in terms of z_t we have for the cosmic time t_t of the expansion transition from the present

$$t_t = \tau \, \frac{(1 + z_t)^2 - 1}{(1 + z_t)^2 + 1}. \qquad (13.8.6)$$

For the above value of z_t and for the age of Universe $\tau = 13.58\,\mathrm{Gyr}$ we have

$$t_t = 8.54^{+0.903}_{-0.662}\,\mathrm{Gyr}. \qquad (13.8.7)$$

Since the Big Bang ($t^* = 0$), the transition cosmic time is $t_t^* = \tau - t_t$,

$$t_t^* = 5.04^{+0.662}_{-0.903}\,\mathrm{Gyr}. \qquad (13.8.8)$$

In Figure 13.8.2 is a plot of the density for $\Omega_m = 0.04$. More than $8.54\,\mathrm{Gyr}$ ago the density was higher than the critical value ($\Omega > 1$). Since the transition the density has become less than critical ($\Omega < 1$). The

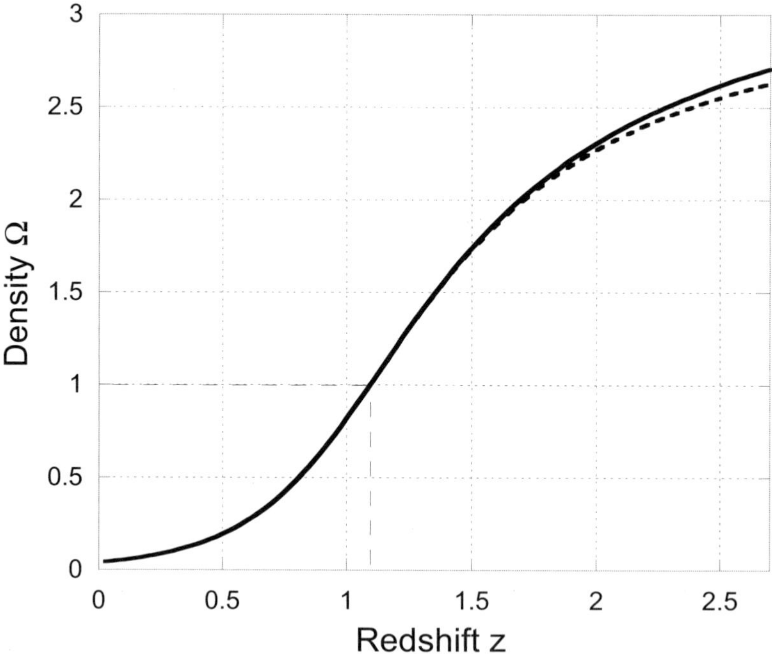

Fig. 13.8.2:   Density model shown as function of redshift for both approximated (broken line) and exact (solid line) with the same value of $\Omega_m = 0.04$. The transition redshift $z_t = 1.095$ where $\Omega = 1$ is indicated by the dashed lines. (Source: Oliveira & Hartnett, 2006)

fit to the SNe Ia data was accomplished without the need for any dark energy, usually associated with the cosmological constant. In CGR there is no cosmological constant although a value for it may be obtained by a comparison study (Carmeli 2002a, pp. 170-172; Hartnett 2006).

13.9 Conclusion

The surface brightness is the same as in standard cosmology, though angular size is smaller by a factor of $(1 - t^2/\tau^2)^{1/2}$.

The analysis in this chapter has shown that the most probable value of the local density of the Universe is $\Omega_m = 0.0401 \pm 0.0199$ the best fit from a combined data set of two totaling 302 data. The fits used a density function with limited range and validity and did not take into account the published errors on the individual magnitude data. The fits to the data are consistent over the entire range of the available redshift data, from $0.1 < z < 2.0$, a result of the more accurate relation for Ω, as well as the proper accounting of the increase in the source luminosity due to the factor $\left(1 - \beta^2\right)^{-1/2}$.

Since Ω_m is within the baryonic matter density budget, there is no need for any dark matter to account for the SNe Ia redshift magnitude data. Furthermore, since the predicted transition redshift $z_t = 1.095\,^{+0.264}_{-0.155}$ is well within the redshift range of the data, the expansion rate evolution from deceleration to acceleration, which occurred about 8.54 Gyr ago, is explained without the need for any dark energy.

The density $\Omega_m < 1$ and the determination of the transition redshift z_t within the data support the conclusion that the expansion is now accelerating and that the Universe is, and will remain open.

13.10 Approximation of Ω

The form for Ω in Eq. (13.5.7) is transcendental, which is not convenient for fitting. A second order approximation can be made by taking $\sinh(x) \approx x + x^3/3!$. With this approximation Eq. (13.5.7) becomes

$$\Omega \approx \Omega_2 = \frac{\Omega_m/\sqrt{1 - \beta^2}}{\left\{1 - \left[\beta\sqrt{1 - \Omega_2} + \beta^3\left(\sqrt{1 - \Omega_2}\right)^3/3!\right]/\sqrt{1 - \Omega_2}\right\}^3}, \quad (13.10.1)$$

which simplifies to

$$\Omega_2 \left[1 - \beta - \frac{\beta^3}{3!} + \frac{\beta^3}{3!}\Omega_2 \right]^3 - \left(\Omega_m / \sqrt{1 - \beta^2} \right) = 0. \qquad (13.10.2)$$

This is a quartic equation in Ω_2 and can be solved for Ω_2 as a function of β by standard methods. Ω_2 is shown in Figure 13.8.2 as the broken line where a matter density $\Omega_m = 0.04$ was assumed. It is compared with Ω given by the exact form (13.5.7).

13.11   Suggested References

P. Astier *et al.*, The Supernova Legacy Survey: Measurement of Ω_M, Ω_Λ and w from the first year data set, *Astron. Astrophys.* **447**, 31-48 (2006); astro-ph/0510447 .

M. Carmeli, Cosmological relativity: Determining the Universe by the cosmological redshift as infinite and curved, *Int. J. Theor. Phys.* **40**, 1871-1874 (2001).

M. Carmeli, *Cosmological Special Relativity,* Second Edition (World Scientific, Singapore, 2002a).

M. Carmeli, Accelerating Universe: Theory versus Experiment, astro-ph/0205396 (2002b).

M. Carmeli, J.G. Hartnett and F.J. Oliveira, The cosmic time in terms of the redshift, *Found. Phys. Lett.* **19**(3), 277-283 (2006); gr-qc/0506079.

W.L. Freedman, *et al.*, Distance to the Virgo cluster galaxy M100 from Hubble Space Telescope observations of Cepheids, *Nature* **371**, 757-762 (1994).

W.L. Freedman *et al.*, Final results from the Hubble Space Telescope Key Project to measure the Hubble constant, *Astrophys. J.* **553**, 47-72 (2001).

M. Fukugita, C.J. Hogan and P.J.E. Peebles, The cosmic baryon budget, *Astrophys. J.* **503**, 518-530 (1998).

R.A. Knop *et al.*, New constraints on Ω_M, Ω_Λ and w from an independent set of 11 high-redshift supernovae observed with the Hubble Space Telescope, *Astrophys. J.* **598**: 102-137 (2003).

J.G. Hartnett, The distance modulus determined from Carmeli's cosmology fits the accelerating Universe data of the high-redshift type Ia supernovae without dark matter, *Found. Phys.* **36**(6), 839-861 (2006); astro-ph/0501526.

J.G. Hartnett and F.J. Oliveira, Luminosity distance, angular size and surface brightness in cosmological general relativity, *Found. Phys.* **37**(3): 446-454 (2007a); astro-ph/0603500.

J.G. Hartnett and F.J. Oliveira, Testing cosmological general relativity against high redshift observations, preprint (2007b); astro-ph/0603500.

L.M. Krauss, The end of the age problem, and the case for a cosmological constant revisited, *Astrophys. J.* **501**, 461-466 (1998).

J.V. Narlikar, *An Introduction to Cosmology,* 3rd Ed., (Cambridge University Press, Cambridge, 2002).

F.J. Oliveira and J.G. Hartnett, Carmeli's cosmology fits data for an accelerating and decelerating Universe without dark matter or dark energy, *Found. Phys. Lett.* **19**(6), 519-535 (2006); astro-ph/0603500.

S. Perlmutter *et al.*, Measurements of the cosmological parameters Ω and Λ from the first seven supernovae at $z > 0.35$, *Astrophys. J.* **483**, 565-581 (1997).

A.G. Riess, A.V. Filippenko, P. Challis, A. Clocchiatti and A. Diercks, Observational evidence from supernovae for an accelerating Universe and a cosmological constant, *Astron. J.* **116**, 1009-1038 (1998).

D.N. Spergel *et al.*, Wilkinson Microwave Anisotropy Probe (WMAP) three year results: Implications for cosmology, astro-ph/0603449.

A.G. Riess *et al.*, Type Ia supernovae discoveries at $z > 1$ from the Hubble Space Telescope: Evidence for past deceleration and constraints on dark energy evolution, *Astrophys. J.* **607**, 665-687 (2004).

R.C. Tolman, On the estimation of distances in a curved Universe with non-static line element, *Proc. Nat. Acad. Sci.* **16**, 515-520 (1930).

Y. Tutui *et al.*, *PASJ* **53**, 701, (2001); astro-ph/0108462.

E.L. Wright, Homogeneity and isotropy; many distances; scale factor, http://www.astro.ucla.edu/~wright/cosmo_02.htm

Chapter 14

Extending the Hubble Diagram to Higher Redshifts in CGR

John Hartnett

The redshift-distance modulus relation, the Hubble Diagram, derived from Cosmological General Relativity are extended to arbitrarily large redshifts. Numerical methods are employed and a density function is found that results in a valid solution of the field equations at all redshifts. The extension is compared to 302 type Ia supernova data as well as to 69 Gamma-ray burst data. The latter however do not truly represent a 'standard candle' as the derived distance modulii are not independent of the cosmology used. Nevertheless the analysis shows a good fit can be achieved without the need to assume the existence of dark matter.

The Carmelian theory is also shown to describe a Universe that is always spatially flat. This results from the underlying assumption of the energy density of a cosmological constant $\Omega_\Lambda = 1$, the result of vacuum energy. The curvature of the Universe is described by a spacevelocity metric where the energy content of the curvature at any epoch is $\Omega_K = \Omega_\Lambda - \Omega = 1 - \Omega$, where Ω is the matter density of the Universe. Hence the total density is always $\Omega_K + \Omega = 1$. The following is based on Hartnett 2008.

14.1   Introduction

Carmeli's cosmology, also referred to as cosmological general relativity (CGR), is a space-velocity theory of the expanding Universe. It is a description of the Universe at a particular fixed epoch of cosmic time t. In CGR time is measured from the present back toward the beginning. At the present epoch $t = 0$, the Universe can be described by its *spacevelocity* coordinates (v, r, θ, ϕ). It is based on the Hubble law which says that the observed redshift z in the light emitted from a distant source of atoms is directly proportional to the distance r to the source, viz. $v = H_0 r$, where H_0 is Hubble's constant. CGR incorporates this basic law, where $H_0^{-1} \approx \tau = 13.56 \pm 0.48$ Gyr, a universal time constant, into a general 4D Riemannian geometrical theory satisfying the Einstein field equations (see Chapter 4).

14.1.1 *Spacevelocity equations*

The line element in CGR

$$ds^2 = \tau^2 dv^2 - e^\mu dr^2 - R^2(d\theta^2 + \sin^2\theta d\phi^2), \qquad (14.1.1)$$

represents a spherically symmetrical isotropic Universe, that is not necessarily homogeneous.

It is fundamental to the theory that $ds = 0$. In the case of Cosmological Special Relativity (see Chapter 2), which is very useful pedagogically, we can write the line element as

$$ds^2 = \tau^2 dv^2 - dr^2, \qquad (14.1.2)$$

ignoring θ and ϕ coordinates for the moment. By equating $ds = 0$ it follows from Eq. (14.1.2) that $\tau dv = dr$ assuming the positive sign for an expanding Universe. This is then the Hubble law in the small v limit. Hence, in general, this theory requires that $ds = 0$.

Using spherical coordinates (r, θ, ϕ) and the isotropy condition $d\theta = d\phi = 0$ in Eq. (14.1.1) then dr represents the radial coordinate distance to the source and it follows from Eq. (14.1.1) that

$$\tau^2 dv^2 - e^\mu dr^2 = 0, \qquad (14.1.3)$$

where μ is a function of v and r alone. This results in

$$\frac{dr}{dv} = \tau e^{-\mu/2}, \qquad (14.1.4)$$

where the positive sign has been chosen for an expanding Universe.

Carmeli found a solution, Eq. (4.3.16), to his field equations modified from Einstein's, of the form

$$e^\mu = \frac{R'^2}{1 + f(r)}, \qquad (14.1.5)$$

with $R = r$ and hence $R' = 1$, which must be positive. From the field equations and Eq. (14.1.5) we get a differential equation

$$f' + \frac{f}{r} = -\kappa\tau^2 \rho_{eff} r, \qquad (14.1.6)$$

where $f(r)$ is a function of r and satisfies the condition $f(r) + 1 > 0$. The prime is the derivative with respect to r. Here $\kappa = 8\pi G/c^2\tau^2$ and $\rho_{eff} = \rho - \rho_c$ where ρ is the averaged matter density of the Universe and $\rho_c = 3/8\pi G\tau^2$ is the critical density. Finally the solution to the inhomogeneous equation (14.1.6) was found to be

$$f(r) = -\frac{\kappa}{3}\tau^2\rho_{eff}r^2 = (1 - \Omega)\frac{r^2}{c^2\tau^2}. \qquad (14.1.7)$$

Therefore this requires that

$$1 + (1 - \Omega)\frac{r^2}{c^2\tau^2} > 0. \qquad (14.1.8)$$

In the above Carmelian theory it was initially assumed that the Universe has expanded over time and at any given epoch it has an averaged density ρ, and hence ρ_{eff}. The solution of the field equations has been sought on this basis. However because the Carmelian metric is solved in an instant of time (on a cosmological scale) any time dependence is neglected. In fact, the general time dependent solution has not yet been found. But since we observe the expanding Universe with the coordinates of Hubble at each epoch (or redshift z) we see the Universe with a different density $\rho(z)$ and an effective density $\rho_{eff}(z)$. In Section 14.2.1 this fact is taken into account to extend the validity of the solution.

14.2   Comparison with Observation

In order to compare CGR theory with observations one requires a relationship that gives proper distance as a function of redshift. From Eqs. (14.1.4), (14.1.5) and (14.1.7) we get the differential equation

$$\frac{dr}{dv} = \tau\sqrt{1 + \left(\frac{1 - \Omega}{c^2\tau^2}\right)r^2}, \qquad (14.2.1)$$

where $\Omega = \rho/\rho_c$. Equation (14.2.1) relates the velocity of the receding sources to their distance. Carmeli integrated Eq. (14.2.1) and found the redshift-distance relation

$$\frac{r}{c\tau} = \frac{\sinh\left(\beta\sqrt{1 - \Omega}\right)}{\sqrt{1 - \Omega}}, \qquad (14.2.2)$$

where,

$$\beta = \frac{v}{c} = \frac{(1 + z)^2 - 1}{(1 + z)^2 + 1}, \qquad (14.2.3)$$

and Ω the average matter density of the Universe, a function of redshift z, must be found.

This equation was used in both Hartnett 2006a, which forms the basis of the work presented here, and Oliveira and Hartnett 2006 where comparisons were made to the high redshift SNe Ia data from Riess *et al.* (2004), Astier *et al.* (2006) and Knop *et al.* (2003). The proper distance is converted to magnitude as follows.

$$m(z) = \mathcal{M} + 5 \log\left[\mathcal{D}_L(z; \Omega)\right], \qquad (14.2.4)$$

where $\mathcal{D}_L$ is the dimensionless "Hubble constant free" luminosity distance. (See Section 13.7.) Here

$$\mathcal{M} = 5 \log\left(\frac{c\tau}{\mathrm{Mpc}}\right) + 25 + M_B + a. \qquad (14.2.5)$$

The units of $c\tau$ are Mpc. The constant 25 results from the luminosity distance expressed in Mpc. However, $\mathcal{M}$ represents a scale offset for the distance modulus (m-M_B). It is sufficient to treat it as a single constant chosen from the fit. In practice we use a, a small free parameter, to optimize the fits.

In CGR the luminosity distance was found to be slightly different from the expression used in the FRW theory and hence $\mathcal{D}_L$ is given by

$$\mathcal{D}_L(z; \Omega) = \frac{r}{c\tau}\frac{1+z}{\sqrt{1-\beta^2}}, \qquad (14.2.6)$$

using Eq. (14.2.2), which is a function of Ω and z.

The parameter $\mathcal{M}$ incorporates the various parameters that are independent of the redshift z. The parameter M_B is the absolute magnitude of the supernova at the peak of its light-curve and the parameter a allows for any uncompensated extinction or offset in the mean of absolute magnitudes or an arbitrary zero point. The absolute magnitude then acts as a "standard candle" from which the luminosity and hence distance can be estimated.

The value of M_B need not be known, neither any other component in $\mathcal{M}$, as $\mathcal{M}$ has the effect of merely shifting the fit curve (14.2.6) along the magnitude axis. However by choosing the value of the Hubble-Carmeli constant $\tau = 4.28 \times 10^{17}\, s = 13.58$ Gyr, which is the reciprocal of the chosen value of the Hubble constant in the gravity-free limit $h = 72.17 \pm 0.84$ (statistical) km s^{-1} Mpc^{-1}, $\mathcal{M} = 43.09 + M_B + a$.

14.2.1 *Extended redshift range*

The proper distance from Eq. (14.2.2) however has limited application because it has essentially been found by integrating Eq. (14.2.1) assuming constant density then using Eq. (14.2.3). The result being only valid over a limited redshift range. In order to improve on this, because the observables are distance (magnitude) and redshift not velocity, the differential equation (14.2.1) must first be converted to dr/dz using the chain rule as follows.

$$\frac{1}{\tau^2}\left(\frac{dr}{dz}\frac{dz}{dv}\right)^2 = 1 + \frac{1 - \Omega(z)}{c^2\tau^2}r^2, \tag{14.2.7}$$

where it must be remembered that the matter density Ω is also a function of redshift. The derivative dv/dz is then calculated from Eq. (14.2.3) and substituted into Eq. (14.2.7), which becomes

$$\frac{1}{c^2\tau^2}\left(\frac{dr}{dz}\right)^2 = \frac{16(1+z)^2}{\left(1 + (1+z)^2\right)^4}\left(1 + (1 - \Omega(z))\frac{r^2}{c^2\tau^2}\right). \tag{14.2.8}$$

This differential equation is then solved for $r/c\tau$, but because it involves mixed terms on the rhs, it must be numerically solved. But we can employ a 'bootstrapping' technique. By substituting the limited redshift solution from Eq. (14.2.2) into the right hand side of Eq. (14.2.8) and assuming the averaged matter density in the Universe

$$\Omega(z) = \Omega_m (1 + z)^3, \tag{14.2.9}$$

Eq. (14.2.8) was numerically solved for $r/c\tau$. The result is shown in curve 1 of Figure 14.2.1. This is compared with $r/c\tau$ resulting from the initial equation (14.2.2) represented by curve 2.

Then $r/c\tau$ (of curve 1) was taken as input on the right hand side of Eq. (14.2.8) and iterated again to numerically solve for $r/c\tau$. However because of the requirement (14.1.8) on the solution from the field equations the result is only valid to about $z = 2.5$ with the assumed density dependence (14.2.9) on the present epoch density $\Omega_m = 0.04$. Dimensionally the matter density depends on the length cubed if we assume constant mass, but the precise density dependence is unknown at high redshifts. Therefore curve 1 of Figure 14.2.1 plotted beyond $z = 2.5$ can only be considered approximate.

Now in CGR there is the scale radius

$$R_0 = \frac{c\tau}{\sqrt{|1 - \Omega|}}. \tag{14.2.10}$$

From Eq. (14.2.10) and the condition (14.1.8) the range of validity of the solution (14.2.1) is $0 \leq r/c\tau < 1 \leq R_0$ if $\Omega \leq 1$ and $0 \leq r/c\tau < R_0$ if

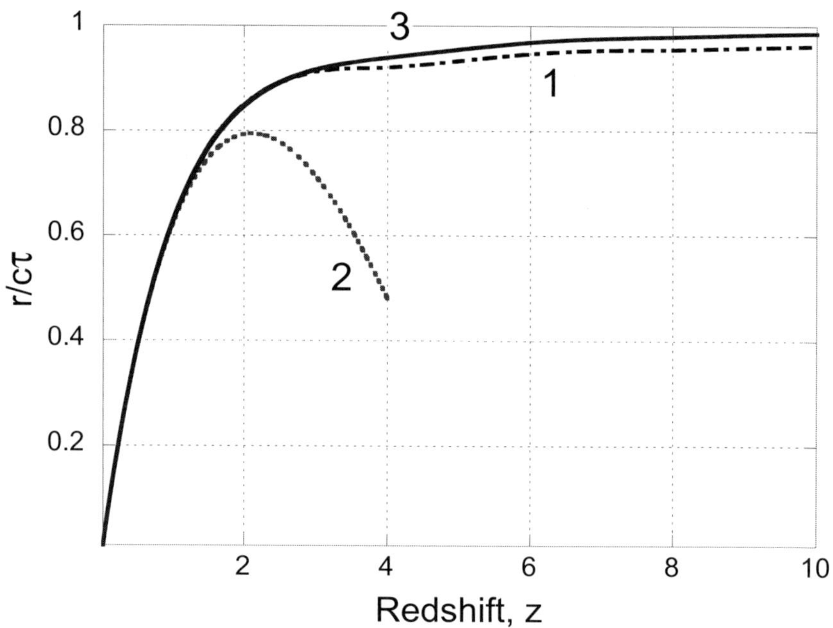

Fig. 14.2.1: Redshift-distance relation: Curve 1 represents $r/c\tau$ from the solution of Eq. (14.2.8) with $\Omega(z) = \Omega_m(1 + z)^3$, curve 2 represents $r/c\tau$ from Eq. (14.2.2) with $\Omega(z) = \Omega_m(1 + z)^3$ and curve 3 represents $r/c\tau$ from the solution of Eq. (14.2.8) but with the density taken from curve 1 of Figure 14.2.2. (Source: Hartnett, 2008)

$\Omega > 1$. This means the solution is valid for all values of the matter density $\Omega(z) < 2$.

Now if we again use the solution (curve 1 of Figure 14.2.1) and solve Eq. (14.2.8) for matter density Ω, it results in a density function such that Eq. (14.1.8) is always true. The density function is shown as curve 1 in Figure 14.2.2 and is compared with the initial density function (14.2.9) shown as curve 2. Finally using this as the density function in Eq. (14.2.8) and numerically solving for $r/c\tau$ results in curve 3 in Figure 14.2.1.

The new density function has some unexpected features, but it produces a smooth monotonically increasing function of distance on redshift as expected. See Figure 14.2.3 where it has been plotted to $z = 100$ for the case where $\Omega_m = 0.04$. The fluctuating part of $\Omega(z)$ for $z > 4$ is approximately

$$1 + 1.1 \cos^2\left[\beta\sqrt{0.04\,(1 + z)^3 - 1} + 1.6\right], \tag{14.2.11}$$

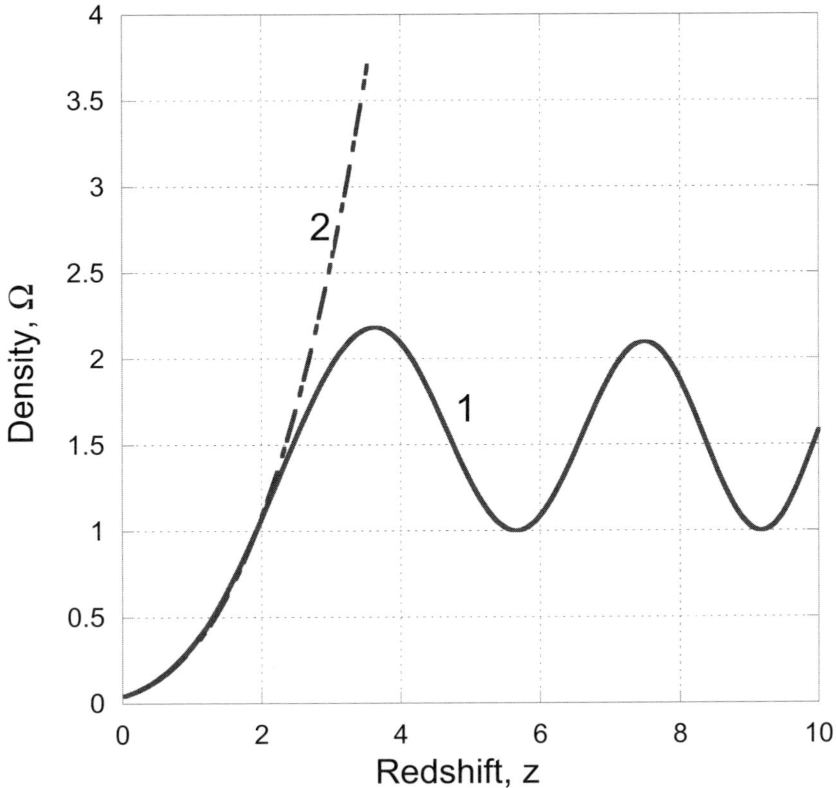

Fig. 14.2.2: Density as a function of redshift: Curve 1 represents density derived from Eq. (14.2.8) and curve 2 represents the density function $\Omega(z) = \Omega_m(1 + z)^3$ with $\Omega_m = 0.04$. (Source: Hartnett, 2008)

where β is determined from Eq. (14.2.3). The function (14.2.11) has a decreasing period as a function of z, a minimum at $\Omega = 1$ and a maximum slightly greater than $\Omega = 2$. Any density function that remains within the range of this function for $z > 4$, even a constant density, will yield a smooth monotonic redshift-distance relation. However the oscillating density as a function of redshift could represent episodic creation from the vacuum as the Universe expands. Particle production at those epochs is a byproduct of the expansion process. Gemelli (see Chapter 8) found from a hydrodynamic solution of the 5D problem that particle production must occur from the expansion provided the Universe is not isentropic.

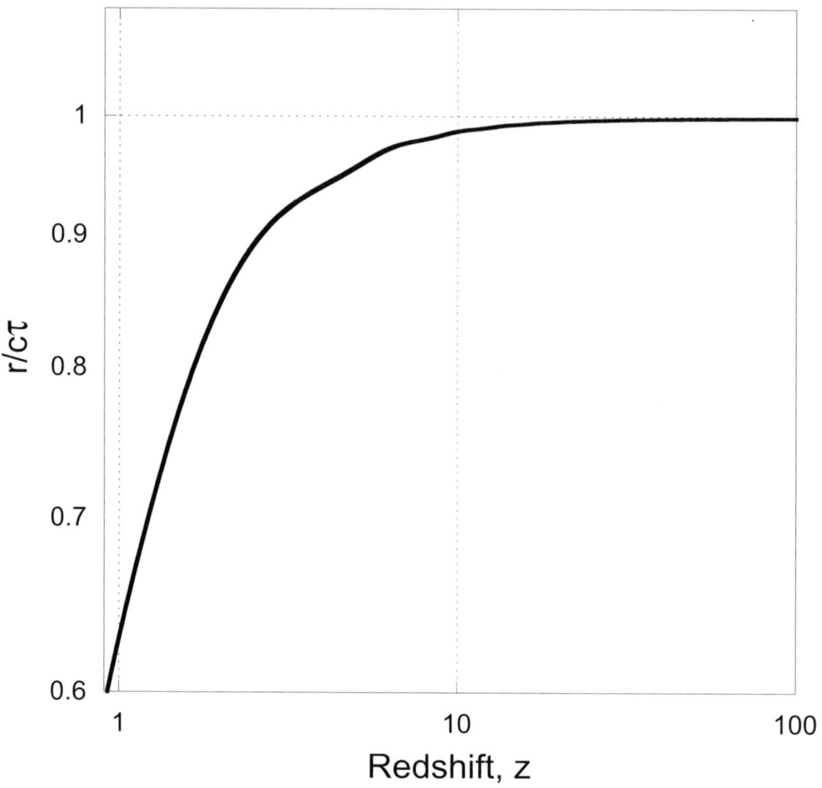

Fig. 14.2.3:   Redshift-distance relation to $z = 100$ using density function of curve 1 from Figure 14.2.2. (Source: Hartnett, 2008)

14.2.2   *Quality of curve fits*

The initial equation (14.2.2) has been curve fitted to SNe Ia data (Oliveira and Hartnett 2006) as described in Section 13.7 and excellent fits resulted without any dark matter component. The best statistical fit resulted in a value of $\Omega_m = 0.0401 \pm 0.0199$, which is consistent with the observed baryonic matter density (Fukugita *et al.*, 1998). The fit was accomplished with the combined Gold and Silver SNe Ia data sets of Riess *et al.* (2004) with that of Astier *et al.* (2006) that extended to $z = 1.75$, the current limit of observational data. Inspection of the $r/c\tau$ curve (curves 1 or 3 shown in Figure 14.2.1) and the initial curve (curve 2) for the region $1.5 < z < 1.75$ indicates that only the last two SNe Ia data would be sensitive to the

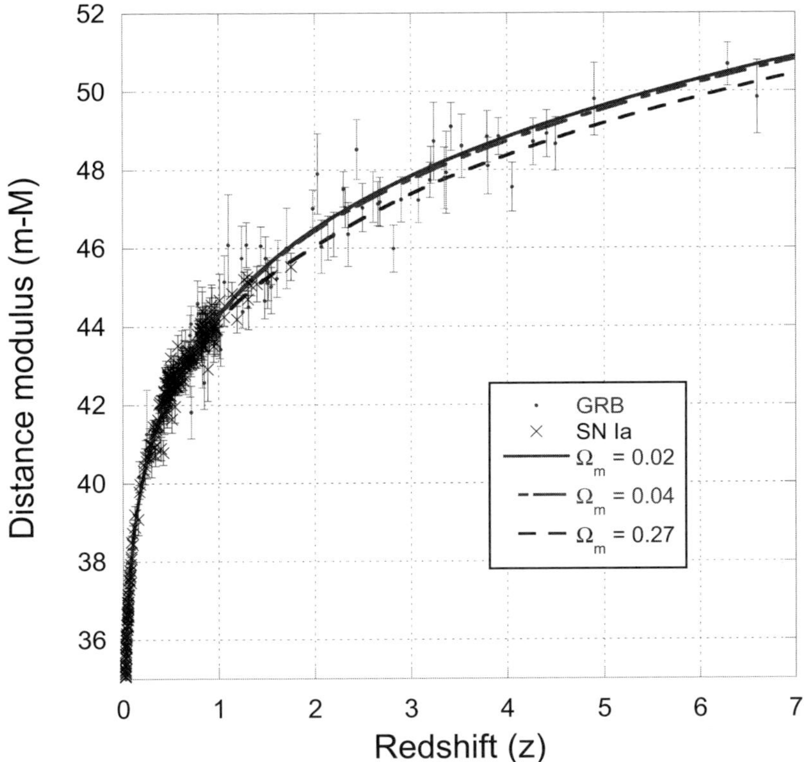

Fig. 14.2.4:   Distance modulus data from the combined SN Ia data set of Riess *et al.* and Astier *et al.* (crosses) and the GRB data (dots) derived from Schaefer (2007) with curve fits using the extended equation (14.2.8). The top solid curve represents the distance modulus where $\Omega_m = 0.02$, the broken curve below it is where $\Omega_m = 0.04$ and the dashed curve below the latter is where $\Omega_m = 0.27$. (Source: Hartnett, 2008)

numerically determined extension here. See crosses in Figure 14.2.4.

Using $r/c\tau$ from the extended equation (14.2.8) we have curve fitted to the data set of Riess *et al.* and also to combined data set of Riess *et al.* and Astier *et al.* However before using Astier *et al.* data we added 0.14 magnitudes to their quoted errors to align their errors with those of Riess *et al.* Otherwise Astier *et al.* seem to have underestimated their magnitude errors, which unfairly bias the χ^2 calculations. The resulting χ^2s for the model here are shown in Table 14.1.

Clearly more SNe Ia data, at higher redshifts, are needed to test this prediction. Consequently GRBs have been used to get distance modulus

Table 14.1:   Curve fit χ^2 parameter.

Data set(s)	Ω_m	χ^2/N
SNe Ia	0.00	1.4120
Riess et al. Gold & Silver	0.02	1.4047
$N = 185$	0.04	1.3987
$a = 0.2400$	0.06	1.3941
(a from $z < 0.23$)	0.10	1.3888
	0.27	1.4258
SNe Ia	0.00	1.2366
Riess et al. + Astier et al. SNLS	0.02	1.2341
$N = 302$	0.04	1.2326
$a = 0.1894$	0.06	1.2321
(a from $z < 0.24$)	0.10	1.2340
	0.27	1.2871
SNe Ia + GRBs	0.00	1.2191
Riess + Astier + Schaefer	0.02	1.2005
$N = 371$	0.04	1.1902
$a = 0.2113$	0.06	1.1860
(a from $z < 2$)	0.10	1.1870
	0.27	1.2538

and redshift as independent parameters (from Schaefer 2007) and are shown as dots in Figure 14.2.4. However the magnitude calculation is not really independent of the chosen cosmology as explained in the latter reference. Nevertheless using the data from columns 2 and 8 of Table 6 (Schaefer 2007), and making the necessary changes to convert the magnitudes from the FRW Concordance Model to that of the Carmeli model, new distance modulii are obtained.

The following corrections were applied to convert the distance modulus of Schaefer (2007) to that of the extended theory of CGR developed here. Since the luminosity distance in CGR is different from that in FRW theory, by the factor $(1 - \beta^2)^{-1/2}$, its effect must be added to the distance modulus. Therefore the luminosity distance equation (10) of Schaefer must be replaced by that in CGR. So equation (14.2.6), with $r/c\tau$ determined from the numerical solution of (14.2.8) and $\Omega_m = 0.04$, was used to calculate a correction, which was added as magnitude to the distance modulus. Equation (10) of Schaefer with $\Omega_m = 0.27$ and $\Omega_\Lambda = 0.73$ for the Concordance Model was numerically integrated and removed as a magnitude from the data. These three corrections have the combined effect on the data as shown in Figure 14.2.6.

Finally by fitting the extended equation (14.2.8) with $\Omega_m = 0.04$ to this data and to the combined set of Riess *et al.* and Astier *et al.* it was

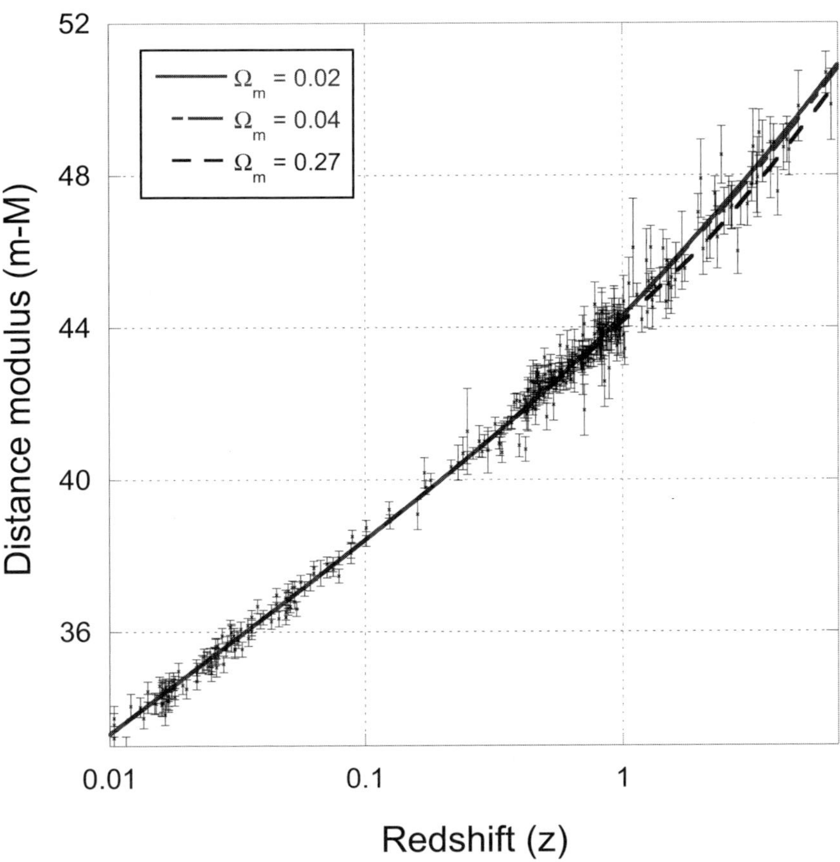

Fig. 14.2.5: All 371 distance modulus data from the combined SN Ia data set of Riess *et al.* and Astier *et al.* and the GRB data derived from Schaefer (2007) with curve fits as shown in Figure 14.2.4, but on a redshift log scale. The top solid curve represents the distance modulus where $\Omega_m = 0.02$, the broken curve below it is where $\Omega_m = 0.04$ and the dashed curve below the latter is where $\Omega_m = 0.27$. (Source: Hartnett, 2008)

determined that 0.1988 magnitudes needed to be added to shift the GRB set so it could be represented on the same plot as the SN Ia data. This was done by determining the best fit value of a from Eq. (14.2.5) for each data set then adding the difference to the GRB distance modulii so a is the same for both. Hence a fit to the total data set can then be accomplished. See Figure 14.2.5. There the data and fits using the extended Carmeli theory with $\Omega_m = 0.02, 0.04$ and 0.27 are shown on a redshift log axis. This better shows the low redshift fit. At high redshifts the fits with the smaller values

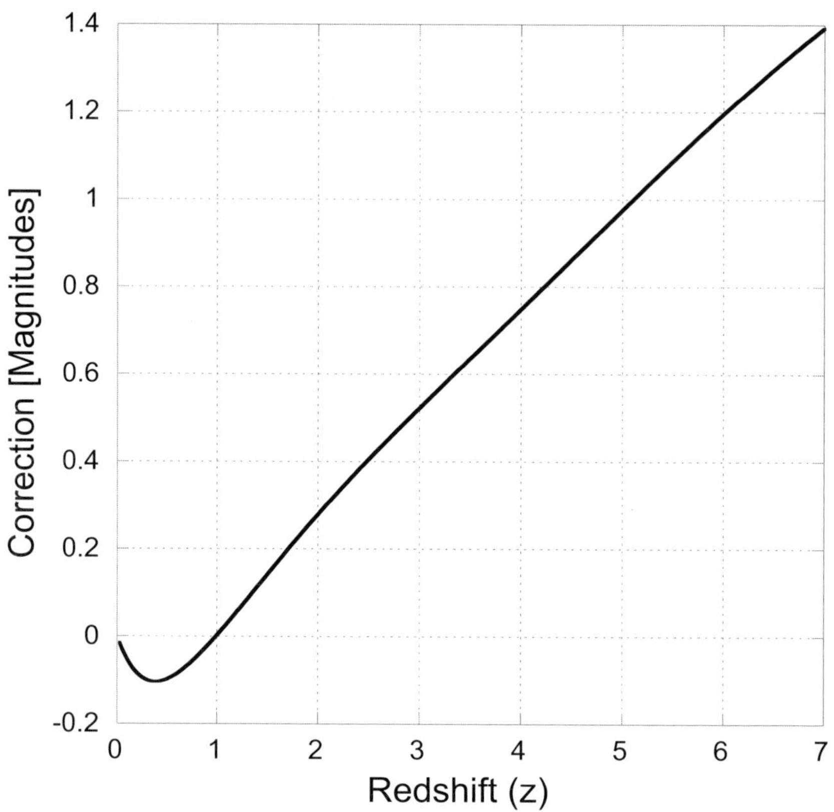

Fig. 14.2.6:   The correction applied to the distance modulus data of Schaefer (2007) to account for the Carmeli cosmology. (Source: Hartnett, 2008)

of Ω_m are favored.

In order to quantify the goodness of the least squares fitting we have used the χ^2 parameter which measures the goodness of the fit between the data and the theoretical curve assuming the two fit parameters a and Ω_m. Hence χ^2 is calculated from

$$\chi^2 = \sum_{i=1}^{N} \frac{1}{\sigma_i^2} \left[(m - M)(z)_i - (m - M)(z_{obs})_i \right]^2 , \qquad (14.2.12)$$

where N are the number of data; $(m - M)(z)$ are determined from Eq. (14.2.4) with fit values of a and Ω_m; $(m - M)(z_{obs})$ are the observed distance modulus data at measured redshifts z_{obs}; σ_i are the published magnitude errors. The values of χ^2/N $(\approx \chi^2_{d.o.f})$ are shown in Table 14.1, calculated using published errors on the distance modulus data. The best fit value of a is found at low redshift and then applied to all fits with various values of Ω_m.

From Table 14.1 the smallest χ^2/N value are generally obtained where $0.04 < \Omega_m < 0.06$, especially when the GRBs are included. But the quality of the fits are not strongly dependent on the matter density. The locally measured baryonic matter budget has determined that $0.007 < \Omega_b < 0.041$ (Fukugita *et al.* 1998), where a Hubble constant of 70 km s^{-1}Mpc^{-1} was assumed. This region is consistent with the most probable values of Ω_m from this analysis. Definitely a matter density of $\Omega_m = 0.27$ is not necessary for the best fit. Therefore no exotic dark matter need be assumed. However to properly test this model much more SNe Ia data are needed where $z > 2$ and much better constrained magnitude errors. The GRB data are not determined independently of the cosmology tested and therefore by themselves don't provide a 'standard candle' test of the theory.

14.3 Spatially Flat Universe

In Section 14.2 the time part of the metric $g_{\mu\nu}$ was ignored. Here we use coordinates $x^\mu = (ct, r, \theta, \phi, \tau v)$ where $\mu = 0, \ldots, 4$, but the full 5D description of cosmos has never been determined in the Carmelian theory. So the question must be answered, "What is the g_{00} metric component for the large scale structure of the Universe in CGR?"

First note from Eqs. (14.1.5) and (14.1.6) the g_{11} metric component

$$g_{11} = -\left(1 + \frac{1-\Omega}{c^2\tau^2}r^2\right)^{-1}. \tag{14.3.1}$$

Using the scale radius we can define an energy density from the curvature

$$\Omega_K = \frac{c^2}{h^2 R_0^2} = \frac{c^2\tau^2}{R_0^2}, \tag{14.3.2}$$

which, when Eq. (14.2.10) is used, becomes

$$\Omega_K = 1 - \Omega. \tag{14.3.3}$$

This quantifies the energy in the curved *spacevelocity*.

In the FRW theory the energy density of the cosmological constant is defined $\rho_\Lambda = \Lambda/8\pi G$ hence

$$\Omega_\Lambda = \frac{\Lambda}{3H_0^2}. \tag{14.3.4}$$

Even though the cosmological constant is not explicitly used in CGR, it follows from the definition of the critical density that

$$\rho_c = \frac{3}{8\pi G \tau^2} = \frac{\Lambda}{8\pi G}, \tag{14.3.5}$$

when the cosmological constant Λ is identified with $3/\tau^2$. Therefore in CGR it follows that

$$\Omega_\Lambda = \frac{\Lambda}{3h^2} = \Lambda\left(\frac{\tau^2}{3}\right) = 1. \tag{14.3.6}$$

In CGR $h = \tau^{-1}$. This means that in CGR the vacuum energy $\rho_{vac} = \Lambda/8\pi G$ is encoded in the metric via the critical density since $\rho_{eff} = \rho - \rho_c$ principally defines the physics. So $\Omega_\Lambda = 1$ identically and at all epochs of time. (The determination of Ω_Λ in Hartnett (2006a) was flawed due to an incorrect definition.) Also we can relate Ω_Λ to the curvature density by

$$\Omega_K = \Omega_\Lambda - \Omega, \tag{14.3.7}$$

which becomes

$$\Omega_k = \Omega_\Lambda - \Omega_m, \tag{14.3.8}$$

at the present epoch ($z \approx 0$). Here $\Omega = \Omega_m(1 + z)^3$ ($z \ll 1$) and hence $\Omega_K \to \Omega_k$ as $z \to 0$.

Finally we can write for the total energy density, the sum of the matter density and the curvature density,

$$\Omega_t = \Omega + \Omega_K = \Omega + 1 - \Omega = 1, \tag{14.3.9}$$

which means the present epoch value is trivially

$$\Omega_0 = \Omega_m + \Omega_k = \Omega_m + 1 - \Omega_m = 1. \tag{14.3.10}$$

This means that the 3D spatial part of the Universe is always flat as it expands. This explains why we live in a Universe that we observe to be identically geometrically spatially flat. The curvature is due to the velocity dimension. Only at some past epoch, in a radiation dominated Universe, would the total mass/energy density depart from unity.

We therefore write

$$g_{00}(r) = 1 + (1 - \Omega_t)r^2, \tag{14.3.11}$$

where r is expressed in units of $c\tau$. Equation (14.3.11) follows from $g_{00} = 1 - 4\Phi/c^2$, where Φ is taken from the gravitational potential but with effective density, which in turn involves the total energy density because we are now considering *spacetime*. The factor 4 results from a comparison of the CGR theory with Newtonian theory.

Clearly from Eq. (14.3.9) it follows that $g_{00}(r) = 1$ regardless of epoch. Thus from the usual relativistic expression

$$1 + z_{grav} = \sqrt{\frac{g_{00}(0)}{g_{00}(r)}} = 1, \tag{14.3.12}$$

and the gravitational redshift is zero regardless of epoch. As expected if the emission and reception of a photon both occur in flat space then we'd expect no gravitational effects.

Therefore we can write down the full 5D line element for CGR in any dynamic spherically symmetrical isotropic Universe,

$$ds^2 = c^2 dt^2 - \frac{dr^2}{1 + \dfrac{1 - \Omega}{c^2 \tau^2} r^2} + \tau^2 dv^2. \tag{14.3.13}$$

The θ and ϕ coordinates do not appear due to the isotropy condition $d\theta = d\phi = 0$. Due to the Hubble law the 2nd and 3rd terms sum to zero leaving $dt = ds/c$, the proper time. Clocks, co-moving with the galaxies in the Hubble expansion, would measure the same proper time.

However inside the Galaxy we expect the matter density to be much higher than critical, i.e. $\Omega_{galaxy} \gg 1$ and the total mass/energy density can be written

$$\Omega_0|_{galaxy} = \Omega_{galaxy} + \Omega_k \approx \Omega_{galaxy}, \tag{14.3.14}$$

because $\Omega_k \approx 1$, since it is cosmologically determined. Therefore this explains why the galaxy matter density only is appropriate when considering the Poisson equation for galaxies (Hartnett 2006b).

As a result inside a galaxy we can write

$$g_{00}(r) = 1 + \Omega_K \frac{r^2}{c^2 \tau^2} + \Omega_{galaxy} \frac{r^2}{c^2 \tau^2}, \tag{14.3.15}$$

in terms of densities at some past epoch. Depending on the mass density of the galaxy, or cluster of galaxies, the value of g_{00} here changes. As we approach larger and larger structures their mass densities approach that of the Universe as a whole and $g_{00} \to 1$ as we approach the largest scales of the Universe. Galaxies in the cosmos then act only as local perturbations but have no effect on Ω_K. That depends only on the average mass density of the whole Universe, which depends on epoch (z).

Equation (14.3.15) is in essence the same expression used on page 173 of Carmeli (2002a) in his gravitational redshift formula rewritten here as

$$\frac{\lambda_2}{\lambda_1} = \sqrt{\frac{1 + \dfrac{\Omega_K r_2^2}{c^2 \tau^2} - \dfrac{R_S}{r_2}}{1 + \dfrac{\Omega_K r_1^2}{c^2 \tau^2} - \dfrac{R_S}{r_1}}}, \qquad (14.3.16)$$

involving a cosmological contribution $(\Omega_K r^2/c^2\tau^2)$ and $R_S = 2GM/c^2$, a local contribution where the mass M is that of a compact object. The curvature (Ω_K) results from the averaged mass/energy density of the whole cosmos, which determines how the galaxies 'move' but the motion of particles within galaxies is dominated by the mass of the galaxy and the masses of the compact objects within. Therefore when considering the gravitational redshifts due to compact objects we can neglect any cosmological effects, only the usual Schwarzschild radius of the object need be considered. The cosmological contributions in Eq. (14.3.16) are generally negligible. This then leads back to the realm of general relativity.

14.4   Conclusion

The solution of the Einstein field equations using the *spacevelocity* of Moshe Carmeli requires the condition (14.1.8). This means that the redshift-distance relation can only be valid, in general, where $\Omega \leq 2$, in a matter dominated Universe. If follows then if the Universe underwent periodic particle creation at high redshift, while the expansion progressed, then we can construct a Hubble Diagram that is valid over arbitrary redshift. This approach extends the Carmeli theory to where it makes a clear prediction and can be tested with future space-based telescopes, like the James Webb.

Also using the Carmelian metric we are able to answer the question, 'Why do we observe the Universe at this present epoch to be Euclidean?' Carmeli's choice of metric implicitly involves a vacuum energy density with an energy content that requires the Universe to be spatially flat at all epochs. This is because the total curvature of the Universe ($\Omega_k = 1 - \Omega_m$ at the present epoch) is the result of the average matter content warping *spacevelocity* away from flatness whereas the 3D spatial part remains flat. That is, $\Omega_0 = \Omega_k + \Omega_m = 1$ at the present epoch.

In the next chapter homogeneous spaces are discussed mathematically, using the concepts of Lie derivative and Killing equations and vectors.

14.5 Suggested References

P. Astier *et al.*, The Supernova Legacy Survey: Measurement of Ω_M, Ω_Λ and w from the first year data set, *Astron. Astrophys.* **447**, 31-48 (2006), astro-ph/0510447 .

M. Carmeli, *Cosmological Special Relativity,* Second Edition (World Scientific, Singapore, 2002a).

M. Carmeli, Accelerating Universe: Theory versus Experiment, astro-ph/0205396 (2002b).

M. Carmeli, *Cosmological Relativity,* (World Scientific, Singapore, 2006).

M. Carmeli, J.G. Hartnett and F.J. Oliveira, The cosmic time in terms of the redshift, *Found. Phys. Lett.* **19**(3), 277-283 (2006), gr-qc/0506079.

M. Fukugita, C.J. Hogan and P.J.E. Peebles, The cosmic baryon budget, *Astrophys. J.* **503**, 518-530 (1998).

G. Gemelli, Particle production in 5-dimensional Cosmological Relativity, *Int. J. Theor. Phys.* **45**(12), 2261–2269 (2006).

J.G. Hartnett, The distance modulus determined from Carmeli's cosmology fits the accelerating Universe data of the high-redshift type Ia supernovae without dark matter, *Found. Phys.* **36**(6), 839-861 (2006a), astro-ph/0501526.

J.G. Hartnett, Spiral galaxy rotation curves determined from Carmelian general relativity, *Int. J. Theor. Phys.* **45**(11), 2147-2165 (2006b), astro-ph/0511756.

J.G. Hartnett, Extending the redshift-distance relation in cosmological general relativity to higher redshifts, *Found. Phys.* **38**(3), 201-215 (2008), gr-qc/07053097.

J.G. Hartnett and F.J. Oliveira, Luminosity distance, angular size and surface brightness in Cosmological General Relativity, *Found. Phys.* **37**(3), 446-454 (2007a), astro-ph/0603500.

R.A. Knop *et al.*, New constraints on Ω_M, Ω_Λ and w from an independent set of 11 high-redshift supernovae observed with the Hubble Space Telescope, *Astrophys. J.* **598**, 102-137 (2003).

F.J. Oliveira and J.G. Hartnett, Carmeli's cosmology fits data for an accelerating and decelerating Universe without dark matter or dark energy, *Found. Phys. Lett.* **19**(6), 519-535 (2006), astro-ph/0603500.

S. Perlmutter *et al.*, Measurements of the cosmological parameters Ω and Λ from the first seven supernovae at $z > 0.35$, *Astrophys. J.* **483**, 565-581 (1997).

A.G. Riess *et al.*, Type Ia supernovae discoveries at $z > 1$ from the Hubble Space Telescope: Evidence for past deceleration and constraints on dark energy evolution, *Astrophys. J.* **607**, 665-687 (2004).

A.G. Riess, A.V. Filippenko, P. Challis, A. Clocchiatti and A. Diercks, Observational evidence from supernovae for an accelerating Universe and a cosmological constant, *Astron. J.* **116**, 1009-1038 (1998).

B.E. Schaefer, The Hubble Diagram to redshift > 6 from 69 Gamma-Ray Bursts, *Astrophys. J.* **660**, 16-46 (2007).

Chapter 15

Homogeneous Spaces and Bianchi Classification

Moshe Carmeli

In this chapter homogeneous spaces will be discussed in some details that will include the concepts of Lie derivative and the Killing equation. Following that the Bianchi classification of homogeneous spaces, frequently mentioned in discussing cosmology, will be given.

15.1   Lie Derivative

In this section we will deal with homogeneous spaces that are not in particular isotropic as discussed in Chapter 11. Our starting point is with the concept of Lie derivative[1].

---

[1] (Marius) Sophus Lie (Born: 17 Dec 1842 in Nordfjordeide, Norway; Died: 18 Feb 1899 in Kristiana (now Oslo), Norway)

In 1857 Sophus entered Nissen's Private Latin School in Christiania (later became Kristiania, then Oslo in 1925). He decided to take up a military carrier but he could not implement his plan because of his eyesight problems. At the University Lie attended lectures by Sylow, which included Abel's and Galois' work, and also lectures by Carl Bjerknes. In 1867 Lie had his first brilliant new mathematical idea, that came to him in the middle of the night. He rushed to his friend Ernst Motzfeldt, woke him up and shouted: "I have found it, it is quite simple." He became interested in geometry by reading Plücker and Poncelet's papers. Lie wrote a short mathematical paper in 1869 which he published at his own expense. After long delays, in 1869, *Crelle's Journal* accepted his work for publication. On the basis of that paper, Lie was awarded a scholarship to meet leading mathematicians. He went to Prussia and visited Göttingen and then Berlin, where he met Kronicker, Kummer and Weierstrass. Most important to Lie was that in Berlin he met Felix Klein. In 1870 Lie and Klein were in Paris where they met Darboux, Chasles and Camille Jordan. Group theory was now realized to be an impor-

In Chapter 3 the geometry of curved spacetime was discussed, where use was made of coordinate systems. No description was given, however, of how the coordinate system should be chosen, since the discussion was intended for general spaces rather than for particular ones with special symmetries. We have also discussed coordinate transformations between two or more coordinate systems. We recall that different coordinates describing the same spacetime meant that the same spacetime point could be described by two sets of four coordinates, x^μ and x'^μ, for instance. A simple example

---

tant vehicle for Lie's theory. Because of political arguments France and Prussia declared war against each other, and Klein had to return fast to Berlin. In August the German army trapped part of the French army in Metz and Lie decided it was time for him to leave, to hike to Italy. He reached Fontainebleau but there he was arrested as a German spy, being accused that his mathematics notes are top secret coded messages. Only after the intervention of Darboux was he released from prison. On 1 September the French army surrendered and on 19 September the German army began to blockade Paris. Lie fled again to Italy, then to Christiania via Germany so he could meet Klein. In 1871 Lie became an assistant at Christiania. He submitted a dissertation for his doctorate which was approved in 1872. The University of Christiania recognized Lie's ability and created a chair for him in 1872. The famous Norwegian mathematician Abel had died more than 40 years earlier but his complete works had not been published at that time. Between 1873 and 1881 Sylow and Lie published Abel's complete works. At that time Lie married Anna Birch and the family had three children, a daughter and two sons.

Lie began examining partial differential equations hoping to find an analogous theory to the Galois theory of equations. He worked on combining the transformations, which today is called the Lie algebra. Later Killing was to examine the Lie algebras associated with Lie groups, but it was Elie Cartan who completed the classification of semisimple Lie algebras in 1900.

In later years Lie became depressed by the feeling of not getting enough recognition from other mathematicians. One reason for that was his isolation in Christiania, a second reason was his papers were not easily understood. Klein understood the problem and suggested sending Friedrich Engel to Christiania to help Lie. Engel went to work with Lie in Christiania in 1884 for nine months, leaving in 1885. Engel was then appointed to Leipzig, and when Klein left the chair at Leipzig in 1886 Lie was appointed to succeed him. The collaboration between Engel and Lie continued for nine years during which they produced their major publication *Theorie der Transformationsgruppen* in three volumes. In Leipzig Lie was rather active and students from many countries came to study under him, he also had a heavier teaching load. Lie was also troubled by constant homesickness. Towards the end of 1880s Lie's relationship with Engel broke down, and so did the friendship between Lie and Klein. Lie returned to Christiania in 1898 to assume a position specially created for him. With so much disagreement between Lie and Klein, an indication of Lie's love for his homeland is the fact that he continued to hold his chair in Christiania from his first appointment in 1872, being officially on leave while holding the chair in Leipzig. His health was already deteriorating when he returned to his chair in Christiania in 1898. He died of pernicious anemia in February 1899 soon after taking up the post.

The above report on Sophus Lie is based on an article by J J O'Connor and E F Robertson (February 2000).

of this designation of two sets of coordinates to the same spacetime is the use of Cartesian coordinates $x^\mu(ct, x, y, z)$ and of spherical coordinates $x'^\mu(ct, r, \theta, \phi)$ in the Minkowskian flat spacetime. Thus each point of the Minkowskian spacetime is described by either one of the two sets of four coordinates at the same time.

In this section we introduce and discuss an essentially different kind of coordinates transformation. This different transformation will lead us, among other things, to a general prescription of assigning particular pre-ferred coordinates to spaces with particular symmetries. Thus this method involves discussing the symmetry properties of the spacetime itself. It also involves the introduction of a new kind of derivative. The discussion will be followed, in the next section, by introducing a new kind of transformation, a mapping of the spacetime onto itself, and a certain differential equation which indicates if a given spacetime has any symmetries or not.

We start our discussion by considering the coordinate transformation

$$\tilde{x}^\mu = \tilde{x}^\mu(\varepsilon; x^\nu), \qquad (15.1.1)$$

where

$$x^\mu = \tilde{x}^\mu(0; x^\nu), \qquad (15.1.2)$$

and ε is a parameter. Equation (15.1.1) describes a one-parameter set of transformations $x^\mu \to \tilde{x}^\mu$ and is interpreted as follows. Let us suppose a point P, which is labeled by the set of four coordinates x^μ, is given. We assign to the point P *another* point Q of the same spacetime, which is labeled by the four coordinates $\tilde{x}^\mu$, in the *same* coordinate system that was used to label the first point P. In this way to *each* point of our spacetime we assign another point of the same spacetime using the same system of coordinates. As a consequence, the transformation (15.1.1) describes a *mapping of the spacetime onto itself.*

15.1.1 *Infinitesimal transformation*

Let us now concentrate on *infinitesimal transformations.* The transforma-tions (15.1.1) may then be written in the form

$$\tilde{x}^\mu = x^\mu + \varepsilon \xi^\mu(x), \qquad (15.1.3)$$

and are called *infinitesimal mapping.* Here ε is an infinitesimal parameter, and $\xi^\mu(x)$ is a contravariant vector field which, in general, might be defined by

$$\xi^\mu(x) = \left[\frac{\partial \tilde{x}^\mu}{\partial \varepsilon}\right]_{\varepsilon=0}. \qquad (15.1.4)$$

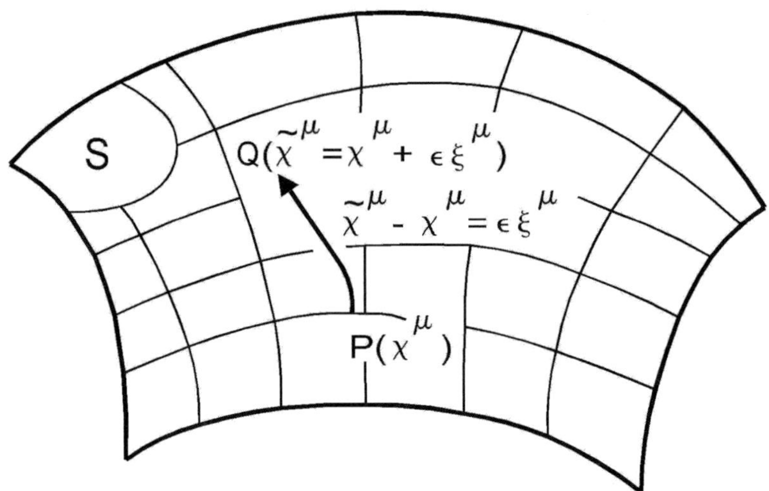

Fig. 15.1.1: Two points P and Q of spacetime S, labeled x^{μ} and $\tilde{x}^{\mu} = x^{\mu} + \varepsilon \xi^{\mu}$, in the same coordinate system. The transformation $x^{\mu} \to \tilde{x}^{\mu}$ describes an infinitesimal mapping of the spacetime onto itself, namely, to each point P of S there is an image point Q also in S.

The meaning of the infinitesimal mapping (15.1.3) is as follows (see Figure 15.1.1). To each point P, whose coordinates are x^{μ}, of the spacetime there corresponds another point Q, whose coordinates are $x^{\mu} + \varepsilon \xi^{\mu}(x)$, using the same coordinate system.

Assume now that there is some tensor field $T(x)$ which is defined in our spacetime. At the point Q we can evaluate the tensor $T(x)$ in two different ways. First we have the value of T at the point Q, namely $T(\tilde{x})$. Then we have the value of $\tilde{T}(\tilde{x})$, namely, the transformed tensor $\tilde{T}$ using the usual coordinate transformations for tensors at the point Q. The difference between these two values of the tensor T evaluated at the point Q with the coordinates $\tilde{x}^{\mu}$ leads to the possibility of defining the concept of *Lie derivative* of the tensor T.

We now illustrate the above procedure by defining the Lie derivative of a tensor of order zero, namely, a scalar field $\phi(x)$. At the point Q the value of ϕ is $\phi(\tilde{x})$, which can be written in terms of its value at the point x^{μ} by means of the infinitesimal expansion,

$$\phi(\tilde{x}) = \phi(x + \varepsilon\xi) = \phi(x) + \varepsilon \frac{\partial \phi(x)}{\partial x^{\alpha}} \xi^{\alpha}. \tag{15.1.5}$$

Under the infinitesimal coordinate transformation (15.1.3), on the other hand, the scalar function ϕ is, of course, unchanged,

$$\tilde{\phi}(\tilde{x}) = \phi(x). \tag{15.1.6}$$

Here $\tilde{\phi}$ is a function evaluated at the point Q whose coordinates are $\tilde{x}^{\mu}$, whereas ϕ is a function that is evaluated at the original point P whose coordinates are x^{μ}.

The Lie derivative of the scalar function $\phi(x)$, denoted by $\mathcal{L}_{\xi}\phi(x)$, is then defined by

$$\mathcal{L}_{\xi}\phi(x) = \lim_{\varepsilon \to 0} \frac{\phi(\tilde{x}) - \tilde{\phi}(\tilde{x})}{\varepsilon} = \xi^{\alpha}(x)\frac{\partial \phi(x)}{\partial x^{\alpha}}. \tag{15.1.7}$$

Thus the Lie derivative of the function ϕ is just the scalar product of the vector field ξ^{α} with the gradient of ϕ. We may also present the definition of the Lie derivative in a somewhat different way. This is achieved if we evaluate all functions involved at the same point P. Hence the function $\tilde{\phi}(\tilde{x})$ is expanded as

$$\tilde{\phi}(\tilde{x}) = \tilde{\phi}(x + \varepsilon\xi) = \tilde{\phi}(x) + \varepsilon\xi^{\alpha}(x)\frac{\partial \tilde{\phi}(x)}{\partial x^{\alpha}}. \tag{15.1.8}$$

Furthermore, neglecting terms of second and higher order in ε, we may replace $\tilde{\phi}(x)$ by $\phi(x)$ in all terms containing ε. Hence Eq. (15.1.8) may be written as

$$\tilde{\phi}(\tilde{x}) = \tilde{\phi}(x) + \varepsilon\xi^{\alpha}(x)\frac{\partial \phi(x)}{\partial x^{\alpha}}, \tag{15.1.9}$$

or, using Eq. (15.1.6),

$$\tilde{\phi}(x) = \phi(x) - \varepsilon\xi^{\alpha}(x)\frac{\partial \phi(x)}{\partial x^{\alpha}}. \tag{15.1.10}$$

Accordingly we have

$$\mathcal{L}_{\xi}\phi(x) = \lim_{\varepsilon \to 0} \frac{\phi(x) - \tilde{\phi}(x)}{\varepsilon} = \xi^{\alpha}(x)\frac{\partial \phi(x)}{\partial x^{\alpha}} \tag{15.1.11}$$

for the Lie derivative of the scalar function $\phi(x)$.

It will be noted that, since ϕ is a scalar function, we may replace the partial derivative in Eq. (15.1.11) by a covariant derivative, thus getting

$$\mathcal{L}_{\xi}\phi(x) = \xi^{\alpha}(x)\nabla_{\alpha}\phi(x) \tag{15.1.12}$$

for the Lie derivative of a scalar function.

For a general tensor field T the Lie derivative is defined, following Eq. (15.1.11) for the scalar field, by

$$\mathcal{L}_{\xi}T(x) = \lim_{\varepsilon \to 0} \frac{T(x) - \tilde{T}(x)}{\varepsilon}. \tag{15.1.13}$$

In the following the Lie derivatives of some fields are calculated.

In the next section we discuss the Killing equation.

15.1.2 *Problems*

P 15.1.1. Calculate the Lie derivative of a contravariant vector V^α.

Solution: Under the infinitesimal coordinate transformation (15.1.3) the vector V^α is transformed into

$$\tilde{V}^\alpha (\tilde{x}) = \frac{\partial \tilde{x}^\alpha}{\partial x^\beta} V^\beta (x) . \tag{1}$$

From the coordinate transformation (15.1.3) we find that

$$\frac{\partial \tilde{x}^\alpha}{\partial x^\beta} = \delta_\beta^\alpha + \varepsilon \frac{\partial \xi^\alpha}{\partial x^\beta} . \tag{2}$$

We therefore obtain for the transformed components of the vector V^α the following:

$$\tilde{V}^\alpha (\tilde{x}) = V^\alpha (x) + \varepsilon V^\beta (x) \frac{\partial \xi^\alpha (x)}{\partial x^\beta} . \tag{3}$$

The left-hand side of this equation may also be written as a function of x^μ if we expand $\tilde{V}^\alpha$ around the point x^μ:

$$\tilde{V}^\alpha (\tilde{x}) = \tilde{V}^\alpha (x) + \varepsilon \xi^\beta (x) \frac{\partial V^\alpha (x)}{\partial x^\beta} . \tag{4}$$

Hence, comparing Eqs. (3) and (4), we obtain

$$\tilde{V}^\alpha = V^\alpha + \varepsilon \left(V^\beta \frac{\partial \xi^\alpha}{\partial x^\beta} - \xi^\beta \frac{\partial V^\alpha}{\partial x^\beta} \right) , \tag{5}$$

where all functions now are evaluated at point P.

Therefore, using the definition (15.1.13), the Lie derivative of the contravariant vector V^α is given by

$$\mathcal{L}_\xi V^\alpha = \lim_{\varepsilon \to 0} \frac{V^\alpha (x) - \tilde{V}^\alpha (x)}{\varepsilon}$$

$$= \xi^\beta \frac{\partial V^\alpha}{\partial x^\beta} - V^\beta \frac{\partial \xi^\alpha}{\partial x^\beta} . \tag{6}$$

As for the scalar function case, the partial derivatives in the above equation can be replaced by covariant derivatives, giving

$$\mathcal{L}_\xi V^\alpha = \xi^\beta \nabla_\beta V^\alpha - V^\beta \nabla_\beta \xi^\alpha , \tag{7}$$

for the Lie derivative of a contravariant vector V^α.

P 15.1.2. Calculate the Lie derivative of a covariant vector V_α.

Solution: Under the infinitesimal coordinate transformation (15.1.3) the vector V_α is transformed into

$$\tilde{V}_\alpha\left(\tilde{x}\right) = \frac{\partial x^\mu}{\partial \tilde{x}^\alpha} V_\mu\left(x\right). \tag{1}$$

The coordinate transformation (15.1.3) gives, by taking its partial derivative with respect to the coordinate $\tilde{x}^\nu$,

$$\delta^\mu_\nu = \frac{\partial x^\mu}{\partial \tilde{x}^\nu} + \varepsilon \frac{\partial \xi^\mu}{\partial \tilde{x}^\nu}, \tag{2}$$

or, to the first order in ε,

$$\frac{\partial x^\mu}{\partial \tilde{x}^\nu} = \delta^\mu_\nu - \varepsilon \frac{\partial \xi^\mu}{\partial x^\nu}. \tag{3}$$

Using the latter relation in the above law of transformation for the vector V_α, we get

$$\tilde{V}_\alpha\left(\tilde{x}\right) = V_\alpha\left(x\right) - \varepsilon V_\mu\left(x\right) \frac{\partial \xi^\mu\left(x\right)}{\partial x^\alpha}. \tag{4}$$

Expanding $\tilde{V}_\alpha(\tilde{x})$ around the point P, and neglecting terms of higher order than 1 in ε, gives

$$\tilde{V}_\alpha\left(\tilde{x}\right) = \tilde{V}_\alpha\left(x\right) + \varepsilon \xi^\mu\left(x\right) \frac{\partial V_\alpha\left(x\right)}{\partial x^\mu}. \tag{5}$$

Comparing the latter two expressions for $\tilde{V}_\alpha\left(\tilde{x}\right)$, we obtain

$$\tilde{V}_\alpha = V_\alpha - \varepsilon \left(V_\mu \frac{\partial \xi^\mu}{\partial x^\alpha} + \xi^\mu \frac{\partial V_\alpha}{\partial x^\mu} \right). \tag{6}$$

Using the definition (15.1.13), we therefore obtain

$$\mathcal{L}_\xi V_\alpha = \xi^\mu \frac{\partial V_\alpha}{\partial x^\mu} + V_\mu \frac{\partial \xi^\mu}{\partial x^\alpha} \tag{7}$$

for the Lie derivative of a covariant vector V_α. Once again we may replace the partial derivatives in the above formula by covariant derivatives, thus getting

$$\mathcal{L}_\xi V_\alpha = \xi^\mu \nabla_\mu V_\alpha - V_\mu \nabla_\alpha \xi^\mu, \tag{8}$$

for the Lie derivative of V_α.

P 15.1.3. Find the Lie derivatives of covariant and contravariant tensors of order 2. Apply the results in particular to the contravariant metric tensor.

Solution: The transformed components of a covariant tensor of order 2, $T_{\mu\nu}$, are given by

$$\tilde{T}_{\alpha\beta}\left(\tilde{x}\right) = \frac{\partial x^\mu}{\partial \tilde{x}^\alpha} \frac{\partial x^\nu}{\partial \tilde{x}^\beta} T_{\mu\nu}\left(x\right). \tag{1}$$

Using Eq. (3) of Problem 15.1.2 in the above equation, we obtain

$$\tilde{T}_{\alpha\beta}\left(\tilde{x}\right) = T_{\alpha\beta}\left(x\right) - \varepsilon\left(T_{\alpha\nu}\frac{\partial\xi^{\nu}}{\partial x^{\beta}} + T_{\mu\beta}\frac{\partial\xi^{\mu}}{\partial x^{\alpha}}\right), \tag{2}$$

where terms of order higher than first in ε are neglected. Expanding now the tensor $\tilde{T}_{\alpha\beta}(\tilde{x})$ around the point x^{α}, on the other hand, we obtain, to the first order in ε,

$$\tilde{T}_{\alpha\beta}\left(\tilde{x}\right) = \tilde{T}_{\alpha\beta}\left(x + \varepsilon\xi\right) = \tilde{T}_{\alpha\beta}\left(x\right) + \varepsilon\xi^{\rho}\frac{\partial T_{\alpha\beta}}{\partial x^{\rho}}. \tag{3}$$

Comparing Eqs. (2) and (3) we obtain

$$\tilde{T}_{\alpha\beta}\left(x\right) = T_{\alpha\beta}\left(x\right) - \varepsilon\left(\xi^{\rho}\frac{\partial T_{\alpha\beta}}{\partial x^{\rho}} + T_{\alpha\nu}\frac{\partial\xi^{\nu}}{\partial x^{\beta}} + T_{\mu\beta}\frac{\partial\xi^{\mu}}{\partial x^{\alpha}}\right). \tag{4}$$

Accordingly we obtain, using the definition (15.1.13),

$$\mathcal{L}_{\xi}T_{\alpha\beta} = \lim_{\varepsilon\to 0}\frac{T_{\alpha\beta}\left(x\right) - \tilde{T}_{\alpha\beta}\left(x\right)}{\varepsilon} = \xi^{\rho}\frac{\partial T_{\alpha\beta}}{\partial x^{\rho}} + T_{\alpha\nu}\frac{\partial\xi^{\nu}}{\partial x^{\beta}} + T_{\nu\beta}\frac{\partial\xi^{\nu}}{\partial x^{\alpha}} \tag{5}$$

for the Lie derivative of the covariant tensor $T_{\alpha\beta}$.

Using the same method we obtain for the Lie derivative of the contravariant tensor of order 2 the following formula:

$$\mathcal{L}_{\xi}T^{\alpha\beta} = \xi^{\rho}\frac{\partial T^{\alpha\beta}}{\partial x^{\rho}} - T^{\alpha\nu}\frac{\partial\xi^{\beta}}{\partial x^{\nu}} - T_{\mu\beta}\frac{\partial\xi^{\alpha}}{\partial x^{\mu}}. \tag{6}$$

We notice that both Eqs. (5) and (6) can be written in terms of covariant derivatives instead of partial derivatives:

$$\mathcal{L}_{\xi}T_{\alpha\beta} = \xi^{\rho}\nabla_{\rho}T_{\alpha\beta} + T_{\alpha\nu}\nabla_{\beta}\xi^{\nu} + T_{\mu\beta}\nabla_{\alpha}\xi^{\mu}, \tag{7}$$

$$\mathcal{L}_{\xi}T^{\alpha\beta} = \xi^{\rho}\nabla_{\rho}T^{\alpha\beta} - T^{\alpha\nu}\nabla_{\nu}\xi^{\beta} - T^{\mu\beta}\nabla_{\mu}\xi^{\alpha}. \tag{8}$$

Equations (7) and (8) may be applied to the covariant and contravariant components respectively of the metric tensor. Remembering that the covariant derivative of the metric tensor vanishes, we obtain for its covariant part:

$$\mathcal{L}_{\xi}g_{\alpha\beta} = g_{\alpha\nu}\nabla_{\beta}\xi^{\nu} + g_{\mu\beta}\nabla_{\alpha}\xi^{\mu}, \tag{9}$$

or

$$\mathcal{L}_{\xi}g_{\alpha\beta} = \nabla_{\alpha}\xi_{\beta} + \nabla_{\beta}\xi_{\alpha} = 2\nabla_{(\alpha}\xi_{\beta)}, \tag{10}$$

Of course, the above formula is completely equivalent to

$$\mathcal{L}_{\xi}g_{\alpha\beta} = \xi^{\rho}\frac{\partial g_{\alpha\beta}}{\partial x^{\rho}} + g_{\alpha\nu}\frac{\partial\xi^{\nu}}{\partial x^{\beta}} + g_{\nu\beta}\frac{\partial\xi^{\nu}}{\partial x^{\alpha}}, \tag{11}$$

obtained from the general formula (5) for the Lie derivative of a covariant tensor of order 2. The Lie derivative of the contravariant metric tensor is given by

$$\mathcal{L}_\xi g^{\alpha\beta} = \xi^\rho \frac{\partial g^{\alpha\beta}}{\partial x^\rho} - g_{\rho\beta}\frac{\partial \xi^\alpha}{\partial x^\beta} - g_{\alpha\rho}\frac{\partial \xi^\beta}{\partial x^\rho}, \tag{12}$$

which may also be written in the form

$$\mathcal{L}_\xi g^{\alpha\beta} = -\left(\nabla^\alpha \xi^\beta + \nabla^\beta \xi^\alpha\right) = -2\nabla^{(\alpha}\xi^{\beta)} \tag{13}$$

if covariant derivatives are used.

P 15.1.4. Show that the Lie derivative of a product of two tensors satisfies the usual rule of derivative of products, namely

$$\mathcal{L}_\xi (VT) = V\mathcal{L}_\xi T + T\mathcal{L}_\xi V. \tag{1}$$

Prove the above formula for a vector V and a tensor T.

Solution: Let us prove, for instance, that

$$\mathcal{L}_\xi (V^\alpha T_{\alpha\beta}) = V^\alpha \mathcal{L}_\xi T_{\alpha\beta} + T_{\alpha\beta}\mathcal{L}_\xi V^\alpha. \tag{2}$$

From the rules of algebra of tensors we know that the quantity $V^\alpha T_{\alpha\beta}$ is a covariant vector. Hence we have, using Eq. (8) of Problem 15.1.2 for the Lie derivative of a covariant vector, the following:

$$\mathcal{L}_\xi (V^\alpha T_{\alpha\beta}) = \xi^\rho \nabla_\rho (V^\alpha T_{\alpha\beta}) + V^\alpha T_{\alpha\rho}\nabla_\beta \xi^\rho$$

$$= \xi^\rho (V^\alpha \nabla_\rho T_{\alpha\beta} + T_{\alpha\beta}\nabla_\rho V^\alpha) + V^\alpha T_{\alpha\rho}\nabla_\beta \xi^\rho$$

$$= V^\alpha (\xi^\rho \nabla_\rho T_{\alpha\beta} + T_{\alpha\rho}\nabla_\beta \xi^\rho + T_{\rho\beta}\nabla_\alpha \xi^\rho)$$

$$+ T_{\alpha\beta}(\xi^\rho \nabla_\rho V^\alpha - V^\rho \nabla_\rho \xi^\alpha). \tag{3}$$

Using now Eq. (7) of Problem 15.1.3 and Eq. (15.1.7), we finally obtain

$$\mathcal{L}_\xi (V^\alpha T_{\alpha\beta}) = V^\alpha \mathcal{L}_\xi T_{\alpha\beta} + T_{\alpha\beta}\mathcal{L}_\xi V^\alpha. \tag{4}$$

P 15.1.5. Find the formula for the Lie derivative of a scalar density of weight $W = +1$.

Solution: Let us denote such a scalar density by ψ. Then ψ may be represented as the product

$$\psi = \sqrt{-g}\,\phi, \tag{1}$$

where ϕ is an ordinary scalar function. Expanding $\tilde{\phi}\left(\tilde{x}\right)$ around the point x^{α}, we obtain

$$\tilde{\psi}\left(\tilde{x}\right) = \tilde{\psi}\left(x + \varepsilon\xi\right) = \tilde{\psi}\left(x\right) + \varepsilon\xi^{\alpha}\frac{\partial\psi}{\partial x^{\alpha}}, \tag{2}$$

where terms of order higher than the first in ε have been neglected.

Under the infinitesimal coordinate transformation (15.1.3), on the other hand, the scalar density ψ is transformed into

$$\tilde{\psi}\left(\tilde{x}\right) = \sqrt{-\tilde{g}\left(\tilde{x}\right)}\tilde{\phi}\left(\tilde{x}\right) = \sqrt{-\tilde{g}\left(\tilde{x}\right)}\phi\left(x\right), \tag{3}$$

since ϕ is an ordinary scalar function. The latter equation can be written in the form

$$\tilde{\psi}\left(\tilde{x}\right) = \left|\frac{\partial x}{\partial\tilde{x}}\right|\sqrt{-g\left(x\right)}\phi\left(x\right) = \left|\frac{\partial x}{\partial\tilde{x}}\right|\psi\left(x\right). \tag{4}$$

We now calculate the Jacobian $\left|\partial x/\partial\tilde{x}\right|$. From Eq. (15.1.3) we have

$$\frac{\partial x^{\mu}}{\partial\tilde{x}^{\nu}} = \delta^{\mu}_{\nu} - \frac{\partial\xi^{\mu}}{\partial x^{\nu}}. \tag{5}$$

Hence we get for the Jacobian, to the first order in ε, the following expression:

$$\left|\frac{\partial x}{\partial\tilde{x}}\right| = 1 - \varepsilon\frac{\partial\xi^{\mu}}{\partial x^{\mu}}, \tag{6}$$

and, accordingly, Eq. (4) becomes

$$\tilde{\psi}\left(\tilde{x}\right) = \psi\left(x\right) - \varepsilon\psi\left(x\right)\frac{\partial\xi^{\mu}}{\partial x^{\mu}}. \tag{7}$$

Comparing now Eq. (2) with Eq. (7), we find

$$\tilde{\psi} = \psi - \varepsilon\left(\xi^{\alpha}\frac{\partial\psi}{\partial x^{\alpha}} + \psi\frac{\partial\xi^{\alpha}}{\partial x^{\alpha}}\right), \tag{8}$$

where all functions are now evaluated at point x^{μ}. Hence we finally obtain

$$\mathcal{L}_{\xi}\psi = \lim_{\varepsilon\to 0}\frac{\psi\left(x\right) - \tilde{\psi}\left(x\right)}{\varepsilon} = \xi^{\alpha}\frac{\partial\psi}{\partial x^{\alpha}} + \psi\frac{\partial\xi^{\alpha}}{\partial x^{\alpha}}, \tag{9}$$

for the Lie derivative of the scalar density ψ of weight $W = +1$.

Finally we may now rewrite the above result in terms of covariant derivatives instead of the partial derivatives. The result we obtain is

$$\mathcal{L}_{\xi}\psi = \xi^{\alpha}\nabla_{\alpha}\psi + \psi\nabla_{\alpha}\xi^{\alpha}. \tag{10}$$

15.2 The Killing Equation

In the last section the concept of Lie derivative was defined and applied to an arbitrary tensor. In this section we make more applications of this concept in order to get an insight into the structure of the metric tensor, namely, the structure of the spacetime itself. For clarity we confine our illustrations in this section to the Minkowskian flat spacetime.

Suppose that a spacetime with a metric tensor $g_{\mu\nu}$ is given. The question naturally arises as to whether or not the given metric tensor changes its value under the infinitesimal coordinate transformation (15.1.3),

$$\tilde{x}^\mu = x^\mu + \varepsilon\xi^\mu(x). \tag{15.2.1}$$

We recall that, under such an infinitesimal transformation, the metric tensor $g_{\mu\nu}(x)$ goes over into its transformed value $\tilde{g}_{\mu\nu}(\tilde{x})$. The tensor $\tilde{g}_{\mu\nu}(\tilde{x})$ usually differs from the value of the metric tensor at the point $\tilde{x}^\mu$, namely, $g_{\mu\nu}(\tilde{x})$.

The expression for the Lie derivative of $g_{\mu\nu}$, however, is based on exactly this difference between the two different values of the metric tensor , namely,

$$\mathcal{L}_\xi g_{\mu\nu}(x) = \lim_{\varepsilon\to 0} \frac{g_{\mu\nu}(\tilde{x}) - \tilde{g}_{\mu\nu}(\tilde{x})}{\varepsilon}. \tag{15.2.2}$$

As a consequence, the condition for the "constancy" of the metric tensor is exactly the vanishing of its Lie derivative.

15.2.1 *Isometric mapping*

A mapping of the spacetime onto itself of the form (15.2.1) is called an *isometric mapping* if the Lie derivative of the metric tensor associated with it vanishes,

$$\mathcal{L}_\xi g_{\mu\nu}(x) = 0. \tag{15.2.3}$$

From Eq. (10) of Problem 15.1.3 we see that the latter condition on the metric tensor is equivalent to the following formula:

$$\nabla_\mu \xi_\nu(x) + \nabla_\nu \xi_\mu(x) = 0. \tag{15.2.4}$$

Accordingly, the conditions for the existence of isometric mappings in a given spacetime is the existence of solutions $\xi^\mu(x)$ of the differential equation (15.2.4).

15.2.2 *Killing equation. Killing vector*

Equation (15.2.4) is called the *Killing equation*[2]. The solutions $\xi_\mu\,(x)$ of the Killing equation are called the *Killing vectors*. Of course, a given spacetime

---

[2]Wilhelm (Karl Joseph) Killing (Born: 10 May 1847 in Burbach (near Siegen), Westphalia, Germany; Died: 11 Feb 1923 in Münster, Germany)

Wilhelm Killing was one of three children of Josef Killing, legal clerk, and Anna Catharina Kortenbach, the daughter of the pharmacist Wilhelm Kortenbach. As a child Wilhelm's health was not good and he was described as "... quite weakly and besides very awkward ..., always excited, but a completely unpractical bookworm."

Wilhelm was given a conservative education, with a great love of his country. Killing attended elementary school and was given private lessons to enter the Gymnasium in Brilon. At the Gymnasium Killing especially liked the classical languages: Greek, Latin and Hebrew. His teacher Harnischmacher first gave Killing his love of mathematics, and later Killing dedicated his thesis to him. At the Gymnasium Killing became convinced that he should become a mathematician. He graduated from the Gymnasium in 1865 and in the autumn of the same year began his university studies at the Royal Academy at Münster, which became a full university in 1902. However, at the Academy mathematics was not taught to a high level, so Killing learnt his mathematics from studying books. He read Gauss's "Disquisitiones Arithmeticae", works by Hesse and also Plücker's works on geometry, whose results he tried to extend.

At Münster Killing was having to educate himself, and although he greatly appreciated the genius of the authors whose works he read, he felt that his self-education was not enough and that he needed expert teaching. After four terms he moved to Berlin, where he found the highest quality of teaching and was particularly influenced by Kummer, Weierstrass and Helmholtz. In 1871 he began working towards his doctorate under the supervision of Weierstrass. His doctoral thesis, which applied Weierstrass's theory of elementary divisors of a matrix to surfaces, was entitled "Der Flächenbüschel zweiter Ordnung" (Bundles of surfaces of the second degree). It was presented in March 1872. After completing his doctorate and until 1878 Killing taught at schools in Berlin: the Frdr Werder Gymnasium and St Hedwig's Catholic school. In 1875 he married Anna Commer, daughter of a lecturer in music. They had four sons, the first two of whom died as infants, and two daughters Maria and Anka.

In 1878 Killing taught at the Gymnasium in Brilon where he had been a pupil. In spite of a heavy teaching load (he spend around 36 hours each week either teaching in the classroom or tutoring pupils), he published his first papers: "Über zwei Raumformen mit konstanter positiver Krümmung" (1879), "Die Rechnung in den Nicht-Euklidischen Raumformen" (1880) and "Die Mechanik in den Nicht-Euklidischen Raumformeni" (1885) on non-euclidean geometry in n-dimensions in Crelle's Journal. He also published the book "Die nichteuklidischen Raumformen in analytischer Behandlung" on non-euclidean geometry in Leipzig in 1885.

In 1882 on Weierstrass's recommendation Killing was appointed to a chair of mathematics at the Lyceum Hosianum in Braunsberg where for ten years he was isolated mathematically, but during this period he developed some of the most original mathematics ever produced: Killing introduced Lie algebras independently from Lie with a different purpose, being interested in non-euclidean geometry. He classified the semisimple Lie algebras, and it was a fine mathematical research. The main tools in this classification, Cartan subalgebras and the Cartan matrix, were first introduced by Killing. He also introduced the idea of a root system which appears a lot in algebra today. We will now follow the development of Killing's ideas.

might not have even one solution to the Killing equation. This is the case when the spacetime has no symmetry whatsoever.

In general, however, the existence of a Killing vector, namely, the existence of a solution to the Killing equation for a given spacetime metric tensor, means the existence of an isometric mapping of the spacetime onto itself. The latter statement, in turn, means the existence of a certain intrinsic symmetry in that spacetime.

15.2.3 *Example: The Poincaré group*

As an illustration of the above discussion, let us assume that our spacetime is the Minkowskian flat spacetime. Thus $g_{\mu\nu} = \eta_{\mu\nu}$, where $\eta_{\mu\nu}$ is the

Killing introduced Lie algebras in "Programmschrift" (1884) published by the Lyceum Hosianum in Braunsberg. His aim was to systematically study all geometries with specific properties relating to infinitesimal motions. He translated this geometrical aim into the problem of classifying all finite dimensional real Lie algebras. At this stage Killing was not aware of Lie's work and therefore his definition of a Lie algebra was made quite independently of Lie. He was examining conditions on the Lie algebra which he studied for their geometrical significance.

Killing sent Klein a copy of "Programmschrift" in July 1884, and Klein wrote him that his work was closely related to algebras on which Sophus Lie had published a number of papers over the preceding ten years. In October 1885 Killing wrote to Lie, requesting copies of Lie's papers, and Lie sent copies of his papers to Killing, who returned them in March 1886. He had not had time to fully appreciate all that they contained. However Killing had also written to Engel in November 1885 and they started a long scientific correspondence which was helpful to them both.

When Killing wrote to Engel in April 1887 he had come up with the definition of a semisimple Lie algebra, and by 18 October he had discovered the complete list of simple algebras. However, he did not have concrete representations of these algebras.

Finally, it is worth noting that Killing introduced the term "characteristic equation" of a matrix.

It was Cartan, in his doctoral thesis submitted in 1894, who found concrete representations of all the exceptional simple Lie algebras. He also reworked Killing's proofs to make them more easily understood.

In 1892 Killing returned to Münster as professor of mathematics and he spent the rest of his life there submerged in teaching, administration and charitable work. He was rector of the University of Münster in 1897-98. He always upheld tradition and disliked change.

Killing was honored with the award of the Lobachevsky Prize by the Kazan Physico-mathematical Society in 1900. This was the second award made of the Prize, the first in 1897 going to Lie.

The collapse of social cohesion in Germany after 1918 caused Killing much pain in his last years as he was a great patriot. He had already suffered the loss of two infant sons, but even more devastating was the loss of his remaining two sons, one of whom died in 1910 while working for his habilitation on a topic on the history of music, the other became ill in an army camp and died briefly before end of World War I in 1918.

The above description is based on the article by J J O'Connor and E F Robertson.

Minkowskian metric,

$$\eta_{\mu\nu} = \begin{pmatrix} +1 & 0 & 0 & 0 \\ 0 & -1 & 0 & 0 \\ 0 & 0 & -1 & 0 \\ 0 & 0 & 0 & -1 \end{pmatrix}. \tag{15.2.5}$$

Using the Minkowskian metric in the Killing equation (15.2.4), the latter gives

$$\frac{\partial \xi_\mu}{\partial x^\nu} + \frac{\partial \xi_\nu}{\partial x^\mu} = 0. \tag{15.2.6}$$

For $\mu = \nu = 0, 1, 2, 3$ the above equation yields the following conditions:

$$\frac{\partial \xi_0}{\partial x^0} = \frac{\partial \xi_1}{\partial x^1} = \frac{\partial \xi_2}{\partial x^2} = \frac{\partial \xi_3}{\partial x^3} = 0, \tag{15.2.7}$$

and therefore

$$\xi_0 = \xi_0\left(x^1, x^2, x^3\right), \qquad \xi_1 = \xi_1\left(x^0, x^2, x^3\right),$$

$$\xi_2 = \xi_2\left(x^0, x^1, x^3\right), \qquad \xi_3 = \xi_3\left(x^0, x^1, x^2\right). \tag{15.2.8}$$

In addition to Eqs. (15.2.8), the components of Killing vector ξ_μ satisfy the following six relations:

$$\frac{\partial \xi_0}{\partial x^n} + \frac{\partial \xi_n}{\partial x^0} = 0, \qquad \frac{\partial \xi_m}{\partial x^n} + \frac{\partial \xi_n}{\partial x^m} = 0, \tag{15.2.9}$$

where $m, n = 1, 2, 3$ and $m \neq n$.

The solution of the above system of equations is then given by

$$\xi_\mu\left(x\right) = \varepsilon_{\mu\nu} x^\nu + \zeta_\mu, \tag{15.2.10}$$

where $\varepsilon_{\mu\nu}$ and ζ_μ are some constants, with $\varepsilon_{\mu\nu}$ being antisymmetric, $\varepsilon_{\mu\nu} = -\varepsilon_{\nu\mu}$. Using matrix notation, the above formula gives

$$\begin{pmatrix} \xi_0 \\ \xi_1 \\ \xi_2 \\ \xi_3 \end{pmatrix} = \begin{pmatrix} 0 & \varepsilon_{01} & \varepsilon_{02} & \varepsilon_{03} \\ -\varepsilon_{01} & 0 & \varepsilon_{12} & \varepsilon_{13} \\ -\varepsilon_{02} & -\varepsilon_{12} & 0 & \varepsilon_{23} \\ -\varepsilon_{03} & -\varepsilon_{13} & -\varepsilon_{23} & 0 \end{pmatrix} \begin{pmatrix} x^0 \\ x^1 \\ x^2 \\ x^3 \end{pmatrix} + \begin{pmatrix} \zeta_0 \\ \zeta_1 \\ \zeta_2 \\ \zeta_3 \end{pmatrix} \tag{15.2.11}$$

for the covariant components of the Killing vector.

The contravariant components of the Killing vector ξ^μ may be obtained, as usual, by

$$\xi^\mu\left(x\right) = \eta^{\mu\alpha} \xi_\alpha\left(x\right) = \varepsilon^\mu_\alpha x^\alpha + \zeta^\mu, \tag{15.2.12}$$

where

$$\varepsilon^\mu_\alpha = \eta^{\mu\nu} \varepsilon_{\nu\alpha}, \qquad \zeta^\mu = \eta^{\mu\nu} \zeta_\nu. \tag{15.2.13}$$

Hence, using matrix notation, we obtain

$$
\begin{pmatrix} \xi^0 \\ \xi^1 \\ \xi^2 \\ \xi^3 \end{pmatrix} = \begin{pmatrix} 0 & \varepsilon_{01} & \varepsilon_{02} & \varepsilon_{03} \\ \varepsilon_{01} & 0 & -\varepsilon_{12} & \varepsilon_{31} \\ \varepsilon_{02} & \varepsilon_{12} & 0 & -\varepsilon_{23} \\ \varepsilon_{03} & -\varepsilon_{31} & \varepsilon_{23} & 0 \end{pmatrix} \begin{pmatrix} x^0 \\ x^1 \\ x^2 \\ x^3 \end{pmatrix} + \begin{pmatrix} \zeta^0 \\ \zeta^1 \\ \zeta^2 \\ \zeta^3 \end{pmatrix} \tag{15.2.14}
$$

The geometrical meaning of the above solution for the Killing equation is as follows.

The set of the four parameters ζ^μ describes the four *translations* in the Minkowskian spacetime along the axes $0x^0$, $0x^1$, $0x^2$, and $0x^3$. They are the *infinitesimal generators* of the *translational subgroup* of the Poincaré group (inhomogeneous Lorentz group), the symmetry group of the Minkowskian flat spacetime.

The other set of the six parameters $\varepsilon_{\mu\nu}$, on the other hand, describes the six *Lorentz rotations* in the flat spacetime. Each one of the six parameters describes either a three-dimensional *rotation* or a homogeneous Lorentz transformation (boost). The set of ε_{23}, ε_{31}, and ε_{12} describes the three-dimensional rotations around the axes $0x^1$, $0x^2$, and $0x^3$, respectively. The set ε_{01}, ε_{02}, and ε_{03}, on the other hand, describes the Lorentz boosts along the axes $0x^1$, $0x^2$, and $0x^3$, respectively.

We can also write the *infinitesimal matrices* corresponding to each one of the above six Lorentz rotations in the Minkowskian space. The infinitesimal matrices for the three-dimensional rotations are then given by the three matrices

$$
a_1 = \begin{pmatrix} 0 & 0 & 0 & 0 \\ 0 & 0 & 0 & 0 \\ 0 & 0 & 0 & -1 \\ 0 & 0 & 1 & 0 \end{pmatrix}, \quad a_2 = \begin{pmatrix} 0 & 0 & 0 & 0 \\ 0 & 0 & 0 & 1 \\ 0 & 0 & 0 & 0 \\ 0 & -1 & 0 & 0 \end{pmatrix},
$$

$$
a_3 = \begin{pmatrix} 0 & 0 & 0 & 0 \\ 0 & 0 & -1 & 0 \\ 0 & 1 & 0 & 0 \\ 0 & 0 & 0 & 0 \end{pmatrix}, \tag{15.2.15}
$$

while those for the Lorentz transformations are given by

$$
b_1 = \begin{pmatrix} 0 & 1 & 0 & 0 \\ 1 & 0 & 0 & 0 \\ 0 & 0 & 0 & 0 \\ 0 & 0 & 0 & 0 \end{pmatrix}, \quad b_2 = \begin{pmatrix} 0 & 0 & 1 & 0 \\ 0 & 0 & 0 & 0 \\ 1 & 0 & 0 & 0 \\ 0 & 0 & 0 & 0 \end{pmatrix},
$$

$$b_3 = \begin{pmatrix} 0 & 0 & 0 & 1 \\ 0 & 0 & 0 & 0 \\ 0 & 0 & 0 & 0 \\ 1 & 0 & 0 & 0 \end{pmatrix}, \qquad (15.2.16)$$

around and along the axes $0x^1$, $0x^2$, and $0x^3$, respectively.

The 4×4 matrix, appearing in the Killing vector (15.2.14), may then be written as a linear combination of the above infinitesimal matrices as follows:

$$(\varepsilon_{23}a_1 + \varepsilon_{31}a_2 + \varepsilon_{12}a_3) + (\varepsilon_{01}b_1 + \varepsilon_{02}b_2 + \varepsilon_{03}b_3). \qquad (15.2.17)$$

Furthermore we may also verify that the infinitesimal Lorentz matrices indeed satisfy the usual commutation relations of the homogeneous Lorentz group. For if we define the matrics J_m and K_n by means of

$$J_m = i a_m, \qquad K_n = i b_n, \qquad (15.2.18)$$

with m, n=1,2,3, then we find that

$$[J_i, \ J_j] = i\varepsilon_{ijk}J_k, \qquad (15.2.19a)$$

$$[K_i, \ K_j] = -i\varepsilon_{ijk}J_k, \qquad (15.2.19b)$$

$$[J_i, \ K_j] = i\varepsilon_{ijk}K_k. \qquad (15.2.19c)$$

In the above formulas $[A, \ B] = AB - BA$, and ε_{ijk} is the three-dimensional skew-symmetric tensor with values $+1$ or -1, depending upon whether ijk is an even or an odd permutation of 123, and zero otherwise.

The above commutation relations may also be simplified into the single formula

$$[J_{\kappa\lambda}, \ J_{\mu\nu}] = i\left(\delta_{\kappa\mu}J_{\lambda\nu} + \delta_{\lambda\nu}J_{\kappa\mu} - \delta_{\kappa\nu}J_{\lambda\mu} - \delta_{\lambda\mu}J_{\kappa\nu}\right), \qquad (15.2.20)$$

where $J_{\mu\nu}$ is skew-symmetric in the indices μ and ν, and is related to J_k and K_m by means of

$$J_k = \frac{1}{2}\varepsilon_{klm}J_{lm}, \qquad K_n = iJ_{0n}. \qquad (15.2.21)$$

The *Lorentz matrices*, describing the finite three-dimensional rotations and the finite Lorentz transformations around and along the axes $0x^1$, $0x^2$, and $0x^3$ are then obtained from the infinitesimal matrices a_k and b_k by exponentiation. We obtain

$$a_k\left(\psi\right) = \exp\left(\psi a_k\right), \qquad b_k\left(\psi\right) = \exp\left(\psi b_k\right), \qquad (15.2.22)$$

with k=1, 2, 3.

We obtain, for instance, for $a_1(\psi)$ the following expression:

$$a_1(\psi) = \exp(\psi a_1) = I + \psi a_1 + \frac{\psi^2}{2!} a_1^2 + \frac{\psi^3}{3!} a_1^3 + \ldots,$$

where I is the 4×4 unit matrix. A simple calculation then shows that

$$a_1^{2m+1} = (-1)^m a_1, \qquad a_1^{2m} = (-1)^{m+1} a_1^2,$$

for $m=1,\ 2,\ 3,\ldots$. The matrix a_1 is given by Eq. (15.2.15), whereas a_1^2 is given by

$$a_1^2 = \begin{pmatrix} 0 & 0 & 0 & 0 \\ 0 & 0 & 0 & 0 \\ 0 & 0 & -1 & 0 \\ 0 & 0 & 0 & -1 \end{pmatrix}.$$

As a result one obtains

$$a_1(\psi) = I + \left(\psi - \frac{\psi^3}{3!} + \cdots\right) a_1 + \left(\frac{\psi^2}{2!} - \frac{\psi^4}{4!} + \cdots\right) a_1^2$$

$$= I + \sin\psi\, a_1 + (1 - \cos\psi)\, a_1^2$$

or, using an expression for the matrices a_1 and a_1^2,

$$a_1(\psi) = \begin{pmatrix} 1 & 0 & 0 & 0 \\ 0 & 1 & 0 & 0 \\ 0 & 0 & \cos\psi & -\sin\psi \\ 0 & 0 & \sin\psi & \cos\psi \end{pmatrix}. \tag{15.2.23a}$$

In the same way we obtain

$$a_2(\psi) = \begin{pmatrix} 1 & 0 & 0 & 0 \\ 0 & \cos\psi & 0 & \sin\psi \\ 0 & 0 & 1 & 0 \\ 0 & -\sin\psi & 0 & \cos\psi \end{pmatrix}, \tag{15.2.23b}$$

$$a_3(\psi) = \begin{pmatrix} 1 & 0 & 0 & 0 \\ 0 & \cos\psi & -\sin\psi & 0 \\ 0 & \sin\psi & \cos\psi & 0 \\ 0 & 0 & 0 & 1 \end{pmatrix}. \tag{15.2.23c}$$

Likewise, for the Lorentz transformations we obtain

$$b_1(\psi) = \begin{pmatrix} \cosh\psi & \sinh\psi & 0 & 0 \\ \sinh\psi & \cosh\psi & 0 & 0 \\ 0 & 0 & 1 & 0 \\ 0 & 0 & 0 & 1 \end{pmatrix}. \tag{15.2.24a}$$

$$b_2(\psi) = \begin{pmatrix} \cosh\psi & 0 & \sinh\psi & 0 \\ 0 & 1 & 0 & 0 \\ \sinh\psi & 0 & \cosh\psi & 0 \\ 0 & 0 & 0 & 1 \end{pmatrix}. \qquad (15.2.24b)$$

$$b_3(\psi) = \begin{pmatrix} \cosh\psi & 0 & 0 & \sinh\psi \\ 0 & 1 & 0 & 0 \\ 0 & 0 & 1 & 0 \\ \sinh\psi & 0 & 0 & \cosh\psi \end{pmatrix}. \qquad (15.2.24c)$$

We note that the infinitesimal matrices a_k and b_k can be obtained from the Lorentz matrices $a_k(\psi)$ and $b_k(\psi)$ by

$$a_k = \frac{da_k(\psi)}{d\psi}\,|_{\psi=0}, \qquad b_k = \frac{db_k(\psi)}{d\psi}\,|_{\psi=0}, \qquad (15.2.25)$$

where $k = 1,\,2,\,3$.

In Eqs (15.2.23) ψ denotes the angle of a three-dimensional rotation between two Lorentz frames. In Eqs. (15.2.24), however, ψ denotes the "angle of rotation" between two Lorentz frames moving with respect to each other. If v is the relative velocity between the two coordinate systems, then ψ is related to v by the following:

$$\sinh\psi = \frac{-v/c}{\sqrt{1 - v^2/c^2}}, \qquad \cosh\psi = \frac{1}{\sqrt{1 - v^2/c^2}}. \qquad (15.2.26)$$

In conclusion, we have obtained 10 independent Killing vectors as solutions of the Killing equation in the Minkowskian space. These solutions represent the 10 parameters of the Poincaré group. This is also the maximum number of solutions to the Killing equation, and therefore the Minkowskian space is the most symmetric spacetime.

In the next section we make a simple classification of the gravitational field in terms of use of the Killing equation.

15.2.4   *Problems*

P 15.2.1. Solve the Killing equation corresponding to the *Euclidean group* in the plane $E(2)$.

Solution: Using Cartesian coordinates, the metric in the plane is given by $g_{AB} = \delta_{AB}$, where A, $B=1,\,2$. The Killing equation then reduces to

$$\frac{\partial\xi^A}{\partial x^B} + \frac{\partial\xi^B}{\partial x^A} = 0, \qquad (1)$$

Table 15.1: Enumeration of, and canonical structure constants for, the Bianchi types.

Class	G_3A						G_3B				
Type	I	II	VI_0	VII_0	VIII	IX	V	IV	III	VI_h	VII_h
Rank (N^{DE})	0	1	2	2	3	3	0	1	2	2	2
Signature (N^{DE})	0	1	0	2	1	3	0	1	0	0	2
A	0	0	0	0	0	0	1	1	1	$\sqrt{-h}$	$\sqrt{h}$
N_1	0	1	0	0	-1	1	0	0	0	0	0
N_2	0	0	-1	1	1	1	0	0	-1	-1	1
N_3	0	0	1	1	1	1	0	1	1	1	1
Dimensions of canonical basis freedom	9	6	4	4	3	3	6	4	4	4 (each $h < 0$)	4 (each $h > 0$)

which gives

$$\frac{\partial \xi^1}{\partial x} = \frac{\partial \xi^2}{\partial y} = 0, \qquad \frac{\partial \xi^1}{\partial y} + \frac{\partial \xi^2}{\partial x} = 0, \tag{2}$$

if we denote $x^1 = x$ and $x^2 = y$. The solutions of Eqs. (2) are given by

$$\xi^1 = Y(y), \qquad \xi^2 = X(x), \tag{3}$$

along with the additional condition

$$\frac{dX(x)}{dx} + \frac{dY(y)}{dy} = 0. \tag{4}$$

Using the method of separation of variables, we obtain

$$X(x) = \phi x + y_0, \qquad Y(y) = -\phi y + x_0, \tag{5}$$

or

$$\xi^1 = -\phi y + x_0, \qquad \xi^2 = \phi x + y_0. \tag{6}$$

In the above equations ϕ is a separation constant, and x_0 and y_0 are some constants, too.

The above solution to the Killing equation shows that we have three degrees of freedom, describing the infinitesimal group of motions of the Euclidean plane in two dimensions. The two parameters x_0 and y_0 correspond to translations in the plane along the x axis and the y axis, respectively, whereas ϕ corresponds to the rotations around the origin $x = y = 0$.

We may also write the above results in terms of the isometric mapping (15.2.1). Assume first that $\phi = 0$ in the solution (6). Hence we obtain, using Eqs. (15.2.1) and (6),

$$\tilde{x} = x + x_0, \qquad \tilde{y} = y + y_0, \tag{7}$$

Table 15.2: Expressions for Killing vectors and reciprocal group generators in canonical coordinates for the Bianchi types.

	ξ_A	η_A	ω_A
I	∂_x ∂_y ∂_z	∂_x ∂_y ∂_z	dx dy dz
II	∂_x ∂_y $\partial_z + z\partial_x$	∂_x $\partial_y + z\partial_x$ ∂_z	$dx - zdy$ dy dz
IV	$\partial_x - y\partial_y - (y+z)\partial_z$ ∂_y ∂_z	∂_x $e^{-x}(\partial_y - x\partial_z)$ $e^{-x}\partial_z$	dx $e^x dy$ $e^x(dz + xdy)$
V	$\partial_x - y\partial_y - z\partial_z$ ∂_y ∂_z	∂_x $e^{-x}\partial_y$ $e^{-x}\partial_z$	dx $e^x dy$ $e^x dz$
VI (including III)	$\partial_x + (z - Ay)\partial_y + (y - Az)\partial_z$ ∂_y ∂_z	∂_x $e^{-Ax}(\cosh x\,\partial_y + \sinh x\,\partial_z)$ $e^{-Ax}(\sinh x\,\partial_y + \cosh x\,\partial_z)$	dx $e^{Ax}(\cosh x\,dy - \sinh x\,dz)$ $e^{Ax}(-\sinh x\,dy + \cosh x\,dz)$
VII	$\partial_x + (z - Ay)\partial_y - (y + Az)\partial_z$ ∂_y ∂_z	∂_x $e^{-Ax}(\cos x\,\partial_y - \sin x\,\partial_z)$ $e^{-Ax}(\sin x\,\partial_y + \cos x\,\partial_z)$	dx $e^{Ax}(\cos x\,dy - \sin x\,dz)$ $e^{Ax}(\sin x\,dy + \cos x\,dz)$

	ξ_A	η_A	ω_A
VIII	∂_x $-\sinh x \tanh y\,\partial_x + \cosh x\,\partial_y - \sinh x \operatorname{sech} y\,\partial_z$ $\cosh x \tanh y\,\partial_x - \sinh x\,\partial_y \cosh x \operatorname{sech} y\,\partial_z$	$\operatorname{sech} y \cos z\,\partial_x - \sin z\,\partial_y - \tanh y \cos z\,\partial_z$ $\operatorname{sech} y \sin z\,\partial_x + \cos z\,\partial_y - \tanh y \sin z\,\partial_z$ ∂_z	$\cosh y \cos z\,dx - \sin z\,dy$ $\cosh y \sin z\,dx + \cos z\,dy$ $\sinh y\,dx + dz$
IX	∂_x $\sin x \tan y\,\partial_x + \cos x\,\partial_y + \sin x \sec y\,\partial_z$ $\cos x \tan y\,\partial_x - \sin x\,\partial_y \cos x \sec y\,\partial_z$	$\sec y \cos z\,\partial_x - \sin z\,\partial_y + \tan y \cos z\,\partial_z$ $\sec y \sin z\,\partial_x + \cos z\,\partial_y + \tan y \sin z\,\partial_z$ ∂_z	$\cos y \cos z\,dx - \sin z\,dy$ $\cos y \sin z\,dx + \cos z\,dy$ $-\sin y\,dx + dz$

where we have substituted x_0 and y_0 for εx_0 and εy_0, for simplicity. Assume now that $x_0 = y_0 = 0$. We then obtain

$$\begin{pmatrix} \tilde{x} \\ \tilde{y} \end{pmatrix} = \left\{ \begin{pmatrix} 1 & 0 \\ 0 & 1 \end{pmatrix} + \phi \begin{pmatrix} 0 & -1 \\ 1 & 0 \end{pmatrix} \right\} \begin{pmatrix} x \\ y \end{pmatrix}$$

$$\approx \begin{pmatrix} \cos\phi & -\sin\phi \\ \sin\phi & \cos\phi \end{pmatrix} \begin{pmatrix} x \\ y \end{pmatrix}, \tag{8}$$

where ϕ stands for $\varepsilon\phi$. The transformation (8) describes an infinitesimal rotation with the angle ϕ around the origin.

P 15.2.2. Find the relationship between the "angles of rotation" between two Lorentz frames and their relative velocity.

Solution: Let us assume a Lorentz transformation between two systems that is given by the matrix (15.2.24a). Denote the two systems by K and K', and assume that their coordinates are given by x^μ and x'^μ. If we denote $x^0 = ct$, $x^1 = x$, $x^2 = y$, $x^3 = z$, and similarly for the primed system, then we may write

$$ct' = ct\cosh\psi + x\sinh\psi, \tag{1a}$$

$$x' = ct\sinh\psi + x\cosh\psi, \tag{1b}$$

$$y' = y, \tag{1c}$$

$$z' = z. \tag{1d}$$

The motion of the origin of the coordinate system K, as seen from K', is found by putting $x = 0$. This gives

$$t' = t\cosh\psi, \qquad x' = ct\sinh\psi, \qquad y' = y, \qquad z' = z. \tag{2}$$

If v is the relative velocity between the two systems, then

$$\frac{x'}{t'} = -v = c\tanh\psi. \tag{3}$$

Thus

$$\sinh\psi = \frac{-v/c}{\sqrt{1 - v^2/c^2}}, \qquad \cosh\psi = \frac{1}{\sqrt{1 - v^2/c^2}}. \tag{4}$$

The Lorentz transformation along the x axis is therefore given by

$$t' = \frac{t - vx/c^2}{\sqrt{1 - v^2/c^2}}, \qquad x' = \frac{x - vt}{\sqrt{1 - v^2/c^2}}, \qquad y' = y, \qquad z' = z. \tag{5}$$

In the same way one finds other Lorentz transformations along the y axis and the z axis.

15.3 Bianchi Types

The dimension of the subgroup of the linear transformations that preserves the canonical form is given in Table 15.1 (for proof see Siklos, 1976).

Using Table 15.1 we obtain Table 15.2 as was originally given by Bianchi (Bianchi 1897).

The Bianchi classification is less useful than the well-known Petrov classification of gravitational fields (Carmeli 1982). The possibility of presenting the gravitational field as a Hilbert space was given by the author (Carmeli 1969).

15.4 Suggested References

L. Bianchi, On three-dimensional spaces which admit a group of motions (Sugli spazii a tre dimensioni che ammettono un gruppo continuo di movimenti), *Soc. Ital. Sci. Mem di Mat.* **11**, 267 (1897).

F. Brickell and R.S. Clark, *Differentiable Manifolds: An Introduction* (Van Nostrand Reinhold, London, 1970).

M. Carmeli, "Hilbert Space Description of the Gravitational Field," *Phys. Lett. A* **28**, 683 (1969).

M. Carmeli, *Classical Fields: General Relativity and Gauge Theory* (John Wiley, New York, 1982; reprinted by World Scientific, Singapore, 2001).

C. Chevalley, *Theory of Lie Groups* (Princeton University Press, Princeton, 1946).

P.M. Cohn, *Lie Groups* (Cambridge University Press, 1957).

L. Defrise, *Groupes d'isotropie et groupes de stabilité conforme dans les espaces lorentziens* (Thése, Université Libre de Bruxelles, 1969).

L.P. Eisenhart, *Continuous Groups of Transformations* (Princeton University Press, 1933).

D. Kramer, H. Stephani, M. Maccallum, E. Herlt and Ernst Schmutzer, (Eds.), *Exact Solutions of Einstein's Field Equations* (VEB Deut-scher Verlag der Wissenschaften, Berlin, 1980).

A. Krasinski, *Inhomogeneous Cosmological Models* (Cambridge University Press, 1997).

L. Landau and E. Lifshitz, *The Classical Theory of Fields*, 4-th Edition (Addison-Wesley Publishing Company, Inc., Reading, Massachusetts, 1975).

F.D. Murnaghan, *The Theory of Group Representations* (Dover Publications, Inc., New York, 1938).

A.Z. Petrov, *Einstein Spaces* (Pergamon Press, London, 1969).

G. Racah, *Group Theory and Spectroscopy* (Notes by Eugen Mer-zbacher and David Park, The Hebrew University, Jerusalem, 1951).

M.P. Ryan, Jr. and L.C. Shepley, *Homogeneous Relativistic Cosmologies* (Princeton University Press, Princeton, New Jersey, 1975).

A. Salam, Lectures on Group Theory (preprint, published by Atomic Energy Agency, 1960).

S.T.C. Siklos, *Singularities, Invariants and Cosmology*, Ph. D. Thesis, Cambridge, see Sect. 5.1, 8.2, 9.2, 10.2, 11.2.

F.W. Warner, *Foundations of Differentiable Manifolds and Lie Groups* (Scott, Foresman, Glenview, Illinois, 1971).

H. Weyl, *The Classical Groups, Their Invariants and Representation* (Princeton University Press, Princeton).

Appendix A

Mathematical Conventions

Moshe Carmeli

Throughout this appendix we use the convention

$$\alpha, \beta, \gamma, \delta, \cdots = 0, 1, 2, 3, 4,$$

$$a, b, c, d, \cdots = 0, 1, 2, 3,$$

$$p, q, r, s, \cdots = 1, 2, 3, 4,$$

$$k, l, m, n, \cdots = 1, 2, 3.$$

The coordinates are $x^0 = ct$, x^1, x^2 and x^3 (spacelike coordinates) $r^2 = (x^1)^2 + (x^2)^2 + (x^3)^2$, and $x^4 = \tau v$. The signature is $(+ - - - +)$. The metric, approximated up to ϕ and ψ, is:

$$g_{\mu\nu} = \begin{pmatrix} 1+\phi & 0 & 0 & 0 & 0 \\ 0 & -1 & 0 & 0 & 0 \\ 0 & 0 & -1 & 0 & 0 \\ 0 & 0 & 0 & -1 & 0 \\ 0 & 0 & 0 & 0 & 1+\psi \end{pmatrix}, \qquad (A.1)$$

$$g^{\mu\nu} = \begin{pmatrix} 1-\phi & 0 & 0 & 0 & 0 \\ 0 & -1 & 0 & 0 & 0 \\ 0 & 0 & -1 & 0 & 0 \\ 0 & 0 & 0 & -1 & 0 \\ 0 & 0 & 0 & 0 & 1-\psi \end{pmatrix}. \qquad (A.2)$$

The nonvanishing Christoffel symbols are (in the linear approximation):

$$\Gamma^0_{0\lambda} = \frac{1}{2}\phi_{,\lambda}, \qquad \Gamma^0_{44} = -\frac{1}{2}\psi_{,0}, \qquad \Gamma^n_{00} = \frac{1}{2}\phi_{,n}, \qquad (A.3a)$$

$$\Gamma^n_{44} = \frac{1}{2}\psi_{,n}, \qquad \Gamma^4_{00} = -\frac{1}{2}\phi_{,4}, \qquad \Gamma^4_{4\lambda} = \frac{1}{2}\psi_{,\lambda}, \tag{A.3b}$$

$$\Gamma^a_{00} = -\frac{1}{2}\eta^{ab}\phi_{,b}, \qquad \Gamma^a_{44} = -\frac{1}{2}\eta^{ab}\psi_{,b}, \tag{A.3c}$$

$$\Gamma^p_{00} = -\frac{1}{2}\eta^{pq}\phi_{,q}, \qquad \Gamma^p_{44} = -\frac{1}{2}\eta^{pq}\psi_{,q}. \tag{A.3d}$$

The Minkowskian metric η in five dimensions is given by

$$\eta = \begin{pmatrix} 1 & 0 & 0 & 0 & 0 \\ 0 & -1 & 0 & 0 & 0 \\ 0 & 0 & -1 & 0 & 0 \\ 0 & 0 & 0 & -1 & 0 \\ 0 & 0 & 0 & 0 & 1 \end{pmatrix}. \tag{A.4}$$

A.1 Components of the Ricci tensor

The elements of the Ricci tensor are:

$$R_{00} = \frac{1}{2}\left(\nabla^2\phi - \phi_{,44} - \psi_{,00}\right), \tag{A.5}$$

$$R_{0n} = -\frac{1}{2}\psi_{,0n}, \qquad R_{04} = 0, \tag{A.6}$$

$$R_{mn} = -\frac{1}{2}\left(\phi_{,mn} + \psi_{,mn}\right), \tag{A.7}$$

$$R_{4n} = -\frac{1}{2}\phi_{,4n}, \tag{A.8}$$

$$R_{44} = \frac{1}{2}\left(\nabla^2\psi - \phi_{,44} - \psi_{,00}\right). \tag{A.9}$$

The Ricci scalar is

$$R = \nabla^2\phi + \nabla^2\psi - \phi_{,44} - \psi_{,00}. \tag{A.10}$$

The mixed Ricci tensor is given by

$$R^0_0 = \frac{1}{2}\left(\nabla^2\phi - \phi_{,44} - \psi_{,00}\right), \tag{A.11}$$

$$R^n_0 = \frac{1}{2}\psi_{,0n}, \qquad R^0_n = -\frac{1}{2}\psi_{,0n}, \tag{A.12}$$

$$R^4_0 = R^0_4 = 0, \tag{A.13}$$

$$R^n_m = \frac{1}{2}\left(\phi_{,mn} + \psi_{,mn}\right), \tag{A.14}$$

$$R^4_n = -\frac{1}{2}\phi_{,n4}, \qquad R^n_4 = \frac{1}{2}\phi_{,n4}, \tag{A.15}$$

$$R^4_4 = \frac{1}{2}\left(\nabla^2\psi - \phi_{,44} - \psi_{,00}\right). \tag{A.16}$$

Appendix B

Integration of the Equation of the Universe Expansion

Moshe Carmeli

The Universe expansion was shown to be given by Eq. (7.3.13),

$$\frac{dr}{dv} = \tau\sqrt{1 + \frac{(1-\Omega)\,r^2}{c^2\tau^2}}. \qquad (B.1)$$

This equation can be integrated exactly by the substitutions

$$\sin\chi = \frac{\alpha r}{c\tau}; \qquad \Omega > 1 \qquad (B.1a)$$

$$\sinh\chi = \frac{\beta r}{c\tau}; \qquad \Omega < 1 \qquad (B.1b)$$

where

$$\alpha = \sqrt{\Omega - 1}, \qquad \beta = \sqrt{1 - \Omega}. \qquad (B.2)$$

For the $\Omega > 1$ case a straightforward calculation using Eq. (B.1a) gives

$$dr = \frac{c\tau}{\alpha}\cos\chi d\chi \qquad (B.3)$$

and the equation of the Universe expansion (B.1) yields

$$d\chi = \frac{\alpha}{c}dv. \qquad (B.4a)$$

The integration of this equation gives

$$\chi = \frac{v}{c}\alpha + \text{const.} \qquad (B.5a)$$

The constant can be determined using Eq. (B.1a). At $\chi = 0$, we have $r = 0$ and $v = 0$, thus

$$\chi = \frac{v}{c}\alpha, \qquad (B.6a)$$

407

or, in terms of the distance, using (B.1a) again,

$$r\left(v\right) = \frac{c\tau}{\alpha}\sin\frac{v}{c}\alpha; \qquad \alpha = \sqrt{\Omega - 1}. \tag{B.7a}$$

This is obviously a decelerating expansion .

For $\Omega < 1$, using Eq. (B.1b), a similar calculation yields for the Universe expansion (B.1)

$$d\chi = \frac{\beta}{c}dv, \tag{B.4b}$$

thus

$$\chi = \frac{v}{c}\beta + \text{const.} \tag{B.5b}$$

Using the same initial conditions as above then gives

$$\chi = \frac{v}{c}\beta \tag{B.6b}$$

and in terms of distances,

$$r\left(v\right) = \frac{c\tau}{\beta}\sinh\frac{v}{c}\beta; \qquad \beta = \sqrt{1 - \Omega}. \tag{B.7b}$$

This is now an accelerating expansion.

For $\Omega = 1$ we have, from Eq. (B.1),

$$\frac{d^2 r}{dv^2} = 0. \tag{B.4c}$$

The solution is, of course,

$$r\left(v\right) = \tau v. \tag{B.7c}$$

This is a constant expansion.

Finally we calculate the Hubble constant H_0 from Eq. (B.1). From the Hubble expansion formula

$$v = H_0 r, \tag{B.8}$$

we obtain, since H_0 is constant at the instant of the observation,

$$\frac{dv}{dr} = H_0. \tag{B.9}$$

Using now Eq. (B.1) we obtain

$$H_0 = \frac{1}{\tau}\frac{1}{\sqrt{1 + \dfrac{\left(1 - \Omega\right)r^2}{c^2\tau^2}}}. \tag{B.10}$$

Appendix C

Spheroidal and Elliptical Galaxy Velocity Dispersion from CGR

John Hartnett

Rotational velocity dispersion in spheroidal and elliptical galaxies, as a function of radial distance from the center of the galaxy, is derived from Cosmological Special Relativity. For velocity dispersions in the outer regions of spherical galaxies, the dynamical mass calculated for a galaxy using Carmelian theory may be 10 to 100 times less than that calculated from standard Newtonian physics. This means there is no need to include halo dark matter. The velocity dispersion is found to be approximately constant across the galaxy after falling from an initial high value at the center. The following is based on Hartnett 2008.

C.1 Introduction

The motion of stars or the motion of gases as characterized by the spectroscopic detection of neutral hydrogen and other gases in spheroidal and elliptical galaxies has caused concern for astronomers for many decades. Newton's law of gravitation predicts much lower radial velocity dispersions than those measured. This has led to the assumption of the existence of halo 'dark matter' surrounding galaxies but transparent to all forms of electromagnetic radiation.

Carmeli (2000 and 2002) formulated an extension of Einstein general relativity theory, in an expanding Universe taking into account the Hubble expansion as a fundamental axiom, which imposes an additional constraint on the dynamics of particles (Carmeli 1982). Carmeli believes the usual assumptions in deriving Newton's gravitational force law from general relativity are insufficient, that gases and stars in galaxies are not immune from

409

Hubble flow. As a consequence a universal constant a_0 is introduced as a characteristic acceleration in the cosmos.

Using spherical coordinates Carmeli (1998) theoretically derived the Tully-Fisher law for galaxies; see Section 7.4. Following Carmeli's lead, Hartnett (2006) showed that the same line of reasoning leads to plausible galaxy rotation curves in spiral galaxies using cylindrical coordinates; see Chapter 10.

In this Appendix we take the analysis further, using spherical coordinates and model the gravitational potential and the resulting forces determining how test particles move in spheroidal and elliptical galaxies with an appropriate density distribution that reflects the observed luminous matter distribution. Similar to what was found in spiral galaxies (Hartnett 2006), two acceleration regimes are apparent here also. In one, normal Newtonian gravitation applies. In that regime the effect of the Hubble expansion is not observed or is extremely weak. It is as if particle accelerations are so great that they slip across the expanding space. In the other, new physics is needed. There the Carmelian metric provides it. In this regime the accelerations of particles are so weak that their motions are dominated by the Hubble expansion and as a result particles move under the combined effect of both the Newtonian force and a post-Newtonian contribution.

C.2   Gravitational Potential

In the weak gravitational limit, where Newtonian gravitation applies, it is sufficient to assume the Carmelian metric with non-zero elements $g_{00} = 1 + 2\Phi/c^2$, $g_{44} = 1 + 2\Psi/\tau^2$, $g_{kk} = -1$, $(k = 1, 2, 3)$ in the lowest approximations in both $1/c$ and $1/\tau$. The potential functions Φ and Ψ are determined by Einstein field equations and from their respective Poisson equations,

$$\nabla^2 \Phi = 4\pi G\rho, \qquad\qquad (C.1)$$

$$\nabla^2 \Psi = \frac{4\pi G\rho}{a_0^2}, \qquad\qquad (C.2)$$

where ρ is the mass density, a_0 a universal characteristic acceleration $a_0 = c/\tau$ and c is the speed of light in vacuum.

A comparison of Φ and Ψ in Eqs. (C.1) and (C.2) leads to $\Psi = \Phi/a_0^2$ within an arbitrary additive constant. Since both potentials are defined with respect to the same coordinate system, in reality, we only need deal with one potential function, the gravitational potential, Φ.

The density function that best describes the luminous distribution of matter in spherical and elliptical galaxies is

$$\rho(r) = \frac{M}{4\pi} \frac{r_c}{r^2(r+r_c)^2}, \tag{C.3}$$

where r_c is the core radius which contains half of the luminous matter. The parameter M is the total mass of the galaxy as measured at $r = \infty$. Therefore the mass at radius r is

$$M(r) = M \frac{r}{r+r_c}. \tag{C.4}$$

Spherical symmetry has been assumed to simplify the problem. To consider fully oblate spheroids would introduce some asymmetry to the problem but in principle the solution found here should broadly apply to any spherical matter distribution and would also apply, to first order, to oblate spheroids.

In spherical coordinates (r, θ, ϕ) the potential Φ that satisfies Eq. (C.1) can be found by integrating over Eq. (C.1) using Eq. (C.3). Only the radial dependence remains.

$$\Phi(r) = \frac{GM}{r_c} \left\{ \ln\left(\frac{1+r_c/\Delta}{1+r_c/r}\right) - \frac{r_c/\Delta}{1+r_c/\Delta} \right\}, \tag{C.5}$$

where Δ is the radial extent of the matter distribution and ln is the natural logarithm. This may be approximated where $\Delta \gg r_c$, which is the usual case. Hence we get

$$\Phi(r) = -\frac{GM}{r_c}\ln\left(1 + \frac{r_c}{r}\right). \tag{C.6}$$

C.3   Equations of Motion

The 5D line element for any two points in the CGR theory is $ds^2 = g_{00}c^2 dt^2 + g_{kk}(dx^k)^2 + g_{44}\tau^2 dv^2$, where $k = 1, 2, 3$. The relative separation in 3 spatial coordinates $r^2 = (x^1)^2 + (x^2)^2 + (x^3)^2$ and the relative velocity between points connected by ds is v. The Hubble-Carmeli constant, τ, is a universal constant for all observers. Equations (8.4.9a) and (8.4.10a) are used in the sequel to derive the appropriate equations of motion to lowest approximation in $1/c$.

C.3.1   *Newtonian*

It follows from Carmeli's equation (8.4.9a) using Eqs. (C.3) and (C.5), and the usual form of the circular motion equation

$$\frac{v^2}{r} = \frac{d\Phi}{dr}, \tag{C.7}$$

that

$$v^2 = \frac{GM}{r}\,\frac{1}{1 + r_c/r},\tag{C.8}$$

where G denotes the gravitational constant.

Equation (C.8) is the usual Newtonian result for the speed of circular motion in a spherical gravitational potential. This equation has been plotted in curve 1 of Figure C.3.1(a) as a function of radial position from the center of a galaxy in kiloparsecs (kpc) where kpc $\approx 3.08 \times 10^{19}$ m. Figure C.3.1(b) shows the corresponding accelerations. Curve 1 is the Newtonian acceleration. Throughout this appendix M is expressed in solar mass units $M_\odot \approx 2 \times 10^{30}$ kg.

C.3.2   *Carmelian*

Using $\Psi = \Phi/a_0^2$ in Carmeli's Eq. (8.4.10a) results in a new equation

$$v = a_0 \int_0^r \frac{dr}{\sqrt{-\Phi}},\tag{C.9}$$

which has been integrated and solved for v as a function of r. Using the potential Φ, determined from Eqs. (C.3) and (C.5), in Eq. (C.9), results in

$$v = \frac{2}{3}a_0\frac{r^{3/2}}{\sqrt{GM}},\tag{C.10}$$

which describes the expansion of space within a galaxy.

In Carmeli (1998 and 2000), using spherical coordinates, it was found that in the limit of large r and where all the matter was interior to the position of a test particle, such a particle is also subject to an additional circular motion described by Eq. (C.10). Apparently this is the result of the expansion of space itself within the galaxy but in an azimuthal direction to the usual center of coordinates of the galaxy.

Carmeli (1998 and 2000) determined a Tully-Fisher type relation using the Newtonian circular velocity equation expressed in spherical coordinates,

$$v^2 = \frac{GM}{r},\tag{C.11}$$

where it is assumed that test particles orbit at radius r outside of a fixed mass M. Then by eliminating r between Eqs. (C.11) and (C.10) we get the result. This is achieved by taking the 3/2 power of Eq. (C.11) and multiplying it with Eq. (C.10), yielding

$$v^4 = GM\frac{2}{3}a_0.\tag{C.12}$$

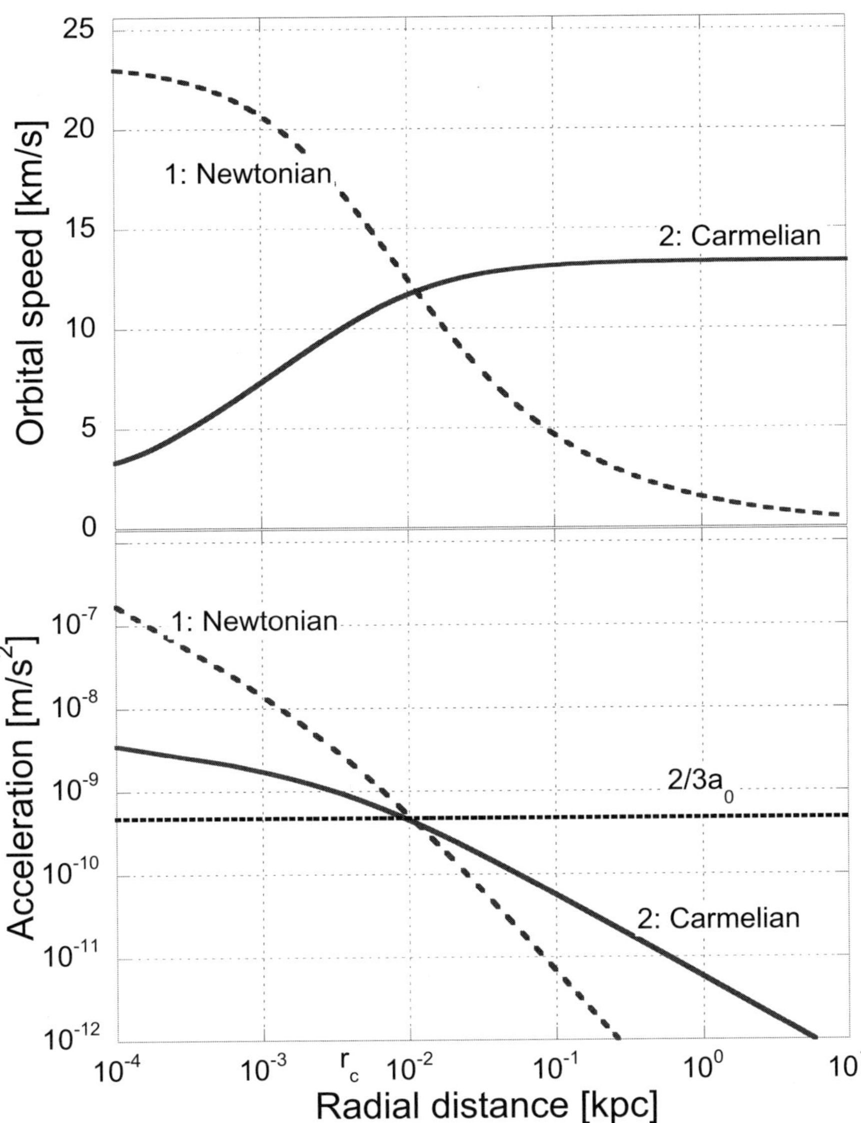

Fig. C.3.1: (a) Orbital velocity of a test particle in a dwarf spheroidal galaxy calculated from Eqs. (C.8) and (C.10) and where $M = 10^6 M_\odot$ and $r_c = 0.004$ kpc, indicated by the black vertical line; (b) Corresponding acceleration of a test particle for both the Newtonian and Carmelian regimes. Note that they coincide at the critical acceleration $2/3a_0$. (Source: Hartnett, 2008)

So by applying the same approach with Eq. (C.8), raising it to the 3/2 power and multiplying it with Eq. (C.10), we can derive an equation describing the circular motion of test particles in spherical galaxies. The result is

$$v^4 = GM\frac{2}{3}a_0 \left(\frac{r}{r+r_c}\right)^{3/2}. \qquad (C.13)$$

But in elliptical and spherical galaxies we generally don't observe any group circular motion . Therefore the trajectories of the individual orbits, though circular around the center of mass, are randomly oriented. Nevertheless we can calculate the orbital speed of a test particle from Eq. (C.13). Curve 2 of Figure C.3.1(a) shows one for the same parameters as is assumed for the Newtonian curve, as a function of radial position from the center of a galaxy. However for an individual particle if $r \gg r_c$ Eq. (C.13) recovers the form of the Tully-Fisher relation (C.12).

The corresponding acceleration may also be calculated from v^2/r using Eq. (C.13) and is shown in curve 2 of Figure C.3.1(b) for the same galaxy parameters. It is also compared with the critical acceleration $2/3a_0$.

C.4   Radial Velocity Dispersion

Assuming a spherically isotropic dynamic cloud of gases and stars, where the radial velocity dispersion is equal to the θ or ϕ dispersion (i.e. $\sigma_r^2 = \sigma_{\theta,\phi}^2$) the hydrostatic Jeans equation becomes

$$\frac{d}{dr}(\rho(r)\sigma_r^2) = -\rho(r)\frac{d\Phi}{dr}. \qquad (C.14)$$

C.4.1   *Newtonian*

Using Eqs. (C.7), (C.8) and (C.3) we solve the differential equation (C.14) for the radial velocity dispersion

$$\sigma_r^2 = \frac{1}{2}\frac{GM}{r_c}\Xi(x), \qquad (C.15)$$

where

$$\Xi(x) = (1 + 2x)(1 - 6x(1 + x)) + 12x^2(1 + x)^2\ln\left(\frac{x}{1+x}\right), \qquad (C.16)$$

and $x = r/r_c$. The dimensionless function $\Xi(r/r_c)$ is shown as curve 1 in Figure C.3.2.

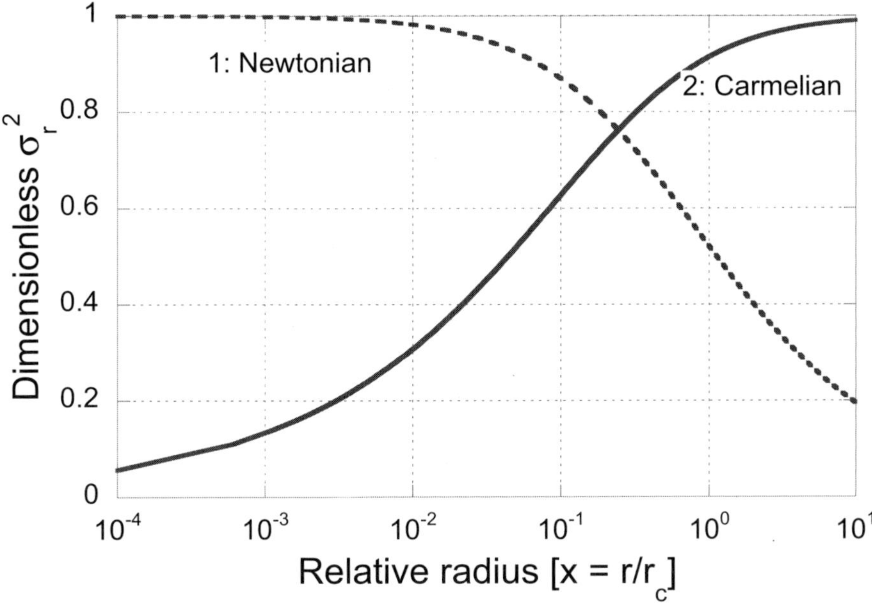

Fig. C.3.2: Dimensionless velocity dispersion squared. Curve 1 is $\Xi(x)$ from Eq. (C.16) and curve 2 is $\Gamma(x)$ from Eq. (C.20) as functions of the relative radius $x = r/r_c$. (Source: Hartnett, 2008)

For $r \gg r_c$ the function $\Xi(x) \approx 2/5x$ which means Eq. (C.15) becomes

$$\sigma_r^2 = \frac{GM}{5r}. \tag{C.17}$$

This is the same result determined from the virial theorem.

So in the limit where $r \to \infty$, the function $\Xi \to 0$ and where $r \to 0$, the function $\Xi \to 1$. From Newtonian theory we expect the radial velocity dispersion in spheroidal and elliptical galaxies to tend to zero where the gravitational acceleration is very weak. And where the gravitational acceleration is strongest $(r = 0)$ the radial dispersion is maximum and equal to the rotation velocity of a particle at twice the core radius.

C.4.2 *Carmelian*

Using Eq. (C.13) we can construct an equivalent gravitational potential which includes the effects of the Hubble flow on test particles in the galaxy. By taking v^2/r and using this in Eq. (C.14) instead of the gradient of the

Newtonian potential we get

$$\sigma_r^2 = \frac{1}{4}\left(GM\frac{2}{3}a_0\right)^{1/2}\Gamma(x), \qquad (C.18)$$

where

$$\Gamma(x) = \frac{16}{35}x^{3/4}(1+x)^{1/4}\{7 - 4x(21 - 32(x(1+x)^3)^{1/4} + ...$$

$$... + 8x(7 + 4x - 4(x(1+x)^3)^{1/4}))\}, \qquad (C.19)$$

and $x = r/r_c$. See curve 2 in Figure C.3.2 for $\Gamma(r/r_c)$.

In the limit where $r \to \infty$, the function $\Gamma \to 1$ and where $r \to 0$, the function $\Gamma \to 0$. From Carmelian theory then we expect the radial dispersion in spherical and elliptical galaxies to tend to zero where the gravitational acceleration is strongest ($r = 0$) and to tend to a constant where the gravitational acceleration is very weak. It also follows from Eq. (C.18) that for $r \gg r_c$

$$\sigma_r^4 = \frac{1}{16}\left(GM\frac{2}{3}a_0\right). \qquad (C.20)$$

C.4.3 *Discussion*

Similarly to what was seen in the case of rotation curves in Chapter 10 (Hartnett 2006), two regimes must be considered. When accelerations are greater than the critical acceleration $2/3a_0$ we have the Newtonian regime and we expect the velocity dispersion to follow Eq. (C.15) but when accelerations are less than $2/3a_0$ we have the Carmelian regime and we expect the velocity dispersion to follow Eq. (C.18). Notice that even though the velocity dispersion from the Newtonian calculation Eq. (C.15) scales with core radius (r_c), the velocity dispersion from the Carmelian calculation (C.18) is independent of core radius.

The Newtonian regime applies where the Newtonian acceleration $a_N > 2/3a_0$ and the Carmelian regime applies where the Carmelian acceleration $a_C < 2/3a_0$. In Figure C.4.3 radial velocity dispersion curves are shown for a dwarf spheroidal (curve 1) with $M = 10^6 \, M_\odot$ and a massive elliptical (curve 2) with $M = 10^{13} \, M_\odot$. Only the regions of the curves which represent valid velocity dispersion are retained. The results indicate that, as a function of radius from the center of the galaxy, the velocity dispersion should fall from a central high value to a constant value as r becomes much greater than r_c. This is commonly observed in these type of galaxies.

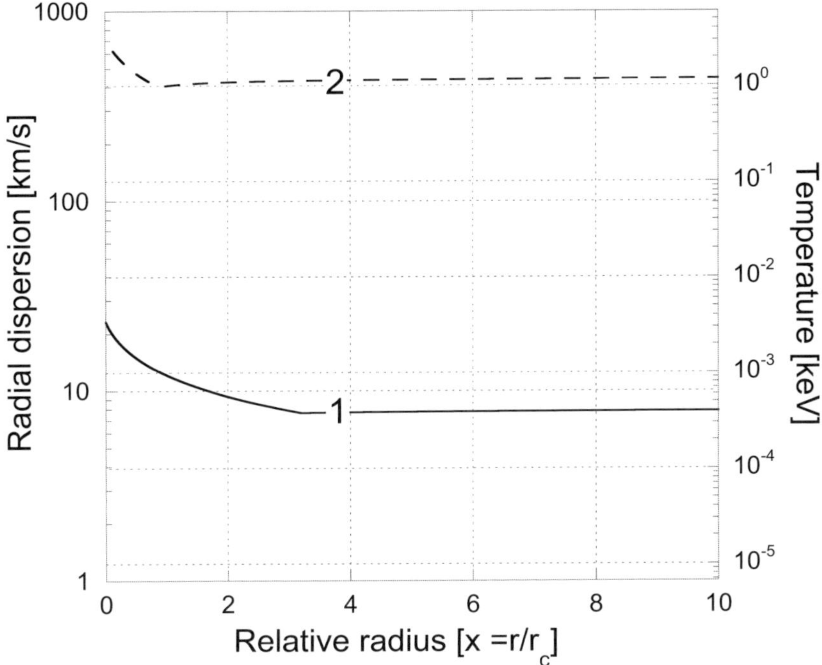

Fig. C.4.3:   Left axis: Radial velocity dispersion as a function of $x = r/r_c$. Right axis: Temperature of gas in units of keV. Curve 1 is $\sigma_r(x)$ for a dwarf spheroidal galaxy with $M = 10^6\ M_\odot$ and curve 2 is $\sigma_r(x)$ for a massive elliptical galaxy with $M = 10^{13}\ M_\odot$. (Source: Hartnett, 2008)

C.4.3.1   *Dwarf spheroidal galaxies*

It follows from Eq. (C.20) that we can write an equation for the radial velocity dispersion that we expect to observe in a typical dwarf galaxy, of mass M (in units of $M_\odot$) as

$$\sigma_r = 4.40 \left(\frac{M}{10^5 M_\odot} \right)^{1/4} \text{km/s}, \qquad (C.21)$$

assuming the observations are obtained from the sources far from the core radius. That is, sources that are in the Carmelian acceleration regime. Here a value of $\tau = 4.276 \times 10^{17}$ s has been assumed (Oliveira 2006), which is equivalent to $\tau^{-1} = 72.17$ km s^{-1}Mpc^{-1}, and hence $a_0 = 4.674 \times 10^{-10}$ m/s^2.

By calculating the mass required for typically observed values of σ_r, in dwarf spheroidal galaxies, according to the both Newtonian (C.15) and

Table C.1: Masses of spheroidal dwarf galaxies from velocity dispersion σ_r at $x = r/r_c = 10$

σ_r [km/s]	Carmelian $M[\times 10^5 M_\odot]$	Newtonian $M[\times 10^5 M_\odot]$	Ratio
4.40	1.00	9.38	9.38
6.58	5.00	20.9	4.20
7.82	10.0	29.7	2.96
11.7	50.0	66.2	1.33

Carmelian (C.18) equations developed here a comparison can be made. The galaxy mass (M) calculated for different radial velocity dispersions σ_r at $x = r/r_c = 10$ (for a typical core radius of $r_c = 0.004$ kpc) are listed in Table C.1.

C.4.3.2   *Bright central elliptical galaxies*

Similarly from Eq. (C.20) we can write an equation for the radial velocity dispersion that we expect to observe in a typical massive elliptical galaxy , of mass M (in units of $M_\odot$) as

$$\sigma_r = 247.4 \left(\frac{M}{10^{12} M_\odot} \right)^{1/4} \text{km/s}. \tag{C.22}$$

also assuming the observations are obtained from the sources that are far from the core radius, in the Carmelian acceleration regime.

Again by calculating the mass required for typically observed values of σ_r, in bright ellipticals using both models a comparison is made results listed in Table C.2. In this case a typical core radius of $r_c = 40$ kpc was assumed.

C.4.3.3   *Temperature*

By considering the random motion of gas particles we can write an expression for the temperature of a cloud of gas in the galaxies that is heated by the gravitational potential the gas feels whether it be in the Newtonian or Carmelian regime. Using $\mu = 0.609$ as the mean atomic weight of the gas we can equate

$$\mu m_p \sigma_r^2 = kT, \tag{C.23}$$

Table C.2: Masses of bright elliptical galaxies from
velocity dispersion σ_r at $x = r/r_c = 10$

| σ_r | Carmelian | Newtonian | |
[km/s]	$M[\times 10^{12} M_\odot]$	$M[\times 10^{12} M_\odot]$	Ratio
139.1	0.1	9.38	93.8
247.4	1.0	29.7	29.7
369.9	5.0	66.3	13.3
439.9	10.0	93.8	9.38

where k is Boltzmann's constant, m_p is the proton mass and T is the temperature of the gas. Using Eq. (C.23) the temperature of the gas in a galaxy heated in this manner can be determined from the radial velocity dispersion. This is shown on the right axis of Figure C.4.3. The result indicates that the temperature of the gas across spherical galaxies is expected to be approximately isothermal. Also, due to the thermal heating of intragalactic gas, massive elliptical and spheroidal galaxies emit radiation in the X-ray part of the spectrum, while dwarf spheroidals emit in the near infrared. This is consistent with observations.

C.5   Conclusion

This theory suggests that it is the Carmelian regime that is applicable at low accelerations where $r \gg r_c$ and hence the masses of galaxies are overestimated from the observed dynamics. In Tables C.1 and C.2 typical dispersion velocities are used but the core radius is fixed at $r_c = 0.004$ kpc and $r_c = 40$ kpc, respectively. So depending on the exact details for a galaxy one may get 10 to 100 times more mass from a Newtonian calculation than from a Carmelian calculation. This brings the masses more in line with estimates from the luminous material. Therefore the need to invoke halo dark matter is avoided. It is also true that σ_r is observed to fall from a central higher value near the center of the spheroidal galaxy and become constant as a function of radius from the center. This is indicated by a Newtonian regime in the center which becomes Carmelian as a function of radius. In massive ellipticals radial velocity dispersion is often observed to be approximately constant as a function of radius. This is indicated by observations where most of the galaxy is dominated by the Carmelian regime.

C.6 Suggested References

M. Carmeli, *Classical Fields, General Relativity and Gauge Fields* (New York, Wiley, 1982).

M. Carmeli, Is galaxy dark matter a property of spacetime? *Int. J. Theor. Phys.* **37**(10), 2621-2625 (1998).

M. Carmeli, Derivation of the Tully-Fisher law: Doubts about the necessity and existence of halo dark matter, *Int. J. Theor. Phys.* **39**(5), 1397-1404 (2000).

M. Carmeli, *Cosmological Special Relativity* (World Scientific, Singapore, 2002).

J.G. Hartnett, Spiral galaxy rotation curves determined from Carmelian general relativity, *Int. J. Theor. Phys.* **45**(11), 2147-2165 (2006).

J.G. Hartnett, Spheroidal and elliptical galaxy rotational velocity dispersion determined from Cosmological General Relativity, *Int. J. Theor. Phys.* **47**(5), 1252-1260 (2008), astro-ph/07072858.

F.J. Oliveira and J.G. Hartnett, Carmeli's cosmology fits data for an accelerating and decelerating universe without dark matter or dark energy *Found. Phys. Lett.* **19**(6), 519-535 (2006).

Appendix D

Bibliography

The following bibliography includes papers and books specifically referred to in the text and many others besides, so that it includes a fairly extended account of the literature related to the theory of relativity at large.

S. Abe, On space-time with a non-symmetric fundamental tensor admitting a six-parameter group of motions, *Tensor* **18**, 289 (1967).

E.S. Abers and B.W. Lee, Gauge theories, *Phys. Lett. C* **9**, 1 (1973).

R.J. Adler, Spinors in a Weyl geometry, *J. Math. Phys.* **11**, 1185 (1970).

R.J. Adler, M. Bazin and M. Schiffer, *Introduction to General Relativity* (McGraw Hill, New York, 1965).

R.J. Adler and C. Sheffield, Classification of space-time in general relativity, *J. Math. Phys.* **14**, 465 (1973).

A.G. Agnese and A. Wataghin, On the Schwarzschild exterior space-time, *Lett. al Nuovo Cimento* **2**, 1041 (1971).

P.C. Aichelburg, High symmetry fields and the homogeneous field in general relativity, *J. Math. Phys.* **11**, 1330 (1970).

P.C. Aichelburg and R.U. Sexl, The gravitational field of a massless particle, *Lett. al Nuovo Cimento* **4**, 1316 (1970).

A.I. Akhiyezer and V.B. Berestetskii, *Quantum Electrodynamics* (Moscow, 1953).

A.I. Akhiyezer and I.M. Glazman, *The Theory of Linear Operators in Hilbert Space* (Moscow, 1950).

J.A. Alcarás and P.L. Ferreira, Algebraic construction of the basis for the irreducible representations of rotation groups and for the homogeneous Lorentz group, *J. Math. Phys.* **6**, 578 (1965).

S.I. Alishauskas, On the Clebsch-Gordon coefficients of the SL(2,C) group, *Litov. Fiz. Sb. (USSR)* **13**, 829 (1973). [In Russian]

G.R. Allcock, A geometrized class of Yang-Mills fields, *Nuovo Cim.* **28**, 858 (1963).

H. Anandan, Ph.D. Thesis (University of Pittsburgh, Pittsburgh, PA, 1978).

J.L. Anderson, *Principles of Relativity Physics* (Academic Press, New York, 1967).

J.L. Anderson and R. Gautreau, Possible causal violations at radii greater than the Schwarzschild radius, *Phys. Lett.* **20**, 24 (1966).

J.L. Anderson and R. Gautreau, Operational formulation of the principle of equivalence, *Phys. Rev.* **185**, 1656 (1969).

A.M. Anile, *Relativistic Fluids and Magneto–Fluids* (Cambridge University Press, Cambridge, New York, 1989).

R. Arnowitt, S. Deser and C.W. Misner, Interior Schwarzschild solutions and interpretation of source terms, *Phys. Rev.* **120**, 321 (1960).

R. Arnowitt and S.I. Fickler, Quantization of the Yang-Mills field, *Phys. Rev.* **127**, 1821 (1962).

B.A. Aronson, *Twisting null congruences in asymptotically flat space* (Thesis, University of Pittsburg, 1971).

B.A. Aronson, R. Lind, J. Messmer and E.T. Newman, A note on asymptotically flat spaces, *J. Math. Phys.* **12**, 2462 (1971).

B.A. Aronson and E.T. Newman, Coordinate systems associated with asymptotically shear-free null congruences, *J. Math. Phys.* **13**, 1847 (1972).

H. Arzeliés, *General Relativity and Gravitation*, Vols. I and II (*Trav. Inst. Sci. Cherifien, Ser. Sci. Phys. (Morocco)*, No. 7, 1961, and No. 8, 1963. [In French])

E. Ascher and A. Janner, Space-time symmetry of transverse electromagnetic plane waves, *Helv. Phys. Acta* **43**, 296 (1970).

P. Astier *et al.*, The Supernova Legacy Survey: Measurement of Ω_M, Ω_Λ and w from the first year data set, *Astron. Astrophys.* **447**, 31-48 (2006), astro-ph/0510447 .

M.F. Atiyah, N.J. Hitchin and I.M. Singer, *Proc. Nat. Acad. Sci. (USA)* **74** (1977).

M.F. Atiyah, V. Patodi and L. Singer, *Math. Proc. Camb. Philos. Soc.* **77**, 43 (1975); **78**, 405 (1975); **79**, 71 (1976).

J. Audretsch and W. Graf, Neutrino radiation in gravitational fields, *Commun. Math. Phys.* **19**, 315 (1970).

L. Auslander and R. Mackenzie, *Introduction to Differential Manifolds* (McGraw-Hill, New York, 1977).

A. Avez, The Schwarzschild ds^2 amongst stationay ds^2, *Ann. Inst. Poincaé (A)* **1**, 291 (1964).

W.L. Bade and J. Jehle, An introduction to spinors, *Rev. Mod. Phys.* **25**, 714-728 (1953).

A. Banerjee, Null electromagnetic fields in general relativity, *J. Phys. A* **3**, 501 (1970).

A. Banerjee, Null electromagnetic fields in general relativity admitting timelike or null killing vectors, *J. Math. Phys.* **11**, 51 (1970).

R. Bairerlain, Testing general relativity with laser ranging to the Moon, *Phys. Rev.* **162**, 1275 (1967).

J.M. Bardeen, Kerr Metric black holes, *Nature* **226**, 64 (1970).

J.M. Bardeen and W.H. Press, Radiation fields in the Schwarzschild background, *J. Math. Phys.* **14**, 7 (1973).

V. Bargmann, Irreducible unitary representations of the Lorentz group, *Ann. Math.* **48**, 568 (1947).

B.M. Barker and R.F. O'Connell, Derivation of the equations of motion of a gyroscope from the quantum theory of gravitation, *Phys. Rev. D* **2**, 1428 (1970).

A. Barnes, On Birkhoff's theorem in general relativity, *Commun. Math. Phys.* **33**, 75 (1973).

I. Bars and F. Gürsey, Operator treatment of the Gelfand-Naimark basis of SL(2,C), *J. Math. Phys.* **13**, 131 (1972).

I. Bars and F. Gürsey, Projective representations of SL (3,C) in the Z-operator formalism, *J. Math. Phys.* **14**, 759 (1973).

A.O. Barut, H. Kleinert and S. Malin, The "anomalous Zitterbewegung" of composite particles, *Nuovo Cimento A* **58**, 835-847 (1968).

A.O. Barut and M. Leiser, Note on gauge transformations in quantum mechanics, *Am. J. Phys.* **29**, 24 (1961).

A.O. Barut and S. Malin, Position operators and localizability of quantum systems described by finite- and infinite-dimensional wave equations, *Rev. Mod. Phys.* **40**, 632-651 (1968).

D. Basu and S.D. Majumdar, Complex angular momenta and the Lorentz group, *J. Phys. A* **6**, 1097 (1973).

S. Bazanski, in: *Recent Developments in General Relativity* (Pergamon, PWN, Warsaw, 1962), p. 137.

S.L. Bazanski, Decomposition of the Lorentz transformation matrix into skew-symmetric tensors, *J. Math. Phys.* **6**, 1201 (1965).

G. Beck and H. Weyl, *Math. Z.* **13**, 142 (1921).

B.L. Beers and R.S. Millman, Analytic vector harmonic expansions on SU(2) and S^2, *J. Math. Phys.* **16**, 11 (1975).

K.G. Begeman, A.H. Broeils *et al.*, Extended rotation curves of spiral galaxies: Dark haloes and modified dynamics, *Mon. Not. R. Astr. Soc.* **249**, 523-537 (1991).

S. Behar and M. Carmeli, Cosmological relativity: A new theory of cosmology, *Int. J. Theor. Phys.* **39**, 1375 (2000), astro-ph/0008352.

W. Beiglböck, Theory of the infinitesimal holonomy group in general relativity theory, *Z. Phys.* **179**, 148 (1964).

W. Beiglböck, *Comm. Math. Phys.* **5**, 106 (1967).

L. Bel, Su la radiation gravitationnelle, *C.R. Acad. Sci., Paris* **247**, 1094 (1958).

L. Bel, Introduction d'un tenseur du quatrieme order, *C.R. Acad. Sci., Paris* **248**, 1297 (1959).

L. Bel, Quelques remarques sur la classification de Petrov, *C.R. Acad. Sci.,*

Paris **248**, 2561 (1959).

L. Bel, Schwarzschild singularity, *J. Math. Phys.* **10**, 1501 (1969).

L. Bel, Les souces des solutions statiques de Schwarzschild et Curson, *Gen. Relat. Grav.* **1**, 337 (1971).

F.J. Belinfant, Kruskal space without wormholes, *Phys. Lett.* **20**, 251 (1966).

V.A. Belinskii, General solution of the gravitational equations with a physical singularity, *Zh. Eksp. Teoret. Fiz.* **57**, 2163 (1969).

V.A. Belinskii and I.M. Khalatnikov, A general solution of the gravitational equations with a simultaneous fictious singularity, *Zh. Eksp. Teoret. Fiz.* **49**, 1000 (1965); *Soviet Physics JETP* **22**, March 1966.

V.A. Belinskii and I.M. Khalatnikov, On the nature of the singularities in the general solutions of the gravitational equations, *Zh. Eksp. Teoret. Fiz.* **56**, 1700 (1969). [In Russian]

V.A. Belinskii and I.M. Khalatnikov, General solution of the gravitational equations with a physical singularity, *Zh. Eksp. Teoret. Fiz.* **57**, 2163 (1969). [In Russian]

V.A. Belinskii and I.M. Khalatnikov, On a generalized solution of the gravitational equations with a physical singularity of an oscillatory nature, *Zh. Eksp. Teoret. Fiz.* **59**, 3141 (1970).

K. Bera, Note on an interesting metric of the field equations in general relativity, *J. Phys. A* **2**, 138 (1969).

F.A. Berezin, Laplace operators on semi-simple Lie groups, *Dokl. Akad. Nauk SSSR* **107**, 9 (1956); *Trud. Mosk. Mat. Ob-va* **6**, 371 (1957).

F.A. Berezin and I.M. Gelfand, Some remarks on the theory of functions on symmetric Riemannian manifolds, *Trud. Mosk. Mat. Ob-va* **5** (1956).

O. Bergmann and R. Leipnik, Space-time structure of a spherical symmetrical field, *Phys. Rev.* **107**, 1157 (1957).

P.G. Bergmann, Gauge-invariant variables in general relativity, *Phys. Rev.* **124**, 274 (1961).

P.G. Bergmann, Foundation problems in general relativity, in: *Problems in the Foundations of Physics* (M. Bunge, Ed.) (Springer-Verlag, Berlin,

1971).

P.G. Bergmann, M. Cahen and A.B. Komar, Spherically symmetric gravitational fields, *J. Math. Phys.* **6**, 1 (1965)

P.G. Bergmann, I. Robinson and E. Schuking, Asymptotic properties of a system with nonzero total mass, *Phys. Rev.* **126**, 1227 (1962).

B. Bertotti, D. Brill and R Krotkov, Experiments on gravitation, in: *Gravitation: An Introduction to Current Research*, L. Witten (Ed.) (John Wiley, New York, 1962).

B. Bertotti and Plebanski, Theory of gravitational pertubations in the fast motion approximation, *Ann. Phys.* (*N.Y.*) **11**, 169 (1960).

N.J. Bhabha, Relativistic wave equations for the elementary particles, *Rev. Mod. Phys.* **17**, 200 (1945).

A. Bialas, On the continuity of the Petrov classification, *Acta Phys. Polon.* **23**, 699 (1963).

E. Bialas and A. Bialas, On the relation between the Penrose classification of gravitational fields and the classification according to group of motions, *Acta Phys. Polon.* **24**, 515 (1963).

I. Bialynicki-Birula, Gauge-invariant variables in the Yang-Mills theory, *Bull. Acad. Polon. Sci., Ser. Math. Astron. Phys.* **11**, 935 (1963).

L. Bianchi, On three-dimensional spaces which admit a group of motions (Sugli spazii a tre dimensioni che ammettono un gruppo continuo di movimenti), it Soc. Ital. Sci. Mem di Mat. **11**, 267 (1897).

G. Birkhoff, *Relativity and Modern Physics* (Harvard University Press, Cambridge, Mass., 1923).

J.D. Bjorken and D. Drell, *Relativistic Quantum Mechanics* (McGraw-Hill Book Co., New York, 1964).

J.M. Blatt and V.F. Weisskopf, *Theoretical Nuclear Physics* (John Wiley, New York, 1952).

H. Boerner, *Representations of Groups*, 2nd Edition (North-Holland, Amsterdam, 1970).

D. Bohm, *The Special Theory of Relativity* (Benjamin, New York, 1965).

J.R. Bond *et al.*, in *Proc. IAU Symposium* 201 (2000), astro-ph/0011378.

H. Bondi, *Mon. Not. R. Astr. Soc.* **107**, 401(1947).

H. Bondi, Brandeis Summer School 1955.

H. Bondi, The space traveller's youth, *Discovery*, p. 505, December 1957.

H. Bondi, Negative mass and general relativity, *Rev. Mod. Phys.* **29**, 423 (1957).

H. Bondi, M.G.J. van der Burg and A.W.K. Metzner, Gravitational waves in general relativity, VII. Waves from axi-symmetric isolated systems. *Proc. R. Soc. Lond. A* **269**, 21 (1962).

H. Bondi, *Relativity and Common Sense: A New Approach to Einstein* (Dover Publications, Inc., New York, 1962).

H. Bondi, Some special solutions of the Einstein equations, in: *Lectures on General Relativity* (Prentice-Hall, Inc., Englewood Cliffs, New Jersey, 1965), pp. 379-406 (Brandeis Summer Institute in Theoretical Physics, Vol. 1, 1964.)

W.B. Bonnor, Certain exact solutions of the equations of general relativity with an electrostatic field, *Proc. Phys. Soc. Lond. A* **66**, 145 (1953).

W.B. Bonnor, *J. Math. Mech.* **9**, 439 (1960).

W.B. Bonnor, Exact solution of the Einstein-Maxwell equations, *Z. Phys.* **161**, 439 (1961).

W.B. Bonnor, On Birkhoff's theorem, in: *Recent Developments in General Relativity* (PWN, Warsaw, 1962).

W.B. Bonnor, Gravitational waves, *Brit. J. Appl. Phys.* **14**, 555, (1963).

W.B. Bonnor, An exact solution of the Einstein-Maxwell equations referring to a magnetic dipol, *Z. Phys.* **190**, 444 (1966).

W.B. Bonnor, Null curves in a Minkowski space-time, *Tensor* **20**, 229 (1969).

W.B. Bonnor, A new interpretation of the NUT metric in general relativity, *Proc. Camb. Phil. Soc.* **66**, 145 (1969).

W.B. Bonnor, On a Robinson-Trautman solution of Einstein's equations, *Phys. Lett. A* **31**, 269 (1970).

W.B. Bonnor, Charge moving with the speed of light in Einstein-Maxwell theory, *Int. J. Theor. Phys.* **3**, 57 (1970).

W.B. Bonnor and M.C. Faulkes, An exact solution for oscillating spheres in general relativity, *Mon. Not. R. Astr. Soc.* **137**, 2391 (1967).

W.B. Bonnor and N.S. Swaminarayan, An exact solution for uniformly accelerating particles in general relativity, *Z. Phys.* **177**, 240 (1964).

W.B. Bonnor and N.S. Swaminarayan, An exact solution of Einstein's equations, *Z. Phys.* **186**, 222 (1965).

W.B. Bonnor and P.C. Vaidya, Spherically symmetric radiation of charge in Einstein-Maxwell theory, *Gen. Relat. Grav.* **1**, 127 (1970).

S. Bonometto, On gauge invariance for a neutral massive vector field, *Nuovo Cimento* **28**, 309 (1963).

M. Born, *Einstein's Theory of Relativity* (Dover, New York, 1962).

H.H.V. Borzeskowski and U. Kasper, Doppler effect and Einstein shift in Schwarzschild space, *Ann. Phys.* **20**, 187 (1967).

S.K. Bose, Remarks on the Yang-Mills gauge transformation, *Lett. al Nuovo Cimento* **4**, 985 (1970).

D.G. Boulvare and S. Deser, External sources in gauge theories, *Nuovo Cimento* **30**, 1009 (1963).

R.H. Boyer and R.W. Linquist, Maximal analytic extension of the Kerr metric, *J. Math. Phys.* **8**, 265 (1967).

R.H. Boyer and T.G. Price, An interpretation of the Kerr metric in general relativity, *Proc. Camb. Phil. Soc.* **61**, 531 (1965).

V.B. Braginsky and A.B. Menukin, *Measurement of Weak Forces in Physics Experiment* (University of Chicago Press, Chicago, 1977).

V.B. Braginsky and V.I. Panov, *Zh. Eksp. Teor. Fiz.* **61**, 873 (1971). [English translation: *Sov. Phys. - JETP* **34**, 463 (1972).]

V.B. Braginsky and V.I. Panov, The equivalence of inertial and passive gravitational mass, *Gen. Relat. Grav.* **3**, 403 (1972).

D. Branch, *Astrophys. J.* **392**, 35 (1992).

C.H. Brans, A unique and invariant representation of all local analytic solutions to the vacuum Einstein equations: Petrov type III, *Lett. al Nuovo Cimento* **1**, 699 (1969).

C.H. Brans, Invariant representation of all analytic Petrov type III solutions of the Einstein equations, *J. Math. Phys.* **11**, 1210 (1970).

C.H. Brans, Complex 2-form representation of the Einstein equation: The Petrov type III solutions, *J. Math. Phys.* **12**, 1616 (1971).

C.H. Brans and R.H. Dicke, Mach's principle and a relativistic theory of gravitation, *Phys. Rev.* **124**, 925 (1961).

R. Brauer and H. Weyl, Spinors in n dimensions, *Am. J. Math.* **57**, 425 (1935).

R.W. Brehme, Geometric representations of the Lorentz transformations, *Am. J. Phys.* **32**, 233 (1964).

R.W. Brehme, Geometrization of the relativistic velocity addition formula, *Am. J. Phys.* **37**, 360 (1969).

G. Breit, E. Condon and R. Present, Theory of scattering of protons by protons, *Phys. Rev.* **50**, 825 (1936).

F. Brickell and R.S. Clark, *Differentiable Manifolds: An Introduction* (Van Nostrand Reinhold, London, 1970).

D.R. Brill, P.L. Chrzanowshi, C.M. Pereira, E.D. Fackerell and J.R. Ipser, Solution of the scalar wave equation in a Kerr background by separation of variables, *Phys. Rev.* D **5**, 1913 (1972).

L.S. Brown, R.D. Carlitz and C. Lee, Massless excitations in pseudoparticle fields, *Phys. Rev.* D **16**, 417-422 (1977).

Y. Bruhat, The Cauchy problem, in: *Gravitation: An Introduction to Current Research*, L.Witten, Editor (Wiley, New York, 1962).

V.A. Brumberg, On the interpretation of coordinates in Schwarzschild's problem, *Astron. Zh.* **45**, 1110 (1968).

H.A. Buchdahl, A special class of solutions of the equations of the gravitational field arising from certain gauge-invariant action principles, *Proc. Natl. Acad. Sci. Wash.* **34**, 66 (1948).

H.A. Buchdahl, Reciprocal static solutions of the equation of gravitational field, *Aust. J. Phys.* **9** (1956); *Quart. J. Math. (Oxford)* **5**, 116 (1954).

H.A. Buchdahl, Reciprocal static solutions of field equations involving an asymmetric fundamental tensor, *Nuovo Cimento* **5**, 1083 (1957).

H.A. Buchdahl, Gauge invariant generalization of the field theories with asymmetric fundamental tensor, *Quart. J. Math. (Oxford)* **8**, 89 (1957), and **9**, 257 (1958).

H.A. Buchdahl, Reciprocal static metrics and scalar fields in the general theory of relativity, *Phys. Rev.* **115**, 1325 (1959).

H.A. Buchdahl, On the non-existance of a class of static Einstein spaces asymptotic at infinity to a space of constant curvature, *J. Math. Phys.* **1**, 537 (1960).

A.M. Buoncistiani, An algebra of the Yang-Mills field, *J. Math. Phys.* **14**, 849 (1973).

P. Burcev, On the theory of gravitational radiation, *Czech. J. Phys.* **11**, 385 (1961). [In Russian]

R. Burghardt, Yang-Mills fields in gravitation theory, III. Group-theoretical foundations of gravitation theory, *Acta Phys. Aust.* **39**, 127 (1974). [In German]

W.L. Burke and K.S. Thorne, Gravitational radiation damping, in: *Relativity*, M. Carmeli, S.I. Fickler and L. Witten (Editors) (Plenum Press, New York, 1970).

M. Cahen, On a class of homogeneous spaces in general relativity, *Bull. Acad. Roy. Belg., Cl. Sci.* **50**, 972 (1964).

M. Cahen and R. Deveber, On Birkhoff's theorem, *C.R. Acad. Sci., Paris* **260**, 815 (1965).

M. Cahen and J. Leroy, Espace-temps du vide admettant une congruence caracteristique simple integrable, *Bull. Acad. Roy. Belg., Cl. Sci.* **54**, 232 (1968).

M. Cahen and J. Sengier-Diels, Espaces de Classe D, admettant un champ electromagnetique, *Bull. Acad. Roy. Belg., Cl. Sci.* **53**, 801 (1967).

M. Cahen and J. Spelkens, Class III spaces, solutions of Maxwell-Einstein equations, *Bull. Acad. Roy. Belg., Cl. Sci.* **53**, 817 (1967).

M.E. Cahill and G.C. McVittie, Spherical symmetry and mass-energy in general relativity, I and II, *J. Math. Phys.* **11**, 1382 and 1392 (1970).

R.R. Caldwell and P.J. Steinhardt, *Phys. Rev. D* **57**, 6057 (1998).

J. Callaway, Gravitational field of a point charge, *Am. J. Phys.* **27**, 469

(1959).

W.B. Campbell, Tensor and spinor spherical harmonics and the spin-s harmonics $_sY_{lm}(\theta,\phi)$, *J. Math. Phys.* **12**, 1763 (1971).

V. Cantoni, A class of representations of the generalized Bondi-Metzner-group, *J. Math. Phys.* **7**, 1361 (1966).

V. Cantoni, Reduction of some representations of the generalized Bondi-Metzner-group, *J. Math. Phys.* **8**, 1700 (1967).

F. Cap, W. Majerotto, W. Raab and P. Unteregger, Spin calculus in Riemannian manifolds, *Fortschr. Phys.* **14**, 205 (1966)

M. Carmeli, The EIH equations of motion up to the 9th order, *Phys. Lett.* **9**, 132 (1964).

M. Carmeli, Has the geodesic postulate any significance for a finite mass? *Phys. Lett.* **11**, 24 (1964).

M. Carmeli, *Ann. Phys. (N.Y.)* **30**, 168 (1964).

M. Carmeli, Motion of a charge in gravitational field, *Phys. Rev. B* **138**, 1003 (1965).

M. Carmeli, The equations of motion of slowly moving particles in general theory of relativity, *Nuovo Cimento* **37**, 842 (1965).

M. Carmeli, Equations of motion without infinite self-action terms in general relativity, *Phys. Rev. B* **140**, 1441 (1965).

M. Carmeli, Semigenerally covariant equations of motion I and II, *Ann. Phys. (N.Y.)* **34**, 465 (1965), and **35**, 250 (1965).

M. Carmeli, *Phys. Rev.* **158**, 1243.

M. Carmeli, *Nuovo Cimento B* **55**, 220 (1968).

M. Carmeli, Representations of the three-dimensional rotation group in terms of direction and angle of rotation, *J. Math. Phys.* **9**, 1987 (1968).

M. Carmeli, Group-theoretic approach to the new conserved quantities in general relativity, *J. Math. Phys.* **10**, 569 (1969).

M. Carmeli, Group Analysis of Maxwell's Equations, *J. Math. Phys.* **10**, 1699 (1969).

M. Carmeli, Hilbert space description of the gravitational field, *Phys. Lett.*

A **28**, 683 (1969).

M. Carmeli, Symmetry of the gravitational field, *Lett. al Nuovo Cimento* **4**, 40 (1970).

M. Carmeli, SL(2,C) symmetry of the gravitational field dynamical variables, *J. Math. Phys.* **11**, 2728 (1970).

M. Carmeli, Constraints in electrodynamics and quantization of the electromagnetic field without charges, *Nuovo Cimento B* **67**, 103 (1970).

M. Carmeli, Infinite-Dimensional Representations of the Lorentz Group, *J. Math. Phys.* **11**, 1917 (1970).

M. Carmeli, Applications of the group SU(2) technique in general relativity, in: *Relativity and Gravitation*, p. 69, C.G. Kupper and A. Peres, Editors (Gordon and Breach, London, 1971).

M. Carmeli, SL(2,C) invariance of the gravitational field, *Gen. Relat. Grav.* **3**, 317 (1972).

M. Carmeli, SL(2,C) invariance of the gravitational field, *Ann. Phys.* **71**, 603 (1972).

M. Carmeli, SL(2,C) symmetry of the gravitational field, *Il Nuovo Cimento, Series 11* **7A**, 9 (1972).

M. Carmeli, Behavior of fast particles in the Schwarzschild field, *Lett. al Nuovo Cimento* **3**, 379 (1972).

M. Carmeli, Gauge fields and gravitational field equations, *Nucl. Phys. B* **38**, 621 (1972).

M. Carmeli, SL(2,C) symmetry of the gravitational field, in: *Group Theory in Nonlinear Problems*, A.O. Barut, Editor (D. Reidel Publishing Co., Dordrecht-Holland/Boston-U.S.A., 1974), p. 59.

M. Carmeli, General relativity and gauge theories, *Gen. Relat. Grav.* **5**, 287 (1974).

M. Carmeli, Modified gravitational Lagrangian, *Phys. Rev. D* **14**, 1727 (1976).

M. Carmeli, Integral formalism of gauge fields and general relativity, *Phys. Rev. Lett.* **36**, 59 (1976).

M. Carmeli, *Group Theory and General Relativity* (McGraw-Hill, New

York, 1977, reprinted by Imperial College Press, London, 2000).

M. Carmeli, *Phys. Lett. B* **68**, 463 (1977).

M. Carmeli, Classification of classical Yang-Mills fields, in: *Differential Geometrical Methods in Mathematical Physics*, K. Bleuler, H.R. Petry and A. Reetz, Editors, 105-149 (Springer-Verlag, Heidelberg, 1978).

M. Carmeli, *Phys. Lett. B* **77**, 188 (1978).

M. Carmeli, *Nuovo Cimento A* **52**, 545 (1979).

M. Carmeli, *Classical Fields: General Relativity and Gauge Theory* (John Wiley, New York, 1982, reprinted by World Scientific, Singapore, 2001).

M. Carmeli, Extension of the principle of minimal coupling to particles with magnetic moments, *Nuovo Cimento Lett.* **37**, 205 (1983).

M. Carmeli, Cosmological relativity: A special relativity for cosmology, *Found. Phys.* **25**, 1029 (1995).

M. Carmeli, Cosmological special relativity, *Found. Phys.* **26**, 413 (1996).

M. Carmeli, Cosmological general relativity, *Commun. Theor. Phys.* **5**, 159 (1996).

M. Carmeli, Space, time and velocity in cosmology, *Int. J. Theor. Phys.* **36**, 757 (1997).

M. Carmeli, Is galaxy dark matter a property of spacetime? *Int. J. Theor. Phys.* **37**, 2621 (1998).

M. Carmeli, Derivation of the Tully-Fisher law: Doubts about the necessity and existence of halo dark matter, *Int. J. Theor. Phys.* **39**, 1397 (2000), astro-ph/9907244.

M. Carmeli, Cosmological relativity: Determining the Universe by the cosmological redshift as infinite and curved, *Int. J. Theor. Phys.* **40**, 1871-1874 (2001).

M. Carmeli, Lengths of the first days of the universe, p.628, in: *Astrophysical Ages and Time Scale*, T. von Hippel, C. Simpson and N. Manset, Editors (Proceeding of a Conference held in Hilo, Hawaii, 5-9 February 2001) The Astronomical Society of the Pacific, Vol. 245 (2001).

M. Carmeli, *Cosmological Special Relativity: The Large-Scale Structure of Space, Time and Velocity*, Second Edition (World Scientific, Singapore,

2002).

M. Carmeli, Accelerating Universe: Theory versus experiment (2002), astro-ph/0205396.

M. Carmeli, The line elements in the Hubble expansion, in: *Gravitation and Cosmology*, Eds. A. Lobo *et al.* (Universitat de Barcelona, 2003), astro-ph/0211043.

M. Carmeli, *Cosmological Relativity: The Special and General Theories for the Structure of the Universe* (World Scientific, Singapore, 2006).

M. Carmeli and S. Behar, Cosmological general relativity, pp. 5–26, in: *Quest for Mathematical Physics*, T.M. Karade, *et al.* Editors, (New Delhi, 2000).

M. Carmeli and S. Behar, Cosmological relativity: A general relativistic theory for the accelerating Universe, Talk given at Dark Matter 2000, Los Angeles, February 2000, pp.182–191, in: *Sources and Detection of Dark Matter/Energy in the Universe*, D. Cline, Ed., (Springer, 2001).

M. Carmeli and Ch. Charach, *Int. J. Theor. Phys.* **16**, 53 (1977).

M. Carmeli, Ch. Charach and M. Kaye, *Phys. Rev. D* **15**, 1501 (1977).

M. Carmeli, Ch. Charach and M. Kaye, *Nuovo Cimento B* **45**, 310 (1979).

M. Carmeli and S.I. Fickler, Gravitaional Lagrangian, *Phys. Rev. D* **5**, 290 (1972).

M. Carmeli, S.I. Fickler and L.Witten, Eds., *Relativity* (Plenum Press, New York, 1970).

M. Carmeli and M. Fischler, Classification of SU(2) gauge fields: Lorentz-invariant versus gauge-invariant schemes, *Phys. Rev. D* **19**, 3653-3659 (1972).

M. Carmeli and M. Fischler, *Phys. Rev. D* **19**, 3653 (1979).

M. Carmeli, J.G. Hartnett and F. Oliveira, The cosmic time in terms of the redshift, *Found. Phys. Lett.* **19**, 276 - 283 (2006); gr-qc/0506079

M. Carmeli and M. Kaye, Metric of a radiating rotating sphere, *Phys. Lett. A* **53**, 14 (1975); **55**, 495 (1976).

M. Carmeli and M. Kaye, Transformation laws of the Newton-Penrose variables, *Ann. Phys. (N.Y.)* **99**, 188 (1976).

M. Carmeli and M. Kaye, *Ann. Phys. (N.Y.)* **103**, 97 (1977).

M. Carmeli and M. Kaye, Einstein-Maxwell equations: Gauge formulation and solutions for radiating bodies, *Fortschr. Phys.* **27**, 261 (1979).

M. Carmeli and T. Kuzmenko, Value of the cosmological constant: Theory versus experiment, in: *Proceedings of the 20th Texas Symposium on Relativistic Astrophysics*, held 10-15 December 2000, Austin, Texas, H. Martel and J.C. Wheeler, Editors (American Institute of Physics, 2001).

M. Carmeli, E. Leibowitz and N. Nissani, *Gravitation: SL(2,C) Gauge Theory and Conservation Laws* (World Scientific, Singapore, 1990).

M. Carmeli and S. Malin, Infinite-dimensional representations of the Lorentz group: Complementary series of representations, *J. Math. Phys.* **12**, 225 (1971).

M. Carmeli and S. Malin, Finite- and infinite-dimensional representations of the Lorentz group, *Fortsrh. Phys.* **21**, 397 (1973).

M. Carmeli and S. Malin, Infinite-dimensional representations of the Lorentz group: The complete series, *Int. J. Theor. Phys.* **9**, 145 (1974).

M. Carmeli and S. Malin, *Representations of the Rotation and Lorentz Groups: An Introduction* (Marcel and Dekker, Inc., New York and Basel, 1976).

M. Carmeli and S. Malin, *Theory of Spinors: An Introduction* (World Scientific, Singapore, 2000).

E. Cartan, *The Theory of Spinors* (The M.I.T. Press, Cambridge, Massachusetts, 1966).

B. Carter, Complete analytic extension of the symmetry axis of Kerr's solutions of Einstein's equations, *Phys. Rev.* **141**, 1242 (1966).

B. Carter, The complete analytic extension of the Reissner-Nordstrom metric in the special case $e^2 = m^2$, *Phys. Lett.* **21**, 423 (1966).

B. Carter, Global structure of the Kerr family of gravitational fields, *Phys. Rev.* **174**, 1559 (1968).

B. Carter, Hamilton-Jacobi and Schrödinger separable solutions of Einstein's equations, *Commun. Math. Phys.* **10**, 280 (1968).

B. Carter, A new family of Einstein spaces, *Phys. Lett. A* **26**, 399 (1968).

B. Carter, Killing horizons and orthogonally transitive groups in space-time, *J. Math. Phys.* **10**, 70 (1969).

B. Carter, The commutation property of a stationary axisymmetric system, *Commun. Math. Phys.* **17**, 233 (1970).

B. Carter, Casual structure in space-time, *Gen. Relat. Grav.* **1**, 349 (1971).

B. Carter, in: *Proceedings in the Marcel Grossmann Meeting*, R. Ruffini, Editor (North-Holland, Amsterdam, 1978).

L.A. Case, Changes of the Petrov type of a space-time, *Phys. Rev. A* **176**, 1554 (1963).

B. Cassen and E. Condon, On nuclear forces, *Phys. Rev.* **50**, 846 (1936).

L. Castillejo, M. Kugler and R. Roskies, *Phys. Rev. D* **19**, 1782 (1979).

S.-J. Chang and L. O'Raifeartaigh, Unitary representations of SL(2,C) in an E(2) basis, *J. Math. Phys.* **10**, 21 (1969).

K.L. Chao and M. Kohler, Vierbein formulation of the gravitational field, *Z. Naturforsch A* **20**, 753 (1965).

J.E. Chase, Event horizons in static scalar-vacuum space-times, *Commun. Math. Phys.* **19**, 276 (1970).

W. Chau, Gravitational radiation from neutron star, *Astrophys. J.* **147**, 664 (1967).

S.F. Chen, Yang-Mills field and its applications, thesis (New York State University, 1971).

M. Chevalier, Goldberg-Sachs theorem and spaces with conformal connection, *C.R. Acad. Sci., Paris A* **261**, 2446 (1965).

C. Chevalley, *The Algebraic Theory of Spinors* (Columbia University Press, New York, 1954).

C. Chevalley, *Theory of Lie Groups* (Princeton University Press, Princeton, 1962).

Y.M. Cho, *J. Math. Phys.* **20**, 2605 (1979).

Y. Choquet-Bruhat, Global theorem of uniqueness for the solutions of Einstein's equations, *C.R. Acad. Sci., Paris A* **226**, 182 (1968).

Y. Choquet-Bruhat, Construction de solutions radiatives approchees des equations d'Einstein, *Commun. Math. Phys.* **12**, 16 (1969).

Y. Choquet-Bruhat, The bearings of global hyperbolicity on existence and uniqueness theorems in general relativity, *Gen. Relat. Grav.* **2**, 1 (1971).

D. Christodoulou, Reversible and irreversible transformations in black-hole physics, *Phys. Rev. Lett.* **25**, 1596 (1970).

Z. Chylinski, The nature of the Lorentz-invariants, *Acta Phys. Polon.* **30**, 293 (1966).

M. Cissoko, Wavefronts in a relativistic cosmic two-component fluid, *Gen. Relat. Grav.* **30**, 521 (1998).

E. Coddington and V. Levinson, *Theory of Ordinary Differential Equations* (McGraw Hill, New York, 1955), p. 103.

J.M. Cohen, Note on the Kerr metric and rotating masses, *J. Math. Phys.* **8**, 1477 (1967).

J.M. Cohen and D.R. Brill, Angular momentum and the Kerr metric, *J. Math. Phys.* **9**, 905 (1968).

J.M. Cohen and R.M. Wald, Point charge in the vicinity of the Schwarzschild black hole, *J. Math. Phys.* **12**, 1845 (1971).

J. Cohn, On a physical interpretation of the Schwarzschild metric, *Int. J. Theor. Phys.* **2**, 267 (1969).

J. Cohn, Additional remarks on the principle of equivalence, electrodynamics and general relativity, *Int. J. Theor. Phys.* **2**, 407 (1969).

P.M. Cohn, *Lie Groups* (Cambridge University Press, 1957).

C.D. Collinson, Symmetry properties of Harrison space-times, *Proc. Camb. Phil. Soc.* **60**, 259 (1964)

C.D. Collinson, Empty space-times algebraically special on a given world line as a hypersurface, *J. Math. Phys.* **8**, 1547 (1967).

C.D. Collinson, Symmetries of type N empty space-time possessing twisting geodesic rays, *J. Phys. A* **2**, 621 (1969).

C.D. Collinson, Conservation laws in general relativity based upon the existence of preferred collineations, *Gen. Relat. Grav.* **1**, 137 (1970).

C.D. Collinson and R.K. Dodd, Symmetries of stationary axisymmetric empty space-times, *Nuovo Cimento B* **3**, 281 (1971).

C.D. Collinson and P.B. Morris, Space-times admitting nonunique neutrino fields, *Nuovo Cimento B* **16**, 273 (1973).

E. Corinaldesi and A. Papapetrou, *Proc. R. Soc. A* **209**, 259 (1951).

E.M. Corson, *Introduction to Tensors, Spinors and Relativistic Wave Equations* (Hafner Publishing Co., New York, 1953).

O. Costa de Beauregard, The principle of equivalence, negative mass and gravitational repulsion, *C.R. Acad. Sci., Paris* **252**, 1737 (1961).

W.E. Couch and W.H. Hallidy, Radiation scattering in Einstein-Maxwell theory, *J. Math. Phys.* **12**, 2170 (1971).

W.E. Couch, W. Kinnersley and R.J. Torrence, Nonlinear radiative interactions in general relativity, *Phys. Lett. A* **31**, 576 (1970).

W.E. Couch and E.T. Newman, Algebraically special perturbations of the Schwarzschild metric, *J. Math. Phys.* **14**, 285 (1973).

W.E. Couch and R.J. Torrence, Self-interaction of gravitational radiation, *J. Math. Phys.* **11**, 2096 (1970).

W.E. Couch and R.J. Torrence, Asymptotic behavior of vacuum space-times, *J. Math. Phys.* **13**, 69 (1972).

W.E. Couch, R.J. Torrence, A.I. Janis and E.T. Newman, Tail of a gravitational wave, *J. Math. Phys.* **9**, 484 (1968).

R. Courant and D. Hilbert, *Methods of Mathematical Physics* (Interscience, New York, 1953).

T.E. Cranshaw, The Mössbauer effect and experiments in relativity, in: *Evidence for Gravitational Theories* (Academic Press, New York, 1962).

T.E. Cranshaw, S.P. Schiffer and A.B. Whitehead, Measurement of the gravitational red shift using the Mössbauer effect in Fe^{57}, *Phys. Rev. Lett.* **4**, 163 (1960).

T.E. Cranshaw and S.P. Schiffer, Measurement of the gravitational red shift with the Mössbauer effect, *Proc. Phys. Soc., Lond.* **84**, 245 (1964).

A. Crumeyrolle, On the representations of finite degree of the special unitary group and the special linear group, *C.R. Acad. Sci., Paris* **260**, 1871 (1965).

A. Crumeyrolle, Spinor structure, *Ann. Inst. Poincaré A* **11** (1969). [In French]

T. Damour, Gravitational radiation and the motion of compact bodies, in *Gravitational Radiation*, N. Deruelle and T. Piran, Eds. (North-Holland, Amsterdam 1983), pp. 59-144.

A. Das, Static electromagnetic fields, II. Ricci rotation coefficients, *J. Math. Phys.* **14**, 1099 (1973).

T. Dass, Gauge groups and Noether's theorem, *Nuovo Cimento A* **42**, 730 (1966).

B.K. Datta, Static electromagnetic fields in general relativity, *Ann. Phys. (N.Y.)* **12**, 295 (1961), and **15**, 304 (1961).

B.K. Datta, Homogeneous nonstatic electromagnetic fields in general relativity, *Nuovo Cimento* **36**, 109 (1965).

B.K. Datta, Nonstatic electromagnetic fields in general relativity, *Nuovo Cimento* **47**, 568 (1967).

B.K. Datta, Stationary electromagnetic fields in general relativity, *J. Math. Phys.* **9**, 1715 (1968).

B.K. Datta, Nonstatic spherically symmetric clusters of particles in general relativity, I, *Gen. Relat. Grav.* **1**, 19 (1970).

B.K. Datta, Spinor fields in general relativity, I: Noether theorem and the conservation laws in Riemann-Cartan space, *Nuovo Cimento B* **6**, 1 (1971).

B.K. Datta, Spinor fields in general relativity, II: Generalized field equations and applications to the Dirac field, *Nuovo Cimento B* **6**, 16 (1971).

B.K. Datta and A.K. Raychaudhuri, Stationary electromagnetic fields in general relativity, *J. Math Phys.* **9**, 1715 (1968).

N. Dauphas, The U/Th production ratio and the age of the Milky Way from meteorites and galactic halo stars, *Nature* **435**, 1203-1205 (2005).

G. Dautcourt, A. Papapetrou and H. Treder, One-dimensional gravitational fields, *Ann. Phys.* **9**, 330 (1962). [In German]

W. Davidson and S.F. Singer, Use of an artificial satellite to test the clock "paradox" and general relativity, *Nature* **188**, 1013 (1960).

M. Davis and R. Ruffini, Gravitational radiation in the presence of a

Schwarzschild black hole: A boundary value search, *Lett. al Nuovo Cimento* **12**, 1165 (1971).

T.M. Davis and J.R. Ray, Neutrinos in cylindrically-symmetric space-times, *J. Math Phys.* **16**, 80 (1975).

W.R. Davis, *Classical Fields, Particles and the Theory of Relativity* (Gordon and Breach, London, 1970).

U.K. De and A. Banerjee, Two classes of solutions of null electromagnetic radiation, *Progr. Theor. Phys.* **47**, 1204 (1972).

P. de Bernardis *et al.*, *Nature* **404**, 955 (2000), astro-ph/0004404.

R. Debever, Space-times of Petrov type III, *C.R. Acad. Sci., Paris* **251**, 1352 (1960).

G.C. Debney, On vacuum space-times admitting null Killing bivectors, *J. Math. Phys.* **12**, 2372 (1971).

G.C. Debney, Null Killing vectors in general relativity, *Lett. al Nuovo Cimento* **5**, 954 (1972).

G.C. Debney, R.P. Kerr and A. Schild, Solutions of the Einstein and Einstein-Maxwell equations, *J. Math. Phys.* **10**, 1842 (1969).

M. de Campos, Tensorial perturbations in an accelerating Universe, *Gen. Relat. Grav.* **34**, 1393 (2002).

L. Defrise, Empty space type D matrices containing an isotropy group G, acting on space like orbits, *Bull. Acad. Roy. Belg., Cl. Sci.* **53**, 827 (1967).

L. Defrise, *Groupes d'isotropie et groupes de stabilité conforme dans les espaces lorentziens* (Thése, Université Libre de Bruxelles, 1969).

V. de la Cruz, J.E. Chase and W. Israel, Gravitational collapse with asymmetries, *Phys. Rev. Lett.* **24**, 423 (1970).

G. Delbourgo, K. Koller and P. Mahanta, On transformations between basis vectors of unitary SL(2,C) representations, *Nuovo Cimento A* **52**, 1254 (1967).

M. Demianski, New Kerr-like space-time, *Phys. Lett. A* **42**, 157 (1972).

M. Demianski and E.T. Newmann, A combined Kerr-NUT solution of the Einstein field equations, *Bull. Acad. Polon. Sci., Ser. Math. Astr. Phys.* **14**, 653 (1966).

P.N. Demmie and A.I. Janis, The characteristic development of trapped surfaces, *J. Math. Phys.* **14**, 793 (1973).

L. Derry, R. Isaacson and J. Winicour, Shear-free gravitational radiation, *Phys. Rev.* **185**, 1647 (1969).

S. Deser, Self-interaction and gauge invariance, *Gen. Relat. Grav.* **1**, 9 (1970).

S. Deser and J. Higbie, Essential singularities in general relativity, *Phys. Rev. Lett.* **23**, 1184 (1969).

B.S. DeWitt, Theory of radiative corrections for non-Abelian gauge fields, *Phys. Rev. Lett.* **12**, 742 (1964).

B.S. DeWitt, *Phys. Rev.* **162**, 1195 and 1230 (1967).

B.S. DeWitt and R.W. Brehme, *Ann. Phys. (N.Y.)* **9**, 220 (1960).

C.M. DeWitt and J.L. Ging, Gravitational radiation damping, *C. R. Acad. Sci., Paris* **251**, 1868 (1960).

R.H. Dicke, New research on old gravitation: Are the observed physical constants independent of the position, epoch and velocity of the laboratory?, *Science, (N.Y.)* **129**, 621 (1959).

R.H. Dicke, *Sci. Am.* **205**, 84 (1961).

R.H. Dicke, *Theoretical Significance of Experimental Relativity* (Gordon and Breach, New York, 1964).

R.H. Dicke, *Relativity, Groups and Topology*, DeWitt *et al.*, Editors (Gordon and Breach, 1964).

R.H. Dicke, The Sun's rotation and relativity, *Nature* **202**, 432 (1964).

R.H. Dicke, Experimental relativity, in: *Relativity, Groups and Topology* (C. DeWitt *et al.*, Eds.) (Gordon and Breach, New York, 1964), p.163.

R.H. Dicke, *Theoretical Significance of Experimental Relativity* (Gordon and Breach, New York, 1964).

R.H. Dicke, The weak and strong principles of equivalence, *Ann. Phys. (N.Y.)* **31**, 235 (1965).

J.O. Dimmock, Representation theory for nonunitary groups, *J. Math. Phys.* **4**, 1307 (1963).

P.A.M. Dirac, *Proc. Roy. Soc. A* **117**, 610 (1928).

P.A.M. Dirac, Unitary representations of the Lorentz group, *Proc. Roy. Soc. (Lond.) A* **183**, 284 (1945).

P.A.M. Dirac, *Spinors in Hilbert Space* (Plenum Press, New York, 1974).

W.G. Dixon, *Nuovo Cimento* **34**, 317 (1964).

W.G. Dixon, Analysis of the Newman-Unti integration procedure for asymptotically flat space-times, *J. Math. Phys.* **11**, 1238 (1970).

W.G. Dixon, *Proc. R. Soc. A* **314**, 499 (1970).

W.G. Dixon, *Gen. Relat. Grav.* **4**, 199 (1973).

W.G. Dixon, *Phil. Trans. R. Soc. A* **277**, 59 (1974).

W.G. Dixon, Extended bodies in general relativity: Their description and motion, in: *Isolating Gravitating Systems in General Relativity*, LXVII Corso (Soc. Italiana di Fisica, Bologna, Italy, 1979).

P. Dolan, A singularity-free solution of the Maxwell-Einstein equations, *Commun. Math. Phys.* **9**, 161 (1968).

G. Domokos, S. Kovesi-Domokos and E. Schonberg, Fully Reggeized scattering amplitudes, *Phys. Rev. D* **2**, 1026 (1970).

S. Doplicher and J.E. Roberts, Fields, statistics and non-abelian gauge groups, *Commun. Math. Phys.* **28**, 331 (1972).

J.S. Dowker, Non-local Yang-Mills theory, *J. Phys. A* **3**, 59 (1970).

J.S. Dowker, The NUT solution as a gravitational dyon, *Gen. Relat. Grav.* **5**, 603 (1974).

J.S. Dowker and M. Goldstone, The geometry and algebra of the representations of the Lorentz group, *Proc. R. Soc. Lond. A* **303**, 381 (1968).

I.M. Dozmorov, Solutions of the Robinson-Trautman type, *Izv. Viss. Ucheb. Zaved. Fiz.* **9**, 43 (1969) [In Russian]

I.M. Dozmorov, Gravitational waves in general relativity: A wave solution of type III, *Izv. Viss. Ucheb. Zaved. Fiz.* **11**, 121 (1969). [In Russian]

D.V. Duc and N.V. Hieu, On the theory of unitary representations of the SL(2,C) group, *Ann. Inst. Poincaré A* **6**, 17 (1967).

M.J. Duff, Covariant gauges and point sources in general relativity, *Ann.*

Phys. (N.Y.) **79**, 261 (1973).

R.J. Duffin, On the characteristic matrices of covariant systems, *Phys. Rev.* **54**, 1114 (1938).

R.L. Duncombe, Relativity effects for the three inner planets, *Astron. J.* **61**, 174 (1956).

H.P. Dürr, Isospin and parity in nonlinear spinor theory, *Z. Naturforsch. A* **16**, 327 (1961).

H.P. Dürr, Poincaré gauge-invariant spinor theories and the gravitational field, *Nuovo Cimento A* **4**, 187 (1971).

D. Eardley and R.K. Sachs, Space-times with a future projective infinity, *J. Math. Phys.* **14**, 209 (1973).

J. Earman, Are spatial and temporal congruence conventional?, *Gen. Relat. Grav.* **1**, 143 (1970).

R. Ebert, in: *Proceedings of the Fifth Marcel Grossman Meeting on General Relativity, Part A & B* (World Scientific, Perth, Teaneck, NJ, 1988).

L.A. Edelstein and C.V. Vishveshwara, Differential equations for perturbations on the Schwarzschild metric, *Phys. Rev. D* **1**, 3514 (1970).

H.J. Efinger, Space-time of electrostatic field in general relativity, *Acta Phys. Austr.* **19**, 361 (1965).

E. Egami, J.-P. Kneib, G. H. Rieke, R. S. Ellis, J. Richard, J. Rigby, C. Papovich, D. Stark, M. R. Santos, J.-S. Huang, H. Dole, E. Le Floch and P. G. Pérez-González, Spitzer and Hubble Space Telescope constraints on the physical properties of the z≈7 galaxy strongly lensed by A2218, *Astrophys. J.* **618**, L5-L8 (2005).

J. Ehlers, Exakte losing der Einstein-Maxwell'schen feldgleichungen fur statischefelder, *Z. Phys.* **140**, 394 (1955).

J. Ehlers, Thesis (University of Hamburg, 1957, unpublished).

J. Ehlers, Transformations of static exterior solutions of Einstein's gravitational field equations into different solutions by means of conformal mappings, *Collection Int. de CNRS* **91**, 275 (1962).

J. Ehlers and W. Kundt, Exact solutions of the gravitational field equations, in: *Gravitation: An Introduction to Current Research*, L. Witten,

Editor (John Wiley, New York, 1962).

J. Ehlers and E. Rudolph, *Gen. Relat. Grav.* **8**, 197 (1977).

A. Einstein, "Zur Electrodynamik bewegter Körper," *Annalen der Physik* **17**, 891-921 (1905); English translation:"On the electrodynamics of moving bodies," in: A. Einstein, H.A. Lorentz, H. Minkowski and H. Weyl, *The Principle of Relativity* (Dover Publications, 1923), pp. 35-65.

A. Einstein, Die Grundlage der allgemeinen Relativitätstheorie, *Ann. Phys.* **49**, 761 (1916); English translation in: *The Principle of Relativity* (Dover, New York, 1923).

A. Einstein, *Ann. Phys.* **55**, 241 (1918).

A. Einstein, *S. B. Preuss. Akad. Wiss.* **414** (1925).

A. Einstein, *Relativity: The Special and General Theory* (Crown Publishers, New York, 1931).

A. Einstein, *The Meaning of Relativity* (Princeton University Press, Princeton, N.J., 1955).

A. Einstein, *Autobiographical Notes*, P.A. Schilpp, Editor (Open Court Publishing Company, La Salle and Chicago, Illinois, 1979).

A. Einstein and J. Grommer, *Preuss. Akad. Wiss., Phys.-Math. Klasse* **1**, 2 (1927).

A. Einstein and J. Grommer, Allgemeine Relativitätstheorie und Bewegungsgesetz, *Sitzungsber. Ber. Akad. Wiss.* **2**, 235 (1928).

A. Einstein, L. Infeld and B. Hoffmann, Gravitational equations and problems of motion, *Ann. Math.* **39**, 65 (1938).

A. Einstein and L. Infeld, On the motion of particles in general relativity theory, *Can. J. Math.* **1**, 209 (1949).

A. Einstein and N. Rosen, The particle problem in the general theory of relativity, *Phys. Rev.* **48**, 73 (1935).

A. Einstein and N. Rosen, On gravitational waves, *J. Franklin Inst.* **223**, 43 (1937).

L.P. Eisenhart, *Continuous Groups of Transformations* (Dover Publications, Inc., New York, 1961).

L.P. Eisenhart, *Riemannian Geometry* (Princeton University Press, New

Jersey, 1949).

L.P. Eisenhart, *Non-Riemannian Geometry* (Americal Mathematical Society, Providence, RI, 1972).

F. Eisenhauer, R. Schödel, R. Genzel, T. Ott, M. Tecza, R. Abuter, A. Eckart and T. Alexander, A geometric determination of the distance to the galactic center, *Astrophys. J.* **597**, L121-L124 (2003).

R.V. Eötvös, *Math. u. Natur. Ber. Ungarn.* **8**, 65 (1890).

R.V. Eötvös, *Beibl. Ann. Phys.* **15**, 688 (1891).

R.V. Eötvös, *Ann. Phys.* **59**, 354 (1896).

R.V. Eötvös, D. Pekar and E. Fekete, *Ann. Phys.* **68**, 11 (1922).

S.T. Epstein, A note on gauge theories of vector particles, *Nuovo Cimento* **25**, 222 (1962).

F. Erdlye, *Asymptotic Expansions* (Dover, New York, 1955).

G. Erez and N. Rosen, The gravitational field of a particle possessing a multipole moment, *Bull. Res. Counc, Israel* **8F**, 47 (1959).

K.-E. Erikson and S. Yngström, A simple approach to the Schwarzschild solution, *Phys. Lett.* **5**, 119 and 327 (1963).

F.J. Ernst, New formulation of the axially symmetric gravitational field I and II, *Phys. Rev.* **167**, 1175 (1968) and **168**, 1415 (1968).

F.J. Ernst, Exterior algebraic derivation of Einstein field equations employing a generalized basis, *J. Math. Phys.* **12**, 2395 (1971).

F.J. Ernst, Charged version of Tomimatsu-Sato spinning-mass field, *Phys. Rev. D* **7**, 2520 (1973).

F.J. Ernst, *J. Math. Phys.* **15**, 1409 (1974).

L. Evans, G. Feldman and P.T. Mathews, Gauge invariance and renormalization constants, *Ann. Phys. (N.Y.)* **13**, 268 (1961).

C.W.F. Everitt, W.M. Fairbank and W.O. Hamilton, General relativity experiments using low temperature techniques, in: *Relativity*, M. Carmeli, S.I. Fickler and L. Witten, Editors (Plenum Press, New York, 1970).

A.A. Evett, Static gravitational fields in general relativity, *Nuovo Cimento* **22**, 779 (1961).

R.A. Exton, E.T. Newman and R. Penrose, Conserved quantities in the Einstein-Maxwell theory, *J. Math. Phys.* **10**, 1566 (1969).

L.D. Faddeev, Quantum theory of gauge fields, in: *Proceedings of the International Seminar on Vector Mesons and Electromagnetic Interactions, Dubna, 1969.*

G. Feinberg, Possibility of faster-than-light particles, *Phys. Rev.* **159**, 1089 (1967).

D.A. Feinblum, Global singularities and the Taub-NUT metric, *J. Math. Phys.* **11**, 2713 (1970).

E. Fermi, Sopra i fenomeni che avvengono in prossimita di una linear oraria, *Atti Accad. Nazl. Lincei* **27**, 21 and 51 (1922).

G. Ferrarese, *Lezioni di Meccanica Relativistica* (Pitagora ed., Bologna, 1985).

R.P Feynman, The quantum theory of gravitation, in: *Warsaw Conference on Relativistic Theories of Gravitation*, L. Infeld, Editor (PWN, Warsaw, 1964); *Acta Phys. Polon.* **24**, 697 (1963).

M. Fierz, Uber die Relativistische Theorie kraftefreier Teilchen mit beliebigen Spin, *Helv. Phys. Acta* **12**, 3 (1938).

M. Fierz and W. Pauli, On relativistic wave equations for particles of arbitrary spin in an electromagnetic field, *Proc. R. Soc. Lond. A* **173**, 211 (1939).

A.V. Filippenko and A.G. Riess, p.227 in: *Particle Physics and Cosmology: Second Tropical Workshop*, J.F. Nieves, Editor (AIP, New York, 2000).

D. Finkelstein, Past-future asymmetry of the gravitational field of a point particle, *Phys. Rev.* **110**, 965 (1958).

F.J. Finkelstein, Electromagnetic interactions of a Yang-Mills field, *Rev. Mod. Phys.* **36**, 632 (1964).

A.E. Fischer and J.E. Marsden, The Einstein equations of evolution - A geometric approach, *J. Math. Phys.* **13**, 546 (1972).

A.E. Fischer and J.E. Marsden, The Einstein evolution equations as a first order quasilinear symmetric hyperbolic system, I, *Commun. Math. Phys.* **28**, 1 (1972).

R. Fitzgerald and S. Schiminovich, Gauge fields with noninvariant interactions, *J. Math. Phys.* **9**, 73 (1968).

P.S. Florides, A rotating sphere as a possible source of the Kerr metric, *Nuovo Cimento B* **13**, 1 (1973).

P.S. Florides and R.L. Jones, On stationary systems with spherical symmetry consisting of many gravitating masses, *Nuovo Cimento B* **69**, 41 (1970).

P.S. Florides and J.L. Synge, Notes on the Schwarzschild line element, *Commun. Dublin Inst. Adv. Stud.* (A), No. 14 1961).

P.S. Florides and J.L. Synge, Stationary gravitational fields due to single bodies, *Proc. R. Soc. Lond.* **A280**, 459 (1964).

V. Fock, *Rev. Mod. Phys.* **29**, 325 (1957).

V. Fock, *The Theory of Space, Time and Gravitation* (Pergamon Press, Oxford, 1959).

V. Fock, Einstein statics in conformal space, in: *Recent Developments in General Relativity* (PWN, Warsaw, 1962).

V. Fock, *Theory of Space, Time and Gravitation* (Pergamon Press, Oxford, 1964).

J. Foster, S-spaces of Robinson and Trautman: Topological identifications and asymptotic symmetry, *Proc. Camb. Phil. Soc.* **66**, 521 (1969).

J. Foster, On Robinson-Trautman solution of Einstein's equations, *Phys. Lett. A* **37**, 313 (1971).

J. Foster and E.T. Newman, Note on the Robinson-Trautman solutions, *J. Math. Phys.* **8**, 169 (1967).

J.M. Foyster and C.B.G. McIntosh, A class of solutions of Einstein's equations which admit a 3-parameter group of isometries, *Commun. Math. Phys.* **27**, 241 (1972).

E.S. Fradkin and I.V. Tyutin, S matrix for Yang-Mills and gravitational fields, *Phys. Rev. D* **2**, 2841 (1970).

B. Frank, Quaternion solution to the Schwarzschild problem and neutrino mass, *Nuovo Cimento A* **61**, 263 (1969).

W.L. Freedman, B.F. Madore, J.R. Mould, R. Hill, L. Ferrarese, R.C. Kennicutt Jr, A. Saha, P.B. Stetson, J.A. Graham, H. Ford, J.G. Hoessel, J.

Huchra, S.M. Hughes and G.D. Illingworth, Distance to the Virgo cluster galaxy M100 from Hubble Space Telescope observations of Cepheids, *Nature* **371**, 757-762 (1994).

W.L. Freedman *et al.*, Final results from the Hubble Space Telescope, Talk given at the 20th Texas Symposium on Relativistic Astrophysics, Austin, Texas 10-15 December 2000. (astro-ph/0012376)

W.L. Freedman, B.F. Madore, B.K. Gibson, L. Ferrarese, D.D. Kelson, S. Sakai, J.R. Mould, R.C. Kennicutt Jr, H.C. Ford, J.A. Graham, J.P. Huchra, S.M.G. Hughes, G.D. Illingworth, L.M. Macri and P.B. Stetson, Final results from the Hubble Space Telescope Key Project to measure the Hubble constant, *Astrophys. J.* **553**, 47-72 (2001).

E. Frehland, The general stationary gravitational vacuum field of cylindrical symmetry, *Commun. Math. Phys.* **23**, 127 (1971).

A.P. French, *Special Relativity* (W.W. Norton, New York and London, 1968).

A. Friedmann, *Z. Phys.* **10**, 377 (1922).

A. Friedmann, *Z. Phys.* **11**, 326 (1924).

C. Fronsdal, Completion and embedding of the Schwarzschild solution, *Phys. Rev.* **116**, 778 (1959).

M. Fukugita, C.J. Hogan and P.J.E. Peebles, The cosmic baryon budget, *Astrophys. J.* **503**, 518-530 (1998).

G. Gamov, Gravity, *Sci. Am.* **204**, 941 (1961).

P. Gard and N.B. Backhouse, Methods for obtaining symmetrized representations of SU(2) and the rotation group, *J. Phys.A* **7**, 1793 (1974).

J. Garecki, On the physical interpretation of the Bel-Robinson tensor, *Acta Phys. Polon. B* **4**, 347 (1973).

P.M. Garnavich *et al.*, *Astrophys. J.* **493**, L53 (1998). [Hi-Z Supernova Team Collaboration (astro-ph/9710123)].

P.M. Garnavich *et al.*, *Astrophys. J.* **509**, 74 (1998). [Hi-Z Supernova Team Collaboration (astro-ph/9806396)].

R. Gautreau, A Kruskal-like extension of the Kerr metric along the symmetry axis, *Nuovo Cimento A* **50**, 120 (1967).

R. Gautreau and R. B. Hoffman, Exact solutions of the Einstein vacuum field equations in Weyl coordinates, *Nuovo Cimento B* **61**, 411 (1969).

R. Gautreau and R. B. Hoffman, Class of exact solutions of the Einstein-Maxwell equations, *Phys. Rev. D* **2**, 271 (1970).

R. Gautreau and R. B. Hoffman, Generating potential for the NUT metric in general relativity, *Phys. Lett. A* **39**, 75 (1972).

R. Gautreau and R. B. Hoffman, The structure of the sources of Weyl-type electrovac fields in general relativity, *Nuovo Cimento B* **16**, 162 (1973).

J. Geheniau, Une classification des espaces einsteiniens, *Compt-Rend. Acad. Sci., Paris* **244**, 723 (1957).

I.M. Gelfand, On one-parameter groups of operators in a normed space, *Dokl. Akad. Nauk SSSR* **25**, 711 (1939).

I.M. Gelfand, *Lectures on Linear Algebra* (Interscience, New York, 1948).

I.M. Gelfand, The center of an infinitesimal group ring, *Mat. Sborn.* **26** (68), 1030 (1950).

I.M. Gelfand, Spherical functions on symmetric Riemann spaces, *Dokl. Akad. Nauk SSSR* **70**, 5 (1950).

I.M. Gelfand and M.I. Graev, On a general method of decomposition of the regular representation of a Lie group into irreducible representations, *Dokl. Akad. Nauk SSSR* **92**, 221 (1953).

I.M. Gelfand and M.I. Graev, Analogue to Plancherel's formula for classical groups, *Trudy Mosk. Matem. Ob-va* **4**, 375 (1955).

I.M. Gelfand, M.I. Graev and I.I. Pyatetskii-Shapiro, *Representation Theory and Automorphic Functions* (W.B. Saunders Co., Philadelphia, 1969).

I.M. Gelfand, M.I. Graev and N. Ya. Vilenkin, *Generalized Functions, Vol. 5: Integral Geometry and Representation Theory* (Academic Press, New York, 1966).

I.M. Gelfand, R.A. Minlos and Z. Ya. Shapiro, *Representations of the Rotation and Lorentz Groups and their Applications* (Pergamon Press, New York, 1963).

I.M. Gelfand and M.A. Naimark, Unitary representations of the Lorentz group, *J. Phys.* **10**, 93 (1946); *Izv. Akad. Nauk SSSR, Ser. Mat.* **11**, 411 (1947).

I.M. Gelfand and M.A. Naimark, Normed rings with involutions and their representation, *Izv. Akad. Nauk SSSR, Ser. Mat.* **12**, 445 (1948).

I.M. Gelfand and M.A. Naimark, The connection between unitary representations of a complex unimodular group and its unitary subgroup, *Izv. Akad. Nauk SSSR, Ser. Mat.* **14**, 239 (1950).

I.M. Gelfand and Z. Ya. Shapiro, Representations of the group of rotation in three-dimensional space and their applications, *Usp. Mat. Nauk* **7**, 3 (1952) [English translation: *Amer. Math. Soc. Translations* **2** (2), 207 (1956)].

I.M. Gelfand and G.E. Shilov, *Generalized Functions, Vol. 1: Properties and Operations* (Academic Press, New York and London, 1964).

I.M. Gelfand and N. Ya. Vilenkin, *Generalized Functions, Vol. 4: Applications of Harmonic Analysis* (Academic Press, New York, 1964).

I.M. Gelfand and A.M. Yaglom, General relativistic invariant equations and infinite-dimensional representations of the Lorentz group, *Zh. Eksp. Teoret. Fiz.* **18**, 703 (1948).

I.M. Gelfand and A.M. Yaglom, Pauli's theorem for general relativistic invariant equations, *Zh. Eksp. Teoret. Fiz.* **18**, 1096 (1948).

G. Gemelli, Five-dimensional special relativistic hydrodynamics and cosmology, gr-qc/0610010.

G. Gemelli, Particle production in 5-dimensional Cosmological Relativity, *Int. J. Theor. Phys.* **45**(12), 2261–2269 (2006).

G. Gemelli, Hydrodynamics in 5-dimensional cosmological special relativity, *Int. J. Theor. Phys.* (2007).

H. Georgi and S.L. Glashow, Gauge theories without anomalies, *Phys. Rev.* D **6**, 429 (1972).

R. Geroch, What is a singularity in general relativity? *Ann. Phys.* **48**, 526 (1968).

R. Geroch, Spinor structure of space-times in general relativity, I & II, *J. Math. Phys.* **9**, 1739 (1968), and **11**, 343 (1970).

R. Geroch, Domain of dependence, *J. Math. Phys.* **11**, 437 (1970).

R. Geroch, General relativity in the large, *Gen. Relat. Grav.* **2**, 61 (1971).

R. Geroch, Einstein algebras, *Commun. Math. Phys.* **26**, 271 (1972).

R. Geroch, A method of generating solutions of Einstein's equation, *J. Math. Phys.* **12**, 918 (1971), and **13**, 394 (1972).

R. Geroch, A. Held and R. Penrose, A space-time calculus based on pairs of null directions, *J. Math. Phys.* **14**, 874 (1973).

R. Geroch and P.S. Jang, Motion of a body in general relativity, *J. Math. Phys.* **16**, 65 (1975).

R. Geroch and E.T. Newman, Application of the semi-direct product of groups, *J. Math. Phys.* **12**, 314 (1971).

C. Gheorghe and E. Muhil, Casual groups of space-time, *Commun. Math. Phys.* **14**, 165 (1969).

N.N. Ghosh, Theory of analytic spinors I, *Proc. Nat. Inst. Sci. India A* **31**, 477 (1965).

N.N. Ghosh and R. Sengupta, The gravitation of a cylindrically symmetric magnetic field, *Nuovo Cimento* **38**, 1579 (1965).

C.H. Gibson, Turbulent mixing, viscosity, diffusion and gravity in the formation of cosmological structures: The fluid mechanics of dark matter, *J. Fluid Eng.* **122**, 830-835 (2000).

J.F. Gille and J. Manuceau, Gauge transformations of second type and their implementation, I. Fermions, *J. Math. Phys.* **13**, 2002 (1972).

V.L. Ginzburg and I.Ye. Tamm, On the theory of spin, *Zh. Eksp. Teoret. Fiz.* **17**, 227 (1947).

L. Girardello and G. Parravicini, Continuous spin in the Bondi-Metzner-Sachs group of asymptotic symmetry in general relativity, *Phys. Rev. Lett.* **32**, 565 (1974).

S.L. Glashow and M. Gell-Mann, Gauge theories of vector particles, *Ann. Phys. (N.Y.)* **15**, 437 (1961).

S.L. Glashow and Iliopaulos, Divergencies of massive Yang-Mills theories: Higher groups, *Phys. Rev. D* **4**, 1918 (1971).

E.N. Glass and J.N. Goldberg, Newman-Penrose constants and their invariant transformations, *J. Math. Phys.* **11**, 3400 (1970).

R. Godement, A theory of spherical functions, I, *Trans. Am. Math. Soc.* **73**, 496 (1952).

B.B. Godfrey, Mach's principle, the Kerr metric and black-hole physics, *Phys. Rev. D* **3**, 2721 (1970).

J.N. Goldberg, Invariant transformations and Newman-Penrose constants, *J. Math. Phys.* **8**, 2161 (1967).

J.N. Goldberg, Conservation of the Newman-Penrose conserved quantities, *Phys. Rev. Lett.* **28**, 1400 (1972).

J.N. Goldberg, Conservation equations and equations of motion in the null formalism, *Gen. Relat. Grav.* **5**, 183 (1974).

J.N. Goldberg and R.P. Kerr, Some applications of the infinitesimal holonomy group to the Petrov clssification of Einstein spaces, *J. Math. Phys.* **2**, 327 (1961).

J.N. Goldberg, A.J. MacFarlane, E.T. Newman, F. Rohrlich and E.A. Sudarshan, Spin-s spherical harmonics and ∂, *J. Math. Phys.* **8**, 2155 (1967).

J.N. Goldberg and R.K. Sachs, A theorem on Petrov types, *Acta Phys. Polon.* **22**, Suppl. 13 (1962).

H. Goldstein, *Classical Mechanics* (Addison-Wesley, Reading, Mass., 1965).

P.A. Goodinson, An electromagnetic interpretation of the Kerr-Vaidya metric, *Int. J. Theor. Phys.* **6**, 47 (1972).

P.A. Goodinson and R.A. Newing, Einstein-Maxwell null fields with nonzero twist, *Int. J. Theor. Phys.* **7**, 25 (1973).

J.H. Grace and A. Young, *Algebra of Invariants* (Cambridge University Press, London, 1903).

M.I. Graev, On a general method of computing traces of infinitedimensional unitary representations of real simple Lie groups, *Dokl. Akad. Nauk SSSR* **103**, 357 (1955).

M. Greenhow and S. Moyo, *Philos. Trans. Roy. Soc. London Ser. A* **355** 551 (1997).

J.B. Griffiths, Gravitational radiation and neutrinos, *Commun. Math. Phys.* **28**, 295 (1972).

J.B. Griffiths and R.A. Newing, The two-component neutrino field in general relativity, *J. Phys. A* **3**, 136 (1970).

J.B. Griffiths and R.A. Newing, Tetrad equations for the two component neutrino field in general relativity, *J. Phys. A* **3**, 269 (1970).

J.B. Griffiths and R.A. Newing, Geometrical aspects of the two-components neutrino field in general relativity, *J. Phys. A* **4**, 208 (1971).

J.B. Griffiths and R.A. Newing, Neutrino radiation fields in general relativity, *J. Phys. A* **4**. 306 (1971).

L.B. Grigor'eva, Chronometrically invariant representation of the classification of Petrov gravitational fields, *Dokl. Akad. Nauk SSSR*, p. 1251 [English translation in: *Sov. Phys. Dokl.* **15**, 579 (1970)].

J.E. Gunn and J.B. Oke, *Astrophy. J.* **195**, 255 (1975).

F. Gürsey, Introduction to group theory, in: *Relativity, Groups and Topology* (Gordon and Breach, New York, 1964).

A.H. Guth, Inflationary Universe: A possible solution to the horizon and flatness problems, *Phys. Rev. D* **23**, 347 (1982).

J.C. Hafele and R.E. Keating, Around the world atomic clocks, *Science* **177**, 166 and 168 (1972).

P. Hajicek, Extensions of the Taub and NUT spaces and extensions of their tangent bundles, *Commun. Math. Phys.* **17**, 109 (1970).

G.S. Hall, On the Petrov classification of gravitational fields, *J. Phys. A* **6**, 619 (1973).

W.H. Hallidy and A.I. Janis, Gravitational radiation: Cutting the tail, *J. Math. Phys.* **11**, 578 (1970).

A. Hamoui, Deux nouvelles classes de solutions non-static a symetrie spherique des equations d'Einstein-Maxwell, *Ann. Inst. Henri Poincaré* **10**, 195 (1969).

O. Hara, Possibility of giving a finite mass to the gauge field, *Nuovo Cimento* **29**, 565 (1963).

Harish-Chandra, Infinite irreducible representations of the Lorentz group, *Proc. R. Soc. A* **189**, 372 (1947).

Harish-Chandra, On relativistic wave equations, *Phys. Rev.* **71**, 793 (1947).

Harish-Chandra, Plancherel formula for complex semi-simple Lie groups, *Proc. Nat. Acad. Sci., Wash.* **37**, 813 (1951).

Harish-Chandra, The Plancherel formula for complex semi-simple Lie groups, *Trans. Am. Math. Soc.* **76**, 485 (1954).

B.K. Harrison, Electromagnetic solutions of the field equations of general relativity, *Phys. Rev. B* **138**, 488 (1965).

B.K. Harrison, New solutions of the Einstein-Maxwell equations from old, *J. Math. Phys.* **9**, 1744 (1968).

J.B. Hartle and S.W. Hawking, Solutions of the Einstein-Maxwell equations with many black holes, *Commun. Math. Phys.* **26**, 87 (1972).

J.G. Hartnett, "Carmeli's cosmology: The Universe is spatially flat without dark matter," invited talk given at the international conference "Frontiers of Fundamental Physics 6," held in Udine, Italy, September 26-29, 2004

J.G. Hartnett, Carmeli's accelerating Universe is spatially flat without dark matter, *Int. J. Theor. Phys.* **44**, 485 (2005); gr-qc/0407083.

J.G. Hartnett, The Carmeli metric correctly describes spiral galaxy rotation curves, *Int. J. Theor. Phys.* **44**, 359 (2005); gr-qc/0407082.

J.G. Hartnett, The distance modulus determined from Carmeli's cosmology fits the accelerating Universe data of the high-redshift type Ia supernovae without dark matter, *Found. Phys.* **36**(6), 839-861 (2006), astro-ph/0501526.

J.G. Hartnett, Spiral galaxy rotation curves determined from Carmelian general relativity, *Int. J. Theor. Phys.* **45**(11), 2118-2136 (2006); astro-ph/0511756.

J.G. Hartnett, Extending the redshift-distance relation in cosmological general relativity to higher redshifts, *Found. Phys.* **38**(3), 201-215, 2008, gr-qc/0705.3097.

J.G. Hartnett, Spheroidal and elliptical galaxy rotational velocity dispersion determined from Cosmological General Relativity, *Int. J. Theor. Phys.* **47**(5), 1252-1260 (2008); astro-ph/0707.2858 .

J.G. Hartnett and F.J. Oliveira, Luminosity distance, angular size and surface brightness in Cosmological General Relativity, *Found. Phys.* **37**(3), 446-454 (2007).

J.G. Hartnett and F.J. Oliveira, Testing cosmological general relativity against high redshift observations, astro-ph/0603500.

J.G. Hartnett and M.E. Tobar, Properties of gravitational waves in cosmological general relativity, *Int. J. Theor. Phys.* **45**, 2213 (2006), gr-qc/0603067.

S. Hatsukade, New representation of ratation and Lorentz groups based on a model of non-Cantorian set theory, *Rep. Math. Phys.* **3**, 173 (1972).

I. Hauser, Type N gravitational field with twist, *Phys. Rev. Lett.* **33**, 1112 (1974).

P. Havas and J.N. Goldberg, *Phys. Rev.* **128**, 398 (1962).

S.W. Hawking, Stable and generic properties in general relativity, *Gen. Relat. Grav.* **1**, 393 (1971).

S.W. Hawking, Black holes in general relativity, *Commun. Math. Phys.* **25**, 152 (1972).

S.W. Hawking and G.F.R. Ellis, *The Large Scale Structure of Space-Time* (Cambridge University Press, Cambridge, England, 1973).

K. Hayashi and A. Bergman, Poincaré gauge invariance and the dynamical role of spin in gravitational theory, *Ann. Phys. (N.Y.)* **75**, 562 (1973).

W. Heine, *Group Theory in Quantum Mechanics* (Pergamon Press, 1970).

H. Heintzmann, New exact static solutions of Einstein's field equations, *Z. Phys.* **228**, 489 (1969).

W. Heisenberg, *Z. Phys.* **77**, 1 (1932).

A. Held, E.T. Newman and R. Posadas, The Lorentz group and the sphere, *J. Math. Phys.* **11**, 3145 (1970).

J. Hely, A new set of states of pure radiation, *Compt. Rend. Acad. Sci., Paris* **257**, 2083 (1963).

W.C. Hernandez, Static axially symmetric, interior solution in general relativity, *Phys. Rev.* **153**, 1359 (1967).

W.C. Hernandez, Material sources for the Kerr metric, *Phys. Rev.* **159**, 1070 (1967).

W.C. Hernandez, Kerr metric, rotating sources and Machian effects, *Phys. Rev.* **167**, 1180 (1968).

L. Hernquist and V. Springel, An analytical model for the history of cosmic star formation, *MNRAS* **341**, 1253-1267 (2003).

L.A. Herrera, A gravitational Lagrangian and some identities in the Newman-Penrose formalism, *Nuovo Cimento A* **17**, 48 (1973).

E. Hiawka, *Acta Phys. Aust.*, Suppl. VII, 265 (1970).

N.J. Hicks, *Notes on Differential Geometry* (Van Nostrand, Princeton, NY, 1965).

R.H. Hildebrand, Neutral meson production in n-p collisions, *Phys. Rev.* **89**, 1090 (1953).

P. Hillion and J.P. Vigier, On the Yang-Mills field, *Compt. Rend. Acad. Sci., Paris* **252**, 1113 (1961).

P. Hillion and J.P. Vigier, An extension of Utiyama's formalism, *Cahiers de Phys.* **15**, 435 (1961); **16**, 2 (1962).

P. Hillion and J.P. Vigier, Gauge group and isospace symmetries associated with the relativistic ratator theory of weak interaction, *Cahiers de Phys.* **16**, 381 (1962). [In French]

E. Hilton, The singularity in the Schwarzschild space-time, *Proc. R. Soc. Lond.* **A283**, 491 (1965).

R.B. Hoffman, Stationary axially symmetric generalizations of the Weyl solutions in general relativity, *Phys. Rev.* **182**, 1361 (1969).

R.B. Hoffman, Stationary "noncanonical" solutions of the Einstein vacuum field equations, *J. Math. Phys.* **10**, 953 (1969).

B. Hoffmann, On the spherically symmetric field in relativity, *Quart. J. Math. Oxford* **3**, 226 (1932), and **4**, 179 (1933).

B. Hoffmann, Aspects of negative mass and the quasers, *Conference on Relativistic Theories and Gravitation*, Vol. II (London, 1965).

B. Hoffmann, Negative mass, *Science Jl.* **1**, 74 (1965).

S. Hojman, Ph.D. Thesis, Joseph Henry Laboratories (Princeton University, Pinceton, NJ, 1975).

M. Honma and Y. Sofue, Rotation curve of the Galaxy, *Publ. Astron. Soc. Jpn.* **49**, 453 (1997).

G. Hootf, Renormalization of massless Yang-Mills fields, *Nucl. Phys. B*

33, 173 (1971).

A. Horak, Analysis of Schwarzschild's solution, *Phys. Lett.* **13**, 318 (1964).

J. Horsky and M. Lenc, A comment on a plane symmetric solution of Einstein equations, *Czech. J. Phys.* B **20**, 1053 (1970).

J. Horsky and J. Novotny, The exact static exterior and interior metric of a thick plane plate, *J. Phys.* A **2**, 251 (1969).

J.P. Hsu and E.C.G. Sudarshan, Theory of massive and massless Yang-Mills fields, *Phys. Rev.* D **9**, 1678 (1974).

X. Hubaut, A class of spaces with metric connection connected to vacuum spaces in relativity, *Bull. Acad. Roy. Belg., Cl. Sci.* **5**, 1094 (1962).

E.P. Hubble, *Proc. Nat. Acad. Sci.* **15**, 168 (1927).

E.P. Hubble, *The Realm of the Nebulae* (Yale University Press, New Haven, 1936); reprinted by Dover Publications, Inc., New York, 1958.

L.P. Hughston, Generalized Vaidya metrics, *Int. J. Theor. Phys.* **4**, 267 (1971).

L.P. Hughston, R. Penrose, P. Sommers and M. Walker, On a quadratic first integral for the charged particle orbits in the charged Kerr solution, *Commun. Math. Phys.* **27**, 303 (1972).

R.A. Hulse and J.H. Taylor, Discovery of a pulsar in a binary system, *Astrophys. J.* **195**, L51 (1975).

D. Husemoller, *Fiber Bundles* (Springer-Verlag, New York, 1975).

M. Huszar, Angular momentum and unitary spinor bases of the Lorentz group, *Acta Phys. Acad. Sci. Hung.* **30**, 241 (1971).

M. Huszar, Unitary representation of the SL (2,C) group in horospheric basis, *J. Math. Phys.* **14**, 1620 (1973).

M. Huszar, Unitary spinors in the Lorentz group, *Acta Phys. Slovaca* **24**, 3 (1974).

M. Huszar, Deformation of the SO(2,C) subgroup of the Lorentz group, *J. Math. Phys.* **15**, 654 (1974).

E. Ihrig, An exact determination of the gravitational potentials g_{ij} in terms of the gravitational fields R_{lmns}, *J. Math. Phys.* **16**, 54 (1975).

M. Ikeda and Y. Miyachi, On the static and spherically symmetric solution of the Yang-Mills field, *Progr. Theor. Phys. (Jpn)* **27**, 474 (1962).

L. Infeld, Equations of motion in general relativity theory and the action principle, *Rev. Mod. Phys.* **29**, 398 (1957).

L. Infeld, Equations of motion and gravitational radiation, *Ann. Phys. (N.Y.)* **6**, 341 (1959).

L. Infeld, The EIH and the k-approximation methods, *Bull. Acad. Polon. Sci., Ser. Math. Astron. Phys.* **9**, 93 (1961).

L. Infeld, Is Planck's constant a constant in a gravitational field? *Bull. Acad. Polon. Sci., Ser. Math. Astron. Phys.* **9**, 617 (1961).

L. Infeld and J. Plebanski, *Motion and Relativity* (Pergamon Press, London, and PWN, Warsaw, 1960).

L. Infeld and A. Schild, On the motion of test particles in general relativity, *Rev. Mod. Phys.* **21**, 408 (1949).

L. Infeld and B.L. van der Waarden, The wave equation of the electron in the general relativity theory, *Sb. preuss. Akad. Wiss., Phys.-mat. Kl.* **380** (1933).

D.R. Inglis, The energy levels and the structure of light nuclei, *Rev. Mod. Phys.* **25**, 390 (1953).

J.R. Ipser, Electromagnitic test fields around a Kerr-metric black hole, *Phys. Rev. Lett.* **27**, 5291 (1971).

C.J. Isham, A. Salam and J. Strathdee, SL(6,C) gauge invariance of Einstein-like Lagrangians, *Lett. al Nuovo Cimento* **5**, 969 (1972).

W. Israel, New representation of the extended Schwarzschild manifold, *Phys. Rev.* **143**, 1016 (1966).

W. Israel, Casual anomalies of the extended Schwarzschild manifold, *Phys. Lett.* **21**, 47 (1966).

W. Israel, Event horisons in static vacuum space-times, *Phys. Rev.* **164**, 1776 (1967).

W. Israel, Event horisons in static electrovac space-times, *Commun. Math. Phys.* **8**, 245 (1968).

W. Israel, Source of the Kerr metric, *Phys. Rev. D* **2**, 641 (1970).

W. Israel and G.A. Wilson, A class of stationary electromagnetic vacuum fields, *J. Math. Phys.* **13**, 865 (1972).

G.J. Iverson and G. Mack, E2 parametrization of SL(2,C), *J. Math. Phys.* **11**, 1581 (1970).

S. Izmailov, On the quantum theory of particles possessing internal rotational degrees of freedom, *Zh. Eksp. Teoret. Fiz.* **17**, 629 (1947).

R. Jackiw, C. Nohl and C. Rebbi, Conformal properties of pseudoparticle configurations, *Phys. Rev.* D **15**, 1642-1646 (1977).

R. Jackiw and C. Rebbi, Conformal properties of a Yang-Mills pseudoparticle, *Phys. Rev.* D **14**, 517-523 (1976).

R. Jackiw and C. Rebbi, Spinor analysis of Yang-Mills theory, *Phys. Rev.* D **16**, 1052-1060 (1977).

J.C. Jackson, Infilling for NUT space, *J. Math. Phys.* **11**, 924 (1970).

J.D. Jackson, *Classical Electrodynamics* (John Wiley, New York, 1975).

J. Jaffe and I.I. Shapiro, Lightlike behavior of particles in a Schwarzschild field, *Phys. Rev.* D **6**, 405 (1972).

A.I. Janis and E.T. Newman, Structure of gravitational forces, *J. Math. Phys.* **6**, 902 (1965).

A.I. Janis, E.T. Newman and J. Winicour, Reality of the Schwarzschild singularity, *Phys. Rev. Lett.* **166**, 878 (1968).

A.I. Janis, D.C. Robinson and J. Winicour, Comments on Einstein scalar solutions, *Phys. Rev.* **186**, 1729 (1969).

L. Janossy, Two subgroups of the Lorentz group and their physical significance, *Acta Phys. Hung.* **20**, 115 (1966).

R.T. Jantzen, P. Carini and D. Bini, The many faces of gravito - electromagnetism, *Ann. Phys.* **215**, 1 (1992).

J.M. Jauch, Gauge invariance as a consequence of Galilei invariance for elementary particles, *Helvet. Phys. Acta* **37**, 284 (1964).

P. Jordan, J. Ehlers and W. Kundt, Exact solutions of the field equations of general relativity, *Akad. Wiss. Lit. Mainz Abhandl., Mat.-Nat. Kl.* (No. 2), 3 (1960). [In German]

P. Jordan, J. Ehlers and R.K. Sachs, Exact solutions of the field equations of general relativity, II. Contributions to the theory of pure gravitational radiation, *Akad. Wiss. Lit. Mainz Abhandl., Mat.-Nat. Kl.* (No. 1), 3 (1961).

P. Jordan and W. Kundt, Exact solutions of the field equations of general relativity, III. , *Akad. Wiss. Lit. Mainz Abhandl., Mat.-Nat. Kl.* (No. 3), 3 (1961).

V. Joseph, A spatially homogeneous gravitational field, *Proc. Camb. Phil. Soc.* **62**, 87 (1966).

R. Jost, The Normal form of a complex Lorentz transformaion, *Helvet. Phys. Acta* **33**, 773 (1960).

B. Jouvet, Gauge invariance, the mass of the photon, and the asymptotic form of the photom, *Compt. Rend. Acad. Sci., Paris* **251**, 1119 (1960)

B. Jouvet, Classical gauge invariance of quantum-electrodynamics, *Nuovo Cimento* **20**, 28 (1961).

K. Just, Gauge invariance and quantum measurements, *Nuovo Cimento* **36**, 294 (1965).

T. Kahan (Editor), *Theory of Groups in Classical and Quantum Theory*, Vol. I (Oliver and Boyd, London, 1965).

R. Kantowski and R.K. Sachs, Some spatially homogeneous anisotropic relativistic cosmological models, *J. Math. Phys.* **7**, 443 (1966).

L. Karlov, How to get the quasi-uniform gravitational field from the Schwarzschild metric, *Am. J. Phys.* **37**, 667 (1969).

L. Karlov and W. Rindler, Torsion-free world lines in curved space-time, *Phys. Rev. D* **13**, 1035 (1971).

E. Kasner, Finite representation of the solar gravitational field in flat space in six dimensions, *Am. J. Math.* **43**, 130 (1921).

H.A. Kastrup, Position operators, gauge transformations and the conformal group, *Phys. Rev.* **143**, 1021 (1966).

A. Katz, Derivation of Newton's law of gravitation from general relativity, *J. Math. Phys.* **9**, 983 (1968).

J. Katz and D. Lyndel-Bell, *Class. Quant. Grav.* **8** 2231 (1991).

R. Katz, *An Introduction to the Special Theory of Relativity* (Van Nostrand,

Princeton, New York and London, 1964).

H. Kerbrat-Lunc, Mathematical introduction to the study of the Yang-Mills field on a space-time curve, *Compt. Rend. Acad. Sci., Paris* **259**, 3449 (1964).

H. Kerbrat-Lunc, The Yang and Mills field on a curved space-time, *Compt. Rend. Acad. Sci., Paris* **260**, 6809 (1965).

H. Kerbrat-Lunc, Conformal group on a space-time curve and associated Yang-Mills field, *Compt. Rend. Acad. Sci., Paris* **266**, 29 (1968).

H. Kerbrat-Lunc, Varied Yang-Mills field, *Compt. Rend. Acad. Sci., Paris* **276**, 711 (1973). [In French]

R. Kerner, On the Cauchy problem for the Yang-Mills field equation, *Ann. Inst. Henri Poincaré* **A20**, 279 (1974).

R.P. Kerr, Gravitational field of a spinning mass as an example of algebraically special metric, *Phys. Rev. Lett.* **11**, 237 (1963).

R.P. Kerr and G.C. Debney, Jr., Einstein spaces with symmetry groups, *J. Math. Phys.* **11**, 2807 (1970).

R.P. Kerr and J.N. Goldberg, Einstein spaces with four-parameter holonomy groups,*J. Math. Phys.* **2**, 332 (1961).

R.P. Kerr and A. Schild, A new class of vacuum solutions of the Einstein field equations, *Centenario della nascita di Galileo, Firenze*, Vol. II, Tome 1, p. 22 (1965).

R.P. Kerr and A. Schild, *Proceedings of Symposia in Applied Mathematics*, vol. XVII: *Applications of Nonlinear Partial Differential Equations in Mathematical Physics*, R. Finn, Editor (American Physical Society, Providence, RI, 1965).

G.H. Keswani, Pulsars and possible new test of general relativity, *Nature* **220**, 148 (1968).

R.I. Khrapko, On the uniqueness of the central-symmetric gravitational field in vacuum, *Zh. Eksp. Teoret. Fiz.* (USSR) **49**, 1884 (1965).

R.I. Khrapko, Propagation of light through a singular Schwazschild sphere, *Soviet Phys. JETP* **23**, 4 (1966).

T.W.B. Kibble, Lorentz invariance and the gravitational field, *J. Math.*

Phys. **2**, 212 (1961).

A. Kihlberg, On a class of explicit representations of the homogeneous Lorentz group, *Ark. Fys.* **27**, 373 (1964).

T. Kimura, On a fixaion of gauge and true observables of Yang-Mills field, *Prog. Theor. Phys. (Jpn)* **26**, 790 (1961).

W. Kinnersley, Type D vacuum metrics, *J. Math. Phys.* **10**, 1195 (1969).

W. Kinnersley, Field of an arbitrary accelerating point mass, *Phys. Rev.* **186**, 1335 (1969).

W. Kinnersley, Generation of stationary Einstein-Maxwell fields, *J. Math. Phys.* **14**, 651 (1973).

W. Kinnersley, Recent progress in exact solutions, *Proceedings of Tel Aviv Conference* (1974).

W. Kinnersley and M. Walker, Uniformly accelerating charged mass in general relativity, *Phys. Rev. D* **2**, 1359 (1970).

O. Klein, Some remarks on general relativity and the divergence problem of quantum field theory, *Nuovo Cimento Suppl.* **6**, 344 (1957).

W.H. Klink, Mixed D-functions and Clebsch-Gordon coefficients of non-compact group, *J. Math. Phys.* **11**, 3210 (1970).

R.A. Knop *et al.*, New constraints on Ω_M, Ω_Λ and w from an independent set of 11 high-redshift supernovae observed with the Hubble Space Telescope, *Ap. J.* **598**: 102-137 (2003).

S. Kobayashi and K. Nomizu, *Foundations of Differential Geometry*, Vol. 1 (Wiley, New York, 1969).

A. Komar, On the necessity of singularities in the solutions of the field equations of general relativity, *Phys. Rev.* **104**, 544 (1956).

M. Konuma, H. Umezawa and M. Wada, General theory of the gauge transformation for vector meson fields, *Nucl. Phys.* **31**, 507 (1962).

A. Koppel, Some exact axially symmetric solutions of Einstein's field equations, *Füüsika ja Astron. Inst. Uurimused (Tartu)* **20**, 58 (1963).

A. Koppel, On the exact axial-symmetric static one-coordinate solutions of the Einstein equations, *Füüsika ja Astron. Inst. Uurimused (Tartu)* **22**, 53 (1963).

M.P. Korkina, The solution of the Einsten equations in the Gaussian coordinates, *Ukrayin Fiz. Zh. (USSR)* **13**, 647 (1968). [In Russian]

J. Kota and Z. Perjes, All stationary vacuum metrics with shearing geodesic eigenrays, *J. Math. Phys.* **13**, 1695 (1972).

B. Kozarzewski, A theorem concerning the gravitational field with shear-free, rotation-free and expansion-free geodesic rays, *Acta Phys. Polon.* **25**, 437 (1964).

D. Kramer, A class of non-stationary solutions of the Einstein-Maxwell equations, *Wiss. Z. Friedrich-Schiller Univ. Jena, Math. Naturwiss. Reihe* **21**, 45 (1972).

D. Kramer, Group structure and field equations in Einstein universe, *Acta Phys. Polon.* **B4**, 11 (1973). [In German]

D. Kramer and G. Neugebauer, Zu axialsymmetrischen stationärer lösungen der Einstein'schen feldgleichungen für das vakuum, *Commun. Math. Phys.* **10**, 132 (1968).

D. Kramer and G. Neugebauer, Eine exakte stationare lösung der Einstein-Maxwell gleichungen, *Ann. Phys.* **24**, 59 (1969).

D. Kramer and G. Neugebauer, Intenal Reissner-Weyl solution, *Ann. Phys.* **27**, 129 (1971).

D. Kramer, H. Stephani, M. Maccallum, E. Herlt and Ernst Schmutzer, Editor, *Exact Solutions of Einstein's Field Equations* (VEB Deut-scher Verlag der Wissenschaften, Berlin, 1980).

A. Krasinski, *Inhomogeneous Cosmological Models* (Cambridge University Press, 1997).

K. Kraus, Remarks on a class of almost solutions of the gravitaional field equations, *Ann. Phys. (N.Y.)* **50**, 102 (1968).

L.M. Krauss, The end of the age problem, and the case for a cosmological constant revisited, *Astrophys. J.* **501**, 461-466 (1998).

E. Kretschmann, *Ann. Phys.* **53**, 575 (1917).

M.P. Krokina, The solution of the Einstein equations in the Gaussian coordinates, *Ukrayin. Fiz. Zh.* **13**, 647 (1968). [In Russian]

K.D. Krori, On motions of test particles in Kerr metric, *Ind. J. Phys.* **44**,

227 (1970).

M.D. Kruskal, Maximal extension of Schwarzschild metric, *Phys. Rev.* **119**, 1743 (1960).

B. Kuchowicz, A simplified method of obtaining the infinitesimal operators of the homogeneous Lorentz group I, *Bull. Acad. Polon. Sci., Ser. Math. Aston. Phys.* **10**, 221 (1962).

B. Kuchowicz, Neutrinos in general relativity: Four(?) levels of approach, *Gen. Relat. Grav.* **5**, 201 (1974).

H.P. Kuenzle, Construction of singularity-free spherically symmetic spactime manifolds, *Proc. R. Soc. Lond. A* **297**, 244 (1967).

W. Kummer, Gauge-transformation and quantization of free electromagnetic field, *Acta Phys. Austr.* **14**, 149 (1961).

W. Kundt, The plane-fronted gravitational waves, *Z. Phys.* **163**, 77 (1961).

W. Kundt and .T. Newman, Hyperbolic differential equations in two dimensions, *J. Math. Phys.* **9**, 2193 (1969).

W. Kundt and A. Thompson, Weyl tensor and an associated shear-free geodesic congruence, *Compt. Rend. Acad. Sci., Paris* **254**, 4257 (1962).

W. Kundt and M. Trümper, Contributions to the theory of gravitational waves, *Akad. Wiss. Lit. Mainz Abhandl., Mat.-Nat. Kl.* **12**, 36 (1962).

W. Kundt and M. Trümper, Orthogonal decomposition of axisymmetric stationary spacetimes, *Z. Phys.* **192**, 419 (1966).

H.P. Künzle, *Proc. Roy. Soc. Ser. A* **297** 244 (1967).

Kuroda *et al.*, *Int. J. Mod. Physics D* **8**, 557 (1999).

A.G. Kurosh, *Theory of Groups* (Chelsea, London, 1955).

B. Kursunoglu, Complex orthogonal and antiorthogonal representation of Lorentz group, *J. Math. Phys.* **2**, 22 (1961).

B. Kursunoglu, *Modern Quantum Theory* (W.H. Freeman, San Francisco, 1962).

P. Kustaanheimo, A note on the transformability of spherically symmetric metrics, *Proc. Edinb. Math. Soc.* **9**, 13 (1953).

K.B. Lal and H. Prasad, Cylindrical wave solutions of field equations of general relativity containing electromagnetic fields, *Tensor* **20**, 45 (1969).

L.D. Landau, in: *Niels Bohr and the Development of Physics*, W. Pauli, Editor (McGraw Hill, New York, 1955).

L.D. Landau and E.M. Lifshitz, *The Classical Theory of Fields* (Addison-Wesley, Reading, Massachusetts, 1959).

L.D. Landau and E.M. Lifshitz, *Quantum Mechanics* (Pergamon Press, London, 1960).

T. Lauritsen, Energy levels of light nuclei, *Ann. Rev. Nucl. Sci.* **1**, 67 (1952).

B. Léauté, A study of the Kerr metric, *Ann. Inst. Henri Poincaré A* **8**, 93 (1968).

K.K. Lee, Global spinor fields in space-time, *Gen. Relat. Grav.* **4**, 421 (1973).

E. Lehman, The Bel-Petrov classification of a pure electromagnetic field in general relativity, *Compt. Rend. Acad. Sci., Paris* **262**, 806 (1966).

C. Leibovitz, Spherically symmetric static solutions of Einstein's equations, *Phys. Rev.* **185**, 1664 (1969).

G. Lemaitre, *Ann. Soc. Sci. Bruxelles* **A53**, 51 (1933).

J. Lense and H. Thirring, Über den einflüss der eigenrotation der zentralkörper auf die bewegung der planeten und monde nach der Einsteinschen gravitationstheorie, *Physik. Z.* **19**, 156 (1918).

T. Levi-Civita, *Atti Accad. Naz. Lincei* **5**, 26 (1917).

T. Levi-Civita, *Rend. Acc. Lincei*, Several Notes (1918 - 19).

T. Levi-Civita, *Math. Ann.* **97**, 291 (1927).

N. Levinson, The asymptotic behavior of a system of linear differential equations, *Am. J. Math.* **68**, 1 (1946).

H. Levy, Classification of stationary axisymmetric gravitational fields, *Nuovo Cimento B* **56**, 253 (1968).

H. Levy, Stationary gravitational fields from static fields, *Phys. Lett. A* **42**, 176 (1972).

T. Lewis, *Proc. R. Soc. (London A)* **136**, 176 (1932).

A. Lichnerowicz, *Theories Relativistes de la Gravitation et de l'Electromagnetisme* (Masson, Paris,1955).

A. Lichnerowicz, *Relativistic Hydrodynamics and Magneto– Hydrodynamics* (Benjamin, New York, 1967).

A. Lichnerowicz, Espaces fibres et espace-temps, *Gen. Relat. Grav.* **1**, 235 (1971).

A. Lichnerowicz, Magnetohydrodynamics: Waves and shock waves in curved space-time, *Mathematical Physics Studies*, Vol. 14 (Kluwer Academic Publishers, Dordrecht, Boston, London, 1994).

A. Lifschitz, A nonlinear spectral problem with periodic coefficients occurring in magnetohydrodynamic stability theory, in: *Differential and integral operators* (Regensburg, 1995), *Oper. Theory Adv. Appl.* **102**, p.97, Birkhuser, Basel (1998).

LIGO, http://www.ligo.caltech.edu/

R.W. Lind, Algebraically special Einstein-Maxwell fields in the spin coefficient formalism, Thesis (University of Pittsburgh, 1970).

R.W. Lind, Shear-free, twisting Einstein-Maxwell metrics in the Newman-Penrose formalism, *Gen. Relat. Grav.* **5**, 251 (1974).

R.W. Lind, Gravitational and electromagnetic radiation in Kerr-Maxwell spaces, *J. Math. Phys.* **16**, 34 (1975).

R.W. Lind, Stationary Kerr-Maxwell spaces,*J. Math. Phys.* **16**, 39 (1975).

R.W. Lind and E.T. Newman, Complexification of the algebraically special gravitational fields, *J. Math. Phys.* **15**, 1103 (1974).

R.W. Lind, J. Messmer and E.T. Newman, Applications of the intertwining operators for representations of the restricted Lorentz group, *J. Math. Phys.* **13**, 1879 (1972).

R.W. Lind, J. Messmer and E.T. Newman, Equations of motion for the sources of asymptotically flat spaces, *J. Math. Phys.* **13**, 1884 (1972).

A.D. Linde, Scalar field fluctuations in the expanding Universe and the new inflationary Universe scenario, *Phys. Lett. B* **116**, 335 (1982).

R.W. Lindquist, R.A. Schwarz and C.W. Misner, Vaidya's radiating Schwarzschild metric, *Phys. Rev. B* **137**, 1364 (1965).

J.S. Lomont and H.E. Moses, Exponential representation of complex Lorentz matrices, *Nuovo Cimento* **29**, 1059 (1963).

K.H. Look, The Yang-Mills fields and the connections of principle fibre bundles, *Acta Phys. Sin.* **23**, 249 (1974).

L. Loomis, *An Introduction to Abstract Harmonic Analysis* (Van Nostrand, New York, 1953).

H.G. Loos, The range of gauge fields, *Nucl. Phys.* **72**, 677 (1965).

H.G. Loos, Rudimentary geometric particle theory, *Ann. Phys.* **36**, 486 (1966).

H.G. Loos, Free ghost gauge fields, *Nuovo Cimento A* **52**, 1085 (1967).

H.G. Loos, Gauge field of a point charge, *J. Math. Phys.* **8**, 1870 (1967).

H.G. Loos, Internal holonomy groups of Yang-Mills fields, *J. Math. Phys.* **8**, 2114 (1967).

H.G. Loos, Nonexistence of classes of Yang-Mills fields, *Nuovo Cimento A* **53**, 201 (1968).

H.G. Loos, Canonical gauge- and Lorentz-invariant quantization of the Yang-Mills field, *Phys. Rev.* **188**, 2342 (1969).

H.G. Loos, Existence of charged states of the Yang-Mills field, *J. Math. Phys.* **11**, 3258 (1970).

E.A. Lord, General relativity from gauge invariance, *Proc. Camb. Phil. Soc.* **69**, 423 (1971).

M. Lorente, P.L. Huddleston and P. Roman, Some properties of matrices occuring in SL(2,C) invariant wave equations, *J. Math. Phys.* **14**, 1495 (1973).

D. Lovelock, A spherically symmetric solution of the Maxwell-Einstein equations, *Commun. Math. Phys.* **5**, 257 (1967).

E. Lubkin, Geometric definition of gauge invariance, *Ann. Phys.* **23**, 233 (1963).

G. Ludwig, Classification of electromagnetic and gravitational fields, *Am. J. Phys.* **37**, 1225 (1969).

G. Ludwig, Geometrodynamics of electromagnetic fields in the Newman-Penrose formalism, *Commun. Math. Phys.* **17**, 98 (1970).

G. Ludwig, On event horisons in static space-times, *Commun. Math. Phys.* **23**, 255 (1971).

L.A. Lugiato and V. Gorini, On the structure of relativity groups, *J. Math. Phys.* **13**, 665 (1972).

J. Lukierski, Gauge transformation in quantum field theory, *Nuovo Cimento* **29**, 561 (1963).

L.A. Lyusternik and V.I. Sobolev, *Elements of Functional Analysis* (Moscow, 1951).

P.J. McCarthy, Asymptotically flat space-times and elementary particles, *Phys. Rev. Lett.* **29**, 817 (1972).

P.J. McCarthy, Representations of the Bondi-Metzner-Sachs group I. Determination of the representations, *Proc. R. Soc. Lond. A* **330**, 517 (1972).

P.J. McCarthy, Structure of the Bondi-Metzner-Sachs group, *J. Math. Phys.* **13**, 1837 (1972).

P.J. McCarthy, Representations of the Bondi-Metzner-Sachs group II. Properties and classification of the representations, *Proc. R. Soc. Lond. A* **333**, 317 (1973).

W.H. McCrea, The interpretation of the Schwarzschild metric and the release of gravitational energy, *Astrophys. Norvegica* **9**, 89 (1964).

A.J. MacFarlane, On the restricted Lorentz group and groups homomorphically related to it, *J. Math. Phys.* **3**, 1116 (1962).

G.C. McVittie and R. Stabell, Spherically symmetric nonstatical solutions of Einstein's equations, *Ann. Inst. Henri Poincaré* **7**, 103 (1967).

J. Madore, Yang-Mills fields with non-zero mass, *Nuovo Cimento A* **4**, 599 (1971).

A. Maheshwari, Schwinger equations for the Yang-Mills and the gravitational fields, *Progr. Theor. Phys.* **45**, 1662 (1971).

M.D. Maia, Isospinors, *J. Math. Phys.* **14**, 882 (1973).

S. Malin, Gauge-free quantization of the linearized equations of general relativity, *Phys. Rev. D* **10**, 2338 (1974).

S. Malin, The Weyl and Dirac equations in terms of functions over the group SU_2, *J. Math. Phys.* **16**, 679 (1975).

M. Manarini, *Atti Accad. Naz. Lincei. Rend. Cl. Sci. Fis. Mat. Nat.* **4**, 427 (1948).

K.B. Marathe, Spaces admitting gravitational fields, *J. Math. Phys.* **14**, 228 (1973).

L. Marder, On uniform acceleration in special and general relativity, *Proc. Camb. Phil. Soc., Math. Phys. Sci.* **53**, 194 (1957).

L. Marder, Gravitational waves in general relativity, V. An exact spherical wave, *Proc. R. Soc. Lond. A* **261**, 91 (1961).

L. Marder, On space-times with bounded empty regions, *Proc. Camb. Phil. Soc.* **60**, 97 (1974).

L. Marder, Gravitational waves in general relativity, XI. Cylindrical-spherical waves, *Proc. R. Soc. Lond. A* **313**, 83 (1969).

L. Marder, Sources for Kerr type space-times in general relativity, *J. Phys. A* **5**, 605 (1972).

J.J. Marek, Static axisymmetric interior solution in general relativity, *Phys. Rev.* **163**, 1373 (1967).

J.J. Marek, Some solutions of Einstein's equations in general relativity, *Proc. Camb. Phil. Soc.* **64**, 167 (1968).

M.S. Marinov and V.I. Roginsky, On invariants of the rotation group, *Fortschr. Phys.* **19**, 509 (1971).

M. Martellini and P. Sonado, *Phys. Rev. D* **22**, 1325 (1980).

G. Matrucci and M. Modygno, Asymptotic deflection of a light ray in Kerr field, *Lett. al Nuovo Cimento* **3**, 553 (1970).

E. Massa, A simple relation between the restricted Lorentz group L^+ and the unimodular group of order two, *Lett. al Nuovo Cimento* **1**, 806 (1969).

R.A. Matzner, On stationary axisymmetric solutions of the Einstein field equations, *Astrophys. J.* **167**, 149 (1971).

R.A. Matzner and C.W. Misner, Gravitational field equations for sources with axial symmetry and angular momentum, *Phys. Rev.* **154**, 1229 (1967).

A.L. Mehra, P.C. Vaidya and R.S. Kushwaha, Gravitational field of a sphere composed of concentric shells, *Phys. Rev.* **186**, 1333 (1969).

H.J. Meister and A. Papapetrou, Die Schwarzschild'sche lösung in De Don-der'schen Koordinated, *Z. Phys.* **145**, 403 (1956).

M.A. Melvin, Pure magnetic and electric geons, *Phys. Lett.* **8**, 65 (1964).

M.A. Melvin, Dynamics of cylindrical electromagnetic universes, *Phys. Rev. B* **139**, 225 (1965).

M.A. Melvin and J. Wallingford, Orbits in magnetic universe, *J. Math. Phys.* **6**, 333 (1965).

M. Milgrom, A modification of the Newtonian dynamics - Implications for galaxies, *Astrophys. J.* **270**, 371-383 (1983).

M. Milgrom, A Modification of the Newtonian Dynamics - Implications for Galaxy Systems, *Astrophys. J.* **270**, 384-389 (1983).

M. Milgrom, A modification of the Newtonian dynamics as a possible alternative to the hidden mass hypothesis, *Astrophys. J.* **270**, 365-370 (1983).

A.I. Miller, *Albert Einstein's Special Theory of Relativity* (Addison-Wesley, Reading, Massachusetts, 1981).

J.G. Miller, Global analysis of the Kerr-Taub-NUT metric, *J. Math. Phys.* **14**, 486 (1973).

J.G. Miller and M.D. Kruskal, Taub-NUT metric and incompatible extensions, *Phys. Rev. D* **4**, 2945 (1971).

J.G. Miller and M.D. Kruskal, Extension of a compact Lorentz manifold, *J. Math. Phys.* **14**, 484 (1973).

W. Miller, Clebsch-Gordan coefficients and special function identities, II: the rotation and Lorentz groups in 3-space, *J. Math. Phys.* **13**, 827 (1972).

H. Minkowski, Space and time (an address delivered at the 80th Assembly of German Natural Scientists and Physicians, at Cologne, 21 September, 1908); English translation in: *The Principle of Relativity* (Dover, New York, 1923), p.73.

C.W. Misner, The flatter regions of Newman, Unti and Tamburino's generalized Schwarzschild space, *J. Math. Phys.* **4**, 924 (1963).

C.W. Misner, K. Thorne and J.A. Wheeler, *Gravitation* (Freeman, San Francisco, 1973).

C.W. Misner and J.A. Wheeler, Classical physics as geometry: Gravitation, electromagnetism, unquantized charge, and mass as properties of curved empty space, *Ann. Phys. (N.Y.)* **2**, 525 (1957).

M. Misra, Some axially symmetric empty gravitational fields, *Proc. Nat. Inst. Sci., India A* **26**, 673 (1960).

M. Misra, A class of solutions representing plane electromagnetic waves in general relativity, *Proc. Nat. Inst. Sci., India A* **29**, 104 (1963).

M. Misra, Some exact solutions of the field equations of general relativity, *J. Math. Phys.* **7**, 155 (1966).

M. Misra, Type-null vacuum solutions in general relativity, *J. Math. Phys.* **9**, 1052 (1968).

M. Misra, On the generalized Goldberg-Sachs theorem, *Lett. al Nuovo Cimento* **1**, 615 (1969).

M. Misra, A unified treatment of the Kerr and Vaidya solutions in general relativity, *Proc. R. Irish. Acad. A* **69**, 39 (1970).

M. Misra, A new interpretation of the Kerr-Schild metric, *Lett. al Nuovo Cimento* **6**, 715 (1973).

M. Misra, A property of the Schild-Kerr metric, *Curr. Sci. (India)* **42**, 349 (1973).

R.M. Misra, On stationary systems with spherical symmetry, *Nuovo Cimento* **32**, 939 (1964).

R.M. Misra, Axially symmetric fields in general relativity, *Phys. Rev. D* **2**, 410 (1970).

R.M. Misra and D.B. Pandey, Stationary axially symmetric fields and the Kerr metric, *J. Math. Phys.* **13**, 1538 (1972).

R.M. Misra and D.B. Pandey, Class of electromagnetic fields, *J. Phys. A* **6**, 924 (1973).

R.M. Misra, D.B. Pandey, D.C. Srivastava and S.N. Tripathi, New class of solutions of the Einstein-Maxwell fields, *Phys. Rev. D* **7**, 1587 (1973).

R.M. Misra and R.A. Singh, Three-dimensional formulation of gravitational null fields I and II, *J. Math. Phys.* **7**, 1836 (1966), and **8**, 1065 (1967).

R.M. Misra, U. Narain and R.S. Mishra, A solution of the field equations representing a gravitational field of magnetic type, *Tensor* **19**, 303 (1968).

M. Missana, Solutions of Einstein's equations in spherical symmetry, *Mem.*

Soc. Astron. Ital. **39**, 275 (1968).

R.N. Mohapatra, Feynman rules for the Yang-Mills fields: A canonical quantization approach, *Phys. Rev. D* **4**, 378, 1007 and 2215 (1971).

C. Moller, On the possibility of terrestrial test of the general theory of relativity, *Nuovo Cimento Suppl.* **6**, 381 (1957).

C. Moller, Further remarks on the localization of energy in the general theory of relativity,*Ann. Phys. (N.Y.)* **12**, 118 (1961).

C. Moller, Gravitational energy radiation, *Phys. Lett.* **3**, 329 (1963).

C. Moller, *The Theory of Relativity* (Clarendon Press, Oxford, 1969).

I. Moret-Bailly, Structure du champ gravitationnel non-radiatif sans hypothese de symetrie en relativite générale, *Ann. Inst. Poincaré* **9**, 395 (1969).

I. Moret-Bailly, The structure of non-radiative gravitational fields in general relativity using Newman-Penrose formalism, *Ann. Inst. Poincaré A* **11**, 415 (1969). [In French]

H.E. Moses, Irreducible representations of the rotation group in terms of Euler's theorem, *Nuovo Cimento A* **40**, 1120 (1965).

H.E. Moses, Irreducible representations of the rotation group in terms of the axis and angle of rotation, *Ann. Phys.* **37**, 224 (1966), and **42**, 343 (1967).

L. Motz, Gauge invariance and the Lorentz ponderomotive force, *Phys. Rev.* **119**, 1102 (1960).

L. Motz, Gauge invariance and the structure of charged particles, *Nuovo Cimento* **26**, 672 (1962).

L. Motz, Gauge invariance and the Hamilton-Jacobi equation, *Nuovo Cimento B* **69**, 95 (1970).

N. Mukunda, Unitary representations of the homogeneous Lorentz group in an O(2,1) basis, *J. Math. Phys.* **9**, 50 (1968).

N. Mukunda and B. Radhakrishnan, New forms for the representations of the three-dimensional Lorentz group, *J. Math. Phys.* **14**, 254 (1973).

H. Müller zum Hagen, On the analyticity of static vacuum solutions of Einstein's equations, *Proc. Camb. Phil. Soc.* **67**, 415 (1970).

H. Müller zum Hagen and D.C. Robinson, Black holes in static vacuum space-times, *Gen. Relat. Grav.* **4**, 53 (1973).

M. Murenbeeld and J.R. Trollope, Slowly rotating radiation sphere and a Kerr-Vaidya metric, *Phys. Rev. D* **1**, 3220 (1970).

F.D. Murnaghan, *The Theory of Group Representations* (Dover Publications, Inc., New York, 1938).

L. Mysak and G. Szekeres, Behavior of the Schwarzschild singularity in superimposed gravitational fields, *Can. J. Phys.* **44**, 617 (1966).

M.A. Naimark, Rings with involutions, *Usp. Mat. Nauk* **3**, 52 (1948) [English translation: *Am. Math. Soc. Translation*, No. 25 (1950)].

M.A. Naimark, On linear representations of the proper Lorentz group, *Dokl. Akad. Nauk SSSR* **97**, 969 (1954).

M.A. Naimark, On the description of all unitary representations of the complex classical groups I and II, *Mat. Sborn.* **35**(77), 317 (1954); **37**(79), 121 (1955).

M.A. Naimark, Linear representation of the Lorentz group, *Usp. Mat. Nauk* **9**, 19 (1954); English translation: *Amer. Math. Soc. Translation*, Series 2, 6 (1957).

M.A. Naimark, *Normed Rings* (Noordhoff, Groningen, Netherlands, 1959).

M.A. Naimark, On irreducible linear representations of the complete Lorentz group, *Dokl. Akad. Nauk SSSR* **112**, 583 (1957).

M.A. Naimark, *Linear Representations of the Lorentz Group* (Pergamon Press, New York, 1964).

Y. Nambu, Infinite-component wave equations with hydrogenlike spectra, *Phys. Rev.* **160**, 1171 (1967).

J. Namyslowski, The Dirac equation in general relativity in the vierbein formalism, *Acta Phys. Polon.* **20**, 927 (1961).

H. Nariai, On some static solutions of Einstein's gravitational field equations in a spherically symmetric case, *Sci. Rep. Tohoku Univ.* **34**, 160 (1950).

J.V. Narlikar, *An Introduction to Cosmology*, 3rd Ed., (Cambridge University Press, Cambridge, 2002).

Y. Ne'eman, Derivation of strong interactions from a gauge invariance, *Nucl. Phys.* **26**, 222 (1961).

Y. Ne'eman, Yukawa terms in the unitary gauge theory, *Nucl. Phys.* **26**, 230 (1961).

V. Nemytskii, M. Sludskaya and A. Cherkasov, *A Course in Mathematical Analysis* (Moscow, 1944).

E.T. Newman, Some properties of empty space-time, *J. Math. Phys.* **2**, 324 (1961).

E.T. Newman, New approach to the Einstein and Maxwell-Einstein field equations, *J. Math. Phys.* **2**, 674 (1961).

E.T. Newman, A generalization of the Schwarzschild metric, *Conference on Relativistic Theories of Gravitation* (PWN, Warsaw, 1962).

E.T. Newman, The nature of gravitational sources, Gravity Research Foundations, 1964.

E.T. Newman, A possible connection between the gravitational field and elementary particle physics, *Nature* **206**, 811 (1965).

E.T. Newman, Singularities: Their structure and motion, *Gen. Relat. Grav.* **1**, 401 (1971).

E.T. Newman, Complex coordinate transformation and the Schwarzschild-Kerr metrics, *J. Math. Phys.* **14**, 774 (1973).

E.T. Newman, Lienard-Wiechert fields and general relativity, *J. Math. Phys.* **15**, 44 (1974).

E.T. Newman, Complex Maxwell and Einstein fields, IAU Symposium No. 64: *Gravitational Radiation and Gravitational Collapse*, Warsaw,Poland, 5 - 8 September 1974 (Reidel Dordrecht, Netherlands, 1974), p. 105.

E.T. Newman, *Phys. Rev. D* **18**, 2901 (1978).

E.T. Newman, W.E. Couch, K. Chinnapared, A. Exton, A. Prakash and R. Torrence, Metric of a rotating chaged mass, *J. Math. Phys.* **6**, 918 (1965).

E.T. Newman and A. Janis, Note on the Kerr spinning particle metric, *J. Math. Phys.* **6**, 915 (1965).

E.T. Newman and R. Penrose, An approach to gravitational radiation by a method of spin coefficients, *J. Math. Phys.* **3**, 566 (1962); **4**, 998 (1963).

E.T. Newman and R. Penrose, 10 exact gravitationally conserved quantities, *Phys. Rev. Lett.* **15**, 231 (1965).

E.T. Newman and R. Penrose, Note on the Bondi-Metzner-Sachs group, *J. Math. Phys.* **7**, 863 (1966).

E.T. Newman and R. Penrose, *Some new gravitationally conserved quantities*, Gravity Research Foundation (1966).

E.T. Newman and R. Posadas, Equations of motion and the structure of singularities, *Phys. Rev. Lett.* **22**, 1196 (1969).

E.T. Newman and R. Posadas, Motion and the structure of singularities in general relativity, *Phys. Rev.* **187**, 1784 (1969).

E.T. Newman and R. Posadas, Motion and the structure of singularities in general relativity, II, *J. Math. Phys.* **12**, 2319 (1971).

E.T. Newman and L.A. Tamburino, New approach to Einstein's empty space field equations, *J. Math. Phys.* **2**, 667 (1961).

E.T. Newman and L.A. Tamburino, Empty space metrics containing hypersurface orthogonal geodesic rays, *J. Math. Phys.* **3**, 902 (1962).

E.T. Newman, L.A. Tamburino and T. Unti, Empty-space generalization of the Schwarzschild metric, *J. Math. Phys.* **4**, 915 (1963).

E.T. Newman and T.W.J. Unti, Behavior of asymptotically flat empty spaces, *J. Math. Phys.* **3**, 891 (1962).

E.T. Newman and T.W.J. Unti, A class of null flat-space coordinate systems, *J. Math. Phys.* **4**, 1467 (1963).

E.T. Newman and T.W.J. Unti, Note on the dynamics of gravitational sources, *J. Math. Phys.* **6**, 1806 (1965).

E.T. Newman and J. Winicour, The Kerr congruence, *J. Math. Phys.* **15**, 426 (1975).

E.T. Newman and R. Young, A new approach to the motion of a pole-dipole particle, *J. Math. Phys.* **11**, 3154 (1970).

K. Nordtvedt, Jr., Solution of the Einstein-Maxwell equations for a static distribution of massive charged particles, *Gen. Relat. Grav.* **3**, 95 (1972).

K. Nordtvedt, Jr. and H. Pagles, Electromagnetic plane wave solutions in general relativity, *Ann. Phys.* (*N.Y.*) **17**, 426 (1962).

I.D. Novikov, Note on the space-times metric inside the Schwarzschild singular sphere, *Astron. Zh. USSR* **38**, 564 (1961).

I.D. Novikov, Light propagation outside and inside the singular Schwarzschild sphere, *Dokl. Akad. Nauk SSSR* **150**, 1019 (1963).

R.F. O'Connel, Precession of Schiff's proposed gyroscope in an arbitrary force field, *Lett. al Nuovo Cimento* **18**, 933 (1969).

R.F. O'Connel, Present status of the theory of the relativity-gyroscope experiment, *Gen. Relat. Grav.* **3**, 123 (1972).

F. Oektem, On an extension of electromagnetic and gravitational tensors and the Bel-Robinson tensor, *Nuovo Cimento B* **58**, 167 (1968).

V.I. Ogievetskii and I.V. Polubarinov, Spinors in gavitation theory, *Zh. Eksp. Teoret. Fiz.* **48**, 1625 (1965).

V.I. Ogievetskii and I.V. Polubarinov, Group theoretical approach to spinors in the Einstein theory of gravitation, in: *International Conference on Relativistic Theories of Gravitation* (King's College, London, 1965).

H.C. Ohanian, Gravitons as Goldstone Bosons, *Phys. Rev.* **184**, 1305 (1969).

H.C. Ohanian, *Gravitation and Spacetime* (New York London, W.W. Norton, 1976).

H.C. Ohanian and R. Ruffini, *Gravitation and Spacetime*, Second Edition (W.W. Norton, New York and London, 1994).

T. Ohta, H. Okamura, T. Kimura and K. Hiida, Physically acceptable solution of Einstein's equation for many body systems, *Progr. Theor. Phys.* **50**, 492 (1973).

A.I. Oksak, Tilinear Lorentz invariant forms, *Commun. Math. Phys.* **29**, 189 (1973).

F. Oktem, Are the neutrinos an aspect of Riemannian geometry? *Nuovo Cimento A* **1**, 38 (1971).

P. Olijnychenko, On static solution in general relativity, *Nuovo Cimento* **21**, 389 (1961).

C.G. Oliveira and C.M. do Amaral, Spinor curvatures in the theory of gravitation, *Phys. Lett.* **22**, 64 (1966).

C.G. Oliveira and C.M. do Amaral, Spinor formalism in gravitation, *Nuovo Cimento A* **47**, 9 (1967).

F.J. Oliveira, Quantised intrinsic redshift in cosmological general relativity, gr-gc/0508094 (2005).

F.J. Oliveira, Exact solution of a linear wave equation in cosmological general relativity, *Int. J. Mod. Phys. D* **15**(11), 1963-1967 (2006); gr-qc/0509115.

F.J. Oliveira and J.G. Hartnett, Carmeli's cosmology fits data for an accelerating and decelerating Universe without dark matter or dark energy, *Found. Phys. Lett.* **19**(6), 519-535 (2006).

L. O'Raifeartaigh and J.L. Synge, A property of empty space-time, *Proc. R. Soc. Lond. A* **246**, 299 (1958).

F.R. Ore, Jr., Quantum field theory about a Yang-Mills pseudoparticle, *Phys. Rev. D* **15**, 470-479 (1977).

J.M. Overduin and P.S. Wesson, *Kaluza-Klein gravity*, gr-qc/9805018.

I. Ozsvath, Solutions of the Einstein field equations with simply transitive groups of motions, *Akad. Wiss. Lit. Mainz Abhandl., Mat.-Nat. Kl.*, No. 13, 22 (1962). [In German]

I. Ozsvath, Homogeneous solutions of the Einstein-Maxwell equations, *J. Math. Phys.* **6**, 1255 (1965).

A.M. Öztaş and M.L. Smith, Elliptical solutions to the standard cosmology model with realistic values of matter density, *Int. J. Theor. Phys.* **45**, 925-936 (2006).

A. Pais, On spinors in n-dimensions, *J. Math. Phys.* **3**, 1135 (1962).

P. Pajas and P. Winternitz, Representations of the Lorentz group: New integral relations between Legendre functions, *J. Math. Phys.* **11**, 1505 (1970).

D.W. Pajerski and E.T. Newman, Trapped surfaces and the development of singularities, *J. Math. Phys.* **12**, 1929 (1971).

A. Palatini, *Rend. Circ. Mat. Palermo* **43**, 203 (1919).

I.M. Pandya and P.C. Vaidya, Wave solutions in general relativity, I, *Proc. Nat. Inst. Sci. India A* **27**, 620 (1961).

A. Papapetrou, Spinning test-particles in general relativity, *Proc. R. Soc. Lond. A* **209**, 248 (1951).

A. Papapetrou, Some remarks on stationary gravitational fields, *Compt. Rend. Acad. Sci., Paris* **257**, 2797 (1963).

A. Papapetrou, Stationary gravitational fields of axial symmetry, *Compt. Rend. Acad. Sci., Paris* **258**, 90 (1964).

A. Papapetrou, Theorem of radiative electromagnetic and gravitational fields, *J. Math. Phys.* **6**, 1403 (1965).

A. Papapetrou, Spherically symmetric gravitational field with electromagnetic radiation, *Compt. Rend. Acad. Sci., Paris* **258**, 6081 (1964); **262**, 162 (1966).

A. Papapetrou, Stationary gravitational fields of axial symmetry, *Ann. Inst. Henri Poincaré A* **4**, 83 (1966). [In French]

A. Papapetrou, Shock waves in the Newman-Penrose formalism, *Commun. Math. Phys.* **34**, 229 (1973).

A. Papapetrou, *Lectures on General Relativity* (D. Reidel, Dordrecht, Netherlands, 1974).

A. Papapetrou and H. Treder, Non-existence of static, everywhere regular solutions in general relativity, *Compt. Rend. Acad. Sci., Paris* **254**, 4254 (1962).

A. Papapetrou and H. Treder, Problem of everywhere regular stationary solutions in general relativity, *Compt. Rend. Acad. Sci., Paris* **254**, 4431 (1962).

G. Papini and S.. Valluri, *Phys. Rep. C* **33**, 51 (1977). .

L. Parker, R. Ruffini and D. Wilkins, Metric of two spinning charged sources in equilibrium *Phys. Rev. D* **7**, 2874 (1973).

D. Parvizi, Representations of the SL(2, C) Lie algebra, *Compt. Rend. Acad. Sci., Paris A* **276**, 905 (1973).

S. Patnaik, Einstein-Maxwell filds with plane symmetry, I, *Proc. Camb. Phil. Soc.* **67**, 127 (1970).

W. Pauli, Zur Quantenmechanik des magnetischen Electrons, *Z. Physik* **43**, 601 (1927).

M. Pauri and G.M. Prosperi, Canonical realization of the rotation group, *J. Math. Phys.* **8**, 2256 (1967).

R. Pavelle, Yang's gravitational field equations, *Phys. Rev. Lett.* **33**, 1461 (1974).

R. Pavelle, Unphysical solutions of Yang's gravitational field equations, *Phys. Rev. Lett.* **34**, 1114 (1975).

T.G. Pavlopoulos, Breakdown of Lorenta invariance, *Phys. Rev.* **159**, 1106 (1967).

D. Peak, Foundation of gauge theory, Thesis (State University of New York, Albany, 1970).

E. Pechlaner and J.L. Synge, Model of a spinning body withou gravitational radiation, *Proc. R. Irish Acad.* **66**, 93 (1968).

J. Peebles, The Eötvös experiment, spatial isotropy, and generally covariant field theories of gravity, *Ann. Phys. (N.Y.)* **20**, 240 (1962).

P.J.E. Peebles, *Principles of Physical Cosmology*, Princeton Series in Physics (Princeton University Press, Princeton, NJ, 1993).

P.J.E. Peebles, Status of the big bang cosmology, p. 84, in: *Texas/Pascos 92: Relativistic Astrophysics and Particle Cosmology*, Eds. C.W. Akerlof and M.A. Srednicki, Vol. 688 (The New York Academy of Sciences, New York, 1993).

Physics Today, "Search and Discovery," May 2006, pp. 16 - 18.

R. Penney, Generalization of the Reisner-Nordstrom solution to the Einstein-field equations, *Phys. Rev.* **182**, 1383 (1969).

R. Penrose, A spinor approach to general relativity, *Ann. Phys. (N.Y.)* **10**, 171 (1960).

R. Penrose, *Report to the Warsaw Conference* (1962). .

R. Penrose, *Phys. Rev. Lett.* **10**, 66 (1963).

R. Penrose, Zero rest-mass fields including gravitation: Asymptotic behavior, *Proc. R. Soc. Lond. A* **284**, 1397 (1965).

R. Penrose, Relativistic symmetry groups, in: *Group Theory in Nonlinear Problems*, A.O. Barut, Editor (Riedel, Dordrecht, Netherlands; Boston, USA, 1974), p.1.

R. Penrose and M.A.H. MacCallum, Twistor theory: An approach to the quantization of fields and spacetime, *Phys. Lett. C* **6**, 241 (1973).

A. Peres, *Nuovo Cimento* **11**, 644 (1959).

A. Peres, Spinor fields in generally covariant theories, *Nuovo Cimento Suppl.* **24**, 389 (1962).

A. Peres, Non-geodesic motion in general relativity, *Phys. Rev. B* **137**, 1126 (1965).

Z. Perjes, A model method of constracting certain axially-symmetric Einstein-Maxwell fields, *Nuovo Cimento B* **55**, 600 (1968).

Z. Perjes, Three-dimensional relativity for axisymmetric stationary spacetime, *Commun. Math. Phys.* **12**, 275 (1969).

Z. Perjes, Spinor treatment of stationary space-times, *J. Math. Phys.* **11**, 3383 (1970).

Z. Perjes, Solutions of the coupled Einstein-Maxwell equations representing the fields of spinning sources, *Phys. Rev. Lett.* **27**, 1668 (1971).

S. Perlmutter *et al.*, Measurements of the cosmological parameters Ω and Λ from the first seven supernovae at $z > 0.35$, *Astrophys. J.* **483**, 565-581 (1997). [Supernova Cosmology Project Collaboration (astro-ph/9608192)]

S. Perlmutter *et al.*, *Nature* **391**, 51 (1998). [Supernova Cosmology Project Collaboration (astro-ph/9712212)].

S. Perlmutter *et al.*, *Astrophys. J.* **517**, 565 (1999). [Supernova Cosmology Project Collaboration (astro-ph/9812133)].

S. Persides, Asymptotically flat spaces in asymptotically null-spherical co-ordinates, *Proc. R. Soc. Lond. A* **320**, 1542 (1970).

S. Perveen, Renormalization problem of the Yang-Mills theoy with non-zero mass: The spectral approach, *J. Phys. A* **3**, 625 (1970).

A.Z. Petrov, About spaces which determine gravitational fields, *Dokl. Akad. Nauk SSSR* **81**, 149 (1951).

A.Z. Petrov, *Sci. Not.* Kazan State University **114**, No8, 55 (1954). (Trans. No. 29, Jet Propulsion Laboratory, Pasadena, Cal., 1963).

A.Z. Petrov, Classification of gravitational fields of general form by means of group motions I, II and III, *Izv. Viss. Ucheb. Zaned. Fizika, Ser. Mat.*,

No. 6 (1959), No. 1 (1960), and No. 6 (1960).

A.Z. Petrov, Invariant classifications of gravitational field geometry as the geometry of automorphisms, in: *Recent Developments in General Relativity* (PWN, Warsaw, 1962), pp. 371 and 379.

A.Z. Petrov, Classification invaiante des champs de gavitation, *Collection Intene de CNRS* **91**, 107 (1962).

A.Z. Petrov, Centrally-symmetric gravitational fields, *Soviet Phys. JETP* **17**, 1026 (1963).

A.Z. Petrov, Central-symmetric gravitational fields, *Zh. Eksp. Teoret. Fiz.* **44**, 1525 (1963).

A.Z. Petrov, On the gravitational fields with spherical symmetry, *Cahiers Phys.* **18**, 1 (1964).

A.Z. Petrov, *Einstein Spaces* (Pergamon Press, Oxford, 1969).

A.Z. Petrov, Gravitational fields of the third type, *JETP Lett.* **11**, 141 (1970).

I.G. Petrovski, *Lectures on the Theory of Integral Equations* (Graylock, London, 1957).

T.H. Phat, A universal regulator? *Ann. Phys.* **27**, 321 (1972).

P.R. Phillops, Is he graviton a Goldstone boson? *Phys. Rev.* **146**, 966 (1966).

M. Pierce *et al.*, *Nature* **371**, 385 (1994).

F.A.E. Pirani, *Acta Phys. Polon.* **18**, 389 (1956).

F.A.E. Pirani, *Phys. Rev.* **105**, 1089 (1957).

F.A.E. Pirani, A note on Fermi coordinates, *Proc. R. Irish Acad. A* **61**, 9 (1960).

F.A.E. Pirani, *Lectures on General Relativity* (Prentice Hall, Englewood Cliffs, N.J., 1965).

F.A.E. Pirani, Introduction to gravitational radiation theory, in: *Lectures on General Relativity* (1964 Brandeis Summer School) (Prentice-Hall, Englewood Cliffs, New Jersey, 1965).

F.A.E. Pirani and A. Schild, Geometrical and physical interpretation of the Weyl conformal curvature-tensor, *Bull. Acad. Polon. Sci., Ser. Math. Astron. Phys.* **9**, 543 (1961).

P. Pochoda and M. Schwarzschild, Variation of the gravitational constant and the evolution of the Sun, *Astrophys. J.* **139**, 587 (1964).

R. Pons, Integration of the Newman-Penrose equations in the Minkowski space time, *Ann. Inst. Henri Poincaré A* **16**, 51 (1972). [In French]

L.S. Pontrjagin, *Topological Groups* (Princeton University Press, New Jersey, 1946).

R. Potier, Sur les systémes d'equation aux dérivées partielles linéaires et du premier ordre, á quatre variables invariantes dans toute transformation de Lorentz, *Compt. Rend. Acad. Sci., Paris* **222**, 638 (1946).

R. Potier, Sur la définition du vecteur-courant en théorie des corpuscules. Cas du spin demi entier, *Compt. Rend. Acad. Sci., Paris* **222**, 855 (1946).

R. Potier, Sur la définition et les propriétés du vecteur-courant associe a un corpuscule de spin quelconque, *Compt. Rend. Acad. Sci., Paris* **222**, 1076 (1946).

R.V. Pound and G.A. Rebka, Jr., Apparent weight of photons, *Phys. Rev. Lett.* **4**, 337 (1960).

W.H. Press and J.M. Bardeen, Non-conservation of the Newman-Penrose conserved quantities, *Phys. Rev. Lett.* **27**, 1303 (1971).

W.H. Press and S.A. Teukolsky, *Astrophys. J.* **185**, 649 (1973).

I. Prigogine, J. Geheniau, E. Gunzig and P. Nardone, *Gen. Relat. Grav.* **21**, 767 (1989).

W. Raab, On the concept of spinors in general relativity, *Acta Phys. Austr.* **19**, 222 (1965). [In German]

G. Racah, *Group Theory and Spectroscopy* (Notes by Eugen Mer-zbacher and David Park, The Hebrew University, Jerusalem, 1951).

J.K. Rao, Type-N gravitational waves in non-empty space-time, *Curr. Sci. (India)* **32**, 350 (1963).

J.K. Rao, Cylindical waves in general relativity, *Proc. Nat. Inst. Sci., India A* **30**, 439 (1964).

J.K. Rao, Cylindrically symmetric null fields in general relativity, *Indian*

J. Pure Appl. Math. **1**, 367 (1970).

J.K. Rao, Radiating Levi-Civita metric, *J. Phys. A* **4**, 17 (1971).

J.R. Rao, A.R. Roy and R.N. Tiwari, A class of exact solutions for coupled electromagnetic and scalar fields for Einstein-Rosen metric, I, *Ann. Phys. (N.Y.)* **69**, 473 (1972).

J.R. Rao, R.N. Tiwari and A.R. Roy, A class of exact solutions for coupled electromagnetic and scalar fields for Einstein-Rosen metric, Part IA, *Ann. Phys. (N.Y.)* **78**, 553 (1973).

S.N. Rasband, Black holes and spinning test bodies, *Phys. Rev. Lett.* **30**, 111 (1973).

P. Rastall, A solution of the Einstein Field equations, *Can. J. Phys.* **38**, 1661 (1960).

P. Rastall, Rotations and Lorentz transformations, *Nucl. Phys.* **57**, 191 (1964).

P. Rastall, A new spinor calculus, *Gen. Relat. Grav.* **3**, 281 (1972).

A.K. Raychaudhuri, Arbitrary concentrations of matter and the Schwarzschild singularity, *Phys. Rev.* **89**, 417 (1953).

N.N. Razgovorov, Electromagnetic radiation in the general theory of relativity, II, *Izv. Viss. Ucheb. Zaved. Fiz. USSR* **5**, 32 (1973).

T. Regge, General relativity without coordinates, *Nuovo Cimento* **19**, 558 (1961).

C. Reina and A. Treves, *J. Math. Phys.* **16**, 834 (1975).

C. Reina and A. Treves, *Gen. Relat. Grav.* **7**, 817 (1976).

B.L. Reinhart, Maximal foliations of extended Schwarzschild space, *J. Math. Phys.* **14**, 719 (1973).

E.A. Remler, Scattering of Yang-Mills quanta, *Phys. Rev. B* **133**, 1267 (1964).

J. Renner, *Hung. Acad. Sci.* **53**, 542 (1935).

J.D. Reuss, A nonstationary, cylindrical gravitational field, *Compt. Rend. Acad. Sci., Paris* **266**, 794 (1968).

H. Reissner, *Ann. Phys.* **50**, 106 (1916).

A. Riess *et al.*, *Astrophys. J.* **438**, L17 (1995).

A.G. Riess *et al.*, *Astron. J.* **116**, 1009 (1998). [Hi-Z Supernova Team Collaboration (astro-ph/9805201)]

A.G. Riess, A. V. Filippenko, P. Challis, A. Clocchiatti and A. Diercks, Observational evidence from supernovae for an accelerating Universe and a cosmological constant, *Astron. J.* **116**, 1009-1038 (1998).

A.G. Riess, L.-G. Strolger, J. Tonry, S. Casertano, H.C. Ferguson, B. Mobasher, P. Challis, A.V. Filippenko, S. Jha, W. Li, R. Chornock, R.P. Kirchner, B. Leibundgut, M. Dickinson, M. Livio, M. Giavalisco, C.C. Steidel, N. Benitez and Z. Tsvetanov, Type Ia Supernova discoveries at z > 1 from the Hubble Space Telescope: Evidence for past deceleration and constraints on dark energy evolution, *Astrophys. J.* **607**, 665-687 (2004).

R. Rindler, *Special Relativity* (Oliver and Boyd, Edinburgh, 1960).

R. Rindler, Elliptic Kuskal-Schwarzschild space, *Phys. Rev. Lett.* **15**, 1001 (1965).

R. Rindler, What are spinors? *Am. J. Phys.* **34**, 937 (1966).

R. Rindler, Kruskal space and the uniformly accelerated frame, *Am. J. Phys.* **34**, 1174 (1966).

H.P. Robertson, Lecture in Toronto, 1939 (unpublished), cited by J.L. Synge.

D.C. Robinson, Conserved quantities of Newman and Penrose, *J. Math. Phys.* **10**, 1745 (1969).

D.C. Robinson, Uniqueness of the Kerr black hole, *Phys. Rev. Lett.* **34**, 905 (1975).

I. Robinson and J.R. Robinson, Vaccum metrics without symmetry, *Int. J. Theor. Phys.* **2**, 231 (1969).

I. Robinson and A. Schild, Generalization of a theorem by Goldbrg and Sachs, *J. Math. Phys.* **4**, 489 (1963).

I. Robinson, A. Schild and H. Strauss, A generalized Reissner-Nordtröm solution, *Int. J. Theor. Phys.* **2**, 243 (1969).

I. Robinson and A. Trautman, Spherical gravitational waves, *Phys. Rev. Lett.* **4**, 431 (1960).

I. Robinson and A. Trautman, Some spherical gravitational waves in general relativity, *Proc. R. Soc. Lond. A* **265**, 463 (1962).

I. Robinson and J.D. Zund, Degenerate gravitational fields with twisting rays, *J. Math. Mech.* **18**, 881 (1969).

J.A. Roche and J.S. Dowker, The algebraic structure of the vacuum Riemann tensor, *J. Phys. A* **1**, 527 (1968).

V.I. Rodichev and I.M. Dozmorov, Plane waves in the general theory of relativity, *Izv. Viss. Ucheb. Zaved. Fiz.* **4**, 20 (1969). [In Russian]

F. Rohrlich, The principle of equivalence, *Ann. Phys. (N.Y.)* **22**, 169 (1963).

F. Rohrlich, On relativistic theories, *Am. J. Phys.* **34**, 897 (1966).

P.G. Roll, R. Krotkov and R.H. Dicke, The equivalence of inertial and passive gravitational mass, *Ann. Phys. (N.Y.)* **26**, 442 (1964).

H. Rollnik, Operator gauge transformations in quantum electrodynamics, *Z. Phys.* **161**, 370 (1961).

P. Roman, *Theory of Elementary Particles* (North-Holland, Amsterdam, 1960).

G. Rosen, Spatially homogeneous solutions to the Einstein-Maxwell equations, *Phys. Rev. B* **136**, 297 (1964).

G. Rosen, Exact solutions to the Yang-Mills fild equations, *J. Math. Phys.* **13**, 595 (1972).

N. Rosen, Some cylindrical gravitational waves, *Bull. Res. Counc. Israel* **3**, 328 (1954).

N. Rosen, The complete Schwarzschild solution, *Ann. Phys. (N.Y.)* **63**, 127 (1971).

N. Rosen, Theory of Gravitation, *Phys. Rev. D* **3**, 2317 (1971).

N. Rosen and D. Yaakobi, A particle test in a static axially symmetic gravitational field, *Bull. Res. Counc. Israel F* **10**, 49 (1961).

R. Roskies, *Phys. Rev. D* **15**, 1722 (1977).

D.K. Ross and L.I. Schiff, Analysis of the proposed planetary radar reflection experiment, *Phys. Rev.* **141**, 1215 (1966).

R.A. Roy, J.R. Rao and R.N. Tiwari, A class of exact solutions for coupled electromagnetic and scalar fields for Einstein-Rosen metric, II, *Ann. Phys. (N.Y.)* **79**, 276 (1973).

S.R. Roy and V.N. Triphani, On non-static axially symmetric space-time, *Gen. Relat. Grav.* **2**, 121 (1971).

R. Ruffini and J. Tiommo, Electromagnetic field of a particle moving in a spherically symmetric black hole background, *Lett. al Nuovo Cimento* **3**, 211 (1972).

W. Rühl, Multilinear invariant forms built of an abitrary number of unitary representations of SL(n, C), *Nuovo Cimento A* **44**, 659 (1966).

W. Rühl, Tensor operators of rank two for the homogeneous Lorentz group, *Nuovo Cimento A* **68**, 2131 (1969).

W. Rühl, *The Lorentz Group and Harmonic Analysis* (Benjamin, New York, 1970).

A.P. Ryabushko, Centrally symmetric Einstein spaces and Birkhoff's theorem, *Soviet Phys. JETP* **19**, 1379 (1964).

M.P. Ryan, Jr. and L.C. Shepley, *Homogeneous Relativistic Cosmologies* (Princeton University Press, Princeton, New Jersey, 1975).

Yu.A. Rylov, The singularity of the Schwarzschild solution of the gravitational equations, *Zh. Eksp. Teoret. Fiz.* **40**, 1755 (1961).

R.K. Sachs, Gravitational waves in general relativity, VI. The outgoing radiation condition, *Proc. R. Soc. Lond. A* **264**, 309 (1961).

R.K. Sachs, Asymptotic symmetries in gravitational theory, *Phys. Rev.* **128**, 2851 (1962).

R.K. Sachs, On the characteristic initial value problem in gravitational theory, *J. Math. Phys.* **3**, 908 (1962).

R.K. Sachs, Gravitational waves in general relativity, VIII. Waves in asymptotically flat space-time, *Proc. R. Soc. Lond. A* **270**, 103 (1962).

R.K. Sachs, Gravitational radiation, in: *Relativity, Groups and Topology*, C. DeWitt *et al.*, Editors (Gordon and Breach, New York, 1964).

A. Sackfield, Physical interpretation of NUT metric, *Proc. Camb. Phil. Soc.* **70**, 89 (1971).

J.L. Safko and L. Witten, Some properties of cylindrically symmetric Einstein-Maxwell fields, *J. Math. Phys.* **12**, 257 (1971).

J.L. Safko and L. Witten, Models of static, cylindrically symmetric solutions of the Einstein-Maxwell field, *Phys. Rev. D* **5**, 293 (1972).

A. Saha *et al.*, *Astrophys. J.* **438**, 8 (1995).

G.S. Sahakian and Y.L. Vartanian, On Weyl's solution for axial-symmic gravitational fields, *Nuovo Cimento* **27**, 1497 (1963).

A. Salam, Lectures on Group Theory (preprint, published by Atomic Energy Agency, 1960).

A. Salam, *The formalism of Lie groups, Lecture Notes*, 1960.

A. Salam, Renormalizibility of gauge theories, *Phys. Rev.* **127**, 331 (1962).

A. Salam, The formalism of Lie groups, in: *Theoretical Physics* (International Atomic Agency, Vienna, 1963), p. 173.

A. Salam, On SL(6, C) gauge invariance, in: *Proceedings of the Coral Cables Conference on Fundamental Interactions* (Center for Theoretical Studies, Miami, Florida, January 1973).

A. Salam and J.C. Ward, On a gauge theory of elementary interactions, *Nuovo Cimento* **19**, 165 (1961).

A. Sandage *et al.*, *Astrophys. J.* **401**, L7 (1992).

S.S. Sannikov, Representations of the Lorentz group with complex spin, *Yadernaya Fiz. SSSR* **2**, 570 (1965). [In Russian]

S.S. Sannikov, A realization of infinite-dimensional representations of the rotation group with complex spin by entire functions, *Dokl. Akad. Nauk SSSR* **176**, 801 (1967).

B.E. Schaefer, The Hubble Diagram to redshift > 6 from 69 Gamma-Ray Bursts, *Astrophys. J.* **660**, 16-46 (2007).

R. Schattner, The center of mass in general relativity, Diplomarbeit (University of Munich, Munich, Germany, 1977).

J.F. Schell, Classification of four-dimensional Riemannian spaces, *J. Math. Phys.* **2**, 202 (1961).

L.I. Schiff, Sign of the gravitational mass of a positron, *Phys. Rev. Lett.* **1**, 254 (1958).

L.I. Schiff, Gravitational properties of antimatter, *Proc. Nat. Acad. Sci. Wash.* **45**, 69 (1959).

L.I. Schiff, Motion of a gyroscope according to Einstein's theory of gravitation, *Proc. Nat. Acad. Sci. Wash.* **46**, 871 (1960).

L.I. Schiff, On experimental tests of the general theory of relativity, *Am. J. Phys.* **28**, 340 (1960).

L.I. Schiff, *Phys. Rev. Lett.* **4**, 215 (1960).

M.M. Schiffer, R.J. Adler, J. Mark and C. Sheffield, Kerr geometry as complexified Schwarzschild geometry, *J. Math. Phys.* **14**, 52 (1973).

A. Schild, Discrete space-time and integral Lorentz transformations, *Phys. Rev.* **73**, 414 (1948).

A. Schild, Equivalence principle and red-shift measurements, *Am. J. Phys.* **28**, 778 (1960).

A. Schild, On gravitational radiation reaction forces, *Bull. Acad. Polon. Sci., Ser. Math. Astron. Phys.* **9**, 103 (1961).

B. Schmidt *et al.*, *Astrophys. J.* **395**, 366 (1992).

B. Schmidt *et al.*, *Astrophys. J.* **432**, 42 (1994).

B.G. Schmidt, Homogeneous Riemannian spaces and Lie algebras of Killing fields, *Gen. Relat. Grav.* **2**, 105 (1971).

B.P. Schmidt *et al.*, *Astrophys. J.* **507**, 46 (1998). [Hi-Z Supernova Team Collaboration (astro-ph/9805200)].

E. Schmutzer, Some remarks on the theory of spinors and bispinors in curved space-time, *Acta Phys. Hung.* **17**, 57 (1964).

E.M. Schrodinger, *Expanding Universe* (Cambridge University Press, New York, 1956).

K. Schwarzschild, *S.B. Press. Akad. Wiss.*, 189 (1916).

K. Schwarzschild, *Sitzungber. Preuss. Akad. Wiss. Berlin*, p. 424 (1916).

S. Schweber, *Relativistic Field Theory* (Row-Peterson, Evanston, Illinois, 1961).

J. Schwinger, On the charge independence of nuclear forces, *Phys. Rev.* **78**, 135 (1950).

J. Schwinger, Gauge invariance and mass, *Phys. Rev.* **126**, 397 (1962).

J. Schwinger, Non-abelian gauge fields: Commutation relations, *Phys. Rev.* **125**, 1043 (1962).

J. Schwinger, Non-abelian gauge fields: Relativistic invariance, *Phys. Rev.* **126**, 2261 (1962).

J. Schwinger, Gauge theories of vector particles, in: *Theoretical Physics* (International Atomic Energy Agency, Vienna, 1963).

D.W. Sciama, Recurrent radiation in general relativity, *Proc. Camb. Phil. Soc.* **57**, 436 (1961).

D.W. Sciama, Recent developments in the theory of gravitational radiation, *Gen. Relat. Grav.* **3**, 167 (1972).

D.W. Sciama, P.C. Wayler and R.C. Gilman, Generally covariant integral formulation of Einstein's field equations, *Phys. Rev.* **187**, 1762 (1969).

A. Sciarrino and M. Toller, Decomposition of the unitary irreducible representations of the group SL(2, C) restricted to the subgroup SU(1, 1), *J. Math. Phys.* **8**, 1252 (1967).

D.M. Sedrakyan and E.V. Chubaryan, On the internal solution of stationary axial-symmetric gravitational fields, *Astrofizika* **4**, 551 (1968).

D.M. Sedrakyan and E.V. Chubaryan, Stationary axial-symmetric gravitational fields, *Astrofizika* **4**, 239 (1968).

J. Segercrantz, On null tetrads and eigenvectors of Lorentz transformations in spino treatment, *Commun. Phys.-Math. (Finland)* **32**, 8 (1966).

R.U. Sexl. Universal conventionalism and space-time, *Gen. Relat. Grav.* **1**, 159 (1970).

I.I. Shapiro, Fourth test of general relativity, *Phys. Rev. Lett.* **13**, 789 (1964).

I.I. Shapiro, Lincoln Laboratory Tech. Rep. 368 (1964).

I.I. Shapiro, Testing general relativity with radar, *Phys. Rev.* **141**, 1219 (1966).

I.I. Shapiro, *Science* **162**, 352 (1968).

I.I. Shapiro, Testing general relativity: progress, problems, and prospects, *Gen. Relat. Grav.* **3**, 135 (1972).

I.I. Shapiro, M.E. Ash, R.P. Ingalls, W.B. Smith, D.B. Campbell, R.B. Dyce, R.F. Jurgens and G.H. Pettengill, *Phys. Rev. Lett.* **26**, 1132 (1971).

I.I. Shapiro, G.H. Pettengill, M.E. Ash, M.L. Stone, W.B. Smith, R.P. Ingalls and R.A. *Phys. Rev. Lett.* **20**, 1265 (1968).

I.I. Shapiro, W.B. Smith, M.E. Ash, R.P. Ingalls and G.H. Pettengill, Gravitational constant: experimental bound on its time variation, *Phys. Rev. Lett.* **26**, 27 (1971).

K.C. Shekhar, Solutions of Einstein's field equations (singular case), *Nuovo Cimento* **25**, 1205 (1962).

K.C. Shekhar, Particular solution of Einstein's field equations, *Nuovo Cimento* **28**, 902 (1963).

L. Sheldon, S.L. Glashow and M. Gell-Mann, Gauge theories of vector particles, *Ann. Phys. (N.Y.)* **15**, 437 (1961).

K. Shima, Local isotropic gauge transformation, *Progr. Theor. Phys.* **31**, 139 (1964).

M.F. Shirokov, Solutions of the Schwarzschild-Nordstrom type for a point charge without singularities, *Soviet Phys. JETP* **18**, 236 (1948).

J.L. Sievers, J.K. Cartwright, C.D. Contaldi, B.S. Mason, S.T. Myers, S. Padin, T.J. Pearson, U.-L. Pen, D. Pgosyan, S. Prunet, A.C.S. Readhead, M.C. Sheppard, P.S. Udonpraesert, L. Bronfman, W.L. Holzapfel and J. May, Cosmological parameters from cosmic background imager observations and comparisons with BOOMERANG, DASI, and MAXIMA, *Astrophys. J.* **591**, 599-622 (2003).

S.T.C. Siklos, *Singularities, Invariants and Cosmology*, Ph. D. Thesis, Cambridge.

S.F. Singer, Application of an artificial satellite to the measurement of the general relativistic red shift, *Phys. Rev.* **104**, 11 (1956).

K.P. Singh, L. Radhakrishna and R. Sharan, Electromagnetic fields and cylindrical symmetry, *Ann. Phys. (N.Y.)* **32**, 46 (1965).

J. Sinsky and J. Weber, New source for dynamical gravitational fields, *Phys. Rev. Lett.* **18**, 795 (1967).

A.A. Slavnov, Massive gauge fields, *Theor. Math. Phys.* **10**, 201 (1972).

M.L. Smith, A.M. Öztaş and J. Paul, A model of light from ancient blue emissions, *Int. J. Theor. Phys.* **45**, 937-952 (2006).

M.L. Smith, A.M. Öztaş and J. Paul, Estimation of redshifts from early galaxies, *Annales de la Foundation Louis de Broiglie* **32**(1) (2007).

I.N. Sneddon, in: *Handbook der Physik*, Edited by S. Flügge (Springer-Verlag, Berlin, 1955), Vol. 2, p. 299.

Y. Sofue, Y. Tutui, H. Honma, A. Tomita, T. Takamiya, J. Koda and Y. Takeda, Central rotation curves of spiral galaxies, *Astrophys. J.* **523**, 136-146 (1999).

G.A. Sokolik, A spinor formalism of gravitational field, *Dokl. Akad. Nauk SSSR* **154**, 565 (1964).

A. Sommerfeld, *Partial Differential Equations in Physics* (Academic Press, New York, 1949).

P. Sommers, On Killing tensors and constants of motion, *J. Math. Phys.* **14**, 787 (1973).

D. N. Spergel *et al.*, Wilkinson Microwave Anisotropy Probe (WMAP) three year results: Implications for cosmology, astro-ph/0603449.

V. Springle, S.D. White, A. Jenkins, C.S. Frenk, N. Yosida, L. Gao, J. Navarro, R. Thacker, D. Croton, J. Helly, J.A. Peacock, S. Cole, P. Thomas, H. Couchman, A. Evrard, J. Colberg and F. Pearce, Simulations of the formation, evolution and clustering of galaxies and quasars, *Nature* **435**, 629-636 (2005).

J. Stachel, Behavior of Weyl-Levi-Civita coordinates for a class of solutions approximating the Schwarzschild metric, *Nature* **219**, 1346 (1968).

N.E. Steenrod, *Topology of Fiber Bundles* (Princeton University Press, Princeton, NJ, 1951).

E.M. Stein, Analysis in matrix spaces and some new representations of SL(N, C), *Ann. Math.* **86**, 461 (1967).

N.M. Stepanyouk, An exact static axial solution of the Einstein equations, *Zh. Eksp. Teoret. Fiz.* **53**, 570 (1967). [In Russian]

G. Stephenson, Variational principles for the gravitational field, *Lett. al Nuovo Cimento* **1**, 97 (1969).

M.D.H. Strauss, A simple proof of the Lorentz transformation, *Nature* **157**, 516 (1946).

R.F. Streater and A.S. Wightman, *PCT, Spin and Statistics, and All That* (Benjamin, New York, 1964).

F. Strocchi, On the physical implications of gauge invariance, *Nuovo Cimento A* **42**, 9 (1966).

R. Sigal, A symmetry of Petrov type N spaces, *J. Math. Phys.* **14**, 1434 (1973).

S. Ström, On the contraction of representations of the Lorentz group to representations of the Euclidean group, *Ark. Fys.* **30**, 267 (1965).

D.J. Struik, *Lectures on Classical Differential Geometry*, Addison Wesley Press, Cambridge, 1950.

N.S. Swaminarayan, An exact solution of Einstein's equations, *Z. Phys.* **188**, 222 (1965).

N.S. Swaminarayan, An exact solution for uniformly accelerated particles in general relativity, III. Singular expanding (contracting) null surface, *Commun. Math. Phys.* **2**, 59 (1966).

J.L. Synge, The gravitational field of a particle, *Proc. R. Irish Acad. A* **53**, 83 (1950).

J.L. Synge, *Relativity: The General Theory* (North-Holland, Amsterdam, 1960).

J.L. Synge, The Petrov classification of gravitational fields, *Commun. Dublin Inst. Adv. Stud.* **15**, 51 (1964).

J.L. Synge and A. Schild, *Tensor Calculus* (University of Toronto Press, 1949).

K. Szego and K. Toth, SL(2, C) representations in explicitly energy-dependent basis, *J. Math. Phys.* **12**, 846 and 853 (1971).

P. Szekeres, Spaces conformal to a class of spaces in general relativity, *Proc. R. Soc. Lond. A* **274**, 206 (1963).

H. Takeno, On plane wave solutions of field equations in general relativity, *Tensor* **7**, 97 (1957).

H. Takeno, The theory of spherically symmetric space-times, *Sci. Rep.*

Res. Inst. Theor. Phys., Hiroshima Univ., No. 5 (1965).

H. Takeno, On the theory of spherically symmetric space-times, I, II, III and IV, *Tensor* **16**, 198 (1965); **16**, 247 (1965); **17**, 1 (1966); **17**, 56 (1966).

H. Takeno and S. Kitamura, On the theory of spherically symmetric space-times, *Tensor* **17**, 190 and 266 (1966).

C.J. Talbot, Newman-Penrose approach to twisting degenerate metrics, *Commun. Math. Phys.* **13**, 45 (1969).

J.D. Talman, *Special Functions, A Group Theoretic Approach* (Benjamin, New York, 1968).

TAMA, http://tamago.mtk.nao.ac.jp/

L.A. Tamburino, Exact empty space metrics containing geodesic rays, Ph.D. Dissertation (University of Pittsburgh, 1962).

L.A. Tamburino and J.H. Winicour, Gravitational fields in finite and conformal Bondi frames, *Phys. Rev.* **150**, 1039, (1966).

F.R. Tangherlini, Postulational approach to Schwarzschild's exterior solution with application to a class of interior solutions, *Nuovo Cimento* **25**, 1081 (1962).

F.R. Tangherlini, Source of the Schwarzschild field, *Nuovo Cimento* **38**, 153 (1965).

A.H. Taub, Empty space-time admitting a three parameter group of motions, *Ann. Math.* **53**, 472 (1951).

G.E. Tauber, The gravitational fields of electric and magnetic dipoles, *Can. J. Phys.* **35**, 477 (1957).

J.H. Taylor, L.A. Fowler and J.M. Weisberg, Measurements of general relativistic effects in the binary pulsar PSR1913+16, *Nature* **277**, 437 (1979).

S.A. Teukolsly, *Phys. Rev. Lett.* **29**, 1114 (1972).

S.A. Teukolsly, *Astrophys. J.* **185**, 635 (1973).

L. Thanh-Phong, The permanence of the Petrov classification, *Compt. Rend. Acad. Sci., Paris* **254**, 1926 (1962).

A.H. Thompson, Yang's gravitational field equations, *Phys. Rev. Lett.* **34**, 507 (1975).

J. Thompson and G. Schrank, Algebraic classification of four-dimensional Riemannian spaces, *J. Math. Phys.* **10**, 166 (1969).

I.H. Thompson and G.J. Whitrow, Time-dependent internal solutions for sphericallysymmetrical bodies in general relativity, *Mon. Not. R. Astron. Soc.* **136**, 207 (1967), and **139**, 499 (1968).

K.S. Thorne and C.M. Will, High-precision tests of general relativity, *Commun. Astrophys. Space Phys.* **11**, 35 (1970).

J. Tiomno, Electromagnetic field of rotating charged bodies, *Phys. Rev. D* **7**, 992 (1973).

R. Tiwari and D.N. Pant, Non existence of spherically symmetric solution representing vacuum electrostatic field, *Phys. Lett. A* **33**, 505 (1970).

M. Toller, An expansion of the scattering amplitude at vanishing four-momentum transfer using the representations of the Lorentz group, *Nuovo Cimento A* **53**, 671 (1968).

R.C. Tolman, On the estimation of distances in a curved Universe with non-static line element, *Proc. Nat. Acad. Sci.* **16**, 515-520 (1930).

R.C. Tolman, *Proc. Nat. Acad. Sci.* **20**, 169 (1934).

A. Tomimatsu and H. Sato, New exact solution for the gravitational field of a spinning mass, *Phys. Rev. Lett.* **29**, 1344 (1972).

A. Tomimatsu and H. Sato, New series of exact solutions for gravitational fields of a spinning mass, *Progr. Theor. Phys. (Jpn)* **50**, 95 (1973).

M.A. Tonnelat, The recent experiments on the red-shift as a proof of general relativity, *Ann. Inst. Poincare* **17**, 59 (1961).

M.A. Tonnelat, *Experimental Tests in General Relativity* (Masson, Paris, 1964).

A. Toomre, On the distribution of matter within highly flattened galaxies, *Astrophys. J.* **138**, 385-392 (1963).

I.I. Toro, A nonlocal extension of the formalism of Utiyama, *Compt. Rend. Acad. Sci., Paris* **261**, 2593 (1965).

I.I. Toro, A generalization of the gauge transformation, *Compt. Rend. Acad. Sci., Paris* **263**, 1 (1967).

I.I. Toro, The Yang-Mills field, *Compt. Rend. Acad. Sci., Paris* **262**, 1413

(1966).

R. Torrence and A. Janis, An approach to gravitational radiation scattering, *J. Math. Phys.* **8**, 1355 (1967).

A. Trautman, Foundations and current problems of general relativity, in: *Lectures on General Relativity* (1964 Brandeis Summer School) (Prentice-Hall, Englewood Cliffs, New Jersey, 1965).

A. Trautman, *Lectures on General Relativity* (Prentice-Hall, Englewood Cliffs, New Jersey, 1965).

A. Trautman, Fiber bundles associated with space-time, *Rep. Math. Phys.* **1**, 29 (1970).

A. Trautman, On the Einstein-Cartan equations, *Bull. Acad. Polon. Sci., Ser. Math. Astron. Phys.* **20**, 503 and 895 (1972).

R.P. Treat, Possible link between isospace and the geometry of space-time, *Phys. Rev. Lett.* **12**, 407 (1964).

R.P. Treat, Local gauge field in general relativity, *J. Math. Phys.* **11**, 2176 (1970).

R.P. Treat, Plane non-abelian Yang-Mills waves, *Nuovo Cimento A* **6**, 121 (1971).

R.P. Treat, Gauge invariant decomposition of Yang-Mills potentials, *J. Math. Phys.* **13**, 1704 (1972).

H.J. Treder, Einstein and symmetry groups of gravitational theory, *Ann. Phys.* **28**, 333 (1973). [In German]

J.R. Trollope, Null field in Einstein-Maxwell field theory, II, *J. Math. Phys.* **8**, 938 (1967).

M. Trümper, On a special class of type-I gravitational fields, *J. Math. Phys.* **6**, 584 (1965).

P.W. Trim and J. Wainwright, Nonradiative algebrically specila spacetimes, *J. Math. Phys.* **15**, 535 (1974).

W. Tulczyjew, *Acta Phys. Polon.* **18**, 393 (1959).

M.S. Turner, *Science* **262**, 861 (1993).

Y. Tutui *et al.*, *PASJ* **53**, 701, (2001); astro-ph/0108462.

I.V. Tyurin and E.S. Fradkin, On the theory of Yang-Mills fields, *Yadernaya Fiz. (USSR)* **13**, 433 (1971).

K.-H. Tzou, A new gauge invariance, *Compt. Rend. Acad. Sci., Paris* **253**, 2868 (1961).

K.-H. Tzou, Isotopic gauge-invariance of a vector field of zero proper mass, *Compt. Rend. Acad. Sci., Paris* **254**, 1210 (1962).

K.-H. Tzou, Formulations of gauge invariance, *Compt. Rend. Acad. Sci., Paris* **254**, 3819 (1962).

V. Unt, On the connection of news function with source in axi-symmetric case, *Füüsika Mat.* **18**, 421 (1969).

T.W.J. Unti, Note on asymptotically flat empty spaces, *Gen. Relat. Grav.* **3**, 43 (1972).

T.W.J. Unti and R.J. Torrence, Theorem on gravitational fields with geodesic rays, *J. Math. Phys.* **7**, 535 (1966).

H.K. Urbantke, Note on Kerr-Schild type vacuum gravitational fields, *Acta Phys. Austr.* **35**, 396 (1972).

R. Utiyama, Invariant theoretical interpretation of interaction, *Phys. Rev.* **101**, 1597 (1956).

R. Utiyama, Quantum theory and general relativity, *Progr. Theor. Phys.* **33**, 524 (1965).

R. Utiyama, Generalized isospinspace, generalized gauge transformation and derivation of meson, *Progr. Theor. Phys. Suppl.* Extra No. 332 (1965).

C.A. Uzes, Internal charge of the gauge field, *Ann. Phys. (N.Y.)* **50**, 534 (1968).

C.A. Uzes, The long-range self-interaction of the gauge fields, *Nuovo Cimento* **62**, 157 (1969).

C.A. Uzes, Gauge fields with positive-definite energy density, *J. Math. Phys.* **12**, 716 (1971).

P.C. Vaidya, The external field of a radiating star in general relativity, *Curr. Sci.* **12**, 183 (1943).

P.C. Vaidya, Non-static solutions of Einstein's field equations for sphere of fluids radiating energy, *Phys. Rev.* **83**, 10 (1951).

P.C. Vaidya, *Proc. Indian Acad. Sci.* **A33**, 264 (1951).

P.C. Vaidya, *Nature* **171**, 260 (1953).

P.C. Vaidya, The general relativity field of a radiating star, *Bull. Calcutta Math. Soc.* **47**, 77 (1955).

P.C. Vaidya, The gravitational field of a radiating star, *Proc. Indian Acad. Sci.* **33**, 264 (1957).

P.C. Vaidya, Non-static analogs of Schwarzschild interior solution on general relativity, *Phys. Rev.* **174**, 1615 (1968).

P.C. Vaidya and I.M. Pandya, Axially symmetric fields of pure radiation in the general theory of relativity, *Proc. Nat. Inst. Sci. India A* **26**, 459 (1960).

P.C. Vaidya and I.M. Pandya, Gravitational field of a radiating spheroid, *Progr. Theor. Phys.* **35**, 129 (1966).

P.C. Vaidya and L.K. Patel, Radiating Kerr metric, *Phys. Rev. D* **7**, 3590 (1973).

H. van Dam, On the mass of the graviton, *Gen. Relat. Grav.* **3**, 215 (1972).

H. van Dam and M. Veltman, Massive and massless Yang-Mills and gravitational fields, *Nucl. Phys. B* **22**, 397 (1970).

H. Van Dam and E.P. Wigner, *Phys. Rev. B* **138**, 1576 (1965).

M.G.J. van der Burg, Multipole expansions for stationary fields in general relativity, *Proc. R. Soc. Lond. A* **303**, 37 (1968).

B.L. van der Waerden, *Nachr. Wiss. Gottingen, Math.-Physik* **100** (1929).

B.L. van der Waerden, *Die Gruppentheoretische Methode in der Quantenmechanik* (Springer, Berlin, 1932).

B.L. van der Waerden, *Modern Algebra* (Ungar, New York, 1953).

V.I. Vashchuk, Conformal plane solution of the Einstein-Maxwell equations, *Izv. Viss. Ucheb. Zaved., Ser. Fiz.* **8**, 76 (1968). [In Russian]

W.J. Van Stockum, *Proc. R. Soc. (Edinburgh)* **57**, 135 (1937).

O. Veblen and J. von Neumann, *Geometry of Complex Dynamics* (The Institute for Advanced Study, Princeton, New Jersey, 1958).

M. Veltman, *Facts and Mysteries in Elementary Particle Physics* (World Scientific, Singapore, 2003).

M. Verde and I. Ferretti, On the unitary representations of the three-dimensional rotation group, *Atti Accad. Sci. Torino* **I103**, 905 (1969).

J.P. Vigier, Mass of the Yang-Mills vector field, *Nuovo Cimento* **23**, 1171 (1962).

N. Ya. Vilenkin, Bessel functions and representations of the group of Euclidean motions, *Usp. Mat. Nauk* **9**, 69 (1956).

C.V. Vishveshwara, Generaliztion of the Schwarzschild surface to arbitrary static and stationary metrics, *J. Math. Phys.* **9**, 1319 (1968).

C.V. Vishveshwara, Stability of the Schwarzschild metric, *Phys. Rev. D* **1**, 2870 (1970).

B.H. Voorhees, Static axially symmetric gravitational fields, *Phys. Rev. D* **2**, 2119 (1970).

R.V. Wagh, Gravitational field of distant rotating masses, *Indian J. Phys.* **34**, 211 (1960).

J. Wainwright, Geometric properties of neutrino fields in curved spacetimes, *J. Math. Phys.* **12**, 828 (1971).

R. Wald, On uniqueness of the Kerr-Newman black holes, *J. Math. Phys.* **13**, 490 (1972).

R. Wald, On perturbation of a Kerr black hole, *J. Math. Phys.* **14**, 1453 (1973).

L.L. Wang and C.N. Yang, *Phys. Rev. D* **17**, 2687 (1978).

F.W. Warner, *Foundations of Differentiable Manifolds and Lie Groups* (Scott, Foresman, Glenview, Illinois, 1971).

D. Watson, J. Hjorth, P. Jakobsson, D. Xu, J.P.U. Fynbo, J. Sollerman, C.C. Thöne and K. Pedersen, Are short gamma-ray bursts collimated? GRB 050709, a flare but no break, *Astron. Astrophys.* **454**, L123-L126 (2006). G.N. Watson, *A Treatise on the Theory of Bessel Functions* (Cambridge, 1966).

J. Weber, *General Relativity and Gravitational Waves* (Interscience, New

York, 1961).

J. Weber, Gravitational radiation experiments, in: *Relativity*, M. Carmeli, S.I. Fickler and L. Witten, Editors (Plenum Press, N.Y., 1970).

J. Weber, Advances in gravitational radiation detection, *Gen. Relat. Grav.* **3**, 59 (1972).

J. Weber and D. Zipoy, Classification of gravitational radiation, *Nuovo Cimento* **18**, 191 (1960).

A. Weil, *Actual. Sci. Industr.* **869**, Paris (1938). A. Weil, *L'integration dans les groups topologiques et ces applications* (Hermann *et* Cie., Paris, 1940);

S. Weinberg, Photons and gravitons in S-matrix theory: Derivation of charge conservation and equality of gravitational and inertial mass, *Phys. Rev. B* **135**, 1049 (1964).

S. Weinberg, Photons and gravitons in perturbation theory: Derivation of Maxwell's and Einstein's equations, *Phys. Rev. B* **138**, 988 (1965).

S. Weinberg, Infrared photons and gravitons, *Phys. Rev. B* **140**, 516 (1965).

S. Weinberg, A model of leptons, *Phys. Rev. Lett.* **19**, 1264 (1967).

S. Weinberg, *Gravitation and Cosmology* (Wiley, New York, 1973).

S. Weinberg, Problems in gauge field theories, *Proceedings of the 17th International Conference on High Energy Physics, London, England, 1 - 10 July 1974* (Science Research Council, Didcot, Berkshire, 1974).

S. Weinberg, Recent progress in gauge theories of the weak, electromagnetic and strong interactions, *Rev. Mod. Phys.* **46**, 255 (1974).

P.S. Wesson, Comments on a class of similarity solutions of Einstein equations relevant to the early Universe, *Phys. Rev. D* **34**, 3925 (1986).

P.S. Wesson, *Space, Time, Matter: Modern Kaluza-Klein Theory* (World Scientific, Singapore, 1999).

H. Weyl, *Ann. Phys. Leipzig* **54**, 117 (1917).

H. Weyl, *Ann. Phys.* **59**, 185 (1919).

H. Weyl, *The Theory of Groups and Quantum Mechanics* (Dover, New York, 1931).

H. Weyl, *The Classical Groups* (Princeton University Press, New Jersey, 1946).

H. Weyl, *The Classical Groups, Their Invariants and Representation* (Princeton University Press, Princeton).

J.A. Wheeler, Geons, *Phys. Rev.* **97**, 511 (1955).

J.A. Wheeler and T. Regge, Stability of Schwarzschild singularity, *Phys. Rev.* **108**, 1063 (1957).

B.C. Whitemore, Rotation curves of spiral galaxies in clusters, in: *Galactic Models*, J.R. Buchler, S.T. Gottesman, J.H. Hunter. Jr., Eds. (New York Academy Sciences, New York, 1990).

E.P. Wigner, On the consequences of the symmetry of the nuclear Hamiltonian on the spectroscopy of nuclei, *Phys. Rev.* **51**, 106 (1937).

E.P. Wigner, On unitary representations of the inhomogeneous Lorentz group, *Ann. Math.* **40**, 149 (1939).

E.P. Wigner, *Rev. Mod. Phys.* **29**, 255 (1957).

E.P. Wigner, *Group Theory and its Application to the Quantum Mechanics of Atomic Spectra* (Academic Press, New York, 1959).

E.P. Wigner, Geometry of light paths between two material bodies, *J. Math. Phys.* **2**, 207 (1961).

E.P. Wigner, Invariant quantum mechanical equations of motion, in: *Theoretical Physics* (International Atomic Agency, Vienna, 1963), p. 59.

E.P. Wigner, *Group Theory and its Application to the Quantum Mechanics of Atomic Spectra* (Academic Press, New York, 1964).

E. Wild, On the first order wave equations for elemetary particles without subsidiary conditions, *Proc. R. Soc. Lond. A* **191**, 253 (1947).

J.H. Winicour, A.I. Janis and E.T. Newman, Static, axially symmetric point horisons, *Phys. Rev.* **176**, 1507 (1968).

J.M. Wittaker, An interior solution in general relativity, *Proc. R. Soc. Lond. A* **306**, 1 (1968).

L. Witten, Invariants of general relativity and the classifications of spaces, *Phys. Rev.* **113**, 357 (1959).

L. Witten, Static cylindrically symmetric solutions of the Einstein-Maxwell field equations, *Colloque sur la Theorie de la Relativite* (CBRM, University Press, Louvain, 1960).

K. Wodkiewicz, *J. Math. Phys.* **18**, 441 (1977).

S.K. Wong, Gauge invariance and Galilean invariance, *Nuovo Cimento B* **4**, 300 (1971).

M.L. Woolley, The structure of groups of motion admitted by Einstein-Maxwell space-times, *Commun. Math. Phys.* **31**, 75 (1973).

E.L. Wright, Homogeneity and isotropy; many distances; scale factor, http://www.astro.ucla.edu/~wright/cosmo_02.htm A.C.T. Wu, Studies in the Yang-Mills field, *Phys. Rev. D* **2**, 3081 (1970).

T.T. Wu and C.N. Yang, Some solutions of the classical gauge field equation, in: *Properties of Matter under Unusual Conditions*, H. Mark and S. Fernbach, Editors (Interscience, London, 1969).

M. Wyman, Static isotropic solutions of Einstein's field equations, *Phys. Rev.* **66**, 267 (1944).

M. Wyman, Isotropic solutions of Einstein's field equations, *Can. Math. Congress*, p. 90 (1946).

M. Wyman, Schwarzschild interior solution in an isotropic coordinate system, *Phys. Rev.* **70**, 74 (1946).

M. Wyman and R. Trollope, Null fields in Einstein-Maxwell field theory, *J. Math. Phys.* **6**, 1995 (1965).

B.P. Yadav, Distorting metric with vanishing twist and divergence, *Phys. Lett. A* **46**, 253 (1973).

C.N. Yang, Integral formalism for gauge fields, *Phys. Rev. Lett.* **33**, 445 (1974).

C.N. Yang, *Phys. Rev. D* **16**, 330 (1977).

C.N. Yang, in: *Proceedings of the Sixth Hawaii Topical Conference in Particle Physics*, P.N. Dobson, Jr., S. Pakvasa, N.Z. Peterson and S.F. Tuan, Editors (University of Hawaii, Honolulu, 1977).

C.N. Yang and R.L. Mills, Conservation of isotopic spin and isotopic gauge invariance, *Phys. Rev.* **96**, 191 (1954).

P. Yasskin, *Phys. Rev. D* **12**, 2212 (1975).

R.J. Yefimov, *A Short Course in Analytical Geometry* (Moscow, 1954).

H. Yilmaz, New theory of gravitation, *Phys. Rev. Lett.* **27**, 1399 (1971), and *Nuovo Cimento B* **10**, 79 (1972).

V.D. Zakharov, A physical characteristic of Einsteinian spaces of degenerate type II in the classification of Petrov, *Dokl. Akad. Nauk SSSR* **161**, 563 (1965). English translation in: *Soviet Phys. Dokl.* **10**, 242 (1965).

R. Zaykov, Static and spheric solutions of Einstein equations, *Compt. Rend. Acad. Bulg. Sci.* **18**, 725 (1965).

E.C. Zeeman, Causality implies the Lorentz group, *J. Math. Phys.* **5**, 490 (1964).

F.J. Zerilli, Effective potential for even-parity Regge-Wheeler gravitational perturbation equations, *Phys. Rev. Lett.* **24**, 737 (1970).

D.M. Zipoy, Topology of some spheroical metrics, *J. Math. Phys.* **7**, 1137 (1966).

B. Zumino, Theories with gauge groups, *Acta Phys. Austr. Suppl. II* **212** (1965).

J.D. Zund, Degenerate gravitational fields with twisting rays, II, *Lett. al Nuovo Cimento* **4**, 879 (1972).

J.D. Zund and W. Maher, A spinor approach to some problems in Lorentzian geometry, *Tensor* **21**, 70 (1969).

J.D. Zund and W. Maher, A spinor approach to the generalized singular electromagnetic field theory of Bel, Lapiedra and Montserrat, *J. Math. Phys.* **14**, 168 (1973).

Index